Quality in Frozen Food

Quality in Frozen Food

Edited by

Marilyn C. Erickson

Center of Food Safety and Quality Enhancement
Department of Food Science & Technology
University of Georgia

Yen-Con Hung

Center of Food Safety and Quality Enhancement
Department of Food Science & Technology
University of Georgia

CHAPMAN & HALL

I(T)P® INTERNATIONAL THOMSON PUBLISHING

New York • Albany • Bonn • Boston • Cincinnati • Detroit
London • Madrid • Melbourne • Mexico City • Pacific Grove
Paris • San Francisco • Singapore • Tokyo • Toronto • Washington

Cover design: Curtis Tow Graphics

Printed in the United States of America

Chapman & Hall
115 Fifth Avenue
New York, NY 10003

Chapman & Hall
2-6 Boundary Row
London SE1 8HN
England

Thomas Nelson Australia
102 Dodds Street
South Melbourne, 3205
Victoria, Australia

Chapman & Hall GmbH
Postfach 100 263
D-69442 Weinheim
Germany

International Thomson Editores
Campos Eliseos 385, Piso 7
Col. Polanco
11560 Mexico D.F
Mexico

International Thomson Publishing–Japan
Hirakawacho-cho Kyowa Building, 3F
1-2-1 Hirakawacho-cho
Chiyoda-ku, 102 Tokyo
Japan

International Thomson Publishing Asia
221 Henderson Road #05-10
Henderson Building
Singapore 0315

1 2 3 4 5 6 7 8 9 10 XXX 01 00 99 98 97

Library of Congress Cataloging-in-Publication Data

Quality in frozen food / edited by Marilyn C. Erickson & Yen-Con Hung.
p. cm.
Includes bibliographical references and index.
ISBN 0-412-07041-3 (alk. paper)
1. Frozen foods. I. Erickson, Marilyn C. II. Hung, Yen-Con.
TP372.3.Q35 1997
664'.02853--dc21 97-10083
CIP

British Library Cataloguing in Publication Data available

To order this or any other Chapman & Hall book, please contact **International Thomson Publishing, 7625 Empire Drive, Florence, KY 41042.** Phone: (606) 525-6600 or 1-800-842-3636. Fax: (606) 525-7778. e-mail: order@chaphall.com.

For a complete listing of Chapman & Hall titles, send your request to **Chapman & Hall, Dept. BC, 115 Fifth Avenue, New York, NY 10003.**

For Chen-Yei (Heidi), Tony, and Jonathan

In Loving Memory of Lorraine N. Erickson

Contents

Preface

To effectively compete in the marketplace, food companies are having to address quality from a number of perspectives. This book addresses the issues of quality in frozen foods, proceeding from a technological view of quality to one that promotes an integrated view of quality. Technological (mechanical and cryogenic methods of freezing), categorical (classification of quality losses), analytical (measurement of quality), theoretical (model building), applied (preventive treatments), and administrative (policy) views have all been covered. As such, the intent has been to provide a comprehensive integrated view of quality. By focusing on the principles of freezing and the concepts of quality, the book transcends the boundaries of product identity.

The book is divided into five sections. The first section describes the technological and experimental aspects of food freezing in four chapters. In chapter 1, a short review of the types of systems currently available for freezing as well as recent advances in equipment designed for new product types have been presented. Chapter 2 serves to dissect and explain the majority of the physical and chemical phenomena involved in freezing of foods, covering the topics of enthalpy, supercooling, nucleation, crystal growth, volume changes, Mollier diagrams, and thermodynamic methods for analyzing the food-freezing process. Chapter 3 introduces the kinetic phenomena involved in the glass-rubber transition whereby upon freezing, concentration of solutes increases the viscosity to a point where an amorphous solid state or "glass" may form within the frozen food. Formation of this glass phase decreases the rates of molecular diffusion and renders the product stable for much longer periods of time. Consequently, through awareness of the glass formation, appropriate freezing and storage temperatures may be recommended and/or may be instrumental in reformulation of product with a higher glass transition temperature. Also included in chapter 3 are techniques used to measure the glass transition and low-temperature thermal events, including differential scanning calorimetry, thermomechanical analysis, dielectric analysis, and nuclear magnetic resonance. Chapter 4, on the other hand, discusses models of freezing from an engineering perspective. Models to predict freezing time in foods range from simple semitheoretical equations that assume regular shape and homogeneous properties to sophisticated computer programs based on finite differences and finite element techniques capable of handling irregular shape and nonhomogeneous temperature-dependent properties. Methods for measuring and predicting the properties needed for these models have also been discussed.

The second section of the book serves to identify, describe, and explain quality losses associated with frozen foods. Factors affecting thaw drip (irreversibility of water removal, diffusion of water due to differences in osmotic pressure, and ice recrystallization) are addressed in chapter 5. Freeze-cracking or shattering of food products upon rapid freezing is covered in chapter 6. Here, methods for evaluating the physical properties of a food product below freezing, and the incorporation of those measurements into predictive equations for freeze-cracking are also presented. Chapter 7 is concerned with the release of hydrolytic enzymes during the freezing process and the resultant deterioration in quality that occurs during further storage. In chapter 8 the influence of protein denaturation upon textural and functional properties of frozen foods is addressed. Discussion is also included on theorized mechanisms responsible for these alterations, such as protein-protein interactions, protein side chain modifications by low–molecular weight metabolites, and modification in pH of the protein's environment. Presented in chapter 9 are basic mechanisms responsible for lipid oxidation and associated deterioration in flavor and nutritional quality. In addition to examining the relationship of various compositional factors to the kinetics of lipid oxidation, the effects of processing and storage conditions on oxidative stabilities have been discussed. The final chapter in this section discusses the factors affecting microbial survival under freezing conditions, and the relationship of survival on product quality.

Techniques designed to minimize quality losses described in section 2 are covered in section 3. Included among these techniques are the application of cryoprotectants (chapter 11), application of antioxidants (chapter 12), application of edible films/coatings (chapter 13), modification in the raw material (chapter 14), and modification in packaging (chapter 15). In the chapters on cryoprotectants and antioxidants, properties critical to their effectiveness are highlighted. In chapter 12 methods of application are also addressed, as modification in the microenvironment of an antioxidant has been found to alter its effectiveness. Chapters 13 and 15 reinforce the influential nature of environment with edible films and packaging films having the capability to modify the migration of water in the product.

The fourth section of the book details methods by which quality losses may be assessed. Color, texture, and ultrastructural evaluations are covered in chapter 16 while chapter 17 reviews analytical methods used to measure chemical constituents that are associated with quality or the loss of quality. Advantages and limitations of sensory analysis, both consumer affective tests and descriptive analysis tests, are discussed in chapter 18. Ultimately, the information generated from these analyses may be incorporated into shelf-life mathematical models, and the principles behind these models are covered in chapter 19.

The final section of the book is concerned with adminstrative policies or strategies within industrial and academic research communities that could impact a frozen product's quality and ultimate purchase. The basis for any strategy, however, must be the consumer's viewpoint of a "quality" product. Consequently, the section begins with chapter 20 addressing consumer acceptance of frozen foods. Marketing strategies and their impact on perceptions of quality in frozen foods is then addressed in chapter 21. Chapter 22 subsequently describes the benefits to implementation of the concepts of total quality management and continuous improvement to all operations during manufacture of the product. Finally, a case is presented in chapter 23 for implementation of an integrated view to quality in the research arena.

Marilyn C. Erickson
Yen-Con Hung

Contributors

Lillian Arroyo-Gallyoun
Dept. Food Science and Technology
University of Tennessee
Knoxville, TN 37901-1071

Vasudevanallur M. Balasubramaniam
The National Center for Food Safety and Technology
Illinois Institute of Technology
6502 S. Archer Road
Summit-Argo, IL 60501

Fredric Caporaso
Dept. Food Science and Nutrition
Chapman University
333 N. Glassell St.
Orange, CA 92666

Manjeet S. Chinnan
Center for Food Safety and Quality Enhancement
Dept. Food Science and Technology
University of Georgia
Griffin, GA 30223

Marilyn C. Erickson
Center for Food Safety and Quality Enhancement
Dept. Food Science and Technology
University of Georgia
Griffin, GA 30223

Bin Fu
Frito-Lay Co.
7701 Legacy Dr.
Plano, TX 75024

R. Martin George
Campden and Chorleywood Food Research Association
Chipping Campden, Glos. GL 55 6LD

H. Douglas Goff
Dept. of Food Science
University of Guelph
Guelph, Ontario, Canada N1G 2W1

David A. Golden
Dept. Food Science and Technology
University of Tennessee
Knoxville, TN 37901-1071

Norman F. Haard
Dept. Food Science & Technology
University of California, Davis
Davis, CA 95616

John F. Haury
Ore-Ida Foods, Inc.
220 West Park Center Boulevard
P.O. Box 10
Boise, ID 83707-0010

Dennis R. Heldman
Dept. Food Science
University of Missouri–Columbia
214 Agricultural Engineering–MU
Columbia, MO 65211

Yen-Con Hung
Center for Food Safety and Quality Enhancement
Dept. Food Science and Technology
University of Georgia
Griffin, GA 30223

Eugene Jones
Dept. Agricultural Economics
2120 Fyffe Road
Ohio State University
Columbus, OH 43210

John M. Krochta
Dept. Food Science and Technology
University of California, Davis
Davis, CA 95616

Theodore P. Labuza
Dept. Food Science and Nutrition
University of Minnesota
1334 Eckles Ave.
St. Paul, MN 55108

Denise W. Lam
Alta Dena Certified Dairy
17637 E. Valley Boulevard
P.O. Box 388
City of Industry, CA 91747-0388

Tyre C. Lanier
Food Science Dept.
North Carolina State University
Raleigh, NC 27695-7624

Grant A. MacDonald
New Zealand Institute for Crop and Food Research, Ltd.
P.O. Box 5114
Nelson, New Zealand

Parameswarakumar Mallikarjunan
Center for Food Safety and Quality Enhancement
Dept. Food Science and Technology
University of Georgia
Griffin, GA 30223

Therese M. M. Malundo
U.S. Distilled Products
1607 S. 12th St.
Princeton, MN 55371

Raymond F. Mawson
Australian Food Industry Science Centre
Private Bag 16
Werribee, Victoria 3030 Australia

Kay H. McWatters
Center for Food Safety and Quality Enhancement
Dept. Food Science and Technology
University of Georgia
Griffin, GA 30223

Q. Tuan Pham
Dept. Chemical Engineering
University of New South Wales
Sydney, Australia

David S. Reid
Dept. Food Science
University of California
Davis, CA 95616

Anna V. A. Resurreccion
Center for Food Safety and Quality Enhancement
Dept. Food Science and Technology
University of Georgia
Griffin, GA 30223

Robert L. Shewfelt
Dept. Food Science and Technology
University of Georgia
Athens, GA 30602

Ramani V. Sista
Center for Food Safety and Quality Enhancement
Dept. Food Science and Technology
University of Georgia
Griffin, GA 30223

Yvonne M. Stuchell
The NutraSweet Kelco Co.
8225 Aero Dr.
San Diego, CA 92123-1718

John G. Surak
Dept. Food Science
224 Poole Agricultural Center
Clemson University
Clemson, SC 29634-0371

Timothy A. Taylor
Dept. Food Science
University of Missouri–Columbia
214 Agricultural Engineering–MU
Columbia, MO 65211

Youling L. Xiong
Dept. Animal Sciences
206 Garrigus Building
University of Kentucky
Lexington, KY 40546-0215

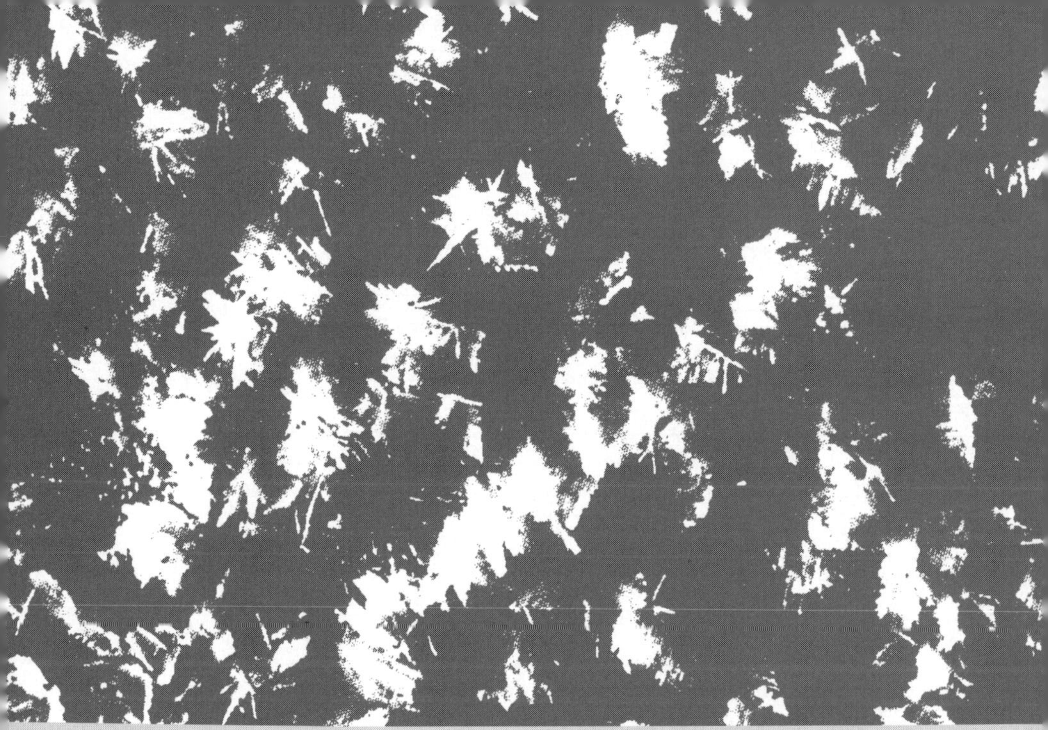

Quality in Frozen Food

SECTION I

Theoretical and Experimental Aspects of Food Freezing

1

Freezing Systems

R. Martin George

1.1 Introduction

Freezing has long been established as an excellent method of preserving high quality in foods. Freezing preserves the taste, texture, and nutritional value of foods better than any other method, and as a result extensive quantities of food are now frozen worldwide. However, frozen foods are experiencing considerable competition from other sectors, including chilled, ambient-stable, and minimally processed foods. The quality and nutritional aspects of frozen foods are critical and depend on the control of the freezing process and careful prefreezing preparation and postfreezing handling of the product.

To obtain high-quality frozen foods, high-quality raw materials are necessary, and each stage of processing, distribution, and storage requires stringent control. Sources of quality losses in frozen foods, that is, diffusion of water, denaturation of proteins, and the oxidation of lipids, challenge the entire frozen-food chain. In terms of the freezing operation, the primary objective is to maintain as much of the original and characteristic product quality as possible. Recent developments in freezing systems can provide a major step toward accomplishing these desirable objectives.

Attention in recent years has focused on freezing the food material as rapidly as possible to promote the growth of large numbers of small ice crystals. However, there is a commercially important trade-off between the speed of freezing and the economy of the freezing operation; the latter may often dictate the choice of operation.

In a general description of the freezing process, two major thermal events occur. These are the formation of the ice crystals (or nucleation) and their subsequent increase in size (crystal growth). Following nucleation, crystal growth occurs at a rate determined by three major factors: rate of reaction at the crystal surface, diffusion rate of water to the growing crystal, and rate of heat removal. The rate of heat removal also determines the number of nucleation sites formed, and crystal size varies inversely with the number of nuclei. At rapid freezing rates, the propagation of the initial ice crystals is insufficient to keep pace with the rate of heat removal, which results in an increased rate of nucleation. As more nucleation sites become active, there is an increase in the number (and a decrease in the size) of ice crystals formed. Conversely, at slow freezing rates, the propagation of ice crystals can better keep pace with the rate of heat removal, resulting in fewer nuclei becoming active and larger ice-crystal formation. Commercial food freezers aim at rapid freezing.

1.2 Food Freezing Systems

Freezing systems may be categorized in many ways, namely by batch or in-line operation, by heat transfer system (air, contact, cryogenic), or by product suitability. A convenient characterization may be made according to the rate of freezing or the rate of movement of the ice front from the product surface to the thermal center of the food. This is related to the properties of the food product and also the efficiency of heat transfer from the freezing medium to the food product surface. The latter may be conveniently quantified by a surface heat transfer coefficient. Table 1.1 shows typical coefficients for the variety of freezing systems described in the following sections.

The slowest rates of freezing are obtained with sharp freezers, which are cold storage rooms with refrigerated air as the cooling medium. These are really more suited solely to the storage of frozen foods. Naturally circulating air has a low surface heat transfer coefficient. Consequently, long freezing times are required to reduce the temperature of the thermal center of the food to desired frozen storage temperatures. Modification of the sharp freezer with extra refrigeration capacity and fans for increased air circulation still does not fulfill the aim to freeze the food rapidly. In addition, there is normally no controlled airflow over the product. Freezing is slow, and product qualities may be compromised. An impor-

Table 1.1 Typical surface heat transfer coefficients (h) for various freezing processes

Type of freezer	Conditions	h(W/m^2 K)
Cold store	Still air	5–10
Air-blast freezer	Air velocity: 2.5 m s^{-1}	17–20
	Air velocity: 5 m s^{-1}	26–30
Tunnel freezer	Counterflow of air and product	15–60
Contact freezer	Contact to cold plate	50–120
Fluidized-bed freezer	Suspending airstream	80–120
Cryogenic freezer (liquid N_2)	Gas zone (prefreezing)	40–60
	Spray zone (freezing)	100–140
Liquid immersion	Circulating brine	60–90
	Specialized refrigerant	500–600
Scraped-surface heat exchanger (e.g, ice-cream freezer)	Contact to cold surface and mixing of food product	1,500–2,000

tant criterion for their use as frozen storage facility is the temperature stability. Constant temperatures during frozen storage help to minimize recrystallization, which is a major cause of quality loss during storage of frozen foods.

More rapid freezing can be obtained with quick freezers, which can use refrigerated air or cryogenic gas as the means of freezing. This class of freezer is undoubtedly the most common in the food industry and offers the processor flexibility, ease of operation, and economy of operation for large production volumes. The air-blast freezer has almost universal application to the food industry and, when the product is loaded onto racks or trolleys and driven through the system, can offer a continuous means of producing many tonnes of product per hour. It is important in such freezers to ensure that the freezing air is allowed to circulate over the food product. Typically, air temperatures are in the range of -20 to -40°C, and air velocities are in the range of 0.5 to 10 m s^{-1}. A typical arrangement for an air-blast tunnel freezer is shown in figure 1.1.

Many modern air-blast freezers use space- and energy-efficient vertical airflow through product lines on an upward-spiraling helical belt. Variation according to product type can be accounted for by using different dwell times within the freezer (belt speed). The vertical airflow forces air through the product layers, but when the product layers are thin or nonexistent the freezing air will concentrate in these areas and bypass the thicker product layers. This effect, known as *channeling*, will result in poor freezing. Advantages in product handling with this type of freezer include the elimination of product transfer points (maintaining the continuity of the line and preventing damage to the product) and the fact that such freezers can be retrofitted into other process lines.

In recent years, there has been a great deal of interest in increasing the value of the surface heat transfer coefficient of the air-blast freezer. A method of substantially increasing this coefficient has been to raise the velocity of the freezing airflow and to direct this airflow onto the surface of the product. This has the effect of removing the insulating membrane of warm air immediately surrounding the product, a process known as *impingement freezing*. Reports suggest that for certain products, such as beef patties, freezing rates comparable with cryogenic freezing can be obtained with such systems.

A fairly recent modification of the air-blast freezer for particular product types has been the fluidized-bed freezer, which is used to provide an individually quick frozen product. The system relies on forced cold air from beneath the conveyor belt causing the products to become suspended or float in the cold airstream. The entire surface area of the product is in intimate thermal contact with the freezing medium, and as a consequence rapid freezing rates can be obtained. Air temperatures similar to those employed for air-blast freezing are used, and air velocity is dependent on the velocity needed to invoke fluidization of the food particles. Products to benefit from this technology include small vegetables, prawns, shrimp, french-fried potatoes, diced meats, and fruits. The high air velocity forces the product to take on the properties of a fluid and increases the rates of heat transfer. A typical fluidized-bed freezer is shown in figure 1.2.

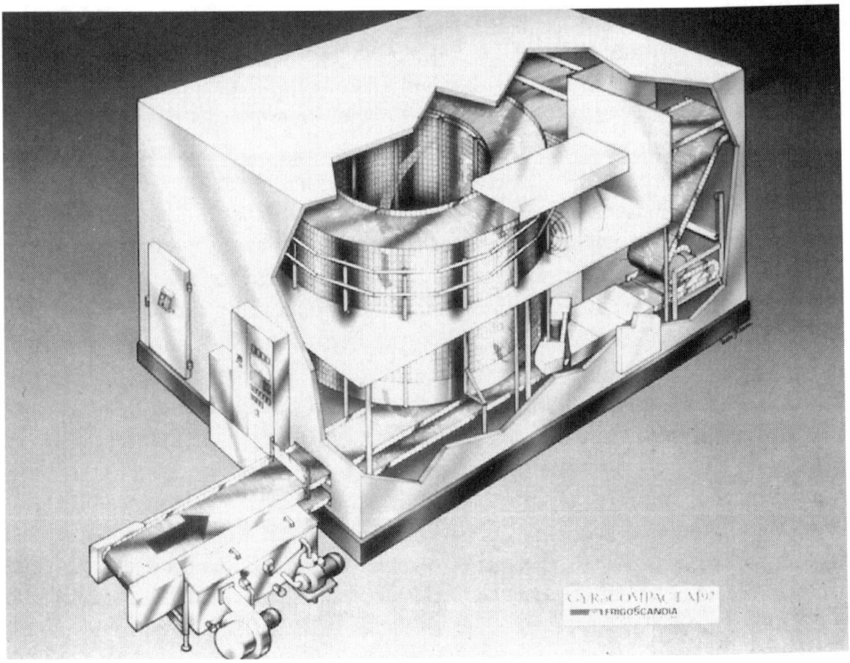

Figure 1.1. A tunnel-type helical air-blast freezer. (Courtesy of Frigoscandia Equipment Ltd., UK.)

An efficient means of heat transfer can also be obtained with the traditional contact freezers. The most recognized form is the well-established plate freezer, which uses a metal plate cooled by a mechanical refrigeration system and placed in direct contact with the food. Heat transfer to the food can be high. The freezer is restricted in the range of product types it can freeze, however; it is more suited to regular-shaped products such as beef patties or block-shaped packaged products. Automation of the plate movement to facilitate a semicontinuous operation is possible. A typical system is shown in figure 1.3.

An alternate form of contact freezer is the immersion freezer, which can be as simple as a tank filled with a suitable refrigerant (for example, salt-sugar-alcohol solution or a cryogenic liquid). Such a system commonly has been used for the shell freezing of large products, which precedes final freezing in an air-blast system. This method has the effect of freezing the outer layers of the product and helps to prevent the dehydration effects sometimes experienced with the blast freezing of large products. Modern systems may have the product enveloped in a flexible membrane, which prevents direct contact of the refrigerant with the food. For regular-shaped food products, this form of freezing offers a uniform and rapid process.

The most rapid form of freezing is obtained with cryogenic systems, which employ expendable, liquefied (cryogenic) gases such as nitrogen or carbon diox-

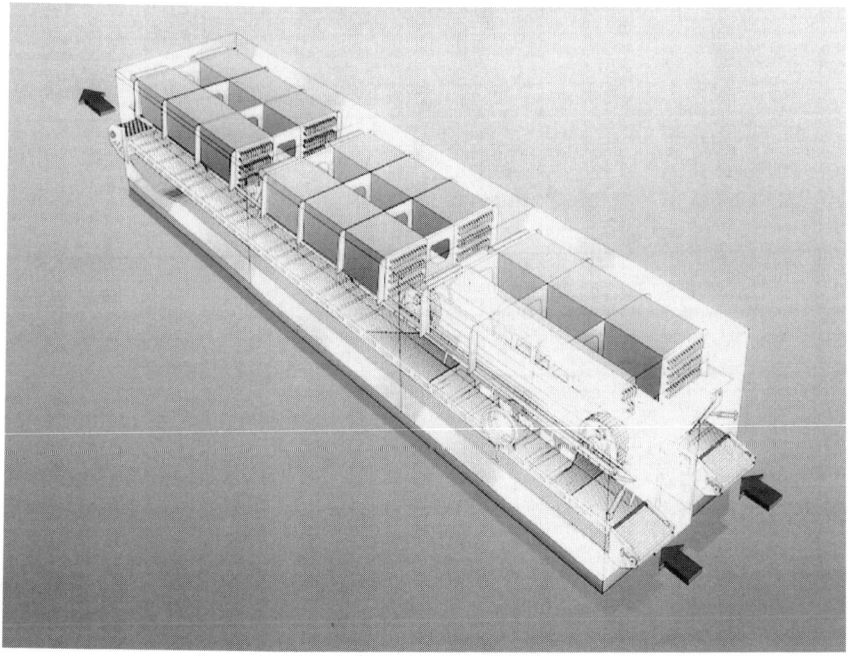

Figure 1.2. A typical fluidized-bed freezer. (Courtesy of Frigoscandia Equipment Ltd., UK.)

ide as the refrigerant. The potential benefits to the food producer include rapid freezing rates, high product throughput, low floor space requirements, flexibility (compatibility with many diverse types of foodstuff), and low capital outlay. Rapid freezing rates at the product surface also minimize the dehydration losses often encountered with air-blast systems. The major disadvantage of the system at present is the high cost of expendable cryogen.

Liquid nitrogen has a boiling point of −196°C at atmospheric pressure and, when sprayed into a food freezer, evaporates in the freezer cavity and at the surface of the food product. The cooling effect is almost instantaneous. This rapid surface freezing also offers advantages to the processor in terms of ease of handling (particularly for fragile or sensitive food products) or further processing. Liquid carbon dioxide (normally stored under pressure) differs from liquid nitrogen in that the release of liquid CO_2 to the atmosphere creates 50% dry ice (solid CO_2) and 50% vapor, both at −70°C. The solid component, dry ice, may appear as a compact frozen solid or as compressed CO_2 snow. Dry ice has an extremely high rate of heat removal from the food product surface, which can often more than compensate for the higher temperatures compared with liquid nitrogen freezing. The use of CO_2 freezers in food processing is dependent on the individual application. Dry ice requires intimate thermal contact with the food material, and the intense localized cooling may not be desirable for all food types because damage to the product may ensue. Carbon dioxide freezers tend to vary widely in

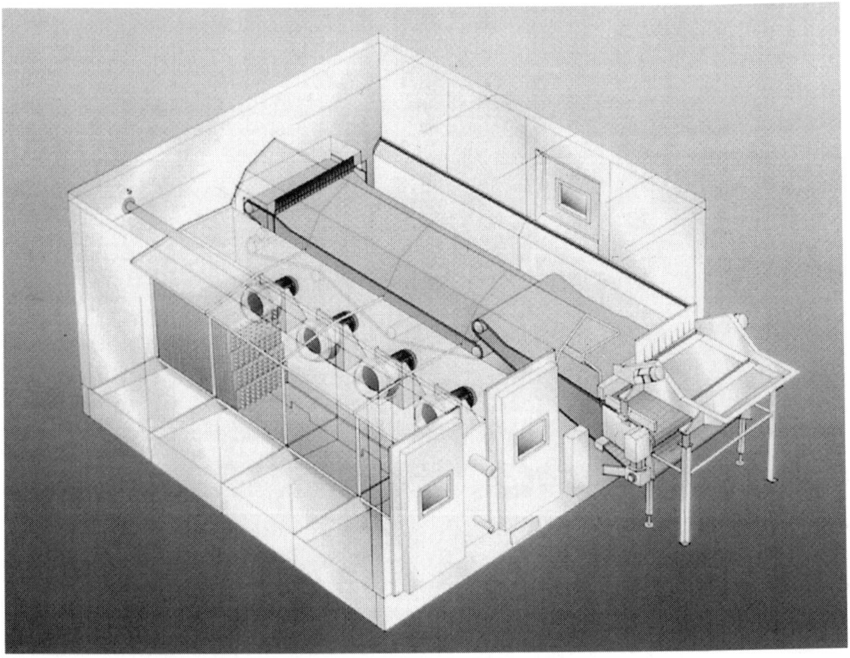

Figure 1.3. A typical plate freezer. (Courtesy of Frigoscandia Equipment Ltd., UK.)

their design and functionality. At present, liquid nitrogen is more commonly used as the cryogen in food freezing systems.

A typical arrangement for a liquid nitrogen cryogenic freezer is shown in figure 1.4. Upon leaving the spray area, liquid nitrogen gas evaporates at the surface of the food, creating a substantial temperature difference between the food surface and core. The equalization of this temperature gradient allows the volumetric freezing of the product. Gas that is not evaporated on the product is circulated to the freezer entrance to allow a precooling of the product entering the freezer. Such freezers are becoming more widespread in their applications to food processing (i.e., in the freezing of fish, shellfish, meat slices, and pizzas). Their advantages are best seen on products that have a large surface-area-to-volume ratio. Their application has to date been on food products that are regarded as high-value-added foods.

The rapid freezing obtained with cryogen gases is able to promote freezing of novel product types. For example, the freezing of liquids and semiliquids into solid frozen pellets is now used, and cryogenic freezing for the initial crust freezing of foods has been widely adopted. The latter example makes best use of the advantages of both cryogenic and mechanical freezing. The cryogenic stage freezes the outer surface of the food material (helping to maintain quality and prevent dehydration), after which the product is transferred to the mechanical freezer where the temperature of the product core is reduced by a more economical means.

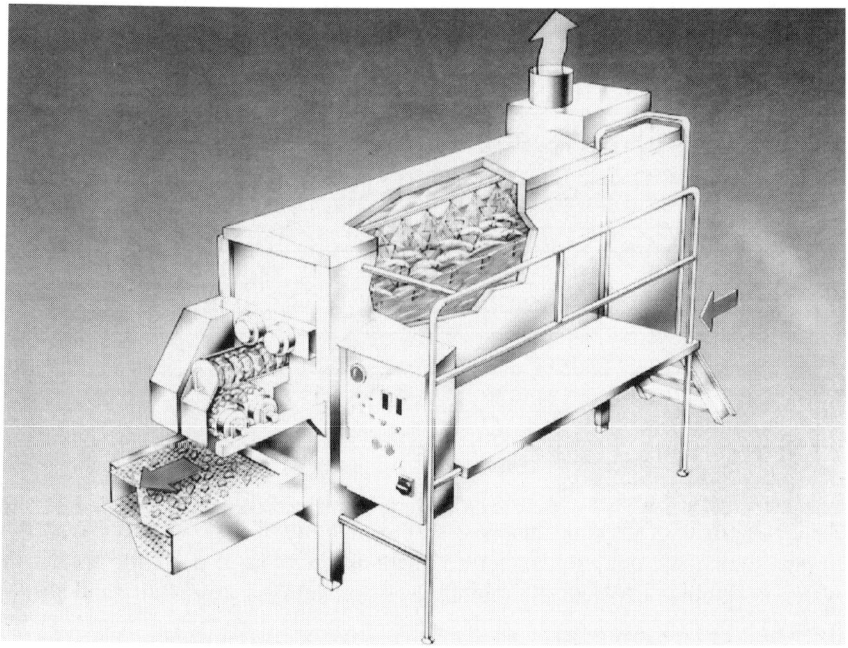

Figure 1.4. A liquid nitrogen cryogenic freezer. (Courtesy of Frigoscandia Equipment Ltd., UK.)

1.3 Future Trends in Food Freezing Processes

Developments in freezing technology over the last few years have been characterized by improvements in mechanical handling and process control to bring about increases in freezing rates and reductions in costs. This has been particularly apparent in fluidized-bed and cryogenic systems. Combination freezers represent a recent development in freezing systems where two or more different methods of freezing are used to ensure easy handling of difficult products such as very wet or sticky foods. Commercial systems of this type are now emerging.

It is likely that in the next few years the freezer manufacturer will increase activity in the development of new freezing systems. This shift will be brought about in part by the need to reduce the use of chlorofluorocarbons (CFCs) and hydrofluorocarbons (HCFCs) in freezing equipment. The search for candidates to replace these refrigerants will undoubtedly cause acceleration of new freezer developments. A key driving force will also be the frozen-food industry, which will need to respond to the market competition from other food sectors and provide the consumer with high-quality and nutritious frozen foods.

2

Overview of Physical/Chemical Aspects of Freezing

David S. Reid

2.1 Introduction

A characteristic of the food supply is that availability of fresh materials does not always match consumption patterns. There is therefore a need for means by which to stabilize perishable commodities to extend their availability and also to make their distribution over large distances much less time critical. A wide range of food-preservation techniques has evolved with the objective of extending the high-quality life of foods and food products. These technologies utilize many scientific principles. Freezing is an appropriate preservation technology because of two primary factors. The first factor is the role of temperature in biosystem stability. In general reactions proceed more rapidly as temperature is increased. The effect of lowered temperature per se is to decrease reaction rates, or rates of change, relating to many important processes. Frozen storage, all other factors being equal, should markedly enhance storage life. A complication results from the potential of the freezing process itself to cause alteration, often detrimental, to the state of a system such that quality is affected adversely by the freeze-thaw cycle. This is a separate issue from the topic of frozen storage stability.

In addition to the consequence of low temperature, which of itself tends to lower reaction rates, the presence of ice in a frozen system implies that the concentration of water in the unfrozen phases is reduced. A well-known principle of preservation is that foods with reduced moisture levels appear to be less susceptible to microbiological deterioration and also often have enhanced chemical storage stability (Labuza, 1980; Leung, 1987). A frozen food with ice has a reduced liquid moisture content, which would be expected to enhance product stability. Thus frozen foods benefit from two stabilizing factors: reduced temperature and reduced effective moisture content.

Since the freezing process is of critical importance to the function of two major preservation methods, this chapter will discuss the scientific basis of the freezing process and then briefly examine the characteristics of a range of processes, systems, and products in light of the underlying mechanisms of the freezing process.

When we consider the freezing process, we have to be concerned particularly with two aspects of the transformation of water into ice. These are the thermodynamic factors that define the positions of equilibrium and the kinetic factors that describe the rates at which these equilibria might be approached. In many systems, kinetics have a profound impact on the evolution of the final frozen system. Kinetics are, however, a modification of the thermodynamic possibilities. That which is thermodynamically prohibited cannot be attained. That which is thermodynamically possible might not be attained. Therefore, it is often best to first describe the thermodynamics of the system and then consider the kinetic constraints that determine the evolutionary routes. Bear in mind that all systems tend at any instant to approach the thermodynamic state that describes equilibrium under the instantaneous applied conditions. The kinetic constraints govern how rapidly, at each instant and under each set of conditions, the system approaches each of these potential equilibria. The time, temperature, and pressure history of a system, and of all individual parts of a system, are important to the attainment of the state in which it is found to exist at any time of interest.

Freezing implies the removal of heat, accompanied by a phase change of water to ice. The rate at which heat can be removed is a function of many factors, most of which can be placed into one of two categories. The first category comprises factors intrinsic to the object to be frozen, including the object's size, surface area, and internal thermal properties. The second category comprises factors characteristic of the freezing system such as the temperature of the heat transfer medium and the heat transfer coefficient between the cooling medium and the object to be frozen.

The rate of removal of heat and the rate of change of internal temperatures within the object under study both affect the phase change process and the spatial patterns of phase change. However, other factors also have to be considered, because heat removal alone is not sufficient to induce a phase change. Additional factors are involved in the initiation and continuation of the growth of ice, such as the rate of nucleation of ice crystals and the rate of propagation of ice. These growth-related factors are influenced by the composition of the material to be frozen and will also change according to the physical state of the system to be frozen. For example, an ice-cream system will have very different freezing characteristics and will require a very different freezing system than will a tissue system. Also, different tissue systems will exhibit characteristically different freezing behavior that will reflect, among other tissue characteristics, the nucleation and growth behavior of ice within the particular tissue (Reid, 1983, 1990, 1994).

2.2 The Physical Chemistry of the Freezing Process

2.2.1 *Effect of Low Temperature*

$$\ln k = -\frac{E_a}{\mathrm{R}T} + \text{const.} \tag{2.1}$$

Temperature is a fundamental descriptor of the state of a system. At high temperatures, molecular motions tend to be more rapid, as described by the Boltzmann relationships. Since molecular motions are more rapid, collision frequencies at the same molecular densities are greater. Thus there is an enhanced kinetics of change. As temperature is reduced, the molecular motions decrease, as do the reaction rates. This situation is described well in the Arrhenius relationship, which relates rates of reaction to temperature. As this relationship indicates, the rate of a reaction will decrease with decreasing temperature while the parameter $\boldsymbol{E_a}$ is negative.

For the temperature to decrease, heat may have to be removed. The removal of heat is in part described by the process known as heat transfer. Heat transfer quantifies the passage of heat down a temperature gradient. The basic equations of heat transfer describe how much heat can flow along the path between two points of known temperature difference for a specific material. Since the removal of heat from a system *may* cause a lowering of temperature (this is discussed in more detail later), there are clearly relationships between the rate of heat flow, the heat content–temperature relationships of the system, and the temperatures at points within the system where heat flow is occurring. In the simplest situation, the heat content of a system is directly related to its temperature. This relationship can be written in the form

$$q = C\,\Delta T_, \qquad \textbf{(2.2)}$$

where q is the change in heat content, C is known as the heat capacity of the system, and ΔT is the associated temperature change. In this case any net heat flow will result in a change in temperature at the points experiencing heat flow. Three primary processes of heat transfer—conduction, convection, and radiation—can be identified. The equations governing these processes are given.

The basic equations that define heat transfer by conduction are of the form

$$\frac{dQ}{dt} = kA\frac{dT}{dL} \qquad \textbf{(2.3)}$$

where dQ/dt is the heat flow, k is the thermal conductivity, A is the cross section, and L is the length, and

$$\frac{dT}{dt} = \frac{k}{C\rho}\frac{d^2T}{dx^2} \qquad \textbf{(2.4)}$$

where ρ is the density, x is the directional coordinate, and t is the time. Equation 2.3 describes the steady state, whereas equation 2.4 describes the dynamic or unsteady-state situation.

The basic equation that defines heat transfer by convection is of the form

$$Q = hA\,(T_b - T_s) \qquad \textbf{(2.5)}$$

where Q is the rate of heat transfer to a surface, h is the heat transfer coefficient, A is the area of the surface, T_b is the bulk fluid temperature, and T_s is the surface temperature. The units of the heat transfer coefficient may be in watts per square meter per kelvin (W m^{-2} K^{-1}).

Heat transfer by radiation is described by an equation of the form

$$\frac{dQ}{dt} = \epsilon A(T_1^{\,4} - T_2^{\,4})\sigma \qquad \textbf{(2.6)}$$

where dQ/dt is the rate of heat flow, ϵ is the emissivity, and σ is Stefan's constant. Since in these simple systems heat content is simply related to temperature, it is fairly easy to link the heat transfer relationships with the heat content relationships to predict the temperature history of such a body placed in an environment that initiates heat transfer. In most frozen systems, the important contributions to heat transfer will be thermal conductivity within the material and perhaps convection at the surface of the object, depending on the nature of the cooling medium. In a few cases, convection may make a significant internal contribution. Before discussing the heat transfer process further, it is necessary to consider the effect of lowered temperature on the state of a system.

A primary characteristic of a frozen system is the coexistence of at least two phases, a liquid phase and a solid phase. The liquid phase, typically, is a melt from the solid phase. It is therefore appropriate to consider the interrelationships between liquid phases and solid phases. In freezing, as the temperature is lowered, there tends to be a reduction in the mobility of the liquid phase. At some point solid phases may be produced. In what way do solid phases differ from the liquid phases from which they originate? First we would recognize that two distinct types of solid phase can exist. The first is the crystalline solid phase, in which there is a regular, defined repeating structure. This defined structure can be deduced from such information as X-ray diffraction patterns. The second form of solid is noncrystalline, or amorphous (sometimes termed glassy), and corresponds to the liquid under conditions in which molecular motion is so constrained that under the timescale of the observation no significant movement takes place. When considering freezing, the characteristics of the transformations from the liquid state to the solid state are of most interest.

Let us first consider the crystalline solid state. For water this is ice, of course. The transformation from liquid water to crystalline ice is a typical first-order phase transition. Figure 2.1 gives typical plots of the thermodynamic functions through the transition zone for pure water. The transition is identified as first order because there is a discontinuity in the derivative of the enthalpy-temperature function. Note that these data are for the transition from liquid water to ice 1h, the stable form of ice at 1 atmosphere pressure and temperatures close to 273 K. Other forms of ice can exist, in particular a high-pressure series of polymorphs (Fletcher, 1970; Hobbs, 1974), but these are of little relevance to frozen foods. One other form of ice that may be relevant is the cubic form of ice 1, that

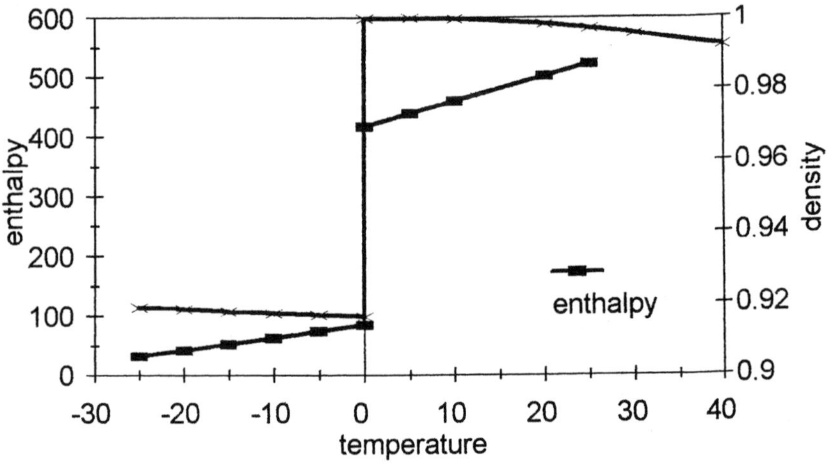

Figure 2.1. The thermodynamic properties of water through the phase transition region

is, ice 1c. This is believed to form if water vapor is cooled rapidly to very low temperatures, entering a glassy state, and then warmed to around 153 K. Evidence for the existence of ice 1c in such systems is not clear-cut (Dowell and Rinfret, 1960; Sugisaki et al., 1968). It is usual to rely on the X-ray diffraction patterns for crystal identification, and unfortunately the primary difference in diffraction pattern for ice 1c and ice 1h is in the presence or absence of two weak reflections. It is difficult, therefore, to demonstrate the confirmed presence of ice 1c in any typical, rapidly frozen food system. Hence in this chapter, ice will refer to ice 1h, the form of ice most likely to be present.

Another fact that should be noted is that the formation of ice is not automatic at temperatures below 273 K. While thermodynamics indicate that here ice is the stable form of water, kinetics may enter the picture. For the crystalline phase to grow, there must be an initiation seed, or nucleus. The processes of nucleation are discussed later. Once nucleation has taken place, the stable phase will grow, or propagate. Here, again, the equilibrium state is not necessarily achieved rapidly. Propagation is affected by rates of heat transfer and by the properties of the system. As we shall see when we discuss solutions, the composition of the system may have a profound effect on propagation in that it greatly modifies the kinetics of molecular motions and their dependence on temperature.

At 1 atmosphere pressure, the transformation from water to ice 1h takes place at 0°C, 273.16 K. Thus in a pure water system, the temperature of the freezing interface might be expected to be at 0°C. This is not quite the case because of heat transfer processes. For the interface to propagate, there must be a temperature gradient down which heat will flow. Also, the velocity of a propagating interface depends on the (slight) extent of supercooling at the interface. For most practical

Table 2.1. The properties of water and ice

Temperature (°C)	Vapor Pressure (kPa)	Density (Kg/M^3)	Enthalpy difference (kJ/kg)	Heat capacity (kJ/kg/K)	Thermal conductivity (W/m K)
−20	0.1034 (ice)	918.4	−375.60 (ice)	2.005	
	0.1254 (water)				
−10	0.2599 (ice)	917.4	−354.53 (ice)	2.083	
	0.2865 (water)				
0	0.6102	916.4 (ice)	−333.47 (ice)		2.219 (ice)
		999.8 (water)	0 (water)	4.217	0.550 (water)
+10	1.2259	999.7	42.03	4.192	0.576
+20	2.3349	998.2	83.86	4.182	0.598

purposes, though, the interface temperature can be assumed to be close to 0°C. The rate of growth of the interface will be controlled by the rate of heat transfer through the ice into the external environment. There will also be some heat transfer from the warmer liquid phase ahead of the freezing interface, but this will be small compared to the heat removal associated with the phase change. Of importance in this process will be the different thermal conductivities of water and ice. Also, the different heat capacities of water and ice will affect the rate of the process. Table 2.1 compares the properties of water and ice.

The presence of a soluble solute alters the phase relationships between ice and liquid. These changing relationships are most clearly illustrated by the use of phase diagrams. In the presence of solute, the solvent vapor pressure above a solution is reduced. Since the equilibrium between solid and liquid solvent occurs at the temperature at which the vapor pressures are equivalent, this lowering of the solvent vapor pressure results in a lowering of the temperature of the equilibrium phase transition. This is illustrated in figure 2.2. This vapor pressure reducing effect and its influence on the equilibrium phase relationships is known as a *colligative effect*. It is due primarily to statistical factors. If this were the only source of vapor pressure lowering, all solutes at the same molar concentration would show the same effect. However, there are specific interactions that influence the effectiveness of a solute in influencing freezing-point depression and so forth. The ideal relationship that describes freezing-point depression is of the form

$$\Delta T = \mathrm{k}\ /\ m \qquad \textbf{(2.7)}$$

where ΔT is the freezing point depression, k is the cryoscopic constant, and m is the solution concentration expressed as molality. This equation derives from Raoult's law and holds well at low concentrations. At higher concentrations, nonideal effects are seen, and the freezing-point depression becomes impossible to

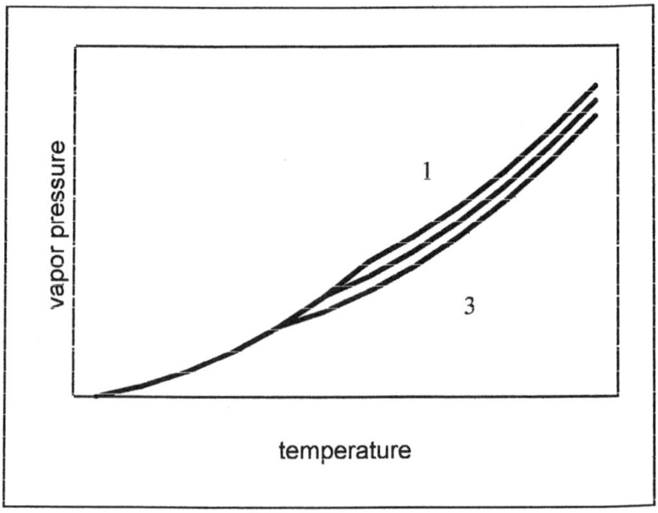

Figure 2.2. The effect of solute on the vapor pressure of water

predict. Since the freezing-point depression–concentration relationship would allow for the calculation of the temperature dependence of the composition of the system, and therefore facilitate enthalpy-temperature prediction, it would be useful to have some way of predicting freezing-point depressions at significant concentrations. Several workers have taken this challenge, and a range of empirical expressions have been derived (Heldman, 1974; Riedel, 1978; Lerici et al., 1983; Succar and Hayakawa, 1983; Chen, 1985, 1987; Chen and Nagy, 1987; Chen, 1988; Chen et al., 1990; Heldman, 1992; Rahman, 1994). Each expression is of value for interpolation of freezing point data but of much less value for extrapolation.

We can now return to the discussion of heat transfer. Earlier, consideration was given only to simple systems in which the temperature changes as a continuous function of the heat content. While this may be an appropriate description of a system without phase change, the fact that phase change is accompanied by a change in heat content must be factored into any discussion of the heat transfer in freezing. In the discussion of phase change, it has been pointed out that for a pure material at the temperature of a first-order phase transition, the heat content is a discontinuous function of temperature. Equation 2.4 adds another term and takes the form

$$q = C\,\Delta T + q_{\mathrm{p}} \tag{2.8}$$

where the additional term q_{p} represents the latent heat of the phase transition, which may or may not exhibit a temperature dependence. For a pure system undergoing the equilibrium phase change, q_{p} is 0 except at the temperature of the transition. In a system where solute results in a temperature range for the phase transition, q_p may be a complex function of temperature over the temperature range of the transition. While the same relationships described previously still govern the heat flow down a temperature gradient, the lack of an analytic function that relates heat content to temperature means that it is no longer a simple task to interrelate heat flows and temperature. Thus, in the real situation, the modeling and prediction of heat transfer become a complex problem. This is discussed further in chapter 4, where modeling of the freezing process is discussed. Here, it is sufficient to discuss qualitatively the consequences of phase change on the heat transfer relationships. Because of the latent heat release associated with the formation of ice, much more heat has to be removed when freezing occurs. Typically the latent heat associated with the formation of ice is equivalent to the heat release that accompanies a temperature change of approximately 80 K in the same mass of liquid water. Thus the heat associated with the phase change tends to dominate the heat transfer process as compared to the sensible heat. If this latent heat is released over a temperature range, it will result in an apparent heat capacity that is significantly larger than the heat capacity only because of sensible heat change. Thus, as heat is removed, the rate of temperature change is slower than it would be if the phase change were not taking place. The rate of heat removal stays higher than it might for the no-phase-change situation, however, because the internal temperatures are higher and so surface heat exchange to the low-temperature medium retains a higher ΔT driving force.

One complication of the phase change, and the existence of a freezing boundary within the object during the freezing process, is that the internal temperature profile becomes quite complex. While in a simple system there tends to be a monotonous change in temperature with position coordinate, in a system with phase change the temperature profile is more complex and can be difficult to predict accurately. Indeed one powerful test of freezing models is to see how close the model can come to correctly describing the temporal and spatial temperature distributions during the freezing process. Even here, the scale of the volume element described can influence the apparent success. If the volume element is small enough to be less than the structural repeat distance of the different phases present (i.e., ice crystals and unfrozen solution in, for example, a dendritic structure), then the spatial temperature profiles may be extremely complicated. This level of detail is beyond current-generation computer models and unnecessary for most purposes.

2.3 Phase Change

As indicated, when considering heat transfer relationships, the heat release as a function of temperature must be known. This is given by the change in heat content of the object as the temperature is changed. There are two main contributions to the heat release as an object is cooled. One is the sensible heat, or specific heat, which is associated with the reducing temperature and reflects the changes in molecular motion and so on. The second source of heat is associated with changes in phase. These changes in heat content are not necessarily associated with change in temperature. Indeed, for a pure material, the phase change may take place at a constant temperature. The change in heat content is associated with changes in the internal bonding and structure of the material. To properly account for the changes associated with removal of heat, and also to quantify the heat release to be expected as accompanying a particular temperature change, the sensible heat change and the latent heat change both must be estimated. The sensible heat change accompanies a change in temperature and can be estimated fairly readily. The latent heat change accompanies a phase change and can be estimated only if the extent of the phase change can be estimated. This is not a trivial problem, and so it is appropriate at this point to discuss the mechanisms of phase change, in particular as they relate to aqueous systems taken below the temperature of 0°C. Two processes must be considered: nucleation and growth.

2.3.1 Nucleation

Thermodynamics describes the equilibrium between a liquid phase and a solid phase. For pure water at 1 atmosphere pressure, the chemical potentials of liquid water and of ice are the same at 0°C. Below this temperature, pure ice has a lower chemical potential than pure liquid water, and above this temperature pure liquid water has a lower chemical potential than pure ice. The effect of added solute will be discussed later. Given that below 0°C the equilibrium state is ice, one might expect cooling below this temperature to produce ice. This would be the case if all processes were under solely thermodynamic control. However, before ice can form, a nucleus, or seed, is required upon which the crystals can grow. The process of producing this seed is termed *nucleation*. Nucleation can be a spontaneous event, on a catalytic center produced by statistical fluctuation within the water, or it can take place on a foreign catalytic center, a heterogeneous nucleation catalyst. Spontaneous nucleation (homogeneous nucleation) of ice from supercooled liquid water does not take place until a temperature approaching −40°C is reached. Heterogeneous nucleation may take place at higher subzero temperatures. Since one active nucleation catalyst is sufficient to initiate crystallization in the whole volume of supercooled water, study of nucleation requires special techniques. One powerful methodology subdivides the water volume into a large number of small drops, often dispersed in a carrier medium. Each drop may nucleate and freeze, but each is independent of all other drops. Thus, the study of the freezing of the individual drops yields information on the statistical probabilities of nucleation (Maki et al., 1974; Charoenrein et al., 1991).

2.3.2 Growth (Propagation)

Once a nucleation catalyst has acted, the crystals can grow. Since there is a release of heat accompanying the phase change, growth may not be instantaneous. Rather, the growth may be controlled by the rate of heat removal and the resulting rate of advance of the freezing-point isotherm. Heat transfer, however, is not the only factor that can influence the rate of propagation of ice. If the ice is crystallizing from a solution, since ice tends to form as a pure crystal, the solute component must be rejected from the spatial region occupied by the ice. Thus there is a mass transfer component to the propagation. Water molecules must diffuse in to add to the growing ice crystal, and at the same time solute molecules must diffuse away from the surface of the growing crystal. It will be important in some processes to determine whether heat transfer or mass transfer is the rate-limiting step. If heat transfer is the rate-limiting step, then the interface probably exists in a state close to the equilibrium state for the temperature that describes that region of the freezing interface. If mass transfer is limiting, the state of the interface region may not be appropriately described by the equilibrium state for the temperature of interest. There may be less ice than the equilibrium amount. This will have consequences for the process and its description. In particular, the state of the system will greatly affect the heat transfer relationships.

To consider the phase change process in solutions rather than in pure water, we must utilize phase and state diagrams, which are convenient methods for describing the state of a system in terms of the compositions of the phases present (Franks, 1985). These diagrams do not indicate the spatial distribution of the phases, only the overall phase composition of the system. In a phase diagram, lines indicate the conditions under which equilibrium phase transformations can occur. While the most familiar phase diagrams are pressure-temperature diagrams, such as that for water (fig. 2.3), in frozen systems the temperature composition phase diagrams are of particular value in describing the phase relationships. It is easiest to illustrate the value of such phase diagrams by describing the features of some typical phase diagrams. Figure 2.4 is the phase diagram for a simple binary system. The curve *a-b-c* follows the composition-temperature relationship of the melting equilibrium. At each temperature, this curve shows the composition for which this is the equilibrium melting point. In region *A* the system is all liquid, a homogeneous solution of composition and temperature as described by the coordinates. In region *B* there is a two-phase system. The temperature of the system is given by the temperature coordinate, and the composition of the system, while described overall by the concentration coordinate, is a mixture of two phases whose compositions are given by the intersection of the temperature line with the opposite boundaries of the region. In the case of a pure crystal, the boundary is the axis. The other boundary describes the composition of the solution that just melts (or just freezes) at the appropriate temperature. This diagram

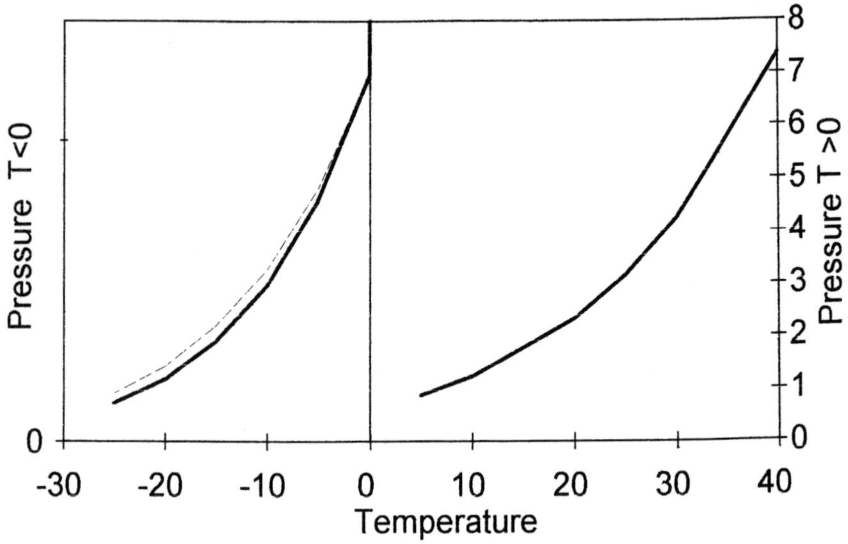

Figure 2.3. The *P*/*T* diagram for water

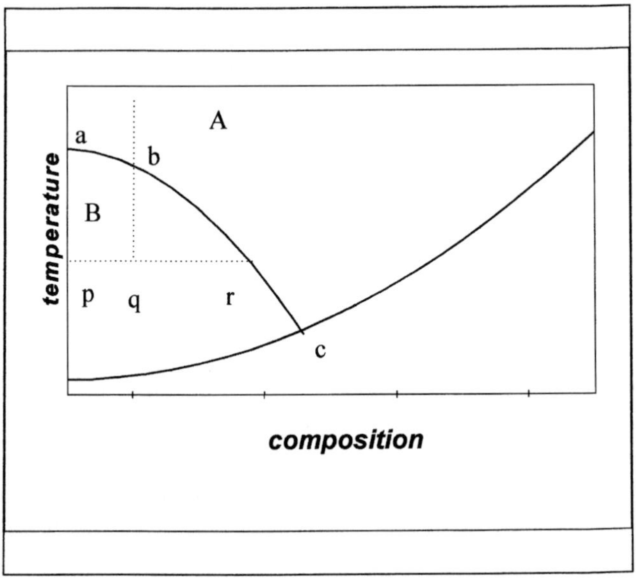

Figure 2.4. A simple binary phase diagram

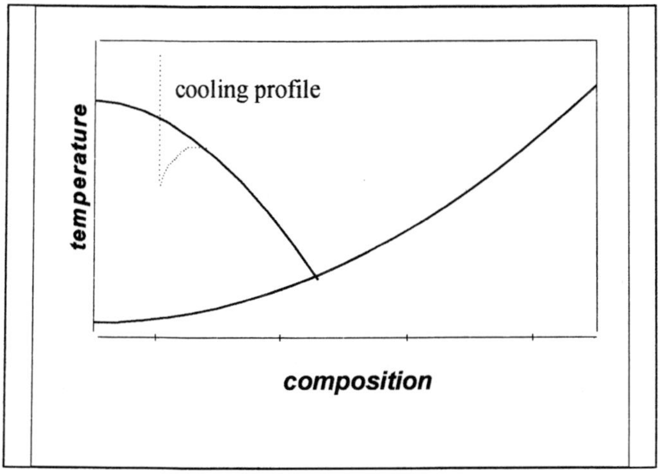

Figure 2.5. The behavior of a freezing system

has some useful properties. For example, if a line is drawn linking the coordinates that describe the two phases coexisting in equilibrium, then the position of the overall system composition on this line allows for an easy calculation of the amounts of each phase present in the equilibrium mixture. In the diagram, the pure solid is at *p*, the solution phase is represented by *r*, the system composition is *q*, and the ratio of solid phase to liquid phase is *qr*/*pq*. This property of *tie-lines*, as they are termed, is of great use.

Consider now the effect of solute on the freezing process, using the phase diagram as a descriptive tool. A freezing path is shown in figure 2.5. At any point, as the system is cooled, the temperature will drop. Depending on the composition of the system, the freezing point is reached at a temperature defined by the liquidus line of the phase diagram. Since nucleation is required for freezing to initiate, the temperature will continue to fall, but no crystallization will occur. At some temperature controlled by a variety of factors that include rate of cooling, system volume, and nature of nucleation catalysts present, nucleation will take place. This will cause a rise in temperature to that described by the liquidus line. Unless the cooling rate is very rapid, the composition of the system will now follow the liquidus line as cooling continues. The time-temperature profile for the point will be governed both by the rate of heat removal, described by the thermal environment, and by the temperature–heat content profile, which is to a large extent governed by the amount of phase change associated with a small temperature change. This can be deduced from the detailed shape of the liquidus if the thermal properties of the unfrozen phase are known or can be estimated. From the shape of the liquidus it is possible to estimate the heat content–temperature relationship. This makes use of the tie-line principle to estimate the amounts of ice and of unfrozen phase that coexist at each temperature. The heat content of ice is a known function of temperature, and the heat content of the unfrozen phase can be estimated from the composition, also as a function of temperature. The reverse calculation is also very useful, allowing the estimation of the amount of ice, and hence the phase diagram, from the changing heat content of a system as a function of temperature. This is described in more detail in Reid et al. (1993) and Kerr et al. (1993).

2.4 Freezing Rate, Definitions, and Effects

While at first sight the term *rate of freezing* would appear to be self-explanatory, a careful analysis indicates that a variety of definitions might serve to express rate of freezing, and through the use of these definitions the terms *fast freezing* and *slow freezing* could mean many things. Indeed it is possible for fast freezing under one definition to be slow freezing under another. It is therefore important to identify more clearly the different concepts of freezing rate. These concepts differ for simple model systems, and for tissue systems. To explain the differences it is best first to discuss the concept of freezing rate in simple models. The first definition of freezing rate is rate of temperature change. Since the phase change influences this rate, the definition is of limited usefulness. Also, rate of temperature change is position dependent. A better definition of freezing rate might be rate of ice formation, which is related to rate of heat removal. Another effective definition equates freezing rate to the rate of advance of the freezing interface. The latter definitions define rates that are related to the appearance of ice in the frozen system. The more rapid the formation of ice, and the more rapid the advance of the freezing interface, the smaller and more numerous are the ice crystals. Fast freezing, visually, is thus often operatively defined as freezing that produces many small ice crystals.

In tissue systems, the existence of permeability barriers, such as cell walls, adds increased complexity to the task of defining freezing rate. Now both the inside and the outside environment must be considered. For the outside connected network the simple definitions remain adequate, but within the cell the permeability of the barrier and the internal properties of the cell must be considered. Two factors influence the internal freezing process (Reid, 1994): The first factor is the loss of water through osmosis to the external environment, since an osmotic gradient exists if there is external ice, while the cell contents are unfrozen. This osmotic transfer increases the internal osmotic concentration, reducing the internal freezing point and hence the degree of supercooling. The nucleation characteristics of the cell contents govern the maximum sustainable degree of supercooling. The interplay of these factors leads to internal cell freezing at different rates of cooling and freezing of the bulk tissue. The membrane permeability, if high, helps prevent internal freezing, as does an ability to sustain a large supercooling. An operational definition of freezing rate in cellular systems might be that fast freezing produces internal freezing in cells and slow freezing produces cellular dehydration and only extracellular ice.

When there is a low rate of heat removal with either low or high water permeability, the rate of change of the unfrozen matrix concentration is slow, and water can transfer from the interior fast enough to minimize internal supercooling. Under these conditions the cell will dehydrate, with the lost water going to external ice crystals. When there is a fast rate of heat removal accompanied by low water permeability, the unfrozen matrix concentration will rise rapidly. Since water cannot transfer rapidly, the cell contents will become increasingly supercooled. At some critical supercooling, the internal contents will freeze. Intracellular freezing will cause structure damage, but there will be little water transfer from the cell interior. When there is a fast rate of heat removal accompanied by high water permeability, then, as the unfrozen matrix concentration increases, the cell exports water and maintains minimal supercooling. In this case, the cell dehydrates but does not freeze. Moisture is transferred from inside the cell.

Note that the critical control factor is the relationship between the rate of concentration increase of the unfrozen external matrix and the maximum rate of water transfer permitted by the cell barrier. A cell with high hydraulic permeability will always tend to dehydrate; a cell with low hydraulic permeability will be easier to freeze internally. The other factor in this process is the maximum degree of supercooling that can be sustained by the cell before internal nucleation of ice will occur. These processes can be visualized easily through use of cryomicroscopy techniques.

2.5 Freezing Damage

The freezing process is often associated with damage (Mazur, 1966; Meryman, 1966; Mazur, 1970; Finkle, 1971; Meryman, 1971; Partman, 1975; Burke et al., 1976; Farrant, 1977; Brown, 1979). After a freeze/thaw cycle, many food products change their characteristics. While usually termed freezing damage, it should be realized that the changes seen may occur either during the freezing or the thawing process. Sources of damage are thought to be associated with ice formation, either directly (e.g., mechanical effects) or indirectly (e.g., changes in solute concentration in the unfrozen phase); through migration of water from cell interior to cell exterior, producing cell shrinkage and membrane damage; through changes in gas solubility; and by phase transformation in nonaqueous membrane components (e.g., lipids). Time of exposure to high solute concentrations during the freezing or thawing process can be significant. In many systems, reaction rates as a function of temperature go through a maximum at some temperature below the initial freezing point. This is a consequence of opposing factors: lowered temperature, which decreases rates, and increasing solute concentration, which may increase rates (Fennema, 1975). If a maximum is seen in the rate-temperature function, it is often around −6°C. Freezing or thawing processes in which the system stays around this temperature for an extended period often show significant solute-induced freezing damage.

2.6 Changes during Frozen Storage

Up to this point we have been discussing the effects of the freezing process. After freezing, but before thawing, products are maintained in frozen storage; thus it is also important to consider the effects of time on the structures and characteristics of frozen products. Consider first the ice crystals. During frozen storage, while the amount of ice in a system will generally remain constant for any given temperature (unless the initial freezing conditions resulted in an amount of ice significantly less than that required by the thermodynamic characteristics of the system), the number of ice crystals will reduce and the average size will increase (Blanshard and Franks, 1987; Reid, 1994). This is a natural consequence of the surface energy between ice and unfrozen matrix and also of the need for a nucleus for growth. Both at constant temperature and as a result of fluctuating temperature there is a trend toward reduced surface area. Fluctuations in temperature will reduce the relative size of small crystals more than that of large crystals as temperature rises, and on the cooling cycle crystals with larger cross sections are more likely to capture the water molecules transferring back to the solid phase. The growth in size of ice crystals can influence the damage during frozen storage and therefore lead to loss in quality. In addition to changes in the size and number of ice crystals, storage conditions can markedly influence the locations of water within a product. The existence of temperature gradients within a product during frozen storage may result in moisture migration, relocating the water within the product. This is a consequence of the temperature dependence of water vapor pressure. Water vapor will tend to transfer from a region of higher vapor pressure to a region of lower vapor pressure. Since temperature fluctuations cannot be avoided in storage, temperature gradients will exist within products. These gradients will reverse in direction in a cyclical pattern. This does not mean, however, that the moisture transfer also reverses; especially at the surface of the product and of internal cavities, when the void volume is colder than the solid volume, moisture easily transfers in. It is more difficult for the moisture to transfer back to the initial location when the void volume is warmer than the solid volume for entropic reasons, and so there is an overall tendency for moisture to move toward surfaces and to leave the denser regions of the product. Should there be void space around product in a package, moisture will transfer into this space and tend to accumulate on the surface of the product and the internal surface of the package.

2.7 Control Factors That Influence Freezing Systems

As indicated previously, the freezing process is influenced by rate of heat removal and by the characteristics of the sample. In practical terms this can be seen in the differences of the freezing behavior of different types of freezers and products. Heat transfer rates are influenced in part by the heat transfer coefficient at the surface of the product, which in some part depends on the thermal contact between the product and the cooling medium. The ability of the cooling medium to carry heat is also important. Hence a boiling cryogen can remove heat rapidly because of both its low temperature and its high heat capacity (a consequence of the latent heat associated with boiling). A cold liquid, being dense and having a fairly high volumetric heat capacity, can also remove heat rapidly. Air has a low heat volumetric capacity; even at a low temperature, heat can be removed rapidly only if a large volume of cold air comes into contact with the surface of the product. The cooling medium can therefore affect the rate of freezing. Product packaging can also influence the effective heat transfer coefficient between cooling medium and product. Product size also enters into the picture. No matter how effective the surface heat transfer, a larger product will take much longer to freeze. This is a function of the surface-to-volume ratio. The heat content is a function of the volume, whereas heat removal is a function of the surface area. Effectively, the freezing rate of a small product is controlled by the external freezing conditions, whereas that of a large product is controlled more by the internal characteristics of the product.

2.8 Factors That Influence Moisture Migration

Moisture migration within a product can occur through two primary mechanisms. During the freezing process, supercooling of cell contents can lead to moisture movement through an osmotic mechanism, as already discussed. During storage, the driving force for moisture migration becomes temperature difference, which translates as vapor-pressure difference. Moisture migrates from high- to low-vapor-pressure regions, that is, from high- to low-temperature regions. To minimize moisture migration, two strategies are particularly important: (a) There should be as little temperature fluctuation and as small internal temperature gradients as possible. At low temperatures the vapor pressure difference per degree change in temperature is also smaller, which results in a lower driving force; and (b) Include internal barriers to moisture migration within the product and within the packaging. A detailed discussion on moisture migration and its prediction can be found in chapter 5.

2.9 Thawing

Thawing is not the reverse of freezing. As a consequence of the different heat capacity, thermal conductivity, and thermal diffusivity of water compared to ice, the temperature profile of a product on thawing is not the reverse of that in freezing. This is effectively demonstrated by an experiment described by Fennema (1973). A can filled with starch gel is immersed in a medium at +80°C. A similar can is immersed in a medium at −80°C. Once equilibrated, the cans are interchanged and the rate of temperature change is followed. It takes much longer for the can at −80°C to reach +80°C than for the can at +80°C to reach −80°C. However, in the thawing process, the gel temperature rapidly rises into the −6°C region, then remains there for an extended time. Since as previously discussed this temperature region is one where accelerated change may take place, the implications of this experiment are serious in terms of product quality. Thawing protocols must therefore be designed to minimize the time spent by the product in this intermediate temperature region, since significant quality loss can be caused by too long a thawing time.

2.10 Conclusions

The freezing process is complex, and there are consequences of both temperature and phase change. This chapter has identified some of those consequences, in particular the effect of ice formation on the progress of heat transfer during both the freezing and thawing processes. Another important effect that has been highlighted is the interaction of temperature and unfrozen phase composition influencing the temperature dependence of reaction kinetics. The need to consider both the thermodynamic characteristics of systems and the kinetic constraints to change has been emphasized.

References

Blanshard, J.M.V. and Franks, F. 1987. Ice crystallization and its control in frozen-food systems. In *Food Structure and Behavior.* (Blanshard, J.M.V. and Lillford, P., eds.) pp. 51–65, Academic Press, London.

Brown, M.S. 1979. Frozen fruits and vegetables: their chemistry, physics, and cryobiology. *Adv. Food Res.* **25**:181–235.

Burke, M.J., Gusta, L.V., Quamme, H.A., Weiser, C.J., and Li, P.H. 1976. Freezing and injury in plants. *Ann. Rev. Plant Physiol.* **27**:507–528.

Charoenrein, S., Goddard, M., and Reid, D.S. 1991. Effect of solute on the nucleation and propagation of ice. In *Water Relationships in Foods: Advances in the 1980s and Trends for the 1990s.* (Levine, H. and Slade, L., eds.) pp. 191–198, Plenum Press, New York.

Chen, C.S. 1985. Thermodynamic analysis of the freezing and thawing of foods: enthalpy and apparent specific heat. *J. Food Sci.* **50**:1158–1162.

Chen, C.S. 1987. Relationship between water activity and freezing point depression of food systems. *J. Food Sci.* **52**:433–435.

Chen, C.S. 1988. Bound water and freezing point depression of concentrated orange juices. *J. Food Sci.* **53**:983–984.

Chen, C.S. and Nagy, S. 1987. Prediction and correlation of freezing point depression of aqueous solutions. *Trans. ASAE* **30**:1176–1180.

Chen, C.S., Nguyen, T.K., and Braddock, R.J. 1990. Relationship between freezing point depression and solute composition of fruit juice systems. *J. Food Sci.* **55**:566–567, 569.

Dowell, L.G. and Rinfret, A.P. 1960. Low-temperature forms of ice as studied by X-ray diffraction. *Nature* **188**:1144–1148.

Farrant, J. 1977. Water transport and cell survival in cryobiological procedures. *Phil. Trans. Royal Soc. Lond. B* **278**:191–205.

Fennema, O.R. 1973. Nature of freezing process. In *Low Temperature Preservation of Foods and Living Matter.* (Fennema, O.R., Powrie, W.D., and Marth, E.H., eds.) p. 153, Marcel Dekker, New York.

Fennema, O.R. 1975. Reaction kinetics in partially frozen aqueous solutions. In *Water Relations of Foods.* (Duckworth, R.B., ed.) pp. 539–558, Academic Press, London.

Finkle, B.J. 1971. Freezing preservation. In *Biochemistry of Fruits and Their Products,* vol. 2. (Hulme, A.C., ed.) pp. 635–685, Academic Press, New York.

Fletcher, N.H. 1970. *The Chemical Physics of Ice*. Cambridge University Press, London.

Franks, F. 1985. Complex aqueous systems at sub-zero temperatures. In *Properties of Water in Foods in Relation to Quality and Stability*. (Simatos, D. and Multon, J.L., eds.) pp, 1–24, Martinus Nijhoff, The Hague.

Heldman, D.R. 1974. Predicting the relationship between unfrozen water fraction and temperature during food freezing using freezing point depression. *Trans. ASAE* **17**:63–66.

Heldman, D.R. 1992. Food freezing. In *Handbook of Food Engineering.* (Heldman, D.R. and Lund, D.B., eds.) pp. 277–315, Marcel Dekker, New York.

Hobbs, P.V. 1974. *Ice Physics.* Oxford University Press, London.

Kerr W.L., Ju, J., and Reid, D.S. 1993. Enthalpy of frozen foods determined by differential compensated calorimetry. *J. Food Sci.* **58**:675–679.

Labuza, T.P. 1980. The effect of water activity on reaction kinetics of food deterioration. *Food Technol.* **34**(4):36–41, 59.

Lerici, C.R., Piva, M., and Dalla Rosa, M. 1983. Water activity and freezing point depression of aqueous solutions and liquid foods. *J. Food Sci.* **48**:1667–1669.

Leung, H.K. 1987. Influence of water activity on chemical reactivity. In *Water Activity: Theory and Applications to Food.* (Rockland, L.B. and Beuchat, L.R., eds.) pp. 27–54, Marcel Dekker, New York.

Maki, L.R., Galyan, E.L., Chang-Chien, M., and Caldwell, D.R. 1974. Ice nucleation induced by *Pseudomonas syringae. Appl. Microbiol.* **28**:456–459.

Mazur, P. 1966. Physical and chemical basis of injury in single-celled microorganisms subjected to freezing and thawing. In *Cryobiology.* (Meryman, H.T., ed.) pp. 214–316, Academic Press, New York.

Mazur, P. 1970. Cryobiology: The freezing of biological systems. *Science* **168**:939–949.

Meryman, H.T. 1966. Review of biological freezing. In *Cryobiology.* (Meryman, H.T., ed.) pp. 1–114, Academic Press, New York.

Meryman, H.T. 1971. Osmotic stress as a mechanism of freezing injury. *Cryobiol.* **8**:489–500.

Partmann, W. 1975. The effects of freezing and thawing on food quality. In *Water Relations of Foods.* (Duckworth, R.B., ed.) pp. 505–537, Academic Press, London.

Rahman, M.S. 1994. The accuracy of prediction of the freezing point of meat from general models. *J. Food Eng.* **21**:127–136.

Reid, D.S. 1983. Fundamental physicochemical aspects of freezing. *Food Technol.* **37**:110–115.

Reid, D.S. 1990. Optimizing the quality of frozen foods. *Food Technol.* **44**(7):78–82.

Reid, D.S. 1994. Basic physical phenomena in the freezing and thawing of plant and animal tissue. In *Frozen Food Technology.* (Mallet, C.P., ed.) pp 1–19, Blackie Academic & Professional, Glasgow.

Reid, D.S., Hsu J., and Kerr, W.L. 1993. Calorimetry. In *The Glassy State in Foods*. (Blanshard, J.M.V. and Lillford, P.J., eds.) Nottingham University Press, Leics, UK.

Riedel, L. 1978. Eine formel zur berechnung der enthalpie fettarmer lebensmitteln in abhangigkeit von wassergehalt und temperature. *Chem. Mikrobiol. Technol. Lebensm.* **5**:129.

Succar, J. and Hayakawa, K. 1983. Empirical formulae for predicting thermal physical properties at freezing or defrosting temperatures. *Food Sci. Technol.* **16**(6):326.

Sugisaki, M., Suga, H., and Seki, S. 1968. Calorimetric study of the glassy state. IV. Heat capacities of glassy water and cubic ice. *Bull. Chem. Soc. Jpn.* **41**:2591–2599.

3

Measurement and Interpretation of the Glass Transition in Frozen Foods

H. D. Goff

3.1 Introduction

The glass transition associated with the freeze-concentrated unfrozen phase (UFP) in frozen foods has received considerable research attention in the last several years, primarily as a result of the efforts of Levine and Slade (1988a,b, 1990; Slade and Levine, 1995) to relate polymer kinetics to food physicochemical processes. The glass transition is of interest for at least three reasons. First, quality losses (e.g., enzymatic reactions and recrystallization) in frozen foods may be greatly reduced or inhibited when the UFP is in the glassy state. Second, the kinetics of reactions associated with quality losses at temperatures close to the glass transition temperature (T_g) can be described by the Williams-Landel-Ferry (WLF) model, in which reaction rates are a function of the temperature difference (ΔT) between the storage temperature (T_s) and T_g. Thus, WLF kinetics may be more applicable than Arrhenius kinetics to studying the rates of detrimental reactions, allowing one to more accurately predict the shelf life of stored frozen foods (Peleg, 1992). Third, knowledge of the influence of environmental and indigenous factors (i.e., temperature, freezing rate, and composition) on the glass transition may make possible the formulation, processing, or storage of frozen foods in a manner that will enhance shelf life. The basic aspects of food freezing have been discussed in the previous chapter, and in detail elsewhere (Goff, 1992; Sahagian and Goff, 1996). This chapter will focus on the physical chemistry and relevant definitions associated with the glass transition and the glassy state, the formation of the glassy state in the UFP (based on the discussion in Sahagian and Goff, 1996), measurement of T_g using several techniques, and the interpretation of the measurements.

3.2 Formation of the Glassy Phase in Frozen Foods

During the freezing of foods, ice is formed as pure water goes through the two-step (nucleation and propagation) crystallization process. As temperature decreases and water is removed from a food in the form of ice, the solutes present in the UFP are freeze concentrated. For each ice/UFP ratio, an equilibrium freezing temperature which is a function of the solute concentration, exists. This equilibrium thermodynamic process can be modeled on a phase diagram as an equilibrium freezing (liquidus) curve (fig. 3.1), which extends from the melting temperature (T_m) of pure water (0°C) to the eutectic temperature (T_e) of the solute, the point at which the solute has been freeze concentrated to its saturation concentration. As temperature is lowered below T_e, it is highly unlikely that solute will crystallize because of high viscosity from concentration of solute and low temperature, so that freeze concentration proceeds beyond T_e in a nonequilibrium state. The highly concentrated UFP can then go through a viscous liquid/glass state transition, driven by the reduction in molecular motion and diffusion kinetics as a result of both the very high concentration and low temperature. A glass is defined as a nonequilibrium, metastable, amorphous, disordered solid of extremely high viscosity (i.e., 10^{10} to 10^{14} Pa s), also a function of temperature and concentration. The glass transition curve extends from the glass transition temperature (T_g) of pure water (−134°C) to the T_g of pure solute. The equilibrium phase diagram and the glass transition can be modeled together on a supplemented state diagram (Roos and Karel, 1991a). The supplementary state diagram showing the solid-liquid coexistence boundaries and glass transition profile for a binary sucrose-water system is shown in figure 3.1. Below and to the right of the glass transition line, the solution is in the amorphous glass state with or without ice present depending on temperature and freezing path followed, whereas above and to the left of the glass transition line, the solution is in the liquid state with or without ice depending on temperature.

As an example, assume a sucrose solution with an initial concentration of 20% at room temperature (point *A*). The initial T_g of this solution at room temperature before phase separation is marked as point *B* (if the solution could be undercooled to this temperature without ice formation). However, upon slow cooling of the system somewhat below its equilibrium freezing point (due to undercooling), nucleation and subsequent crystallization begin at point *C* and initiate the freeze-concentration process, removing water in its pure form as ice. As ice crystallization proceeds, the continual increase in solute concentration (removal of water) further depresses the equilibrium freezing point of the UFP in a manner following the liquidus curve (shown as path *C* in fig. 3.1) while the T_g of the UFP moves up the glass transition line (path *B*; due to increased concentration) with a rapid increase in viscosity in a non-Arrhenius manner, particularly in late stages of the freezing process (Roos and Karel, 1991b,c). Cocrystallization of solute at T_e is unlikely, and thus freeze concentration continues past T_e into a nonequilibrium state since the solute becomes supersaturated. When a critical, solute-dependent concentration is reached, the unfrozen liquid exhibits very resisted mobility and the physical state of the UFP changes from a viscoelastic liquid to a brittle, amorphous solid glass (Ablett et al., 1992; Roos and Karel, 1991b,c; Sahagian and Goff, 1996).

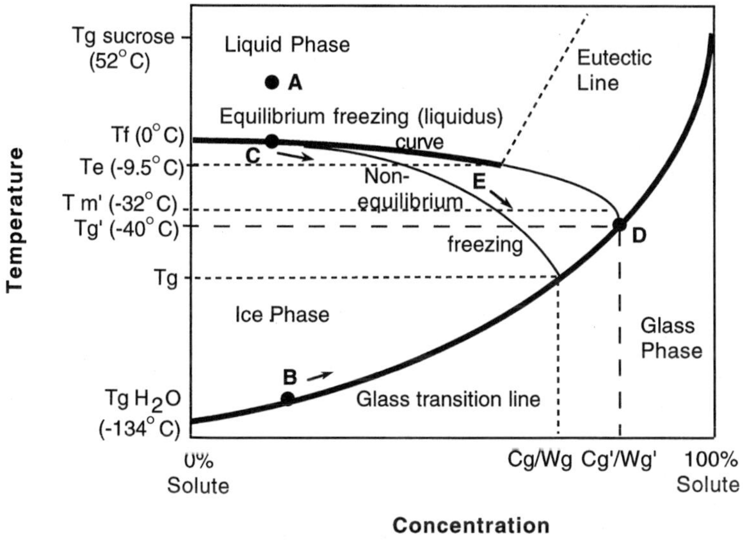

Figure 3.1. A schematic temperature–concentration state diagram for an aqueous carbohydrate solution, showing the glass transition curve, defined by viscosity, which extends from the T_g (glass transition temperature) of pure water (−134°C) to the T_g of pure solute, the equilibrium freezing (liquidus) curve, which extends from the T_m (melting temperature) of pure water (0°C) to the eutectic temperature (T_e) of the solute, and the theoretical eutectic line. Point T_g' represents the glass transition temperature of the maximally freeze-concentrated solution, and W_g' represents the amount of unfrozen water (100% solute, C_g') that becomes trapped in the glass. Point T_m' reflects the temperature needed to cause mechanical collapse and ice melting to occur above the glass transition. Points T_g and W_g represent an example of a temperature concentration relationship in a glass formed as a result of less than maximal ice formation (following the nonequilibrium freezing line). Details of specific lettered points are provided in the text.

The intersection of the nonequilibrium extension of the liquidus curve, beyond T_e, and the glass transition curve, point *D* in figure 3.1, represents the solute-specific, maximally freeze concentrated T_g of the frozen system, denoted T_g', where ice formation ceases within the timescale of normal measurement (Levine and Slade, 1988a,b; 1990). The corresponding maximum concentrations of water and sucrose "trapped" within the glass at T_g' and unable to crystallize are denoted W_g' and C_g', respectively (Hatley et al., 1991). It is worth noting that this unfrozen water is not bound in an "energetic" sense, rather it is unable to freeze within practical time frames (Levine and Slade, 1988a,b, 1990; Franks, 1985). At T_g', the supersaturated solute takes on solid properties because of reduced molecular motion, which is responsible for the tremendous reduction in translational, not rotational, mobility (Roos and Karel, 1991b,c; Ablett et al., 1992; Levine and Slade, 1990; Slade and Levine, 1995). It is this intrinsic slowness of molecular reorganization below T_g' that the food technologist seeks to create within the concentrated phase surrounding constituents of food materials. However, warming from the glassy state to temperatures above T_g' results in a tremendous increase in diffusion, not only from the effects of the amorphous-to-viscous-liquid transition but also from increased dilution as melting of small ice crystals occurs almost simultaneously ($T_g' \leq T_m'$). However, Shalaev and Franks (1995) have suggested that the onset of melting temperature of the maximally freeze-concentrated glass, T_m', may be as much as several degrees above T_g' because of differences in viscosity attributable to molecular motion compared to macromolecular (mechanical) flow. The timescale of molecular rearrangement continually changes as T_g is approached during freezing, so that food technologists can also gain some enhanced stability at temperatures above T_g' by minimizing ΔT between the storage temperature and T_g', either by reduced storage temperatures or enhanced T_g' through freezing methods or formulation. Hence, knowledge of the glass transition provides a clear indication of molecular diffusion and reactivity and therefore shelf stability.

Despite the thermodynamic driving force to achieve the unfrozen water content corresponding to W_g', one must also consider the large kinetic factors that overtake the freezing process. At subzero temperatures, the formation of an amorphous state is time dependent since the limiting factor of the process (water removal in the form of ice) becomes more difficult as concentration increases. The exponential effect of viscosity on mass transfer properties acts as the limiting factor for growth. In addition, under conditions where heat removal is rapid, a high level of undercooling at the interface will only add to a further decrease in propagation rate. The net result is that freezing becomes progressively slower as ice crystallization is hindered, and consequently more time is required for lattice growth at each temperature. Therefore the kinetic restriction imposed on the system can lead to a situation in which nonequilibrium freezing, resulting in a partial dilute glass, can occur (Roos and Karel, 1991b,c). The typical pathway a system may follow during nonequilibrium freezing is shown in figure 3.1 as the line leading to lower T_g (path E) than T_g' with a corresponding lower sucrose concentration in the glass (C_g) and higher water content in the glass (W_g) caused by excess undercooled water plasticized within the glass. This is often referred to as a dilute glass. The magnitude of deviation from the equilibrium curve, and hence the actual path followed, may be regarded as a function of the degree of departure from equilibrium. Systems possessing this undesirable structure may undergo various relaxation-recrystallization mechanisms to maximally freeze concentrate and minimize the unfrozen water content. As a result, during warming, systems formed under these conditions may lead to one or more low temperature transitions, followed by an exothermic devitrification peak caused by crystallization of immobilized water, and finally the onset of ice melting, T_m (Roos an Karel, 1991b,c).

3.3 Measurement of the Glass Transition

Since the glass transition manifests itself as a change in thermal and mechanical properties, thermal analysis techniques are most commonly used to study this phenomenon (Harwalker and Ma, 1990). In particular, measurement of the change in heat capacity (C_p) using differential scanning calorimetry (DSC) has become the established method, largely because of its ease of operation and widespread availability (Ma et al., 1990). Recent advances in modulated (M)DSC, where sinusoidal, oscillating temperature profiles are overlaid onto linear heating gradients, may prove to be more useful than conventional DSC in defining glass transitions. Other techniques, namely dynamic mechanical thermal analysis (DMA or DMTA) (Roos and Karel, 1991b,c; MacInnes, 1993; Kaliachewsky et al., 1992, 1993) and thermomechanical analysis (TMA) (Sahagian and Goff, 1994, 1995a,b; Carrington et al., 1994; Le Meste and Huang, 1992; Maurice et al., 1991) have been used to characterize phase behavior in frozen systems. Because T_g is also associated with a sudden change in the translational diffusion coefficient of molecules or segments of polymers and is both a time- and frequency-dependent phenomena (occurs over a temperature range), dielectric analysis or spectrometry (DEA) and other spectroscopic methods such as nuclear magnetic resonance (NMR) spectroscopy (Sahagian and Goff, 1995a; Chinachoti, 1993; Rubin et al., 1990; Ablett and Lillford, 1991) and electron spin resonance (ESR) spectroscopy (Roozen and Hemminga, 1990; Roozen et al., 1991) are also useful since they allow for the measurement of relaxation times more applicable to time/frequency-dependent materials (Hemminga et al., 1993). However, as with most investigations, the effective approach when studying glass transitions and glasses in food systems is the combination of various thermal analysis techniques (DSC, DEA, TMA, DMA), each complementing the others.

3.3.1 Calorimetric Measurements

Differential Scanning Calorimetry

Differential scanning calorimetry (DSC) is a widely used thermal analysis technique in which the difference in energy inputs into a substance and a reference material is measured as a function of temperature while the substance and reference material are subjected to a controlled temperature program. The substance is usually contained in small sample pans (10–20 μl), and the reference is usually an empty sample pan (Ma et al., 1990). DSC has been used extensively in glass transition research because changes in heat capacity are measured easily and precisely, advanced instrumentation is accurate and available in most research laboratories, and computer analysis of data is well advanced and user-friendly. Examples of recent publications involving DSC research into glass transitions in frozen foods include Ma et al. (1990), Blond and Simatos (1991), Hatley et al. (1991), Izzard et al. (1991), Roos and Karel (1991b,c), Simatos and Blond (1991), Ablett et al. (1992), Hatley and Mant (1993), Reid et al. (1993), Goff (1994, 1995), Sahagian and Goff (1994, 1995a), and Goff and Sahagian (1996).

Second-order (glass) transitions are detected by endothermic step changes in heat flow (baseline shifts) representing discontinuous changes to C_p. To be a true glass transition, it should be possible to scan in and out of the transition (between the vitreous and viscous liquid states) and detect crystallization upon warming above T_g (Wolanczyk, 1989). The transition can be characterized by the onset temperature, the half-height temperature, or the inflection point (peak maximum on the first derivative curve), and all are based on straight-line extrapolations of the baseline before and after the transition. A DSC heat flow versus temperature plot for both rapidly frozen and annealed 40% sucrose solutions is shown in figure 3.2. From such analyses, two distinct transitions are always evident. For discussion purposes, these two transitions are labeled in figure 3.2 as the temperature at transition 1, the lower temperature transition (T_{Tr1}), and the temperature at transition 2, the higher temperature transition (T_{Tr2}), similar to the method adopted by Sahagian and Goff (1994, 1995a). The interpretation of these two transitions will only be presented briefly here and is expanded upon in detail elsewhere (Slade and Levine, 1995; Sahagian and Goff, 1996).

At present, there are three major interpretations regarding the origin of each transition and the assignment of T_g': (1) T_g' is represented by the major transition at T_{Tr2} while the lower temperature T_g at T_{Tr1} is a result of incomplete freeze concentration during freezing (this implies devitrification of the dilute glass if not maximally freeze concentrated with the original freezing protocol and recrystallization to T_g' in the same experiment) (Franks, 1985; Levine and Slade, 1988a,b, 1990; Hatley et al., 1991; Hatley and Mant, 1993); (2) T_g' occurs at the lower transition (T_{Tr1}) while the warmer transition is not a glass transition but rather the onset of ice melting, T_m' (first-order transition), initiated soon after T_g', the separation being attributable to the differences in viscosity required for molecular relaxation/diffusion versus mechanical collapse and flow (this must also imply adequate care to ensure that freeze concentration is maximal) (Izzard et al., 1991; Roos and Karel, 1991b,c; Ablett et al., 1992; Roos, 1992; MacInnes, 1993; Sahagian and Goff, 1994; Shalaev and Franks, 1995); and (3) T_g' is close to T_{Tr2} (Hatley and Mant, 1993) or somewhere between the two transitions (Blond and Simatos, 1991; Simatos and Blond, 1991, 1993), and the complex behavior is partially the result of a relaxation process manifested as an excess enthalpy overshoot close to T_g'. Adding to the difficulty is that some reported values are the midpoint of the transition (Levine and Slade, 1988b, 1990; Hatley and Mant, 1993), whereas others represent the onset (Roos and Karel, 1991b,c; Sahagian and Goff, 1994, 1995a).

Based on the original description of T_g' (Levine and Slade, 1988b, 1990), consistent with the state diagram, the observed T_g will be equal to T_g' if the system is maximally freeze concentrated. However, this stresses the importance of the an-

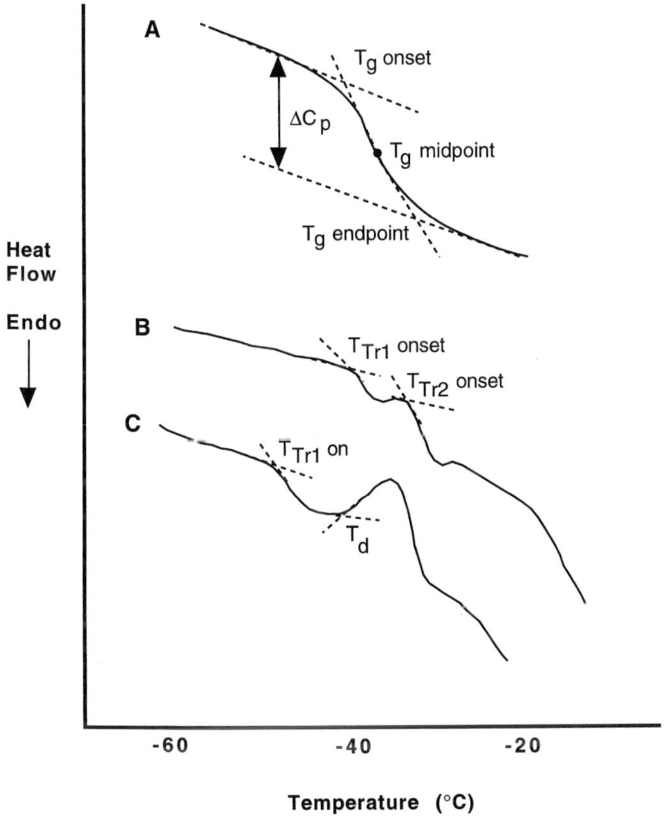

Figure 3.2. A schematic example of a temperature–heat flow profile from DSC analysis of a frozen sucrose solution. Curve *A* (not to temperature scale) illustrates the baseline shift occurring during a glass transition, the measurement of temperature (T_g) of onset, midpoint, and endpoint of the transition, and the ΔC_p. Curve *B* illustrates a typical profile of a maximally freeze-concentrated system achieved through annealing, showing the typical biphasic behavior, the measurement of the onset temperatures of the first (T_{Tr1}) and second (T_{Tr2}) transitions, and the typically larger ΔC_p associated with the second transition. Curve *C* illustrates a typical profile of a more dilute glass formed during rapid freezing, showing a lower T_{Tr1} than the annealed system, followed by a devitrification exotherm and the measurement of its onset (T_d).

nealing process to ensure that maximal freeze concentration has been reached before assignment of a T_g'. However, despite annealing, the two transitions remain evident and do not coincide. The true explanation for this biphasic behavior remains unclear. MacInnes (1993) has suggested that the two transitions would coincide if measured at very low frequency because of their frequency-dependent nature. Also, it should be pointed out that a measurable heat change at T_g may be problematic for some glass-forming materials since ΔC_p can be small, which depends largely on the type of glass, that is, fragile or strong as described by Angell et al. (1994). The relationship between ΔC_p and accompanying T_g is in need of further fundamental understanding (Slade and Levine, 1995).

Modulated (M)DSC

MDSC is a technique in which the temperature during the temperature scan is modulated by a controllable period and amplitude. Hence instantaneous fast heating rates for good sensitivity and slow underlying heating rates for good resolution are obtainable. Separation of the resultant experimental heat flow during this cyclic treatment not only provides the total heat flow signal available from conventional DSC but also separates the total heat flow into its reversing (in-phase) and nonreversing (out-of-phase) components. Typical glass transitions in polymers are reversing phenomena, whereas crystallizations are nonreversing (Sauerbrunn et al., 1992). During warming of frozen solutions, the following thermal events may be evident as discussed previously, depending on their thermal history: a glass transition, a devitrification exotherm, an endothermic event representing an enthalpic relaxation or overshoot after the glass transition (Simatos and Blond, 1991; Sahagian and Goff, 1994), and the melting endotherm. Two transitions representing baseline shifts (changes in heat capacity) always appear despite extensive annealing. Many of these events are overlapping and difficult to separate. MDSC may be the technique to separate and understand these phenomena. Figure 3.3 illustrates the thermal behavior of 20% sucrose solutions following slow freezing, showing the total heat flow, reversing heat flow, and nonreversing heat flow as a function of temperature. It is evident that the devitrification exotherm is present in the nonreversing heat flow curve, while portions of both Tr1 and Tr2 are evident in both the reversing and nonreversing heat flow curves. Research is now underway to determine the significance of variables such as background heating rate, period, and amplitude before much interpretation of this technique is possible. Although there are few publications involving MDSC research directly related to glass transitions in frozen foods, relevant discussion of this technique can be found in Goff and Sahagian (1996).

3.3.2 Mechanical Measurements

Thermomechanical Analysis

Thermomechanical analysis (TMA) measures the mechanical deformation (expansion or contraction) of a solid sample under a constant, unidirectional stress (compression, tension, bending, or twisting) during warming or cooling. The sample can be prepared (frozen) between two parallel plates and then transferred to the TMA instrument. A quartz probe is used to either contact or apply a force

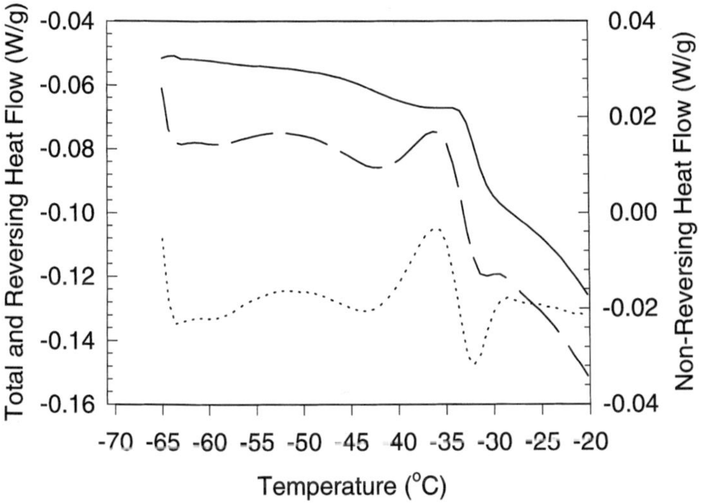

Figure 3.3. Modulated DSC scans of the thermal behavior of 20% sucrose solutions following slow freezing (2°C min^{-1}), showing the total heat flow (- - -), reversing heat flow (—) and nonreversing heat flow (. . . .) as a function of temperature. (From Goff and Sahagian, 1996.)

to the sample, and any movement of the quartz probe as a result of expansion or contraction of the sample is detected by an LVDT and transmitted as an electrical signal (Ma et al., 1990). A glass transition in TMA results from a change to the thermal expansion coefficient. The expansion change at T_{Tr1} is easily measured and seen as an increase in slope on moving from the glass to the viscous liquid state (fig. 3.4), from which the inflection point can be used to determine T_g (Maurice et al., 1991; Le Meste and Huang, 1992; Goff et al., 1993; Carrington et al., 1994; Sahagian and Goff, 1994, 1995a; Goff, 1995). The T_{Tr2} is seen as a subsequent decrease in dimension change as the system begins to flow. Beyond T_{Tr2}, a further drastic change in the slope corresponds to complete solution collapse as the onset of the major ice-melting peak occurs at T_m. In rapidly frozen samples containing a very unstable, dilute glass, sample collapse occurs readily after T_g, making further interpretation difficult (fig. 3.4). TMA also has been used to measure flow properties below and above T_g through stress relaxation measurements (Sahagian and Goff, 1995b).

Dynamic Mechanical Thermal Analysis

Dynamic mechanical thermal analysis (DMA, DMTA) is a technique for measuring the mechanical properties (storage modulus, E' or G', loss modulus, E'' or G'', or tan δ) of materials as they are deformed under periodic stress as a function

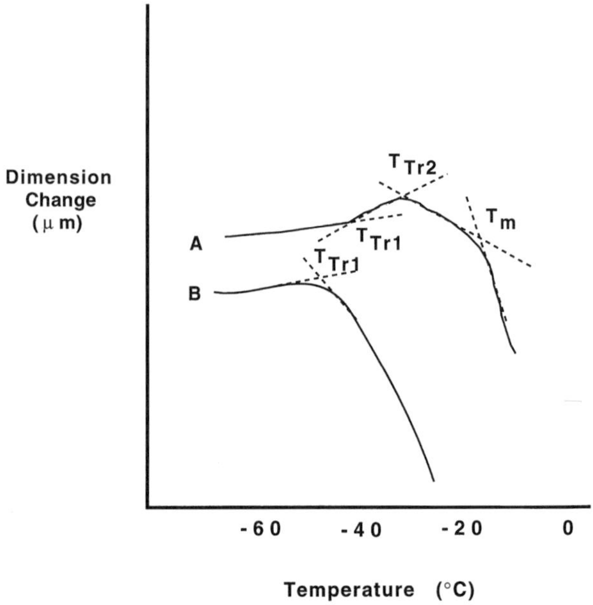

Figure 3.4. A schematic example of a temperature–dimension change profile from thermomechanical (TMA) analysis of a frozen sucrose solution warmed from -80°C at a rate of 2°C min^{-1} under a stress of 140 Pa. Curve *A* illustrates a typical profile of a maximally freeze-concentrated system achieved through annealing, showing the typical biphasic behavior and the measurement of the onset temperatures of the first (T_{Tr1}) and second (T_{Tr2}) transitions, as well as the point at which the sample collapses (T_{m}, melting). Curve *B* illustrates a typical profile of a more dilute glass formed during rapid freezing, showing a lower T_{Tr1} than the annealed system, followed immediately by complete collapse of the system.

of frequency and temperature. Materials can be characterized as solids, demonstrating elastic response such that removal of an applied stress results in a complete recovery to zero strain; liquids, demonstrating viscous response such that removal of an applied stress results in zero recovery; or viscoelastic materials, demonstrating behavior between these two ideals and typical of most foodstuffs. For a linear, viscoelastic material, the strain response lags behind the stress by some phase angle, δ. The elastic component (storage modulus, E' or G') can be determined as the stress amplitude divided by the strain amplitude multiplied by cos δ, while the viscous component (loss modulus, E'' or G'') can be determined as the stress amplitude divided by the strain amplitude multiplied by sin δ. The ratio of energy lost to energy recovered (viscous to elastic components) is then determined as tan δ (Kaliachewsky et al., 1993). The glass transition is a structural mechanical event defined by viscosity. Therefore, DMTA is highly sensitive

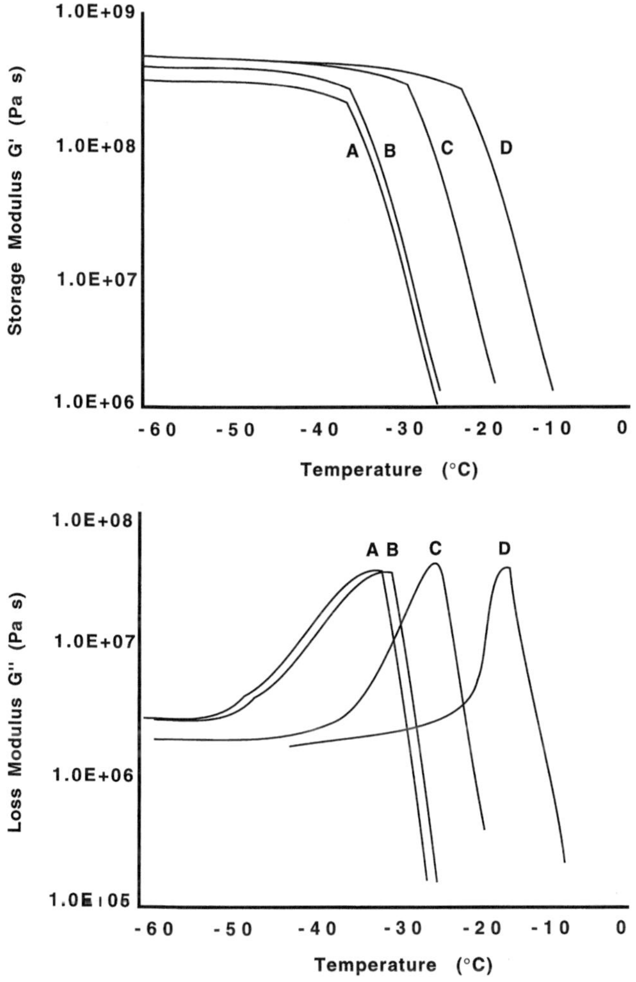

Figure 3.5. Temperature profiles of storage modulus and loss modulus determined by DMTA of 50% T.S. solutions comprised sucrose and dextran (*A* = 50% sucrose; *B* = 40% sucrose, 10% dextran; *C* = 25% sucrose, 25% dextran; *D* = 50% dextran) following slow freezing and annealing. (Unpublished results courtesy of Karen Graham, ENSBANA, France, and the University of Guelph, Canada.)

for detecting second-order transitions caused by the large changes in viscoelastic properties occurring in this temperature region (Ma et al., 1990). Most foods demonstrate a sharp decrease in storage modulus and peak in loss modulus and tan δ as a sample is warmed through T_g (fig. 3.5). DMTA has been reported for glass transition measurements in frozen systems by Kaliachewsky et al. (1992, 1993), MacInnes (1993), and Blond (1994).

3.3.3 Molecular Mobility Measurements

Nuclear Magnetic Resonance Spectroscopy

Nuclear magnetic resonance spectroscopy (NMR) is based on the measurement of resonant radio-frequency (4–600 MHz) energy absorption by nonzero nuclear spins in the presence of an externally applied constant magnetic field. Only a charged atomic nucleus (e.g., ^1H, ^2H, or ^{17}O) possesses a nonzero spin value that can interact with the magnetic field to become either aligned with or opposed to the field. After a radio-frequency pulse generated by a rotating magnetic field has been applied—causing the spins of all nuclei to become aligned—and then shut off, the spin system returns to its equilibrium position through two relaxation processes: longitudinal or spin-lattice (t_1) and transverse or spin-spin relaxation (t_2). After several pulses, the relaxation behavior from all relaxing nuclei can be plotted as voltage intensity or magnetization amplitude against time, called the free induction decay (FID). This is usually then converted by Fourier transform to the frequency domain, resulting in a plot of amplitude versus frequency (Richardson Schmidt, 1990). While the glass transition is defined by viscosity at the macromolecular scale, this is a manifestation of mobility and diffusion at the molecular scale. Thus, NMR is suited ideally for probing directly the molecular mobility in samples either in the glass or viscous liquid states. Figure 3.6 illustrates the change in the slope of t_2 as a function of temperature that occurs when a material passes through T_g. The temperatures corresponding to the change in slope match well to the transition temperatures obtained by DSC. In this example, guar gum was seen to reduce the t_2 at comparable temperatures, unlike DSC analyses, indicating the importance of synthesizing results from several methods in interpreting low-temperature behavior. Recent NMR papers directly related to glass transitions in frozen-food systems include Ablett and Lillford (1991), Kaliachewsky et al. (1992, 1993), Rubin et al. (1990), and Sahagian and Goff (1994, 1995a). The problem with NMR technology is the difficulty of achieving accurate temperature programming capability within existing NMR installations. NMR is also a sophisticated and expensive technique compared to either the calorimetric or mechanical instrumentation described previously and therefore primarily limited to academic or large research institutions.

Electron Spin Resonance Spectroscopy

Electron spin resonance spectroscopy (ESR) is based on the adsorption of microwave radiation by an unpaired electron (whose resonance frequency, 9,500 MHz, is in the microwave region) when it is exposed to a strong magnetic field. Only unpaired electrons will produce an ESR spectra. In biological applications, a spin-label reagent, which contains an odd electron and also reacts with functional groups in organic molecules, is used to provide information, such as struc-

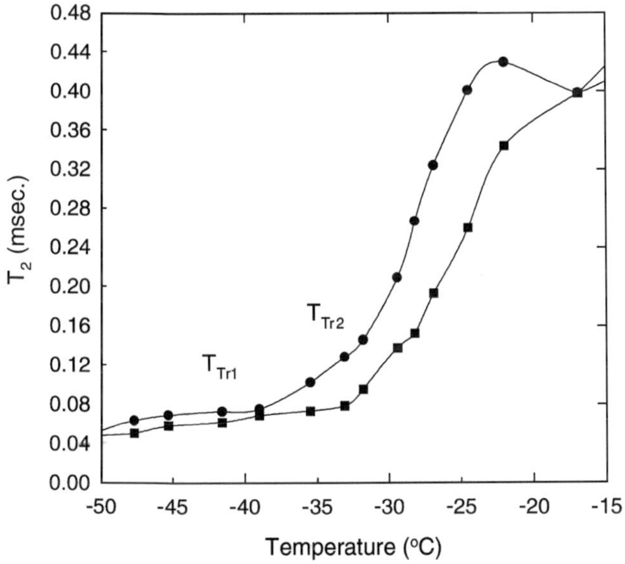

Figure 3.6. NMR transverse relaxation time (t_2) at various subzero temperatures for slowly frozen (approximately 5°C per min) 20% (w/w) sucrose solutions (•) alone and in the presence of 0.5% (w/w) guar gum (▪). The transitions were determined by extrapolating the linear portions of the curves to an intersection point. (From Sahagian and Goff, 1995a.)

tural features or phase changes, about the molecular environment in which the reagent is reacted (Hemminga et al., 1993). ESR has been applied to a study of frozen glass transitions, with the resulting increase in rotational mobility of spin probes around the glass transition temperature on rewarming. ESR also shows that the spin probes in concentrated sucrose-water mixtures above and below the glass transition temperature are present in cavities (Roozen and Hemminga, 1990). In addition, the temperature at which an increase in mobility was observed agreed reasonably well with the glass transition temperature as determined by DSC (Roozen et al., 1991).

Dielectric Analysis

Dielectric analysis (DEA) utilizes a periodic electrical field (sinusoidal voltage) to determine the capacitive and conductive nature of materials (ability to store and transfer electrical charge) and thus molecular relaxations by measuring the resulting current. Permettivity or dielectric constant, e′, is proportional to capacitance and represents the amount of alignment of dipoles to the electric field. The loss factor, e″, is proportional to conductance and measures the amount of energy required to align dipoles or move ions. Temperature and frequency sweeps

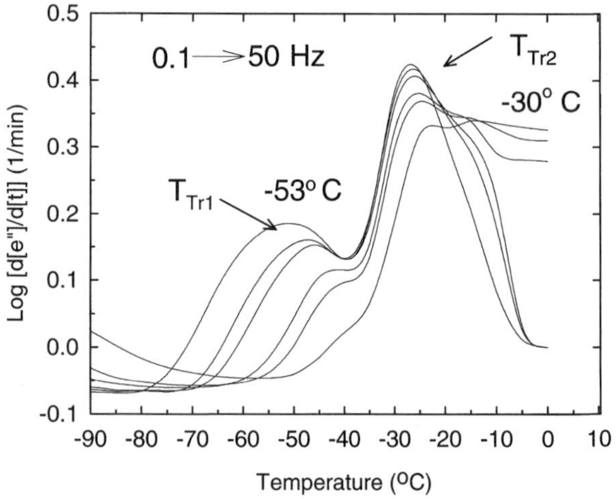

Figure 3.7. First derivative of the loss factor (e″) as a function of temperature at several frequencies from 0.1 to 50 Hz as determined by dielectric analysis (DEA) for a 20% aqueous sucrose solution. Differences in frequency dependence of the two transitions, T_{Tr1} and T_{Tr2}, can be seen. (From Goff, 1995.)

can be conducted simultaneously in the multiplexing mode in advanced instrumentation. First-derivative plots are useful in determining temperature and frequency dependence of each transition. A typical first-derivative DEA plot of the loss factor (e″) as a function temperature at several frequencies (0.1–50 Hz) for a 20% aqueous sucrose solution is shown in figure 3.7 (Goff, 1995). Peaks in the derivative plots represent temperatures at which significant changes in electrical properties of the sample occurred and match fairly well to data obtained by DSC. Much information can also be gained from analyses of the frequency dependence of the various transitions (MacInnes, 1993). Although DEA should provide useful information regarding glass transitions in frozen-food systems, there has been considerably less use of this technique than any of those previously mentioned.

3.3.4 Microscopy

Microscopy techniques, both optical and electron, have been used in exploring physical changes in frozen systems related to both glass and viscous liquid states and transitions in and out of the glass. Although not suited for a direct measurement of transition temperatures, a study of the resultant microstructure of frozen samples based on composition and preparation can provide useful information about the physical state of samples. Optical microscopy is mostly used to either observe crystallization events directly on a cold stage as a function of temperature or to determine the relative amounts of ice and glass formed within a sample (Ablett et al., 1992). Cryoscanning electron microscopy (cryo-SEM) has also been used to determine the relative amounts of recrystallization in samples prepared using different freezing rates and stored at a variety of temperatures (Carrington et al., 1996). Examples of cryo-SEM micrographs of glass formation within frozen 30% fructose and 1% carboxymethyl cellulose (CMC) solutions after ultrarapid (quench in LN_2) and slow freezing are shown in figure 3.8. There was little detectable change in microstructure when rapidly frozen samples were stored at −75°C for 2 wk. However, ice crystals in rapidly frozen samples after storage at −25°C for 2 wk were similar in size to those obtained initially (no storage) by freezing solutions slowly to −25°C. Thus, it was apparent from this work that considerable unfrozen water was trapped in the glass during freezing, and recrystallization of that water following devitrification of the glass resulted in a similar microstructure to slow freezing. Rapid freezing is desirable during food freezing to maximize nucleation, but it appears that rapid freezing to temperatures at or slightly above T_g' are critical to ensure maximal freeze concentration and to reduce possibilities of extensive recrystallization (Carrington et al., 1996). This provides a good example of the use of cryo-SEM for a study of glass transitions because it illustrated the changes in microstructure (in this case ice-crystal size distribution) resulting from storage above and below the glass transition and the effects of dilute versus concentrated glasses on stability.

3.4 Interpretation of the Glass Transition

3.4.1 T_g, T_g', and T_m' in Model Systems

The parameters of T_g and T_g' have distinctly different meanings but have often been confused; there exists a wide variation in published T_g' values (Slade and Levine, 1995), mainly as a result of different interpretations of the same DSC-measured transitions of warming curves of many low–molecular weight (MW) carbohydrates (Reid et al., 1994). T_g' represents the T_g of a maximally freeze-concentrated system, while T_m' represents the onset of ice melting from a maximally freeze-concentrated system, which may coincide with (MacInnes, 1993) or be a few degrees above (Shalaev and Franks, 1995) T_g'. In other words, the term T_{Tr1} used in this chapter would be T_g' if maximally freeze concentrated and T_g if not. Several researchers have demonstrated that proper annealing protocols within a narrow temperature range between the T_g and onset of melting, T_m, are necessary to allow for delayed crystallization and maximal freeze concentration (Roos and Karel, 1991b,c; Sahagian and Goff, 1994; Ablett et al., 1992; Izzard et al., 1991). It is difficult to form ice at concentrations >70% solute without extensive annealing (long times) and/or temperature cycling, and therefore it is virtually impossible to maximally freeze concentrate within realistic time frames (Ablett et al., 1992; Franks, 1985; Hatley et al., 1991; Izzard et al., 1991). Figure 3.2 demonstrates the influence of annealing by showing an increase in T_{Tr1} closer

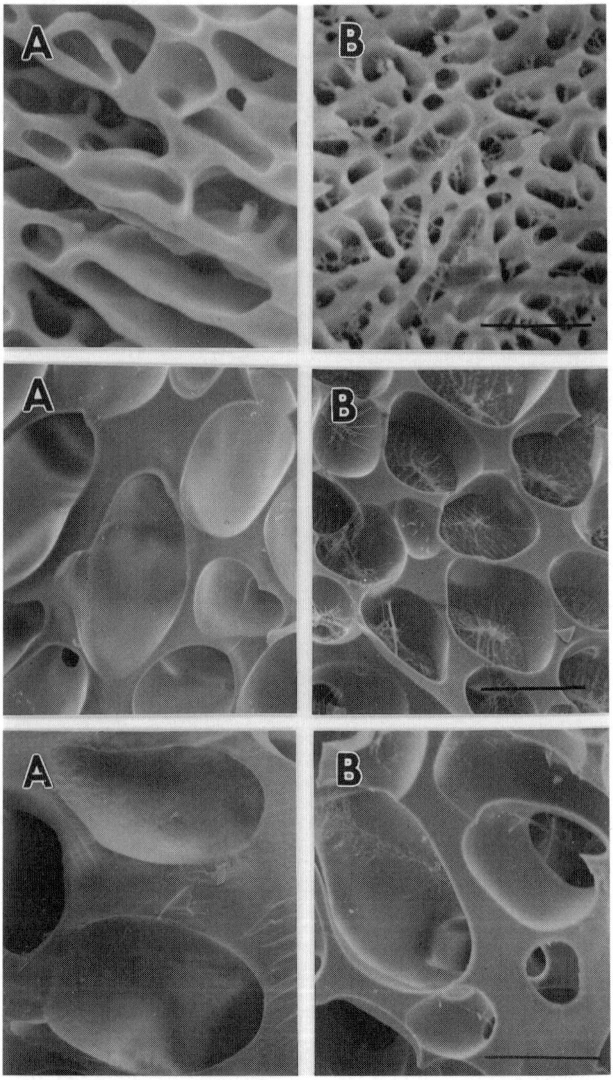

Figure 3.8. Cryo-SEM micrographs of 30% fructose, 1% carboxymethyl cellulose (CMC) solutions after rapid freezing in Freon 22 or slow freezing to −25°C. Top row (same magnification): (A) 30% fructose, rapid freezing, and 2-wk storage at −75°C; (B) 30% fructose + 1% CMC, rapid freezing, and 2-wk storage at −75°C, bar = 3 μm. Middle row (same magnification): (A) 30% fructose, rapid freezing, and 2-wk storage at −25°C; (B) 30% fructose + 1% CMC, rapid freezing, and 2-wk storage at −25°C, *bar* = 30 μm (from Carrington et al., 1996). Bottom row (same magnification): (A) 30% fructose, slow freezing to −25°C, no storage; (B) 30% fructose + 1% CMC, slow freezing to −25°C, no storage, *bar* = 30μm.

to T_{Tr2}, indicative of further promotion of maximal freeze concentration after slow freezing and typical T_g behavior at T_{Tr1}. The same can be seen in figure 3.4, where rapid sample freezing in the TMA led to rapid collapse at temperatures lower than T_{Tr2}, whereas slow freezing and annealing promoted typical T_g response at T_{Tr1}. Thus, in determining T_g' values, great care must be taken to ensure maximal freeze concentration using standardized annealing protocols. Use of the term T_g' should be limited to the theoretical temperature or the measured temperature after rigorous thermal treatment to ensure maximal ice formation and freeze concentration of the solute.

Further confusion results from the typical biphasic behavior of a DSC melting thermogram and the fact that extensive annealing will cause the two transitions to approach but not to coincide. T_{Tr2} thus remains somewhat of an enigma. It has been referred to frequently as T_m', the onset of ice melting after the glass transition (Roos and Karel, 1991b,c; Ablett et al., 1992, Shalaev and Franks, 1995), and its lag after T_g is perhaps simply a reflection of the timescale of the measurement (MacInnes, 1993) or representative of a mechanical relaxation that must occur before sample flow (Shalaev and Franks, 1995). If the latter proposal is correct, then T_m' may be a more technologically significant temperature relating to mechancial collapse and flow in the system. In any case, clarification of these thermal transitions and a consistent, agreed-upon single interpretation in frozen systems is a priority since new technologies and processes that attempt to use this approach must rely on the position of T_g' under specific conditions. It is clear that the present understanding is far from complete, and time-temperature relationships—crucial to the kinetics of ice formation—are of critical importance when dealing with nonequilibrium materials.

3.4.2 Relevance to Frozen-Food Stability and Quality

Although there has been considerable research activity on the formation of glasses in model binary and ternary systems, there has not been as much discussion on the extrapolation of these results to complex food systems. Recent comprehensive reviews on the subject (Levine and Slade, 1991; Blanshard and Lillford, 1993), however, contain numerous references to the application of glass transitions to food systems, and compilations of data, such as that presented by Sá and Sereno (1994) on the glass transition in fruits and vegetables, are beginning to appear. The cryostabilization of frozen foods is dependent on the ability to store the food at temperatures less than its glass transition temperature, which may be very low, or, where appropriate, to modify the food formulations to increase glass transition temperatures to above normal storage temperatures (Levine and Slade, 1988b; Slade and Levine, 1995). However, it may not be practical or economical to maintain foods in the glassy state because of the low temperatures involved, and formulation of foods to raise the T_g' is limited by the sensory properties of many of the large–molecular weight stabilizers. In tissue food systems such as meat and vegetables, the quantity of low–molecular weight solute is much less than in foods such as fruits, ice cream, or sweetened frozen foods (e.g., cakes and toppings), and glass transition temperatures vary accordingly (Levine and Slade, 1990).

Lim and Reid (1991) studied the effect of manipulation of the glass transition temperature using three solutes (sucrose, maltodextrins, and CMC) on three reactions: an enzyme hydrolysis reaction, where both enzyme and substrate must come together by diffusion of large molecules; protein insolubilization during freezing, which occurs by intermolecular bonding; and nonenzymatic oxidation, which is controlled by diffusion of small molecules (oxygen). In the maltodextrin system, all reactions were effectively controlled by storage at temperatures less than T_g' values, which were similar to those reported by Levine and Slade (1990). However, the same was not true for the CMC system, in which none of the reactions was controlled by storage at $T < T_g'$, and only partly true for the sucrose system, in which only the protein insolubilization reaction was effectively controlled. In simulated shelf life studies, Karel and Saguy (1991) reported that shelf life of food products was very dependent on the ΔT between T_g' and the storage temperature, indicative of the importance of knowledge of T_g' and of the applicability of WLF kinetics as a model to predict shelf life. However, it was pointed out by these authors that many of their assumptions may not be met in real food systems.

Simatos and Blond (1991) suggested that, in addition to the glass relaxation, ice melting and its dilution effect at Levine and Slade's glass transition temperature greatly increases the complexity of WLF kinetics determinations and also has a large influence on frozen-food stability. These same authors (Simatos and Blond, 1993) have shown fairly conclusively that in the temperature range above T_g', deviation from WLF kinetics is related directly to ice melting since the chemical composition is not constant over the time and temperature range considered. Therefore, kinetic models of shelf stability must take into account the fact that the reference glass transition temperature should vary according to the dilution of the freeze-concentrated phase.

3.5 Conclusions

There is an increasing awareness about the importance of the state (glass, viscous, or liquid) of the unfrozen phase on the rates of quality loss (e.g., recrystallization and chemical and enzymatic reactions) in stored frozen foods. Hence, the ability to measure this state and the relative temperatures at which the various transitions occur is becoming increasingly important as well. Methods have been limited so far to very sophisticated analytical equipment not often available within the food industry. The interpretations of both the results obtained from various pieces of analytical equipment and the importance of these results to frozen-food quality has also been the subject of intense study by research scientists, without full agreement. Professor David Reid of the University of California has recently stated that, "Whilst, as scientists, we are deeply committed to determining the true nature of the events which lead to 'humps' and 'bumps' in calorimeter traces, for practical purposes it may not be so important. If product stability can be correlated with any particular bump, then this is of practical value, and should not be overlooked" (Reid et al., 1993). Before the measurement of glass transitions becomes a routine part of frozen-food product development, more examples of correlations between state transitions and quality loss, as well as measurement techniques that are both simple and of competitive cost, will be necessary.

References

Ablett, S., Izzard, M.J., and Lillford, P.J. 1992. Differential scanning calorimetric study of frozen sucrose and glycerol solutions. *J. Chem. Soc. Faraday Trans.* **88**:789–794.

Ablett, S. and Lillford, P. 1991. Water in foods. *Chem. Brit.* **11**:1024–1026.

Angell, C.A., Bressel, R.D., Green, J.L., Kanno, H., Oguni, M., and Sare, E.J. 1994. Liquid fragility and the glass transition in water and aqueous solutions. *J. Food Eng.* **22**:115–142.

Blanshard, J.M.V. and Lillford, P.J. (eds.) 1993. *The Glassy State in Foods*, Nottingham University Press, Loughborough, UK.

Blond, G. 1994. Mechanical properties of frozen model solutions. *J. Food Eng.* **22**:253–269.

Blond, G. and Simatos, D. 1991. Glass transition of the amorphous phase in frozen aqueous systems. *Thermochim. Acta* **175**:239–247.

Carrington, A.K., Goff, H.D., and Stanley, D.W. 1996. Structure and stability of the glassy state in rapidly and slowly cooled carbohydrate solutions. *Food Res. Int.* **29**:207–213.

Carrington, A.K., Sahagian, M.E., Goff, H.D., and Stanley, D.W. 1994. Ice crystallization temperatures of sugar/polysaccharide solutions and their relationship to thermal events during warming. *Cryo-Lett.* **15**:235–244.

Chinachoti, P. 1993. Water mobility and its relation to functionality of sucrose-containing food systems. *Food Technol.* **47**:134–140.

Franks, F. 1985. *Biophysics and Biochemistry at Low Temperatures*. p. 21, Cambridge University Press, New York.

Goff, H.D. 1992. Low-temperature stability and the glassy state in frozen foods. *Food Res. Int.* **25**:317–325.

Goff, H.D. 1994. Measuring and interpreting the glass transition in frozen foods and model systems. *Food Res. Int.* **27**:187–189.

Goff, H.D. 1995. The use of thermal analysis in the development of a better understanding of frozen food stability. *Pure Appl. Chem.* **67**:1801–1808.

Goff, H.D., Caldwell, K.B., Stanley, D.W., and Maurice, T.J. 1993. The influence of polysaccharides on the glass transition in frozen sucrose solutions and ice cream. *J. Dairy Sci.* **76**:1268–1277.

Goff, H.D. and Sahagian, M.E. 1996. Glass transitions in aqueous carbohydrate solutions and their relevance to frozen food stability. *Thermochimica Acta* **280**:449–464.

Harwalker, V.R. and Ma, C.Y. (eds.) 1990. *Thermal Analysis of Foods*. Elsevier Applied Science, New York.

Hatley, R.H.M. and Mant, A. 1993. Determination of the unfrozen water content of maximally freeze-concentrated carbohydrate solutions. *Int. J. Biol. Macromol.* **15**:227–232.

Hatley, R.H.M., van den Berg, C., and Franks, F. 1991. The unfrozen water content of maximally freeze concentrated carbohydrate solutions: Validity of the methods used for its determination. *Cryo-Lett.* **12**:113–124.

Hemminga, M.A., Roozen, M.J.G.W., and Walstra, P. 1993. Molecular motions and the glassy state. In *The Glassy State in Foods.* (Blanshard, J.M.V. and Lillford, P.J., eds.) pp. 157–172, Nottingham University Press, Loughborough, UK.

Izzard, M.J., Ablett, S., and Lillford, P.J. 1991. A calorimetric study of the glass transition occurring in sucrose solutions. In *Food Polymers, Gels, and Colloids* (Dickinson, E., ed.) pp. 289–300, The Royal Society of Chemistry, Cambridge.

Kaliachewsky, M.T., Blanshard, J.M.V., and Marsh, R.D.L. 1993. Applications of mechanical spectroscopy to the study of glassy biopolymers and related systems. In *The Glassy State in Foods.* (Blanshard, J.M.V. and Lillford, P.J., eds.) pp. 133–156, Nottingham University Press, Loughborough, UK.

Kaliachewsky, M.T., Jaroszkiewicz, E.M., Ablett, S., Blanshard, J.M.V., and Lillford, P.J. 1992. The glass transition of amylopectin measured by DSC, DMTA, and NMR. *Carbohydr. Polym.* **18**:77–88.

Karel, M. and Saguy, I. 1991. Effects of water on diffusion in food systems. In *Water Relationships in Foods.* (Levine, H. and Slade, L., eds.) pp. 157–173, Plenum Press, New York.

Le Meste, M. and Huang, V. 1992. Thermomechanical properties of frozen sucrose solutions. *J. Food Sci.* **57**:1230–1233.

Levine, H. and Slade, L. 1988a. Thermomechanical properties of small carbohydrate-water glasses and rubbers: kinetically-metastable systems at subzero temperature. *J. Chem. Soc. Faraday Trans. I* **84**:2619–2633.

Levine, H. and Slade, L. 1988b. Principles of cryo-stabilization technology from structure/property relationships of carbohydrate/water systems: a review. *Cryo-Lett.* **9**:21–60.

Levine, H. and Slade L. 1990. Cryostabilization technology: thermoanalytical evaluation of food ingredients and systems. In *Thermal Analysis of Foods.* (Harwalker, V.R. and Ma, C.Y., eds.) pp. 221–305, Elsevier Applied Science, New York.

Levine, H. and Slade, L. (eds.) 1991. *Water Relationships in Foods*. Plenum Press, New York.

Lim, M.H., and Reid, D.S. 1991. Studies of reaction kinetics in relation to the T_g' of polymers in frozen model systems. In *Water Relationships in Foods.* (Levine, H. and Slade, L., eds.) pp. 103–122, Plenum Press, New York.

Ma, C.Y., Harwalker, V.R., and Maurice, T.J. 1990. Instrumentation and techniques of thermal analysis in food research. In *Thermal Analysis of Foods.* (Harwalker, V.R. and Ma, C.Y., eds.) pp. 1–15, Elsevier Applied Science, New York.

MacInnes, W.M. 1993. Dynamic mechanical thermal analysis of sucrose solutions. In *The Glassy State in Foods.* (Blanshard, J.M.V. and Lillford, P.J., eds.) pp. 223–248, Nottingham University Press, Loughborough, UK.

Maurice, T.J., Asher, Y.J., and Thompson, S. 1991. Thermomechanical analysis of frozen aqueous systems. In *Water Relationships in Foods.* (Levine, H. and Slade, L., eds.) pp. 215–224, Plenum Press, New York.

Peleg, M. 1992. On the use of the WLF model in polymers and foods. *Crit. Rev. Food Sci. Nutr.* **32**(1):59–66.

Reid, D.S., Hsu, J., and Kerr, W. 1993. Calorimetry. In *The Glassy State in Foods* (Blanshard, J.M.V. and Lillford, P.J., eds.) pp. 123–132, Nottingham University Press, Loughborough, UK.

Reid, D.S., Kerr, W., and Hsu, J. 1994. The glass transition in the freezing process. *J. Food Eng.* **22**:483–494.

Richardson Schmidt, S.J. 1990. Characterization of water in foods by NMR. In *NMR Applications in Biopolymers.* (Finley, J.W., Schmidt, S.J., and Serianni, A.S., eds.) pp. 415–459, Plenum Press, New York.

Roos, Y. 1992. Phase transitions and transformations in food systems. In *Handbook of Food Engineering.* (Heldman, D.R. and Lund, D.B., eds.) pp. 145–197, Marcel Dekker, New York.

Roos, Y. and Karel, M. 1991a. Applying state diagrams to food processing and development. *Food Technol.* **45**(12):66–71, 107.

Roos, Y. and Karel, M. 1991b. Non-equilibrium ice formation in carbohydrate solutions. *Cryo-Lett.* **12**:367–376.

Roos, Y. and Karel, M. 1991c. Amorphous state and delayed ice formation in sucrose solutions. *Int. J. Food Sci. Tech.* **26**:553–566.

Roozen, M.J.G.W. and Hemminga, M.A. 1990. Molecular motion in sucrose-water mixtures in the liquid and glassy state as studied by spin probe ESR. *J. Phys. Chem.* **94**:7326–7329.

Roozen, M.J.G.W., Hemminga, M.A., and Walstra, P. 1991. Molecular motion in glassy water-malto-oligosaccharide (maltodextrins) mixtures as studied by convention and saturation-transfer spin-probe ESR spectroscopy. *Carbohydr. Res.* **215**:229–237.

Rubin, C.A., Wasylyk, G.M., and Baust, J. 1990. Investigating of vitrification by nuclear magnetic resonance and differential scanning calorimetry in honey: a model carbohydrate system. *J. Agric. Food Chem.* **38**:1824–1827.

Sá, M.M. and Sereno, A.M. 1994. Glass transitions and state diagrams for typical natural fruits and vegetables. *Thermochim. Acta* **246**:285–297.

Sahagian, M.E. and Goff, H.D. 1994. Effect of freezing rate on the thermal, mechanical, and physical aging properties of the glassy state in frozen sucrose solutions. *Thermochim. Acta* **246**:271–283.

Sahagian, M.E. and Goff, H.D. 1995a. Thermal, mechanical, and molecular relaxation properties of stabilized sucrose solutions at sub-zero temperatures. *Food Res. Int.* **28**:1–8.

Sahagian, M.E. and Goff, H.D. 1995b. Influence of stabilizers and freezing rate on the stress relaxation behaviour of freeze-concentrated sucrose solutions at different temperatures. *Food Hydrocoll.* **9**:181–188.

Sahagian, M.E. and Goff, H.D. 1996. Fundamental aspects of the freezing process. In *Freezing Effects on Food Quality.* (Jeremiah, L.E., ed.) pp. 1–50, Marcel Dekker, New York.

Sauerbrunn, S., Crowe, B., and Reading, M. 1992. Modulated differential scanning calorimetry. *Am. Lab.* (Aug.):44–47.

Shalaev, E.Y. and Franks, F. 1995. Structural glass transitions and thermophysical processes in amorphous carbohydrates and their supersaturated solutions. *J. Chem. Soc. Faraday Trans.* **91**:1511–1517.

Simatos, D. and Blond, G. 1991. DSC studies and stability of frozen foods. In *Water Relationships in Foods* (Levine, H. and Slade, L., eds.) pp. 139–156, Plenum Press, New York.

Simatos, D. and Blond, G. 1993. Some aspects of the glassy transition in frozen food systems. In *The Glassy State in Food.* (Blanshard, J.M.V. and Lillford, P.J., eds.) pp. 395–416, Nottingham University Press, Loughborough, UK.

Slade, L. and Levine, H. 1995. Glass transitions and water-food structure interactions. In *Advances in Food and Nutrition Research,* vol. 38 (Kinsella, J.E. and Taylor, S.L., eds.) pp. 103–269, Academic Press, San Diego.

Wolanczyk, J.P. 1989. Differential scanning calorimetry analysis of glass transitions. *Cryo-Lett.* **10**:73–76.

4

Modeling of Food Freezing

D. R. Heldman and T. A. Taylor

4.1 Introduction

Modeling of the freezing process is accomplished in many ways. Computation of refrigeration requirements to achieve the desired temperature reductions and requirements involved in removal of both sensible and latent heats from the product are one component of modeling. In addition, the design of processes for food freezing requires knowledge of time needed to reduce product temperature to desired levels. Ultimately, design of frozen-food storage requires knowledge of changes occurring within the food product as it is exposed to the environmental conditions existing within the storage system. The design of systems requires insight into the changes occurring within the product structure during freezing, the influence of these changes on product properties along with the accepted approaches to computation of freezing times, and the use of these computed times in design calculations.

This chapter will review the following components of modeling as applied to food freezing: (1) the thermodynamics of food freezing and illustration of changes occurring within the product during freezing and the influence of changes to frozen food properties; (2) the impact of temperature-dependent properties on the solution of unsteady-state heat transfer equations; and (3) calculations for freezing time for food and capacity selection of food freezing systems.

4.2 Thermodynamics of Phase Change in Foods

The thermodynamics of food freezing describes the changes in water within a food product as the freezing process proceeds. As illustrated in figure 4.1, the change in product temperature during freezing decreases gradually as the latent heat of fusion is removed from water within the product. The freezing process has two unique characteristics when compared to the freezing of pure water. First, the product temperature at which initial ice-crystals are formed is depressed below the temperature of initial ice crystal formation in pure water. The second difference

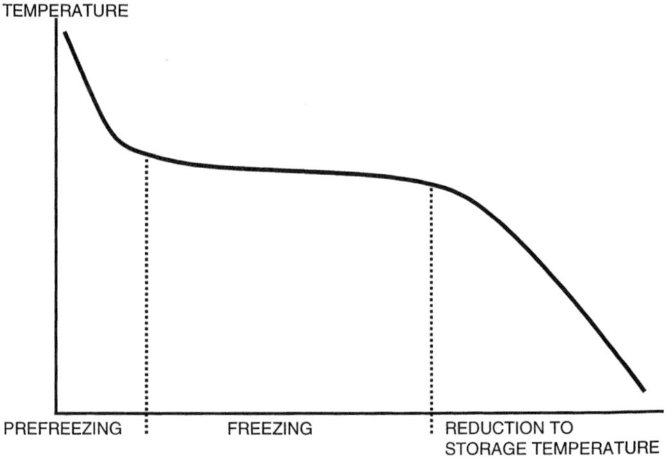

Figure. 4.1. Practical definition of the freezing process

between the food product system and pure water is the gradual removal of latent heat of fusion as the product temperature decreases. The actual shape of the temperature-time curve will vary with product composition and with the location within the product structure.

The relationship between product composition and temperature is most often explained in terms of the freezing temperature depression equation for an ideal solution. This equation is derived by Heldman and Singh (1981) and can be presented as follows:

$$\frac{\lambda}{R}\left[\frac{1}{T_{Ao}} - \frac{1}{T_A}\right] = \ln X_A \tag{4.1}$$

Equation 4.1 illustrates the relationship between the mole fraction (X_A) of water within the product and the absolute temperature (T_A) at which ice-crystal formation occurs as a function of molar latent heat of fusion (λ) and the universal gas constant (R). The mole fraction of water within the product can be defined as follows:

$$X_A = \frac{m_A / M_A}{m_A / M_A + m_s / M_s} \tag{4.2}$$

In equation 4.2, the mole fraction of water in the product is defined in terms of product moisture content expressed as a moisture fraction (m_A), the molecular weight of water (M_A), the mass fraction of product solids (m_s), and the molecular weight of product solids (M_s). The applications of equations 4.1 and 4.2 can be direct in the case of some food products but may be more indirect when product compositions are not well defined. In addition to a complete composition analysis of the food product, knowledge of the product components causing depression of the initial temperature of freezing is needed. More detailed discussion on the thermodynamics of freezing can be found in chapter 2.

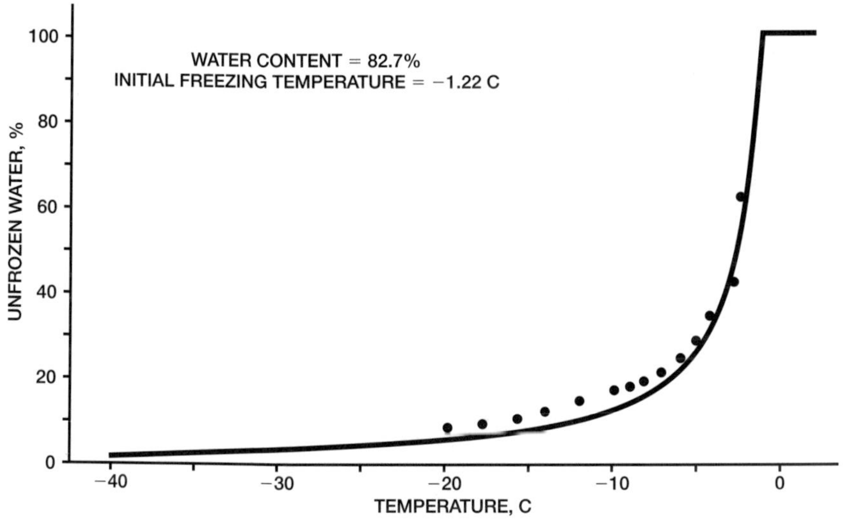

Figure 4.2. Relationship between unfrozen water fraction and temperature of raspberries. (From Heldman, 1974.)

One of the unique characteristics of a frozen food is the relationship between unfrozen water and temperature. This relationship is basic to the design of freezing systems and frozen-food storage facilities. The unfrozen water fraction in a food product decreases gradually as temperature drops below the initial freezing temperature. Since the relationship is similar to freezing an ideal binary solution, the relationship for many foods can be estimated using equation 4.1. The estimation procedure is based on the assumption that pure ice crystals are formed during freezing and that all solute is concentrated in the unfrozen water fraction.

The relationship between the unfrozen water fraction and temperature for raspberries over the temperature range from −40°C to +50°C is illustrated in figure 4.2. The estimated relationship is shown as a solid curve and is compared to experimental data from Riedel (1951). The predicted relationship is ideal for freezing system design. The primary input parameters are initial freezing temperature and moisture content.

4.3 Properties of Frozen Foods

The food product properties of interest when considering the freezing process include density, thermal conductivity, specific heat, enthalpy, and latent heat. These properties must be accounted for in the establishment of refrigeration capacity for the freezing system and in the computation of freezing times needed to assure adequate residence times. The approach suggested in the quantification of property magnitudes during the freezing process depends directly on the relationship between the unfrozen water fraction and temperature.

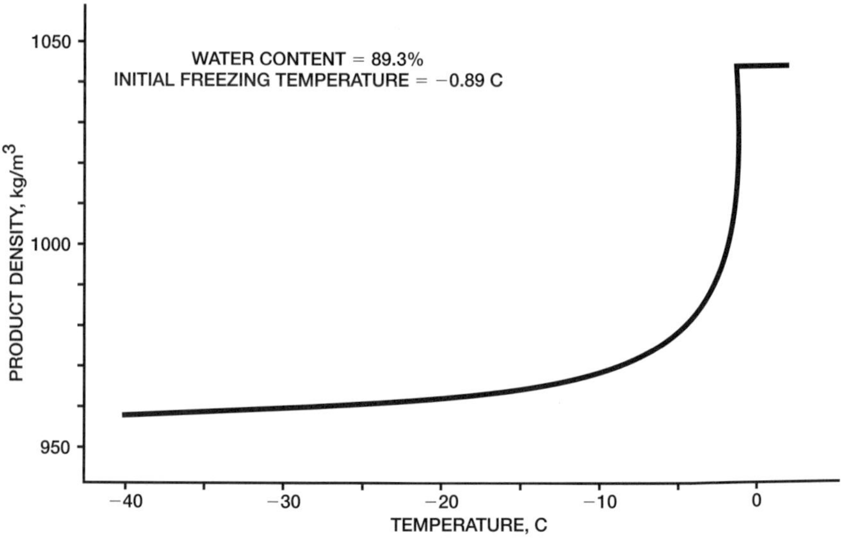

Figure 4.3. Influence of freezing on the predicted density of strawberries. (From Heldman, 1982.)

The research literature contains numerous references to experimental data for properties of unfrozen and frozen foods. These data have been reviewed in several studies by Dickerson (1969), Woodams and Nowry (1968), Reidy (1968), Qashou et al. (1972), Mohsenin (1980), and Heldman and Singh (1983).

The overall influence of freezing on food product density is relatively small, but a dramatic change in density does occur just below the initial freezing temperature. A model to predict the relationship between frozen-food density and temperature was presented by Heldman (1992). Using the prediction model, the influence of freezing on the density of strawberries was obtained and is presented in figure 4.3. Above the initial freezing temperature, the product density is relatively constant. At the initial freezing temperature, the product density decreases rapidly as the fraction of frozen water in the product increases. At temperatures below −20°C, the density of frozen strawberries becomes relatively constant. It should be noted that the overall change in product density between +5°C and −40°C is less than 10%.

The thermal conductivity of a food product is a function of water content and structure. Most models for calculation of thermal conductivity are based on moisture and do not account for structural orientation. The models proposed by Kopelman (1966) have been developed for three different types of structure. Heldman (1992) presented these models and applied them to frozen foods.

The use and implications of these models, when applied to food systems, are presented by Heldman and Singh (1981). The application to frozen foods was first demonstrated by Heldman and Gorby (1975). Frozen foods contain three

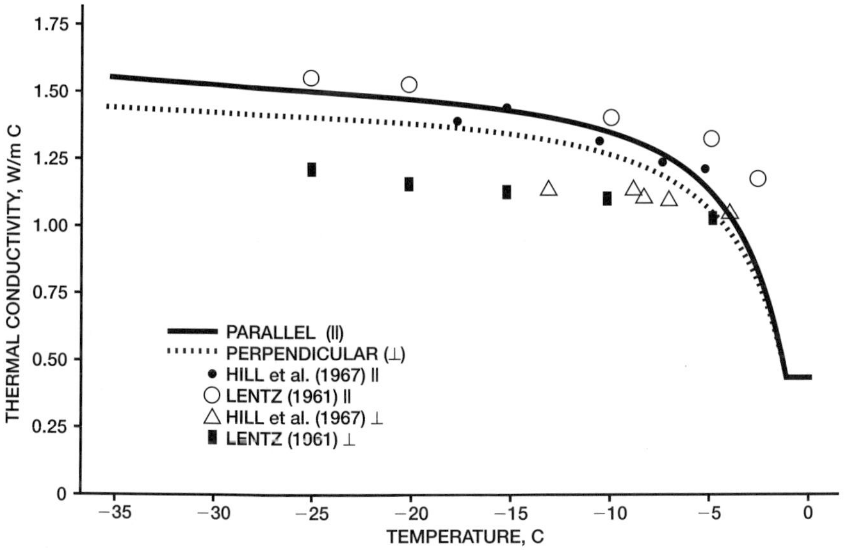

Figure 4.4. Thermal conductivity of frozen lean beef as a function of temperature. (From Heldman and Gorby, 1975.)

distinct phases (solids, liquid water, and solid water); all models are derived for two phases. The equations are usually applied in sequence, first to the water phase with solid water (ice) representing a discontinuous phase within the unfrozen water, which is considered a continuous phase. The second computational step applies the appropriate equations, and the water phase (liquid and solid) represents the continuous phase while the product solids represent the discontinuous phase. An illustration of the relationship between thermal conductivity and temperature for frozen beef is presented in figure 4.4.

To calculate the thermal conductivity of a product, the product solids can be considered a completely homogeneous and disperse mixture of basic components (protein, fat, carbohydrate, and ash). This assumption allows for the computation of thermal conductivity for the product solids to be estimated from the weighted average of the component fractions and the thermal conductivity of each component. Thermal conductivities for each basic component of a food product were published by Choi and Okos (1983).

As illustrated by Heldman and Singh (1981), the specific heat of a food product can be calculated based on composition. Specific heat magnitudes for the product solids can be used in calculations of product enthalpy and apparent specific heat, as illustrated in sections to follow.

The refrigeration requirements for food product freezing are directly dependent on the heat content or enthalpy of the product. Magnitudes of enthalpy for frozen foods were measured by Riedel (1951, 1956, 1957a, 1957b). As is evident, the enthalpy or heat contact is zero at −40°C, which is the reference temperature for refrigerants.

Using the predicted mass fractions of unfrozen and frozen water obtained previously, along with the specific heat of product solids (c_{ps}) and the specific heats of liquid and solid water, the enthalpy of a frozen food is predicted using procedures demonstrated by Heldman (1992). A typical result for the enthalpy of sweet cherries as a function of temperature is illustrated in figure 4.5.

By using the thermodynamic definition of specific heat, the derivative of enthalpy with respect to temperature produces an apparent specific heat function for frozen foods. By evaluating the derivative of enthalpy for sweet cherries over the temperature range from −40°C to +50°C, the relationship in figure 4.6 is obtained. At all temperatures below the initial freezing temperature, the apparent specific heat decreases with decreasing temperature. The function decreases rapidly just below the initial freezing temperature, indicating the region where major portions of the latent heat of fusion are removed from the product. At the initial freezing temperature, the apparent specific heat function reaches a near discontinuity and decreases to a similar magnitude for the unfrozen product at temperatures above the initial freezing temperature.

The thermal property most often incorporated into heat transfer equations is thermal diffusivity. The apparent thermal diffusivity values as a function of temperature are illustrated in figure 4.7. As is evident, the function shape at temperatures below the initial freezing temperature is similar to the shape of the thermal conductivity versus temperature relationship. There is a near discontinuity at the initial freezing temperature before the thermal diffusivity becomes constant at temperatures above freezing.

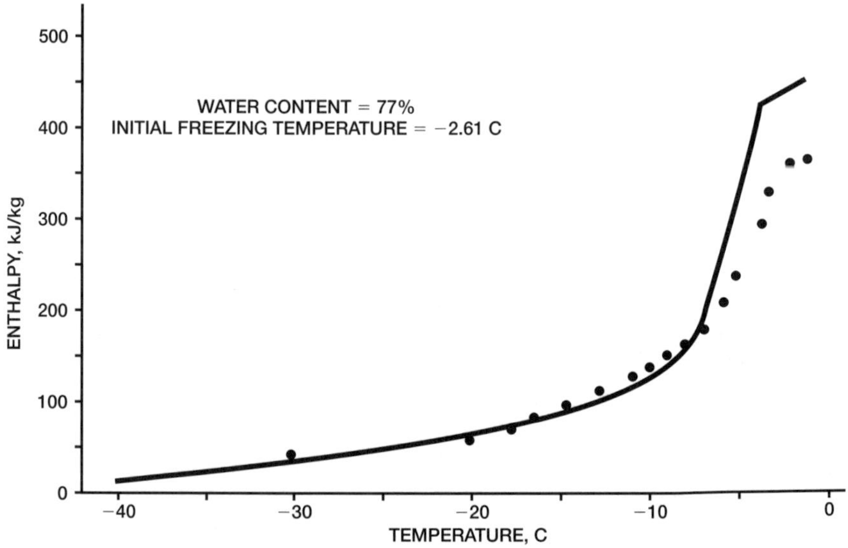

Figure 4.5. Enthalpy of sweet cherries as a function of temperature. (From Heldman, 1982.)

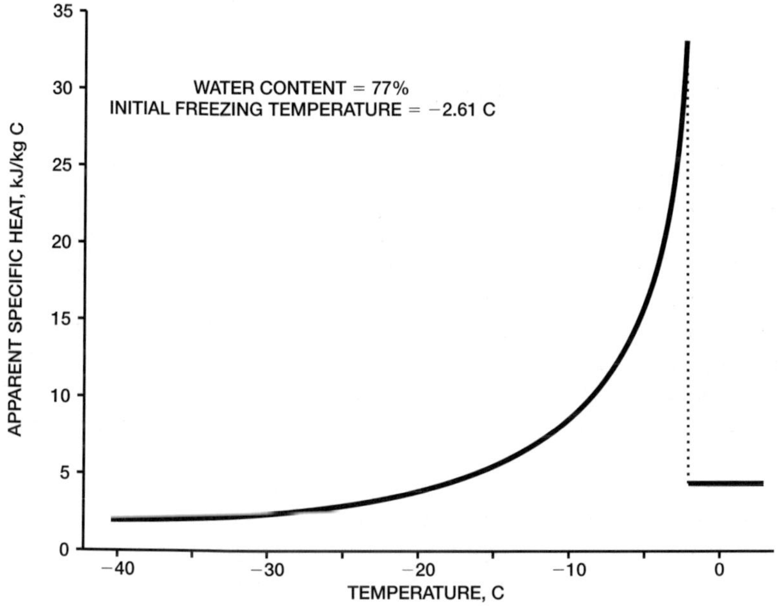

Figure 4.6. Predicted apparent specific heat of frozen sweet cherries as a function of temperature. (From Heldman, 1982.)

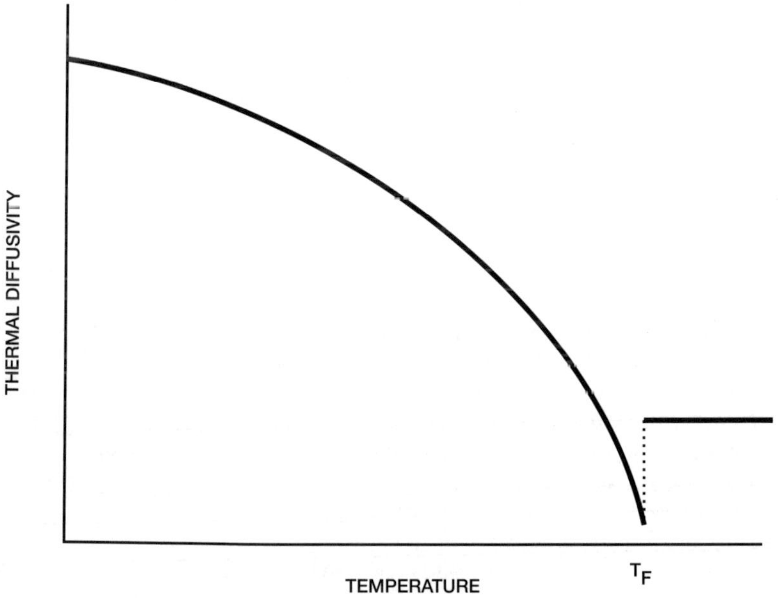

Figure 4.7. Relationship between thermal diffusivity and temperature during freezing of food product as predicted from initial freezing temperature assumption. (From Heldman, 1983.)

4.4 Prediction of Freezing Time in Foods

Freezing time, time to freeze, and freezing rate are the basic design criteria for freezing systems and determine the conditions a food product is exposed to in the freezing process to achieve the desired level of freezing and finished product quality. The most widely accepted definition of freezing time is the time required to reduce the initial product temperature to some established final temperature at the slowest cooling location. An alternative definition changes the end point to the mass average enthalpy equivalent to the desired final temperature for the product. In this case the product leaves the freezer with the outside colder than the inside. As the product equilibrates in storage, the internal temperature cools to the desired final temperature.

Freezing time calculations are completed in order to design the freezing system. The important components of the system design related to freezing time are the residence time in the system and the rate of freezing to achieve the desired final quality. The residence time must be sufficient to ensure that the product temperature is reduced to some desired final temperature or mass average enthalpy. The final temperature is established by the level required to establish and maintain product quality. The residence time is a function of the rate at which the product moves through the system, the length of the system, and the ambient conditions within the cooling zone(s). The specific characteristics of the design will be dependent on the type of freezing system being considered.

The freezing process can be described by several distinct stages (fig. 4.1). The initial stage, referred to as the prefreezing stage, occurs when the product is cooled from the initial temperature to the point that ice crystallization begins to occur. The freezing stage is the plateau during which the majority of the water in the system changes phase into ice. After the majority of the water has changed phase, the rate of cooling increases again during the reduction to storage temperature.

4.4.1 Prefreezing

During the prefreezing stage no phase transitions occur as the product is cooled from an initial temperature to a temperature just above the freezing point. During the cooling and freezing process the temperature changes within the food product. This change in temperature with respect to time and location in the product is referred to as unsteady-state heat transfer. The governing mathematical equation that describes the change in temperature during the prefreezing stage is

$$\frac{\partial T}{\partial t} = \frac{k}{\rho C_p} \frac{\partial^2 T}{\partial x^2} \tag{4.3}$$

where T is temperature (K), t is time (s), x is the location coordinate (m), k is the thermal conductivity (W/m K), ρ is the density (kg/m^3) and C_p is the heat capacity (kJ/kg K).

Equation 4.3 shows that the cooling rate of the product is proportional to the thermal conductivity but inversely proportional to the density and the heat capacity. The rate of cooling is effected to the greatest degree by the temperature differential within the product.

The analytical solution to equation 4.3 requires the use of advanced mathematics and is limited to the simple geometries of an infinite slab (rectangle), infinite cylinder, and sphere. Charts and computer spreadsheets have been developed for simple geometrical shapes. Heisler charts (Heisler, 1947) can be used to find the temperature at the center of a sphere, cylinder, or slab. Computer spreadsheets have been developed to calculate the same information (Singh and Heldman, 1993).

4.4.2 Freezing

The freezing process begins once the temperature of the food has been lowered to the point that ice crystallization first occurs. The rate of freezing or time to freeze has been defined in several ways. Many early researchers tried to relate quality to the freezing time, which was defined as the total time it took for a product to transition from initial temperature to final storage temperature. This value was influenced greatly by initial product temperature and final storage temperature, yet was difficult to relate to product quality.

Freezing rate was the change in temperature divided by the freezing time. Others have tried to evaluate the freezing rate from the speed of movement of the ice front (cm/h) through a product. This definition is difficult to use in practical terms because of limitations in the ability to measure ice front movement in whole products.

The time to freeze can best be defined as the time required for a point in the sample to decrease from the initial freezing temperature to a temperature 5°C lower (Taylor et al., 1994). This is a point source or single point adaptation of the International Institute of Refrigeration (1971) definition for freezing rate. With this definition the freezing time is in seconds and freezing rate is 5°C divided by the time to freeze (°C/s).

The time to freeze definition allows more practical application of current technology, such as thermocouples and temperature recorders, to quantify freezing processes and relate the product quality to the conditions during the freezing process. Studies have shown that product quality is related to the freezing rate more than to traditional freezing time (Añón and Calvelo, 1980; Grujić et al., 1993; Taylor et al., 1994).

4.4.3 Modeling Freezing of Foods

The freezing process involves the removal of sensible and latent heat (Singh, 1995). The latent heat is the energy given up by the water as it is converted into ice. Modeling the freezing process involves accounting for the physical property changes and percentage of water converted into ice.

The best known and most used of the prediction methods is based on Plank's equation (1913)

$$t_F = \frac{\rho L}{T_F - T_\infty}\left(\frac{Pa}{h_c} + \frac{Ra^2}{k}\right) \tag{4.4}$$

where P and R are constants that vary depending on product geometry. The dimension a is product thickness for an infinite slab, diameter for an infinite cylinder, and diameter for a sphere.

	P	R
Infinite slab	0.5	0.125
Infinite cylinder	0.24	0.0625
Sphere	0.167	0.04167

Analytical equations such as Plank's equation are limited by specific geometries and assumptions used to derive these solutions. Plank's equation included assumptions of steady-state conditions and that the sensible heat removed in precooling is negligible compared to the removal of latent heat (Cleland, 1990). The limitations to the use of Plank's equation to compute freezing times have been discussed in reviews by Heldman and Singh (1981) and Ramaswami and Tung (1984). The most obvious difficulties are selection of a latent heat magnitude (L) and a value for the thermal conductivity (k). In addition, the basic equation does not account for the time required to cool the product from initial temperature to the freezing point and the time to cool to the final temperature after freezing has been completed. Analytical equations can generally be used to quickly estimate a rough approximation of the freezing time for products that have near-standard geometries. The values for a given product are generally found to be within 30% of values found experimentally (Cleland, 1990).

With the availability of modern high-speed computers, numerical techniques have been developed that provide greater accuracy and flexibility for modeling the time required for phase transition as well as the entire freezing process. Numerical techniques can provide solutions that include nonstandard geometries and variable thermal properties and can avoid the assumptions required to produce simple analytical equations. While more complex to implement than analytical solutions, numerical techniques are often the only realistic tool for solving a complex problem. Extensive reviews of these methods are provided in Cleland (1990) and Singh and Mannapperuma (1990). These techniques are generally based on enthalpy formulation of heat conduction with gradual phase change. Mannapperuma and Singh (1989) provide a method and software for prediction of freezing and thawing of foods.

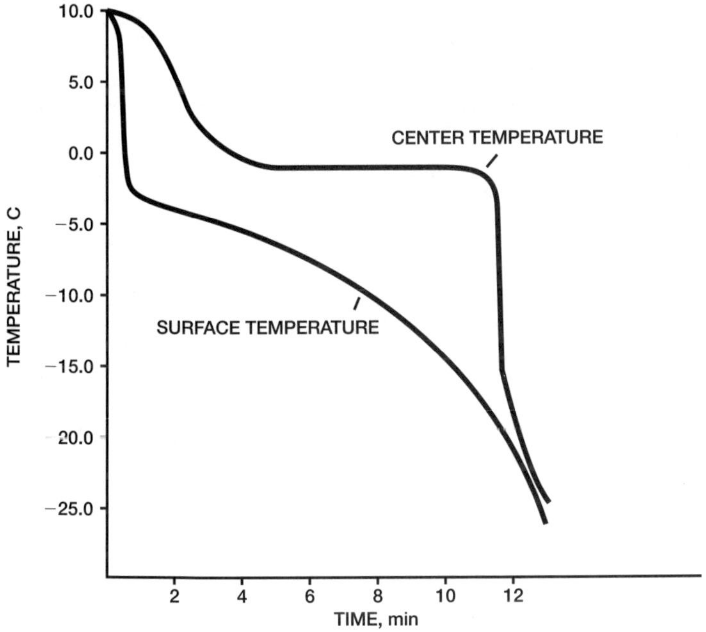

Figure 4.8. Predicted temperature history at the center and surface during freezing of a 2-cm strawberry (T_i = 10°C; T_∞ = −35°C; h_c = 70 W/m^2 K; a = 2 cm). (From Heldman, 1983.)

4.4.4 Freezing Rate Distribution

Freezing processes result in temperature history distributions, as illustrated in figure 4.8. As indicated, the temperature at the center of the product decreases to a plateau at the initial freezing temperature of the product before decreasing to lower temperatures. At all locations away from the center, the temperature decreases gradually during the time period when latent heat is removed. The distribution of freezing rates within a product can also lead to a distribution of quality within the product. To minimize the difference within the product, multiple-stage freezers or multiple-step processes can be used. The product is cooled slowly to just above the freezing point, allowing the temperature to equilibrate, then frozen quickly to minimize the temperature differential and freezing rate within the product.

4.4.5 System Design

The ultimate use of the physical properties and modeling systems discussed in this section is to aid in designing refrigeration systems used for food freezing.

The evaporator on the refrigeration system must be designed to accommodate the heat transfer rate associated with the thermal energy removed from the product as it passes through the freezing system. By computation of the refrigeration requirement or difference between initial and final product enthalpies, the first step in establishing load on the refrigeration system is completed. The second step involves the rate at which the thermal energy is removed from product based on freezing times, freezing rate, and residence time for product within the freezing system. The freezing time was defined as the time required to reduce the product temperature from an initial value to the desired final temperature. Since the freezing time must occur within the freezing system, the residence time must be equal to or greater than the freezing time.

The establishment of residence time is a function of the linear distance that product moves through the freezing system and the speed of product movement through the system. Freezing system design requires the matching of refrigeration requirements with dimensions of the freezing system and speed of product movement in system.

4.5 Summary

The freezing process in foods is complex in terms of temperature history distribution during the removal of latent heat. Due to the complexity of the process, the models for prediction of freezing time are based on assumptions needed to solve appropriate heat transfer equations. Relationships to describe the influence of temperature on physical properties of food during freezing assist in prediction of temperature history distribution. In most situations the operational equations used to predict freezing time are based on assumptions that account for product shape but require estimation of product physical properties.

References

Añón, M.C. and Calvelo, A. 1980. *Meat Science*. Applied Science Publishers, England.

Choi, Y. and Okos, M.R. 1983. Thermal properties of liquid foods: review. Presented at the 1983 Winter Meeting of the American Society of Agricultural Engineers, Chicago, Ill.; paper 83–6516.

Cleland, A.C. 1990. *Food Refrigeration Processes Analysis, Design, and Simulation*. Elsevier Science Publishers, Essex, UK.

Dickerson, R.W., Jr. 1969. Thermal properties of food. In *The Freezing Preservation of Foods*, 4th ed., vol. 2. (Tressler, D.K., Van Arsdel, W.B., and Copley, M.J., eds.) AVI Publishing, Westport, Conn.

Grujić, R., Petrovic, L., Pikula, B., and Amidzic, L. 1993. *Meat Science*. Elsevier Science Publishers, England.

Heisler, M.P. 1947. Temperature charts for induction heating and constant temperature heating. *Trans. ASME* **69**:227.

Heldman, D.R. 1974. Predicting the relationship between unfrozen water fraction and temperature during food freezing, using freezing point depression. *Trans. ASAE* **17**:63.

Heldman, D.R. 1982. Food properties during freezing. *Food Technol.* **36**(2):92–96.

Heldman, D.R. 1992. Food Freezing. In *Handbook of Food Engineering*. (Heldman, D.R. and Lund, D.B., eds.) chapter 6, Marcel Dekker, New York.

Heldman, D.R. and Gorby, D.P. 1975. Prediction of thermal conductivity in frozen food. *Trans. ASAE* **18**:156.

Heldman, D.R. and Singh, R.P. 1981. *Food Process Engineering*, 2d ed. AVI Publishing, Westport, Conn.

Heldman, D.R. and Singh, R.P. 1983. Thermal properties of frozen foods. Presented at the 1983 Winter Meeting of the American Society of Agricultural Engineers, Chicago, Ill., paper 83–6518.

Hill, J.E., Litman, J.D. and Sunderland, J.E. 1967. Thermal conductivity of various meats. Food Technol. 21:1143.

International Institute of Refrigeration. 1971. *Recommendation for the Processing and Handling for Frozen Foods,* 2d ed. International Institute of Refrigeration, Paris.

Lentz, C.P. 1961. Thermal conductivity of meats, fats, gelatin gels and ice. Food Technol. 15:243.

Kopelman, I.J. 1966. Transient heat transfer and thermal properties in food systems. Ph.D. Thesis. Michigan State University, East Lansing.

Mannapperuma, J. and Singh, R.P. 1989. A computer-aided method for the prediction of properties and freezing/thawing times of foods. *J. Food Eng.* **9**:275–304.

Mohsenin, N.N. 1980. *Thermal Properties of Foods and Agricultural Materials*. Gordon and Breach Science Publishers, New York.

Plank, R. 1913. Die Gerfrierdauer von Eisblocken. *Z. ges. Kalte-Ind.* **20**(6):109–114.

Qashou, M.S., Vachon, R.I., and Touloukian, V.A. 1972. Thermal conductivity of foods. *ASHRAE Trans.* **78**:165.

Reidy, G.A. 1968. Thermal properties of foods and methods of their determination. M.S. Thesis. Michigan State University, East Lansing.

Riedel, L. 1951. The refrigeration required to freeze fruits and vegetables. *Refrig. Eng.* **59**:670–673.

Riedel, L. 1956. Calorimetric investigations of the freezing of fresh meat. *Kaltetechnik* **8**(12):374–377.

Riedel, L. 1957a. Calorimetric investigations of the meat freezing process. *Kaltetechnik* **9**:38–40.

Riedel, L. 1957b. Calorimetric investigations of the freezing of egg whites and yolks. *Kaltetechnik* **9**(11):342–345.

Singh, R.P. 1995. Principles of heat transfer. In *Frozen and Refrigerated Doughs and Batters*. (Kulp, K., Lory, K., and Brümmer, J., eds.) American Association of Cereal Chemists, St. Paul, Minn.

Singh, R.P. and Heldman, D.R. 1993. *Introduction to Food Engineering.* Academic Press, Orlando, Fla.

Singh, R.P. and Mannapperuma, J.D. 1990. Developments in food freezing. In *Biotechnology and Food Process Engineering.* (Schwartzburgh, H.G. and Rao, M.A., eds.) Marcel Dekker, New York.

Taylor, T.A., Heldman, D.R., Chao, R.R., and Kramer, H.L. 1994. Measurement of kinetic parameters for quality change during food freezing. In *Advances in Heat and Mass Transfer in Biological Systems,* Heat Transfer Division, vol. 288. ASME.

Woodams, E.E. and Nowry, J.E. 1968. Literature values of thermal conductivities of foods. *Food Technol.* **22**(4):150.

SECTION II

Quality Losses Associated with Frozen Foods

5

Moisture Migration and Ice Recrystallization in Frozen Foods

Q. T. Pham and R. F. Mawson

5.1 Introduction

Moisture migration is the principal physical change occurring in frozen foods and has major effects on the chemical and biochemical properties of frozen foods. It manifests itself in several forms: moisture loss by sublimation, moisture absorption and redistribution in foods or food components, recrystallization of ice, and drip loss during thawing.

Moisture loss from foods has the most important economic consequences and has received the most attention. It is a major factor and sometimes the limiting factor of the shelf life of foodstuffs. For example, Winger (1984) stated that the main limiting factor in the storage life of frozen lamb was desiccation which produces unpleasant cooking odors, but microbial growth and development of rancidity were not important factors for this product. Moisture loss can also lead to freezer burn, a glassy appearance in some meat products produced by the presence of tiny cavities left behind by sublimated ice (Kaess, 1961), or to a whitish, woodlike appearance in desiccated meat. Sometimes even small patches of desiccation due to damaged or imperfect wrapping lead to unacceptable appearance. These various forms of loss of visual quality are also sometimes referred to as freezer burn. Moisture migration in wrapped frozen product results in unattractive formation of ice inside the wrap. Moisture migration during thawing results in drip, which reduces visual attraction and causes nutrient loss. Apart from appearance, loss of moisture will also affect the food's juiciness and texture.

Even when appearance and other quality traits are not affected, moisture migration has a significant quantitative effect in the form of weight loss. The water that is lost has the same economic value as the product it evaporates from and can greatly affect a company's profitability. Typical weight losses during conventional meat processing, for example, amount to about 1–2% during chilling, 1% during freezing, and about 0.5% per month during storage and transport unless the product is wrapped in an impervious film.

On the other hand, some degree of moisture loss is sometimes relied on to ensure food quality and safety. During the chilling of meat, some surface drying is necessary to prevent or retard microbial growth by reducing the surface water activity, as well as to avoid a glassy appearance when the meat is frozen.

Moisture absorption by foods and redistribution among food components has important consequences in certain manufactured foods such as frozen pies and pizzas where the components differ widely in water activity. A normally dry, crisp crust can become moist and soggy by absorbing moisture from the wet filling.

Recrystallization is the process in which ice crystals increase in size and decrease in number through redistribution of water from small to large crystals. In ice cream, frozen desserts, and delicate plant tissues (such as strawberries), recrystallization can have an important effect on appearance and texture. Every consumer who has left ice cream in the home freezer for too long will know this effect. Even in meat there is evidence that recrystallization may affect drip loss during thawing. Recrystallization is receiving increasing attention as a factor in quality losses, and the recent discovery and synthesis of crystal growth inhibitors (antifreeze proteins) has stimulated much research.

Drip loss during thawing is caused by irreversible tissue damage during the freezing, storage (recrystallization), and thawing processes. It will be considered separately from moisture loss because it is driven by quite different mechanisms.

5.2 Water in Food at Low Temperature

The freezing behavior of water-rich foods is often discussed in terms of a simplified model consisting of two or three components: water, a solute, and, in the case of solid foods, an inert solid matrix. If the food is cooled under conditions of thermodynamic equilibrium, the water starts to freeze when the food reaches the initial freezing temperature T_f. Pure ice separates out from the solution, causing the solute concentration to increase and the temperature to gradually decrease. The process continues until the eutectic temperature is reached, at which point the solute begins to solidify. As long as the three phases (ice, crystallized solute, and unfrozen solution) coexist, the mixture temperature remains constant according to the Gibbs phase rule. Once the unfrozen phase has completely disappeared, the temperature is free to fall again.

In practice, equilibrium cooling is never achieved. Some degree of supercooling always takes place, followed by a sudden rise to a value close to the initial freezing temperature. Usually this takes place in the near-surface layers of the food. In the inner parts, which cool more slowly, freezing proceeds by the growing of ice dendrites from the surface inward, or nucleation is facilitated by distortions caused by freezing; consequently, supercooling is much less pronounced.

Another deviation from equilibrium behavior is caused by glass transition. Due to decreased molecular mobility at low temperature, nucleation of the solute does not take place when the food reaches the eutectic temperature; instead, the temperature continues to fall, and the unfrozen solution eventually transforms into an amorphous solid or glass that can be considered as a liquid with very high viscosity. The lowest temperature at which this occurs is the glass transition temperature, T_g'. This has important implications for frozen foods: the rates of diffusion-dependent degradative processes are drastically slowed down in the vicinity of T_g', obeying the Williams-Landel-Ferry (WLF) equation (Williams et al, 1955)

$$\ln k = a + \frac{b}{T - T_g'} \tag{5.1}$$

rather than the Arrhenius equation

$$\ln k = a - \frac{\Delta E_a}{RT} \tag{5.2}$$

In frozen foods, moisture sublimation is likely to follow an Arrhenius-type behavior since it is a vapor-phase process, but recrystallization will obey the WLF equation since it depends on water diffusion in the dense phase (Fennema, 1993).

Because of the complex behavior of frozen food, an empirical bound water model has been proposed by Schwartzberg (1976) in which water is divided into a free phase that can freeze and a bound phase that will not freeze under any condition. The bound water model has proved to be very useful in modeling the calorimetric properties of water-rich foods. Bound water (in the sense of unfreezable water) can be physically interpreted as the water in the monolayer adjacent to the surface (Fennema, 1996).

Some rough quantitative comparisons lend support to this hypothesis. Fennema stated that for water-protein systems, the BET monolayer water content (water bound to the most hydrophilic sites on the protein) is 0.07 g water g^{-1} protein, while the "true monolayer hydration" value (water with properties that may differ significantly from bulk water) is about 0.38 g water g^{-1} protein. The latter value is not far from the values of empirical bound water to protein measured by calorimetric means, which are in the range of about 0.3–0.5 g water g^{-1} protein (Schwartzberg, 1976; Pham, 1987b). The difference in behavior between BET water, true monolayer water, and bulk water is reflected in the sorption isotherm of figure 5.1, in which the water activity a_w is plotted against moisture content. A low value for a_w (low water vapor pressure) indicates a high degree of bonding.

5.3 Moisture Loss by Evaporation or Sublimation

5.3.1 Mathematics of Moisture Loss

Moisture loss during refrigeration and cold storage is a mass transfer phenomenon driven by concentration and temperature gradients. In complex multicomponent

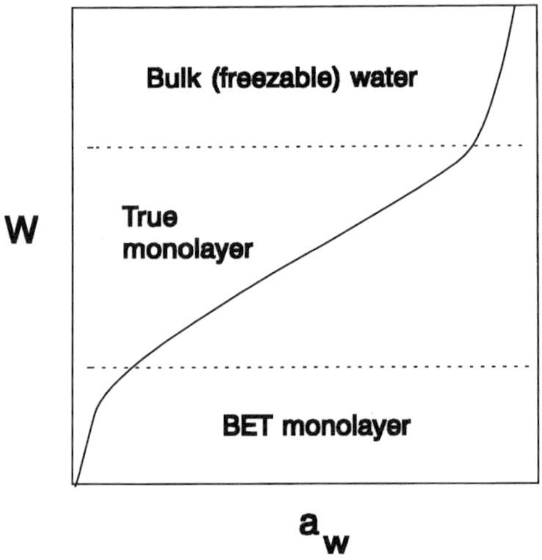

Figure 5.1. Typical moisture sorption isotherm

systems such as foods, the phase rule indicates that moisture can simultaneously exist in several phases (vapor, free liquid, and various types of bound moisture), each of which has its own diffusion rate. However, because the phases are in intimate contact, all must be in thermodynamic equilibrium with each other. Due to lack of data, it is commonly assumed that moisture movement in the (wet) food can be described by a single-phase diffusion equation with effective diffusivity D_W:

$$\frac{\partial W}{\partial t} = \frac{\partial}{\partial x}\left(D_W \frac{\partial W}{\partial x}\right) \quad \textbf{(5.3)}$$

A similar equation applies to heat conduction in the food:

$$\rho_s c_s \frac{\partial T}{\partial t} = \frac{\partial}{\partial x}\left(K \frac{\partial T}{\partial x}\right) \quad \textbf{(5.4)}$$

To solve these equations the boundary conditions must be defined. Two models are in common use, depending on the situation.

The Receding Front Model

This model assumes an evaporation-sublimation interface that gradually moves inward from the food surface. At this interface, the mass flux can be expressed as

$$\dot{m} = \frac{\rho_a (Y_i - Y_a)}{1 / k_c + \delta / D_Y} \quad \textbf{(5.5)}$$

At the same time, heat is transferred from the environment to the evaporating interface

$$q = \frac{(T_a - T_i)}{1/h + \delta/K} \tag{5.6}$$

The mass and heat fluxes are related by an energy balance

$$q + K\left(\frac{\partial T}{\partial x}\right)_{\delta+} = \lambda \dot{m} \tag{5.7}$$

The difference between moisture diffusion from inside the food and the evaporation flux causes the evaporation surface to recede into the food at a speed $d\delta/dt$ given by

$$\rho_s W_i \frac{d\delta}{dt} = \dot{m} - \rho_s D_W \left(\frac{\partial W}{\partial x}\right)_{\delta^+} \tag{5.8}$$

Below the freezing point, water vapor exists in equilibrium with a pure ice phase (Fennema, 1981), and the absolute humidity at the evaporating surface (Y_i) is a function of temperature only

$$Y_i = Y^*_{w,T_i} \tag{5.9}$$

The Partially Wetted Surface Model

This model assumes that evaporation occurs at the food surface ($\delta = 0$). However, as the surface moisture content (W_s) decreases, the humidity (Y_s) at the surface is only a fraction a_{ws} of the free water value, due to partial drying of the surface and possibly to concentration of solutes. The surface water activity (a_{ws}) depends on surface moisture (W_s)

$$Y_s = a_{ws,X_s} Y^*_{w,T_s} \tag{5.10}$$

(In this chapter we shall assume that $a_w = Y/Y^*_w$, which is a reasonable assumption at low temperature.) The other boundary conditions are obtainable from equations 5.5–5.8 by putting $\delta = 0$ and $\partial\delta/\partial t = 0$.

Wet-Bulb Conditions

When the surface is completely wetted ($\delta = 0$ in the receding front model), the object is at uniform temperature ($\partial T/\partial x = 0$) and steady state has been reached ($\partial/\partial t = 0$), the energy balance at the surface is simplified to

$$\frac{Y_s - Y_a}{T_a - T_s} = \frac{h}{\lambda \rho_a \beta} \tag{5.11}$$

This is the well-known psychrometric equation giving the slope dY/dT of the wet-bulb line from which the surface conditions Y_s, T_s can be easily calculated because the ratio h_c/β can be calculated from the Lewis relationship

$$\frac{h_c}{\beta} = \rho_a c_a \left(\frac{Sc}{Pr}\right)^{2/3} = 0.90\ \rho_a c_a \tag{5.12}$$

where c_a is the specific heat of air including the moisture it contains (measured in Joules per degree Kelvin per kilogram dry air), Sc the Schmidt number, and Pr the Prandtl number of air.

5.3.2 Moisture Losses during Chilling

Radford (1976) measured the diffusivity (D_W) of water in meat using tritiated water. He obtained the following Arrhenius relationship

$$D_W = 1.1 \times 10^{-6} \exp(-2300 / T)$$

For surface adipose tissue, values ranging from $8.8 \pm 4.3 \times 10^{-11}$ (for dried tissue) to $34 \pm 21 \times 10^{-11} m^2\ s^{-1}$ (for wet tissue) were found. Using the measured D_W values, Radford et al. (1976) solved the heat and mass transfer equations of the partially wetted surface model by the finite differences method. The results agreed well with measurement. They indicated that evaporation from the meat occurred in three stages. Initially there was a brief period of a few minutes when the evaporation was the same as that from free water. In the second stage the evaporation rate progressively decreased below the free water value as the surface dried. In stage three the rate was still below that of free water but gradually caught up as the evaporation rate slowed down (due to falling surface temperature) and diffusion from inside the meat rewet the surface. In a small-scale experiment on lean meat performed by Radford et al., the minimum relative evaporation rate (maximum surface dryness) occured between 30 and 75 min. from the start of cooling. However, for beef carcasses, data showed that this minimum occured at 8–10 h after cooling started.

Data for D_W in the chilling temperature range are hard to come by. Apart from meat, the D_W for some cheeses has been measured by Luna and Chavez (1992) and is about $1.23 \times 10^{-14}\ m^2\ s^{-1}$ at $T = 12.6°C$.

Moisture Losses during Chilling: Lumped Parameter Model

Califano and Calvelo (1981) made the assumptions that the temperature was uniform throughout the product (hence the term "lumped parameter model"), and the surface of the product was completely wetted ($a_{ws} = 1$). Equations 5.5–5.8 can then be simplified as follows:

- mass flux

$$\dot{m} = \rho_a \beta (Y_s - Y_a) \tag{5.13}$$

- heat transfer between surface and air

$$q = h(T_a - T_s) \tag{5.14}$$

- energy balance at the surface (heat of evaporation = heat from air + heat from product)

$$\lambda \dot{m} = q + \frac{W c_s}{A} \frac{\partial T}{\partial t} \tag{5.15}$$

Solution of this set of equation leads to an equation for the fractional weight loss m_{12}/M when the product is cooled from T_1 to T_2

$$\frac{m_{12}}{M} = c_s \int_{T_1}^{T_2} \frac{Y^*_{w,T} - Y_a}{\frac{h}{\rho_a \beta}(T - T_a) + \lambda[Y^*_{w,T} - Y_a]} dT \tag{5.16}$$

Or, by applying Lewis's heat and mass transfer analogy (equation 5.12)

$$\frac{m_{12}}{M} = c_s \int_{T_1}^{T_2} \frac{Y^*_{w,T} - Y_a}{0.9\, c_a(T - T_a) + \lambda[Y^*_{w,T} - Y_a]} dT \tag{5.17}$$

Pham (1987a) simplified this equation to

$$\frac{m_{12}}{M} = 1.1\, c_s \int_{T_1}^{T_2} \frac{Y^*_{w,T} - Y_a}{H^*_T - H_a} dT \tag{5.18}$$

The humid enthalpy, H (enthalpy of air and the water vapor it contains, per kilogram dry air), is readily available from psychrometric charts. Still another simplification is obtained by assuming that the integrand does not change appreciably over the range considered, so that

$$\frac{m_{12}}{M} = 1.1 \frac{c_s[Y^*_{w,T_m} - Y_a](T_2 - T_1)}{H^*_{T_m} - H_a} \tag{5.19}$$

which can be readily evaluated from a psychrometric chart. In practice, both surface drying and temperature gradients in the product contribute to decrease the weight loss, and hence, values from the above formulas can be considered as theoretical maximal values. The discrepancy is larger for bigger products because of the steeper temperature profile in those products. Califano and Calvelo (1981) suggested the following correction factor for the chilling of lamb and beef carcass:

$$\frac{m_{12\,(actual)}}{m_{12\,(predicted)}} = -\ 0.0542\, M^{1/3} \tag{5.20}$$

Moisture Losses during Chilling: General Observations

During the initial stages of chilling, there is a large temperature difference between the product surface and the air, and $Y_i - Y_a$ will be very large. Variations in the relative humidity of the air is unlikely to significantly affect this difference and, hence, the rate of weight loss. It is advantageous to get through this stage as quickly as possible by using low air temperature and high air velocities to rapidly cool the surface. A high cooling rate has the further advantage of drying the surface and reducing Y_i. Spray chilling is particularly useful during this stage in reducing moisture loss. Once the surface is cooled, it is important to maintain a high relative humidity in the air (as long as there is no risk of condensation) to prevent moisture loss. Finally, when the product has equilibrated, air velocity should be reduced to a minimum to avoid further losses.

5.3.3 Moisture Losses during Freezing

The equations for weight loss during chilling (equations 5.18 and 5.19) can also be applied to the freezing of unwrapped products, but because of the harsher conditions (high air velocities and low air temperatures) found during commercial freezing, the ratio of actual to predicted moisture losses will be smaller. Pham (1987a) found experimental weight losses of around 50% of the theoretical maximum during the freezing of lamb carcasses in commercial blast freezers that operated at -25 to -30°C and 1–2 m s^{-1} air velocity. Gentle freezing by natural convection led to better agreement (about 80% of prediction).

Wrapping the product in an impervious film will reduce moisture loss. With loose wraps, evaporating moisture will form ice inside the wrap. This is minimized if close contact is ensured between film and product by using vacuum or heat-shrink wrapping.

During freezing, some products are susceptible to an appearance defect known as freezer burn. Freezer burn is not necessarily correlated with moisture loss in a simple manner. Kaess (1961) showed that the worst appearance for frozen meat and liver occurred when small cavities were left just under the surface by sublimating ice which refracted light. This tended to happen when the product was initially wet (so that ice formed just under the surface) and freezing was very fast (creating small crystals). Paradoxically, both conditions would tend to minimize overall moisture losses.

5.3.4 Moisture Losses during Frozen Storage

A receding front model can be used for studying moisture losses during frozen storage. Since there is no diffusion of ice, $D_W = 0$. Furthermore pseudo–steady state can be assumed, resulting in uniform product temperature ($\partial T/\partial x = 0$). From these assumptions, Pham and Willix (1984) derived a relationship between humidity and temperature differences that was similar to the wet-bulb relationship (equation 5.11)

$$\frac{Y_i - Y_a}{T_i - T_a} = -s' \tag{5.21}$$

where

$$s' = \left(\frac{1 + \beta\delta / D_Y}{1 + h_c\delta / k_d}\right)s \tag{5.22}$$

and s was the slope of the wet-bulb line such that

$$s = \frac{h_c}{\lambda\rho_a\beta} \tag{5.23}$$

Still further simplifications can be obtained by neglecting δ/k_d on the ground that the thermal conductivity, k_d, of the desiccated food layers is much higher than that of air, which leads to the equation (Pham and Willix, 1984)

$$Y_i - Y_a = \frac{Y^*_{T_a} - Y_a}{1 + \dfrac{1}{1 + \beta\delta / D_Y} \cdot \dfrac{r}{s}} \tag{5.24}$$

where r is the local gradient of the water vapor saturation curve. An expression for the rate of sublimation follows:

$$\dot{m} = \frac{\rho_a\beta(Y^*_{T_a} - Y_a)}{1 + r / s + \beta\delta / D_Y} \tag{5.25}$$

If radiation heat transfer is not negligible, it can be taken into account via

$$\dot{m} = \frac{\rho_a\beta(Y^*_{T_a} - Y_a)}{1 + \dfrac{r}{\alpha s} + \dfrac{\beta\delta}{D_Y}} \tag{5.26}$$

where $\alpha = 1 + h_r/h_c$. Furthermore, at subfreezing temperatures r/s is a psychrometric quantity that depends only on the air temperature T_a (in degrees Celsius)

$$r / s \approx \exp(0.085\, T_a) \tag{5.27}$$

Hence the sublimation rate becomes

$$\dot{m} = \frac{\rho_a\beta(Y^*_{T_a} - Y_a)}{1 + \dfrac{\exp(0.085\, T_a)}{\alpha} + \dfrac{\beta}{D_Y}\delta} \tag{5.28}$$

Since $\dot{m}$ is proportional to dd/dt, the above equation can be integrated to give d, from which the weight loss can be found as a function of storage time. This leads to the prediction that the cumulative weight loss w varies as the square root of time when environmental conditions are constant. This has been confirmed ex-

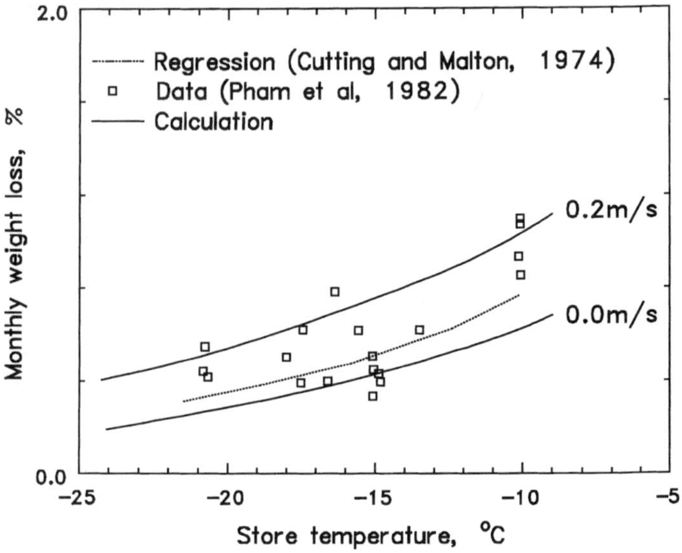

Figure 5.2. Effect of storage temperature on the rate of weight loss from frozen lamb wrapped in netting

perimentally by Storey and Graham (1981). Pham and Willix (1984) found D_Y values ranging from 0.3×10^{-6} to 3.3×10^{-6} m^2 s^{-1} for frozen lamb at -15°C, as compared to about 2.0×10^{-5} m^2 s^{-1} for water vapor in air. Using these values, they successfully predicted the rate of weight loss from lamb carcasses in commercial cold stores (Pham and Willix, 1985) as shown in figure 5.2. Few data have been obtained for other materials.

Experimental weight loss data are subject to very large scatter. Even in the same store, the rate of weight loss can vary by one to two orders of magnitude, due to local variations in the microclimate (Storey and Graham, 1981; Pham et al., 1982). However, from the overall body of data some trends have emerged.

Effect of Storage Temperature on Weight Loss

Typical values for the rate of weight loss from exposed meat carcasses during frozen storage have ranged from nearly zero to about 1% per month. Data from several works (Rutov, 1955; Cutting and Malton, 1974; Pham et al., 1982) indicated that the rate of sublimation doubled for every 10°C rise in temperature. If the relative humidity in cold stores is constant, since the saturation humidity (Y_w^*) doubles approximately every 7°C in the temperature range of -25 to -5°C, one would expect the rate of weight loss to also double for every 7°C. Therefore it seems that relative humidity tends to decrease with decreasing store temperature, for reasons that will become clear later.

Effect of Relative Humidity: Dry vs. Wet Heat

Relative humidity (r.h.) is a major source of variation in determining moisture loss (Pham and Willix, 1985). Relative humidity is calculated from the heat and moisture sources and sinks in the environment and is decreased by dry heat sources. Wet heat is heating accompanied by a high humidity, such as the heat resulting from air interchange through doorways. With wet heat, the driving force for moisture loss is reduced or increased to a lesser extent than in the case of dry heat sources such as lights, electric motors, and heat conduction through walls, floors, and ceilings. An example of the effect of wet heat is the occurrence of frost and ice near doors that are not properly protected or maintained.

Attempts have been made to eliminate dry heat by using jacketed cold stores in which cold air is circulated inside a double-layer wall so as to remove the heat before it reaches the product. Such stores have been reported to be technically successful, reducing weight losses to negligible amounts, but are expensive to construct. Storey and Graham (1981) found that weight loss was almost completely eliminated in an experimental jacketed cold store.

In stores of conventional construction, the floor is the most likely source of dry heat. Heat conducted through walls tends to have only local effects, and heat transferred through the ceiling is quickly removed via air return ducts. The floor, because of its position and large area, can act as a giant radiator. Pham and Willix (1985) found a correlation between the local relative humidity and the neighboring floor temperature. Underfloor temperature should be as close to the freezing point as possible, while avoiding the risk of frost heave (+4°C is a commonly used set point). There was an unreported incident in which a temperature controller broke down, causing underfloor temperature to rise to 40°C or higher (acting as a giant panel dryer). This resulted in a huge (and at the time puzzling) rate of weight loss in unwrapped product.

Heat conduction also explains why lowering the storage temperature does not lead to as much savings in weight loss as could be expected. The increased heat flux through the enclosure results in a lowering of the relative humidity. Another possible source of dry heat is product that is brought inside the store before being fully frozen and continues to radiate heat for some period. In small store rooms, chest freezers, transport containers, and display cabinets, the ratio of area to volume is high and insulation often inadequate, resulting in high conduction heat loads that act as dry heat sources. The relative humidity can therefore be expected to be lower and rates of weight losses higher than in large cold stores.

Lowering the cooling coil (evaporator) temperature will decrease relative humidity because more moisture will be taken out of the air for a given amount of heat. Therefore, whenever exposed product is involved, the evaporator area must be large enough to enable it to do its job at a relatively small temperature difference (about 5°C or less).

5.3.5 Effects of Temperature Fluctuations on Moisture Loss

Temperature fluctuations are unavoidable during frozen storage, particularly in small cold stores. This problem is even more serious in display cabinets, during transport, and in home freezers. The effect of small temperature fluctuations (e.g., 1–2°C) should in theory be negligible. The effects of such small changes are almost linear (i.e., the effect of a negative change is equal and opposite to that of a positive change) and will cancel out. The food will suffer no more than if it was exposed to a steady temperature in the middle of the range.

It is widely believed, however, that temperature fluctuations generally have bad effects. There could be some justification for such beliefs. The effect of large fluctuations will *not* be linear because of the exponential effect of temperature on rate of changes shown by Arrhenius's law or the temperature quotient law. A positive deviation will have a larger effect than a negative deviation. For example, if an increase of 10°C from the mean temperature causes the rate to double, a decrease of 10°C will cause it to halve; the net effect will be an increase of 25% in the mean rate of change. The effect will be even more drastic in the vicinity of the freezing point, the glass transition point, and other phase transition temperatures where serious nonlinearities occur.

Another phenomenon that may magnify the effect of temperature fluctuations is irreversibility. Consider the situation where a frozen capillary porous food is wrapped in a vapor-tight film. A drop in the environment temperature will cool the wrap, causing water vapor to frost on it, and a decrease in the water vapor pressure. Moisture inside the pores will sublime and diffuse to the wrap until the food itself equilibrates to the new temperature. When the environment warms up again, the ice on the wrap will sublime and should in theory diffuse back into the pores where it will deposit. However, the pores might have partially collapsed or shrunk due to loss of the ice that gave them structural strength. Alternatively, ice may bridge some of the pore openings, preventing further resorption. Yet another cause of irreversibility may be the loss of unfrozen water with consequent ionic concentration and chemical denaturation, such that the water binding capacity of the food is lost. Thus, after a cooling-warming cycle, there may be a net migration of moisture from the inside toward the surface and the wrap. This "cavity ice" is a common occurrence in wrapped food after storage and may occur even in foods that are wrapped after freezing.

For unwrapped food, the effect of temperature changes depends on the source of these changes. Earlier, a distinction was made between wet heat sources such as ingress of warm air, and dry heat sources such as lights or heat conduction through walls and floors. Pham (1987a) showed that the effect of temperature changes can be estimated using a psychrometric chart. Figure 5.3 illustrates the situation where air in a cold store is raised from B to D because of ingress of warm air at C. As a result the air near the product surface rises from E to A, where A is the intersection of the wet-bulb line through D with the saturation curve. The

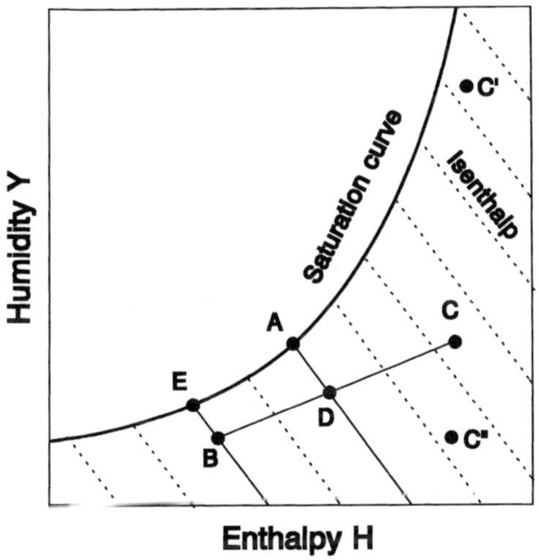

Figure 5.3. Effect of change in air condition on rate of moisture change

ratio *AD*/*EB* then gives the ratio of the new rate of sublimation to the original rate. It can be seen that air ingress may decrease moisture loss or even cause a moisture gain (frosting on the surface) if air conditions are represented by point C' (wet heat). On the other hand, heat gain through conduction (point C'', dry heat) will greatly accelerate moisture loss.

5.3.6 Minimization of Moisture Losses

The unified mathematical model of this chapter shows why the following measures will decrease moisture losses.

During Chilling and Freezing

- Chilling and freezing must be as fast as possible to quickly lower the surface temperature and minimize the moisture diffusing from inside the food. Of course other factors must also be taken into account, such as the effect of cooling rate on sensory qualities. Meat, for example, will become tough when cooled too fast.
- Small pieces of food have a high area: volume ratio and are more susceptible to moisture loss. They should be frozen quickly using low temperature and high heat transfer coefficients.
- As soon as the desired end-point temperature is reached, the product should be taken to storage. The air velocity must be decreased to the greatest degree possible.

- Spray and immersion chilling and freezing, and chilling-freezing in packages are obvious ways of minimizing or reversing moisture loss. If the impervious wrapping is not in close contact with the food, ice will frost inside the package.

During Storage, Transport, and Display

- Moisture loss is least in large, well-insulated, well-stocked cold stores with cold flowing as an envelope around the product to remove heat before it can reach the product. Pay particular attention to the floor temperature and insulation, and minimize other dry heat sources such as lights, motors, and warm products.
- Use large cooling coil areas and low temperature differences between air and coil.
- Certain foods (in particular, seafoods) can be protected by glazing. The frozen food is repeatedly dipped into water near its freezing point to allow the formation of an ice layer that protects the product from desiccation.

5.4 Moisture Absorption and Redistribution

In certain manufactured foods such as frozen pies and pizzas, the components differ widely in water activity. Water will diffuse from the component with highest a_w to that with lower a_w. Since the ice and bound water phases have very low diffusivities, moisture migration proceeds by diffusion of water vapor into the pores of the dry food, followed by adsorption. Moisture absorption is the reverse of drying and both obey the same equations. The physics of both phenomena were reviewed by King (1968).

In the dried food, moisture exists in (at least) two phases: as vapor in the pore and as adsorbed/absorbed (immobilized) moisture. The water vapor diffuses under the influence of a humidity gradient $\partial Y/\partial x$. Moisture accumulates in both vapor and immobilized phases as a result of diffusion; however, the vapor accumulation can be ignored. A material balance can be written as

$$\rho_s \frac{\partial W}{\partial t} = \rho_a \frac{\partial}{\partial x}\left(D_Y \frac{\partial Y}{\partial x}\right) \quad \textbf{(5.29)}$$

where the left-hand side represents the rate of moisture accumulation and the right-hand side the net diffusion. Y and W are related by an equilibrium relationship

$$Y = Y^*_{w,T}\, a_w(T,W) \quad \textbf{(5.30)}$$

where a_w is the water activity as represented by the moisture sorption isotherm. Y can be eliminated to give the standard diffusion equation governed by moisture profile W

$$\frac{\partial W}{\partial T} = \frac{\partial}{\partial x}\left(D_{\mathrm{W}}^{\mathrm{eff}} \frac{\partial W}{\partial x}\right) \tag{5.31}$$

with

$$D_{\mathrm{W}}^{\mathrm{eff}} = \frac{\rho_a}{\rho_s} Y^* \left(\frac{\partial a_w}{\partial W}\right)_T D_Y \tag{5.32}$$

Because of the sigmoidal shape of most sorption isotherms (fig. 5.2), $(\partial a_{\mathrm{w}}/\partial W)_T$ and hence D_{eff} vary in a nearly stepwise manner with W. This behavior has been observed in the drying of foods (Chiriffe, 1984). King (1968) also stated that D_{eff} is very small at low moisture content: this can be understood by looking at the isotherm (fig. 5.2), which shows $(da_{\mathrm{w}}/\partial W)_T$ falling at very low W.

Crisp foods such as potato chips and biscuits lose their crispness when their water activity exceeds about 0.35, which happens at a very low moisture, typically 5% or less (Katz and Labuza, 1981). One theory is that when the moisture content is less than or equal to the BET monolayer value (fig. 5.2), there are strong hydrogen bonding and Van der Waals forces between the molecules or within the carbohydrate matrix. Crispness is lost when water molecules dilute these bonds. Because dried foods need to be maintained at such low water activities, long-term preservation requires a watertight film. In the case of pies and pizzas, edible water tight barriers such as hydrocolloid films or lipid films may be used to separate components that have different a_{w} (Fennema et al., 1993), in order to prevent moisture redistribution affecting the texture of the crust.

5.5 Ice Recrystallization

5.5.1 Mechanisms of Recrystallization

In a frozen food, ice crystals obey an approximately Gaussian size distribution (Bevilacqua and Zaritzky, 1982). Recrystallization causes the mean crystal diameter (d_{m}) to increase, particularly if the temperature fluctuates.

The principal mechanisms of recrystallization in frozen foods are iso-mass recrystallization, migratory recrystallization, and accretion. The driving force behind all forms of recrystallization is the minimization of surface energy. The free energy of a crystal is inversely proportional to its radius of curvature (Kingery et al., 1975). Surfaces with negative curvature (concave surfaces) will have lower energy than plane surfaces, which have lower energy than convex surfaces.

Iso-mass recrystallization happens because crystals gradually take a spherical shape, which for a given mass has the smallest area and therefore lowest surface energy. Accretive recrystallization is also driven by minimization of surface energy, since when two adjacent surfaces merge, the surface area is reduced. Accretion starts with a neck forming between the surfaces. The negative curvature of the neck means that surface energy is minimized there; water molecules will preferentially attach to the neck, which gradually grows until the two crystals become indistinguishable. Migratory recrystallization happens because small crystals have smaller radii of curvature and hence higher specific energy than larger ones. Smaller crystals will gradually lose moisture to the larger ones and eventually disappear, leading to a gradual increase in the mean crystal size.

The rate of crystallization depends on both the activation energy for molecules to move from or to the crystal surface, and the diffusion rate from one crystal to another. With very high moisture foods, in which the crystals are almost contiguous, the process is controlled by the former, while in foods with lower moisture content, such as meat (80% moisture or less), the latter will be the controlling factor and recrystallization will be much slower. The equation for the average crystal size, proposed by Bevilacqua and Zaritzky (1982), is

$$\frac{d\,d_m}{dt} = \frac{k}{d_m} \tag{5.33}$$

where k follows the Arrhenius relationship. Integration of this equation leads to the prediction that d_{m}^2 grows linearly with time

$$d_m^2 = d_0^2 + 2k_1 t \tag{5.34}$$

This equation incorrectly predicts that the mean crystal size keeps growing indefinitely, while empirical evidence has indicated that there is a maximum size that is on the order of 100 μm (Martino and Zaritzky, 1989). Martino and Zaritzky (1989) introduced the following modification to take into account the existence of a limiting diameter d_1:

$$\frac{d\,d_m}{dt} = k_2\left(\frac{1}{d_m} - \frac{1}{d_l}\right) \tag{5.35}$$

Integration led to

$$\ln\left(\frac{d_l - d_0}{d_l - d_m}\right) + \frac{1}{d_l}(d_0 - d_m) = \frac{k_2}{d_l^2}t \tag{5.36}$$

They found the activation energy ΔE of k_2 to be about 11.6×10^4 J mol^{-1} for 0.28 N NaCl and 42.4×10^3 J mol^{-1} for beef muscle. This activation energy leads to a doubling of the recrystallization rate ($d_{\mathrm{m}} << d_1$) approximately every 9°C rise in temperature (as long as the limiting diameter is not approached), and is very close to those for protein denaturation and quality losses, supporting the hypothesis that all these processes are controlled by diffusion processes in the food.

Martino and Zaritzky (1989) also found that the limiting diameter d_1 can be calculated from

$$d_l = \frac{d_s}{\epsilon_s} \quad \textbf{(5.37)}$$

where d_s is the mean diameter of non-ice inclusions and ϵ_s the volume fraction of the inclusions. For beef, d_s (the mean diameter of muscle fibers) ranged from 31 to 44 μm and ϵ_s (volume fraction of fibers) from 0.50 to 0.70, giving limiting diameter values d_1 of about 61 μm, in accord with measurements.

Recrystallization is a particularly important consideration with foods that are frozen quickly, using immersion, IQF, and, especially, cryogenic processes. Usually a fast (and hence expensive) freezing process is justified on the ground that it produces small ice crystals which are less harmful to the product. It can be seen, however, that if d_m^2 grows linearly with time, small ice crystals will grow proportionately much faster than big ones (equation 5.33 or 5.35). For example, Martino and Zaritzky's (1989) data showed that, at −5°C, the crystals grew from 32 to 50 μm in 5 days then from 50 to only 55 μm in another 5 days. The advantages of fast freezing are therefore soon lost unless the food is stored at a very low temperature or at a temperature below its glass transition point.

It has to be said that not all data on the effect of temperature on recrystallization are adequately explained by the combined Arrhenius-WLF models. Although the Arrhenius equation seems to be adequate for Martino and Zaritzky's (1989) data, which covers the temperature range of −16.5 to −5°C, Sy and Fennema (1971) found a sharp break in the rate vs. temperature curve in beef liver at between −7 and −4°C, which is too high to indicate the effect of glass transition. The crystal growth rate, $dD/d(\ln t)$, jumped from 1.5 μm at −7°C to more than 9 μm at −4°C. Sulc (1974) found that recrystallization in cryogenically frozen fruit juices and coffee extract produced noticeable changes in appearance at −18°C but not at −24°C; again this temperature is too high to explain the results as dependent on the glass transition, which occured at around −40°C for fruit juices (Slade and Levine, 1991).

5.5.2 Effect of Temperature Fluctuations on Recrystallization

Temperature fluctuations greatly accelerate the rate of recrystallization. As the temperature rises, some of the ice melts. This melt will come preferentially from the smaller crystals. In fact, many of these crystals will disappear altogether. During recooling, the moisture from these crystals will diffuse toward the remaining large crystals rather than forming new nuclei. This irreversibility will accentuate the effect of temperature fluctuations. Accretive recrystallization is also facilitated by the ice that melts and refreezes, forming bridges between adjacent surfaces. The effect is highly nonlinear because the unfrozen fraction of water rises rapidly as the freezing point is approached. Products with low freezing points are particularly sensitive to this. For example, the frozen fraction of water in ice cream was 89.5% at −30°C, 89% at −25°C, and 86% at −20°C (Everington, 1988). A ±5° fluctuation around −25°C is therefore much more detrimental than a steady storage temperature of −25°C.

5.5.3 Effect of Additives on Recrystallization

Various additives have been tested in an effort to find methods to slow down recrystallization. Harper and Shoemaker (1983) tested various solutions of sweeteners and a polysaccharide stabilizer (sucrose solution, corn syrup, high fructose corn syrup with locust bean gum, and solutions of corn syrup with locust bean gum); they found that the recrystallization rate was affected by sweetener addition but not by the locust bean gum. Smith and Schwartzberg (1985) found that both sucrose and gelatin reduced crystal growth. However, gelatin was found to be ineffective in ice cream at 0.28% w/w (Buyong and Fennema, 1988), although it did have a desirable influence on sensory texture.

Recrystallization inhibitor research received a big boost with the discovery in 1957 of antifreeze peptides found naturally in cold water fish. Similar substances were subsequently found in insects and have now been synthesized chemically or produced by genetic engineering. Both natural and synthetic peptides are extremely effective recrystallization inhibitors (Warren et al., 1993). These substances are covered in more detail in chapter 11.

5.6 Drip Loss

Purge or drip loss from frozen meat is determined by a number of factors that are intrinsic to the food product itself and the freezing and thawing conditions. Thawing drip loss is visually unattractive, soluble nutrients are lost from the food, and it represents a significant economic loss to the processor when the foodstuff is thawed as part of a manufacturing process. The factors that influence thawing drip loss are many so it is not surprising that work that seeks to analyze thawing drip frequently demonstrates conflicting outcomes. For example, Jul (1984) quoted results that showed that strawberries frozen in liquid air had lower drip (53.5%) than those frozen in brine (59%) or air (62%), but mentioned that other works showed no effect or the opposite effect. The best we can do is to combine physical reasoning with selective experimental evidence to build a plausible picture.

Water is retained in foodstuffs by virtue of cellular structure and within hydrocolloid-based tissue components. Freezing is effectively a dehydration process in which water is removed from its original location within the foodstuff and collected in the form of ice crystals. During thawing, the water may or may not be reabsorbed into its original location within the food's microstructure. The underlying factors that determine thawing drip loss include the size and location of ice crystals with respect to tissue microstructure, the rate of thawing, the rate and extent of water reabsorption, the physiological/biochemical status of the tissue prior to freezing, and the intrinsic water binding strength in the tissue prior to freezing.

In tissues of plant origin, water, once it is separated from its original tissue location by freezing, does not reabsorb either into the cells or pectin and plasma hydrocolloid structures. The effects are compounded by blanching prior to freezing; slow freezing generally enhances these effects while rapid freezing does not intensify them. Recrystallization during frozen storage can undo the benefits of rapid freezing in reducing drip losses. The tissues of young plants are more stable to freezing than those of older plants, and some plant tissues contain cryoprotective factors (Spiess, 1979).

Significant reabsorption of water by tissue hydrocolloids can take place in animal tissues. The reabsorption, assessed from electron micrographs of tissues after thawing, by fish hydrocolloids is generally less than by avian and mammalian tissue hydrocolloids (Spiess, 1979). However, measurement of DNA in thawing drip indicates that similar tissue damage occurs in fish and avian tissues (Love, 1955; Crigler and Dawson, 1968).

In muscle tissue prior to freezing, 5–12% of the total water volume is located in spaces between fibers (cells), and 88–95% is within the fibers. Within the fibers, 22–24% is between the myofibrils, and 66–71% is within the myofibrils. Freezing relocates the water from within the myofibrils and within the fibers to an extent determined by the rate of freezing. The slower the freezing rate, the greater the relocation to between the fibers (Voyle, 1974). Añón and Calvelo (1980) found that there was a peak thawing drip loss associated with decreasing freezing rate at $t_c = 10$ min (t_c is the time for the temperature to fall from $-1°C$, the freezing point of meat, to $-7°C$, the point where 80% of the moisture is frozen). At $t_c > 20$ min the thawing drip loss remained constant at about 65% of the peak level (fig. 5.4). At $t_c < 10$ min most of the ice crystals are located intracellularly between the myofibrils; when t_c is in the range of 10–18 min, single large intracellular ice crystals coexist with extracellular ice crystals, and as the freezing rate decreases, water becomes increasingly relocated to extracellular locations, with increasing distortion of the tissues. Drip loss is believed to be associated with the occurrence of a large intracellular crystal, which causes maximal damage to the cell wall. During frozen storage, as previously mentioned, high or fluctuating temperatures and long storage times encourage recrystalization which can lead to tissue damage and drip loss on thawing (Martino and Zaritzky, 1988).

Thawing time and reabsorption of moisture are interrelated in their influence on thawing drip loss. Gonzalez-Sanguinetti et al. (1985) studied this relationship in frozen beef and found that faster thawing rates produced more exudate by virtue of the reduced time for reabsorption, although the effect was small for $t_d > 50$ min (t_d was the time for product temperature to rise from $-5°C$ to $-1°C$). Reabsorption is a slow process and takes 4–5 h to complete; for example, at 1 h and 4°C, 74% of the extracellular moisture was absorbed, and after 3 h this had increased to 80%. At the surface of meat pieces, meat will not have the opportunity

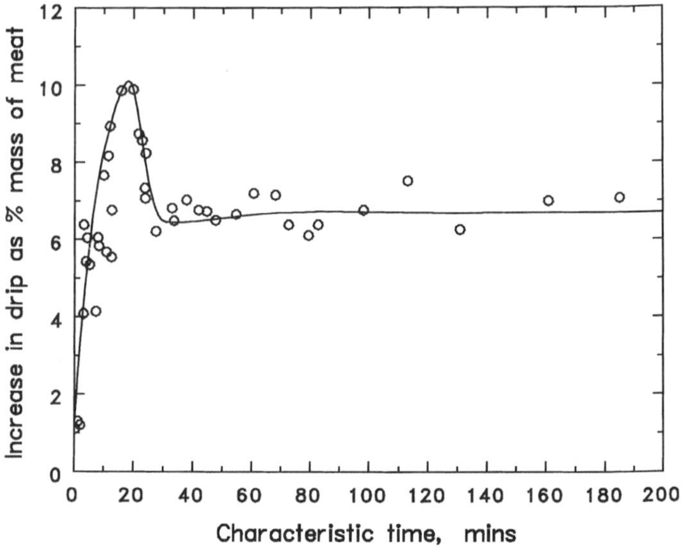

Figure 5.4. Increase in drip loss due to freezing as a function of characteristic freezing time. (Añón and Calvelo, 1980; with kind permission from Elsevier Science, Ltd., Oxford, U.K.)

to reabsorb water; however, surface drip can be minimized by ensuring that $t_d > 50$ min at the surface. Frozen storage will also influence water reabsorption by meat hydrocolloids through mechanisms such as free radical proliferation from oxidizing lipids, and attack on tissue structures and hydrocolloid proteins.

The physiological/biochemical state of meat at the time of freezing can have a profound effect on thawing drip, especially if thawing is rapid. Davey and Gilbert (1976) studied the freezing and thawing of prerigor lamb. If the meat was stored below −12°C the meat went rapidly into rigor during the thawing process with severe contracture (60–70% in boneless meat) such that the meat became very tough and had substantial drip losses. If the storage temperature was above −12°C, enzymic hydrolysis of the adenosinetriphosphate, which provides energy for the contractile process, occurred. After 20–30 days the levels were too low to cause contraction. The meat protein hydrocolloids that bind water have an intrinsic water binding (holding) capacity. Penny (1975) found that in pork the water binding capacity prior to freezing was directly related to the water absorption after freezing.

It must be remembered that the freezing or thawing rate depends on location within the food. Generally, the surface has a smaller t_c or t_d than the material further inside, especially at high heat transfer coefficients (such as in immersion freezing), However, the thermal center will have a smaller t_c or t_d value than the material around it because there is no further latent heat to slow down the rate of

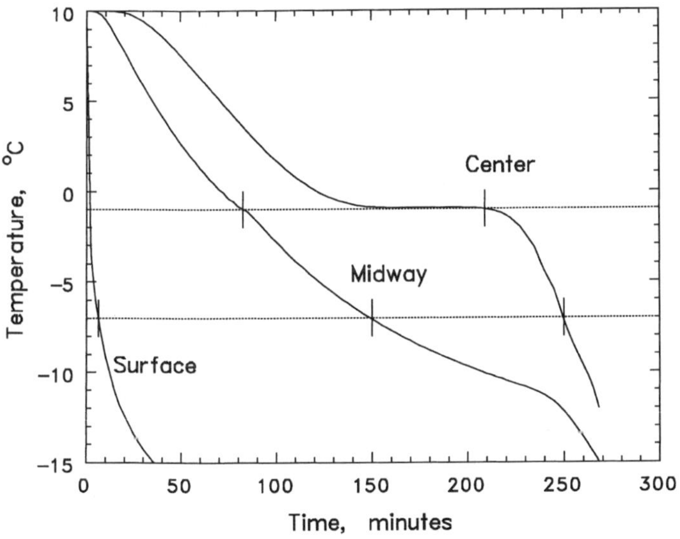

Figure 5.5. Temperatures at the surface, center, and midway to the center in a 10-cm slab of meat frozen by immersion at −25°C, calculated by finite differences

decrease. The information in figure 5.5, obtained by finite difference calculations, illustrates these points. A number of simple methods (see e.g., Plank, 1941; Cleland and Earle, 1982; Hung and Thompson, 1983; Pham, 1986) are also available to quickly estimate the freezing and thawing times, t_c and t_d.

5.7 Conclusion

Since the objective of food freezing is to stop all changes, moisture migration in any form is generally undesirable. An understanding of the driving mechanism of these changes will improve our ability to minimize them. In this chapter we have tried to quantify this understanding by putting it into mathematical form. Much research remains to be done before these equations can be widely used for predictive purposes. Data on moisture diffusivity, sorption curve, and glass transition temperature are still lacking for most foods, and our understanding of recrystallization and especially drip loss remains largely nonquantitative.

5.8 Nomenclature

a_w	Water activity, $p_w/p_w^0 \approx Y/Y_0^*$ at low temperature
a_{ws}	Water activity at food surface

A	Product surface area, m^2
c_a	Specific heat of air and associated moisture, J kg^{-1} dry air K^{-1}
d	Crystal or particle diameter, m
d_0, d_1	Initial and limiting crystal size, m
d_m	Mean crystal size, m
D	Diffusivity, m^2 s^{-1}
D_W, D_Y	Diffusivity of moisture in food and of water vapor through dried food respectively
ΔE	Activation energy, J mol^{-1}
h	Heat transfer coefficient, W m^{-2} K^{-1}
h_c, h_r	Convection and radiation heat transfer coefficients
H	Enthalpy of air, dry basis, J kg^{-1} dry air K^{-1}
H_a	Enthalpy of air in environment
H^*	Saturation enthalpy
k, k_1, k_2	Rate constants
K	Thermal conductivity of food, W m^{-1} K^{-1}
$\dot{m}$	Mass flux, kg m^{-2} s^{-1}
M	Mass of product, kg
m_{12}	Mass loss from time 1 to time 2, kg
p_w	Water partial pressure
q	Heat flux, W m^{-2}
r	Slope of saturation line, K^{-1}
R	Gas constant, J mol^{-1} K^{-1}
s	Slope of wet-bulb line, K^{-1}
s'	Slope of modified wet-bulb line
t	Time, s
T	Absolute temperature, K
T'_g	Glass transition temperature, K
T_i,T_s	Temperatures at evaporating surface and at food surface
x	Distance from food surface to a point in the food, m
W	Moisture content dry basis, kg moisture kg^{-1} dry solid
W_b,W_f	Bound and free moisture contents, dry basis
Y	Absolute humidity dry basis, kg water kg^{-1} dry air
Y_a,Y_i,Y_s	Absolute humidity in environment, at evaporating surface and at food surface

Y^*, Y^*_w	Saturation humidity of an aqueous solution and of pure water separately.
α	$1 + h_r/h_c$
β	Mass transfer coefficient (mass flux/concentration difference), m s^{-1}
δ	Thickness of dried layer
λ	Latent heat of vaporization/sublimation, J kg^{-1}
ρ_a	Air density dry basis, kg dry air m^{-3}
ρ_s	Solid density (dry basis), kg dry matter m^{-3}

References

Añón, M.C. and Calvelo, A. 1980. Freezing rate effects on the drip losses of frozen beef. *Meat Sci.* **4**:1–14.

Bevilacqua, A.E. and Zaritzky, N.E. 1982. Ice recrystallization in frozen beef. *J. Food Sci.* **47**:1410–1414.

Buyong, N. and Fennema, O. 1988. Amount and size of ice crystals in frozen samples as influenced by hydrocolloids. *J. Dairy Sci.* **71**:2630–2639.

Califano, A.N. and Calvelo, A. 1981. Weight loss prediction during meat chilling. *Meat Sci.* **5**:5–15.

Chiriffe, J. 1984. Fundamentals of the drying mechanism during air dehydration of foods. In *Advances in Drying,* vol. 2. (Mujumbar, A.S., ed.) p. 73, Hemisphere, New York.

Cleland, A.C. and Earle, R.L. 1982. Freezing time prediction for foods—a simplified procedure. *Int. J. Refrig.* **5**:134–140.

Crigler, J.C. and Dawson, L.E. 1968. Cell disruption in broiler breast muscle related to freezing time. *J. Food Sci.* **33**:248.

Cutting, C.L. and Malton, R. 1974. Evaporative losses in the commercial freezing and storage of meat. *Meat Res. Inst. Symp.* No. 3, Bristol, England.

Davey, C.L. and Gilbert, K.V. 1976. Thaw contracture and the disappearance of andenosinetriphosphate in frozen lamb. *J. Sci. Food Agric.* **27**:1085 1092.

Everington, D.W. 1988. The special problem of freezing ice cream. In *Food Freezing: Today and Tomorrow.* (Bald, W.B., ed.) pp. 133–142, Springer-Verlag, New York.

Fennema, O. 1981. Water activity at subfreezing temperatures. In *Water Activity: Influences on Food Quality*. (Rockland, L.B. and Stewart, G.F., eds.) pp. 713–733, Academic Press, New York.

Fennema, O. 1993. Frozen foods: challenges for the future. *Food Australia* **45**:374–380.

Fennema, O. 1996. Water and Ice. In *Food Chemistry*, 3d ed. (Fennema, O., ed.) pp. 17–94, Marcel Dekker, New York.

Fennema, O., Donhowe, I.G., and Kester, J.J. 1993. Edible films: barriers to moisture migration in frozen foods. *Food Aust.* **45**:521–525.

Gonzales-Sanguinetti, S., Añón, M.C., and Calvelo, A. 1985. Effect of thawing rate on the exudate production of frozen beef. *J. Food Sci.* **50**:607–700, 706.

Harper, E.K. and Shoemaker, C.F. 1983. Effect of locust bean gum and selected sweetening agents on ice recrystallization rates. *J. Food Sci.* **48**:1801–1803, 1806.

Hung, Y.C. and Thompson, D.R. 1983. Freezing time for slab-shape foodstuff by an improved analytical method. *J. Food Sci.* **48**:555–560.

Jul, M. 1984. *The Quality of Frozen Foods*, p. 31. Academic Press, New York.

Kaess, G. 1961. Freezer burn as a limiting factor in the storage of animal tissue. *Food Technol.* **15**:122–128.

Katz, E.E. and Labuza, T.P. 1981. Effect of water activity on the sensory crispness and mechanical deformation of snack food products. *J. Food Sci.* **46**:403–409.

King, C.J. 1968. Rate of moisture sorption and desorption in porous, dried foodstuffs. *Food Technol.* **22**(4):165–171.

Kingery, W.D., Bowen, H.K., and Uhlmann, D.R. 1975. *Introduction to Ceramics*, 2d ed. p. 452, John Wiley, New York.

Love, R.M. 1955. The expressible fluid of fish fillet. I. Nucleic acid as an index of cell damage in fillets frozen from both sides. *J. Sci. Food Agric.* **6**:30.

Luna, J.A. and Chavez, M.S. 1992. Mathematical model for water diffusion during brining of hard and semi-hard cheese. *J. Food Sci.* **57**:55–58.

Martino, M.N. and Zaritzky, N.E. 1988. Ice crystal size modification during frozen beef storage. *J. Food Sci.* **53**:1631–1637.

Martino, M.N. and Zaritzky, N.E. 1989. Ice recrystallization in a model system and in frozen muscle tissue. *Cryobiology* **26**:138–148.

Penny, I.F. 1975. Use of a centrifuging method to measure the drip of pork *Longissimus dorsi* slices before and after thawing. *J. Sci. Food Agric.* **26**:1593–1602.

Pham, Q.T. 1986. Simplified equation for predicting the freezing time of foodstuffs. *J. Food Technol.* **21**:209–219.

Pham, Q.T. 1987a. Moisture transfer due to temperature changes or fluctuations. *J. Food Eng.* **6**:33–50.

Pham, Q.T. 1987b. Calculation of bound water in frozen foods. *J. Food Sci.* **52**:210–212.

Pham, Q.T., Durbin, J.R., and Willix, J. 1982. Survey of weight loss from lamb in cold storage. *Int. J. Refrig.* **5**:337–342.

Pham, Q.T. and Willix, J. 1984. A model for food desiccation in frozen storage. *J. Food Sci.* **49**:1275–81, 1294.

Pham, Q.T. and Willix, J. 1985. Weight loss from lamb in frozen storage: effect of environmental factors. *Int. J. Refrig.* **8**:231–235.

Plank, R. 1941. Beitrage zur Berichnung und Bewertung der Gefriergeschwindigkeit von Lebensmitteln. *Beih. Z. ges Kalte-Ind.* **3**(10):1–16.

Radford, R.D. 1976. Water transport in meat. *Int. Inst. Refrig. Bull.* Annex **1976**(1):315–552.

Radford, R.D., Herbert, L.S., and Lovett, D.A. 1976. Chilling of meat—a mathematical model for heat and mass transfer. *Int. Inst. Refrig. Bull.* Annex **1976**(1):323–30.

Rutov, D. 1955. The diminishing of weight losses in frozen meat during storage. *Proc. Int. Cong. Refrig.*, **2**:4.153–4.160.

Schwartzberg, H.G. 1976. Effective heat capacities for the freezing and thawing of foods. *J. Food Sci.* **41**:152–156.

Slade, L. and Levine, H. 1991. Beyond water activity: recent advances based on an alternative approach to the assessment of food quality and safety. *Crit. Rev. Food Sci. Nutr.* **30**:115–360.

Smith, C.E. and Schwartzberg, H.G. 1985. Ice crystal size changes during ripening in freeze concentration. *Biotechnol. Prog.* **1**:111–120.

Spiess, W.E.L. 1979. Impact of freezing rates on product quality of deep frozen foods. In *Food Process Engineering 1979. (Proceedings of the Second International Congress on Engineering and Food, Helsinki,* vol. 1) (Linko, P. and Larinkari, J., eds.) pp. 689–694.

Storey R.M. and Graham, J. 1981. Mechanism and measurement of weight loss from frozen fish in cold storage. *Proc. Inst. Refrig.* **78**:28–35.

Sulc, S. 1974. Continuous freezing of liquids with liquid nitrogen. *Proc. Fourth Int. Congr. Food Sci. Technol.* **5**(a):56–58.

Sy, S.H. and Fennema, O. 1971. Rates of recrystallization in beef liver tissue. *Proc. Thirteenth Int. Congr. Refrig.*, **3**:199–204.

Voyle, C.A. 1974. Structural and histological changes associated with the freezing of meat. In *Meat Research Institute, Symposium 3.* pp. 6.1–6.6, Meat Research Institute, Langford.

Warren, G.J., Hague, C.M., Corotto, L.V., and Mueller, G.M. 1993. Properties of engineered antifreeze particles. *FEBS Lett.* **321**(2/3):116–120.

Williams, M.L., Landel, R.F., and Ferry, J.D. 1955. The temperature dependence of relaxation mechanisms in amorphous polymers and other glass-forming liquids. *J. Am. Chem. Soc.* **77**:3701–3707.

Winger, R.J. 1984. Storage and eating-related quality of NZ frozen lamb: a compendium of irrepressible longevity. In *Thermal Processing and Quality of Foods.* (Zeuthen, P. et al., eds.), pp. 541–552, Elsevier, London.

6

Freeze-Cracking

Yen-Con Hung

6.1 Introduction

Freezing has long been recognized as an excellent method for preserving food quality. In general the faster the freezing rates, the higher the end-product quality. The definition of freezing rates has been addressed in chapter 2, and one good working definition is to define the freezing rate as the rate of advance of the freezing interface. In general, the more rapidly the freezing interface advances, the smaller and more numerous the ice crystals. The formation of numerous small ice crystals contributes to very fine and homogeneous food structure. In addition, the amount of water that migrates from the cell to the intercellular space will be minimal and, hence, cause little damage to the tissue structure, minimizing drip losses. Because of the fast freezing rate, food products will spend a minimum amount of time in the freezing process. Not only will the product weight loss be minimized, but the production throughput can be maximized.

However, there is a commercially important trade off between the speed of freezing and the economy of the freezing operation. In general, the costs of a freezing operation increased exponentially with increasing freezing rate; at the same time, the improvement of product quality tended to level off beyond a certain freezing rate. It is also important to note that some products may crack or even shatter if the freezing rates are too high, or products are exposed directly to extremely low-temperature freezing medium. The fundamental aspects of freeze-cracking have been reviewed recently by Hung and Kim (1996). This freeze-crack phenomena has also been reported in the literature for a variety of products (Rasmussen and Olson, 1972; Hoeft et al., 1973; Brennan et al., 1976; Nicholson, 1977). More recently, Ogawa (1987) reported cracking of tuna fish during the $CaCl_2$ brine immersion/spray freezing. Sebok et al. (1991, 1994) also reported cracking incidents in frozen fruits and vegetables. Cracking and shattering of frozen foods is a critical problem and may cause downgrades of frozen-food quality, or even rejection.

6.2 Freeze-Cracking

Two types of freeze-crack exist: (1) surface only cracks and (2) cracks originating from inside the product, which then progress to the surface. The surface only cracks are due primarily to the extreme temperature difference between the product and the freezing medium. This type of the crack is usually very evenly distributed over the product surface and consists of numerous minor cracks. For food products with a nonsmooth surface (e.g., frozen beef patties), or products that are covered following freezing (e.g., application of frosting to frozen cakes), surface cracks may not be economically significant.

Cracks originating from inside the product are small in number but tend to be large. For very uniform products, cracking will occur randomly, but for most food products, cracks will originate from the least structurally resistant location. Some cracks may even be severe enough to break or completely shatter the product. Most of the discussion in the remaining chapter will focus on this type of crack because of its economic significance.

6.3 Mechanisms of Freeze-Cracking

Mechanisms to explain freeze-cracks vary. The most popular explanation has been that mechanical damage induced by cryogenic freezing is due to volumetric changes associated with the water-ice phase transition. Volume changes during freezing were attributed to pure water at 0°C expanding approximately 9% when transformed into ice at the same temperature (Fennema, 1975). While most foods expand during freezing, they expand less than pure water. This may be due to differences in moisture content, composition, or amount of bound (unfreezable) water.

Fennema and Powrie (1964) indicated that, in addition to volume expansion, the amount of empty space in the microstructure affects the degree of mechanical damage on cells during freezing. Reeve and Brown (1968), however, suggested that cracking was probably the result of nonuniform contraction following solidification. Similarly, Wolford and Brown (1965) indicated that freeze-cracking relieves a product's internal stresses from nonuniform contraction during rapid cooling. Sebok et al. (1991), on the other hand, claimed that both contraction and expansion may cause freeze-cracking; however, their recent study (Sebok et al., 1994) supports the hypothesis that freeze-cracking is related mainly to stress build up in response to the volumetric increase; nonuniform contraction has only a secondary effect.

The most recent mechanism postulated for freeze-cracks was made by Kim (1993). He suggested that the crust formed at a product's surface serves as a shell that prevents further volume expansion when the internal portion of the unfrozen material undergoes phase transition. This then contributes to the internal stress build up in the latter part of the freezing process. If the internal stress exceeds the strength of the material at any time point during the freezing process, freeze-cracking will occur. Kim and Hung (1994) suggested that freeze-cracking was affected by both the initiation of stress during phase change, and the distribution of the developed stress. The distribution of the stress would be governed by how the material responds to the stress generated during freezing. Stress could either be absorbed (dissipated) into the structure or reflected to cause build up of internal stress.

6.4 Detection of Freeze-Cracking

Most of the data reported in the literature on degree and severity of freeze-cracking have been obtained by visual observation or through reliance on human hearing to judge the occurrence of cracks. Then the sizes of cracks or the number of cracked pieces were reported. One objective method for monitoring the development of freeze-cracking is to measure the internal pressure of the tested product (Ogawa, 1987). When the internal pressure exceeds the material strength, a crack will develop. However, this method cannot quantify the number and severity of freeze-cracks.

McCambridge et al. (1996) developed an acoustic method to characterize and quantify freeze-cracks. They used the fast Fourier transformation (FFT) technique to transfer time domain sound data to its frequency domain spectrum. Frequency analysis of freeze-cracking sounds indicated that there is a unique frequency distribution associated with cracks in food products. A signal representing a crack possessed peaks at low frequencies, with a tail extending beyond 3,500 Hz. The ratio of the area of signal associated with cracking to background noise in the frequency range of 3,000 to 10,000 Hz was greater than 2. This ratio also increased with increasing crack intensity. Detection of freeze-cracking during liquid nitrogen spray freezing has been tested on several foods using this acoustic method (McCambridge, 1995). The acoustic method developed by McCambridge had an accuracy of 92% in detecting crack signals, and 98% in detecting background signals. By using a sensor to detect sounds in the range of 1,000 to 9,500 Hz and identify the cutoff of the sound intensity for each specific food, an on-line freeze-cracking detection method can be developed.

6.5 Factors Affecting the Development of Freeze-Cracking

A variety of factors may contribute to the development of freeze-cracking. Sebok et al. (1991) evaluated the effects of freezing rate, precooling, size, and maturity on the freeze-cracking frequency of some fruits and vegetables. They found that freeze-cracking can be reduced by precooling and reducing the freezing rate. Physical properties of foods may also affect a food's freeze-cracking susceptibility.

For example, porosity represents the amount of void space inside a food material. The larger the void space, the greater the chance that internal stress will dissipate instead of accumulate. This will reduce the chance of a freeze-cracking incident. Other potential physical factors affecting the development of freeze-cracking are the modulus of elasticity, density, and Poisson's ratio. Modulus of elasticity defines the ratio of stress over strain and indicates the rigidity of the product. Density influences freeze-cracking because it is normally proportional to moisture content and inversely proportional to porosity. Food products with higher moisture content will have a higher volume expansion ratio during freezing and hence higher freeze-cracking susceptibility. Poisson's ratio defines the ratio of lateral strain to axial strain and gives an indication of material expansion perpendicular to compression. In general, material with greater modulus of elasticity and Poisson's ratio are more rigid and more likely to build up stress than those with smaller modulus of elasticity and Poisson's ratio.

Kim and Hung (1994) evaluated the effects of density, modulus of elasticity, Poisson's ratio at both room temperature and −20°C, shape (axial ratio), size (weight), moisture content, and porosity on the freeze-cracking susceptibility of 22 different materials. They found size, moisture content, density (at both temperatures), modulus of elasticity (−20°C), Poisson's ratio (at both temperatures), and porosity all had a significant ($P < 0.05$) influence on freeze-cracking. However, no single property completely explained the development of freeze-cracking. They indicated that if freeze-cracking was caused solely by volume expansion of water during phase transition, moisture content should be the most significant variable contributing to the freeze-cracking incident. Contradicting this notion, however, apples with a high moisture content (87% moisture) did not crack during cryogenic freezing, whereas, margarine (35% moisture) and mozzarella cheese (47% moisture), which had low moisture content, cracked. They suggested that freeze-cracking susceptibility of foods may depend on several interrelated physical properties.

6.6 Predicting a Food's Freeze-Cracking Susceptibility

Kim and Hung (1994) measured physical properties of 22 different foods and then used discriminant analysis (Powers and Ware, 1986) to classify freeze-cracking incidents of these foods during liquid nitrogen (LN_2) immersion freezing into mutually exclusive classes (crack or no-crack). Through a backward elimination process, they found that by using only density at −20°C (DEN20) and porosity (PORO) the freeze-cracking responses of their 500 freezing observations could be successfully classified. The predictive function they developed is presented as follows:

$$PROB(+) = \frac{DJ_l}{DJ_0 + DJ_l} \tag{6.1}$$

$$DJ_0 = \exp(-0.5 \times D_0^2) \tag{6.2}$$

$$DJ_l = \exp(-0.5 \times D_1^2) \tag{6.3}$$

$$D_0^2 = [PORO - 1.350 \quad DEN20 - 1.015] \times \begin{bmatrix} 83.675 & -0.8170 \\ -0.8170 & 0.0138 \end{bmatrix} \times \begin{bmatrix} PORO - 1.350 \\ DEN20 - 1.015 \end{bmatrix} - 0.7213 \tag{6.4}$$

$$D_1^2 = [PORO - 47.254 \quad DEN20 - 0.596] \times \begin{bmatrix} 1401.9 & -14.996 \\ -14.996 & 0.170 \end{bmatrix} \times \begin{bmatrix} PORO - 47.254 \\ DEN20 - 0.596 \end{bmatrix} + 2.5756 \tag{6.5}$$

where *PROB*(+) is the probability that the test material will crack; D_0^2 is the squared Euclidian distance from the no-crack group mean; D_1^2 is the squared Euclidian distance from the crack group mean; 1.350 and 47.254 are *PORO* means of no-crack and crack groups, respectively; 1.015 and 0.5960 are *DEN*20 means of no-crack and crack groups, respectively; the elements in the 2×2 matrices are the variances associated with the no-crack and crack groups; and 0.7213 and 2.5756 in equations 6.4 and 6.5 are the natural logs of the variance of the no-crack and crack groups, respectively. When a food product with a calculated *PROB*(+) of 0.5 or greater is obtained then this product will crack.

The predictive function correctly classified 98.0% (196 out of 200) of the crack samples and 72.3% (217 out of 300) of the no-crack samples. On the prediction side, 70.3% predicted crack samples were confirmed cracked, whereas 98.2% of those predicted not to crack did not.

For example, a fresh apple has determined *DEN*20 and *PORO* values of 0.78 g ml^{-1} and 28%, respectively (Kim and Hung, 1994). These values when used in equations 6.4 and 6.5 resulted in D_0^2 and D_1^2 of 0.2952 and 8.58, respectively. These values were then placed in the right-hand side of equations 6.2 and 6.3 and resulted in DJ_0 and DJ_1 of 0.8628 and 0.01370, respectively. *PROB*(+) can then be calculated as

$$PROB(+) = \frac{0.01370}{0.8628 + 0.01370} = 0.0156$$

A *PROB*(+) of 0.0156 indicates that a fresh apple will not crack during LN_2 immersion freezing.

The study conducted by Kim and Hung (1994) was for LN_2 immersion freezing, and the predictive equations are good only for that freezing process. However, freeze-cracking prediction under LN_2 immersion can be used as the worst-case scenario because of the extremely fast freezing rates. If products will not develop freeze-cracks under that process, they will almost certainly not develop freeze-cracks under any other freezing process.

6.7 Modeling the Development of Freeze-Cracking

One of the main purposes of modeling freeze-cracking is to develop strategies to prevent it. Detailed discussion on modeling of freeze-cracking has been reported in Hung and Kim (1996). As discussed in the section on mechanisms of freeze-cracking, this phenomenon is primarily associated with volume expansion during the water-ice phase transition; however, the modeling process can be divided into two parts: heat transfer and stress analysis. In general, temperature distribution information in food during freezing will be calculated from heat transfer analysis. The development and movement of the freezing interface can then be obtained. Since the outer portion of the frozen material prevents the volume expansion of the internal portion of the food material during phase transition, Hooke's law can be used to model the development of internal stress (Kim, 1993) using nonlinear static stress analysis.

Many simplified assumptions have been made in order to model the development of freeze-cracking; however, modeling of freeze-cracking is very challenging. Chief among these challenges is that, as soon as a crack develops, the internal stress of the product is redistributed. At that time, the stress analysis part of the model must be recalculated.

Shi et al. (1996) calculated the stress development in food during freezing from the equations of mechanics and the measured viscoelastic property of foods. Their calculated tensile stresses show that these can exceed the experimentally measured failure stress and thus cause cracks. Their modeling study demonstrated that large expansion in combination with the viscoelastic property can lead to cracks.

Today, the models proposed do not accurately and precisely model the development of the internal pressure of food during freezing or the exact location and time of the occurrence of freeze-cracking. The main advantage of the modeling, however, is the ability to use the quantitative information generated from the model to develop strategies to reduce or prevent freeze-cracking.

6.8 Prevention of Freeze-Cracking

Although the development of freeze-cracking is undesirable, it is preventable. With a better understanding of the mechanisms and factors affecting the development of freeze-cracking, freezing processes can be modified to prevent the development of freeze-cracking. Several methods have been suggested to prevent or reduce the likelihood of freeze-cracking incidents and are discussed in the following sections.

6.8.1 Precooling

Sebok et al. (1991) reported that precooling can significantly reduce freeze-cracking incidents. In a more recent study, Sebok et al. (1994) suggested that precooling reduced both the temperature difference between product and freezing medium and the time lag between the freezing of the surface and the core. With a smaller temperature difference between product and freezing medium, the driving force for heat exchange will be smaller and, hence, freezing rates will be slower (fewer freeze-cracking incidents). Precooling also reduced the time delay between the freezing of the surface and core so that the expansion of the core during phase change took place at an earlier stage. If the phase change of the core region can occur before the surface becomes brittle, food products can withstand the internal pressure better than when the surface is frozen solidly.

6.8.2 Freezing Rate Optimization

Freezing rate is another important factor that can affect the development of freeze-cracking. In general, the faster the freezing rates the higher the chance of freeze-cracking. However, negative impacts on quality will occur if the freezing rate is too slow.

The purpose of modeling freeze-cracking is to use the model as a tool to develop strategies to prevent freeze-cracking. One approach is to use the model to identify the optimum freezing rate which is the fastest freezing rate of a particular product without causing freeze-cracking (Hung and Kim, 1996). If the predicted internal stress based on a particular set of operating conditions exceeds the frozen material strength, the product will crack during freezing. Freezer operation conditions can then be adjusted in the model to slow down the freezing rate until the predicted product internal stress is less than the material strength. The fastest freezing rate without causing freeze-cracking is the optimum freezing rate for that particular product.

6.8.3 Thermal Equilibration Freezing

Another approach to reducing freeze-cracking is based on the work conducted by Ogawa (1987, 1988). Shortly after immersion into $CaCl_2$ solution, the internal pressure of tuna increased and continued to increase upon further freezing. A sharp reduction in the internal pressure of tuna (from 500 KPa to 100 KPa) was noted if the tuna was equilibrated in a −15°C cold room until the tuna center temperature reached −1 to −2°C but remained unfrozen. Reduction of internal pressure was attributed to the partial softening of the fish surface tissue and, hence, the reduction in the contraction force from the surface frozen crust. Under their proposed equilibration conditions, subsequent reimmersion into the $CaCl_2$ solution to complete the freezing process did not induce freeze-cracking incidents. A similar study has been conducted by Tanaka and Hata (1993) on yellow fin and big eye tuna, in which the thermal equilibration technique resulted in no freeze-cracking.

6.8.4 Combination Method

Hung and Kim (1996) suggested that the heat transfer and stress models developed by Kim (1993) can be used to identify the timing and frequency of tempering periods during fast freezing. Equilibration of the temperature within the product is achieved and accumulation of internal stress is reduced. A similar concept has also been suggested by Shi et al. (1995) in which multistage freezing is used to optimize the process and avoid thermal cracking.

6.9 Summary

In order to retain the best possible quality of foods during freezing, knowledge of the freeze-cracking phenomena and how to prevent them is necessary. A summary of our current understanding and examples regarding how to predict and prevent freeze-cracking has been presented in this chapter. Additional research on physical properties of frozen foods to model the development of freeze-cracking and detect the likelihood of freeze-cracking is needed.

References

Brennan, J.G., Butters, J.R., Cowell, N.D., and Lilly, A.E.V. 1976. *Food Engineering Operations*, 2nd ed. Applied Science Publishers, Essex, England.

Fennema, O.R. 1975. Freezing preservation. In *Principles of Food Science. Part II. Physical Principles of Food Preservation*. (Karel, M., Fennema, O.R., and Lund, D.B. eds.) pp 173–215, Marcel Dekker, New York.

Fennema, O.R. and Powrie, W.D. 1964. Fundamentals of low temperature food preservation. *Adv. Food Res*. **13**:220.

Hoeft, R., Bates, R.P., and Ahmed, E.M. 1973. A research note: cryogenic freezing of tomato slices. *J. Food Sci*. **38**:362.

Hung, Y.-C. and Kim, N.-K. 1996. Fundamental aspects of freeze-cracking. *Food Technol*. **50**(12):59–61.

Kim, N.-K. 1993. Mathematical modeling of cryogenic food freezing. Ph.D. Dissertation. University of Georgia, Athens.

Kim, N.-K. and Hung, Y.-C. 1994. Freeze-cracking in foods as affected by physical properties. *J. Food Sci*. **59**:669–674.

McCambridge, M.A. 1995. Detection of freeze-crack using digital signal processing for the prevention of subsequent freeze-cracks. M.S. Thesis. University of Georgia, Athens.

McCambridge, M.A., Hung, Y.-C., and Mallikarjunan, P. 1996. Detection of freeze-crack using digital signal processing. *Appl. Eng. Agric*. **12**:481–485.

Nicholson, F.J. 1977. Freezing trends and techniques. *Aust. Refrig. Air Cond. Heat.* **5**:48.

Ogawa, Y. 1987. Studies on gape and heave of foodstuffs due to internal pressure during freezing. *Trans. JAR JP.* **4**(2):81–92.

Ogawa, Y. 1988. Freezing system with thermal equalizing. *Jpn. Refrig.* **63**(723):40–46.

Powers, J.J. and Ware, G.O. 1986. Discriminant analysis. In *Statistical Procedures in Food Research.* (Piggot, J.R., ed.) p 125, Elsevier Applied Science, New York.

Rasmussen, C.L. and Olson, R.L. 1972. Freezing methods as related to cost and quality. *Food Technol.* **26**(2):32–47.

Reeve, R.M. and Brown, M.S. 1968. Historical development of the green bean pod as related to culinary texture. 2. Structure and composition at edible maturity. *J. Food Sci.* **33**:326.

Sebok, A., Csepregi, I., and Beke, G. 1991. Cracking of fruits and vegetables during freezing and the influence of precooling. *Proc. 18th Int. Congr. Refrig.* **4**:1754–1758.

Sebok, A., Csepregi, I., and Baar, C. 1994. Causes of freeze cracking in fruits and vegetables. *Food Technol. Int. Eur.* **1994**:66–68.

Shi, X., Data, A.K., and Mukherjee, Y. 1995. Stress analysis of foods during freezing. In *Abstracts of the IFT 1995 Annual Meeting.* Abstract 87-4, p. 260, Institute of Food Technologists, Chicago.

Shi, X., Data, A.K., and Mukherjee, Y. 1996. Cracking in fast freezing of foods: cooling rate effects and crack development. In *Abstract of the IFT 1996 Annual Meeting.* Abstract 74-7, p. 173, Institute of Food Technologists, Chicago.

Tanaka, K. and Hata, M. 1993. Thermal equalize freezing. *Jpn Refrig.* **68**(788): 620–626.

Wolford, E.R. and Brown, M.S. 1965. Liquid-nitrogen freezing of green beans. *Food Technol.* **19**:1133–1135.

7

Quality Deterioration in Frozen Foods Associated with Hydrolytic Enzyme Activities

R. V. Sista, M. C. Erickson, and R. L. Shewfelt

7.1 Introduction

Freezing is an excellent method of preserving the quality of meat, seafood, poultry, vegetables, and fruits. Unfortunately, formation of ice crystals, which is an integral part of the freezing process, can penetrate fragile membranes causing release of enclosed contents and changes in the microenvironment of food tissues. Other unit operations during processing, in addition, can involve comminution of intact tissue that disrupts and releases chemicals and enzymes, potentially affecting the quality of the product during freezing. Storage at low freezing temperatures, though effective in adequately controlling microbial spoilage, can slow but not inactivate these enzymes in the tissue. These enzymatic activities in turn may lead to quality deterioration. Though animal and plant tissues differ structurally as well as functionally, there are many undesirable changes common to both. This chapter focuses on effects of hydrolytic enzymes on quality in frozen plant and animal products.

7.2 Structure and Function of Edible Tissues

Major structural, functional, and metabolic differences exist between plant and animal tissues. The most notable difference of interest to food scientists is that respiration in plants continues even after harvest while cessation of anabolic activities occurs in edible animal tissue. These activities, as well as the cellular organelles and their function, will be discussed in the following sections.

7.2.1 Plant Tissue

On a structural level, edible plant tissue consists of an outer epidermis, parenchymatous cells, and supportive tissue (Schwimmer, 1981a). The epidermis, which is structurally adapted to provide protection against biological or physical stress,

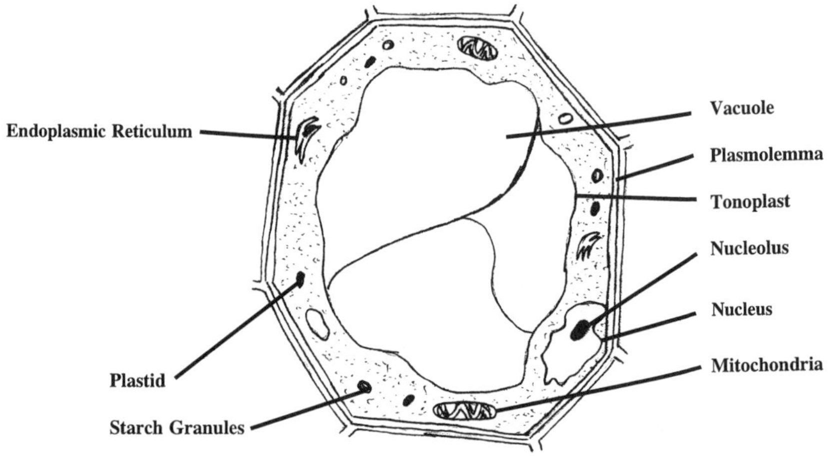

Figure 7.1. Diagrammatic representation of vegetative parenchymal cell and its components

consists of tightly packed cells containing waxy material. The parenchymatous cells, on the other hand, perform much of the metabolic activity of the plant and constitute the bulk of the tissue (Haard and Chism, 1996). Figure 7.1 illustrates the main components of these plant cells. Understanding the function of organelles within the plant cell is important to identifying the source of quality changes during postharvest handling, storage, and processing. For example, firmness and crispness (textural attributes associated with fruits and vegetables) are attributed to the osmotic pressure developed within the cell when pressure is exerted on the rigid cell walls (Schwimmer, 1981a). Exposure of the cell wall to hydrolytic enzymes that attack the pectins, hemicelluloses, and noncellulosic carbohydrate material constituents would dissipate this osmotic pressure.

Another organelle within the cell that has a major functional role in plants is the chloroplast. Responsible for entrapping solar energy into chemical energy both before and after harvest (Nobel, 1974), the chloroplast and its constituent chlorophyll will affect color quality of many plant foods during storage.

The vacuole, which houses hydrolytic enzymes in the cell, may also impact frozen storage quality. If damage to the tonoplastic membrane surrounding the vacuole occurs, hydrolytic enzymes are released in an uncontrolled manner (Boller, 1986). This release, in turn, may lead to a cascade of damaging events, of which further discussion may be found later in this chapter.

7.2.2 Animal Tissue

While a number of different types of animal tissue exist, the major edible tissue is muscle. As such, the major macronutrient is protein compared to carbohydrates in plants. Varying in size, shape, activity, and location among different species, muscles are converted to meat through a series of biochemical reactions after death of the animal. Before this conversion, the major function of contraction and relaxation in muscles is carried out by the cellular machinery. The shutdown of these activities and the basic cellular structure will dictate the quality of the meat.

Figure 7.2 illustrates the cellular structure and organelles found in a muscle cell. While many of the organelles are found in plant cells, there are notable differences in plant and muscle cells, chief among these being the presence of myofibrils and sarcoplasmic reticulum in the muscle cell (Foegeding et al., 1996). Myofibrils are long parallel bundles of the contractile proteins, myosin and actin, that slide and form a cross bridge on contraction of the muscle. The contraction process is initiated by release of calcium from stimulated sarcoplasmic reticulum followed by conformational changes in troponin that in turn activate myosin-ATPase to furnish the chemical energy (ATP) to myosin and actin for sliding. To relax muscle, sequestration of the released calcium by sarcoplasmic reticulum is required. Since this is an energy-dependent process, it does not occur once the ATP is used up. Consequently, meat is left in a state of contraction until hydrolytic enzymes present in the cytoplasm can disrupt the protein complexes and tenderize the meat.

A large portion of the hydrolytic enzymes are housed in the lysosome, an organelle comparable to the vacuole in the plant cell (Lehninger, 1975). Excessive release of these enzymes during comminution or frozen storage is detrimental to the quality of many muscle foods.

7.3 Hydrolytic Enzymes and Their Effect on Quality

Enzymes are substrate-specific biological catalysts that hasten biochemical reactions. A myriad of enzymes exist that aid in reactions specific to a particular organelle. Table 7.1 briefly describes some of these enzymes and their location in plant or animal cells.

Hydrolases or hydrolytic enzymes are catabolic enzymes that catalyze the transfer of groups to water for degradation and turnover of cellular macromolecules. In raw or nonheated products, these hydrolases may remain active during frozen storage. Blanching of vegetables or cooking of meats, on the other hand, generally inactivates these enzymes. If inadequate heating is applied, however, the cellular organelles housing the enzymes may be ruptured and thus accentuate the role of hydrolytic enzymes in quality deterioration.

7.3.1 Lipolytic Enzymes

The two primary categories of lipolytic enzymes are lipases and phospholipases that hydrolyze ester linkages of triacylglycerols and phospholipids, respectively. Physiologically, lipolytic enzymes are important since they hydrolyze fats and oils, giving rise to free fatty acids and partial glycerides, which are essential for

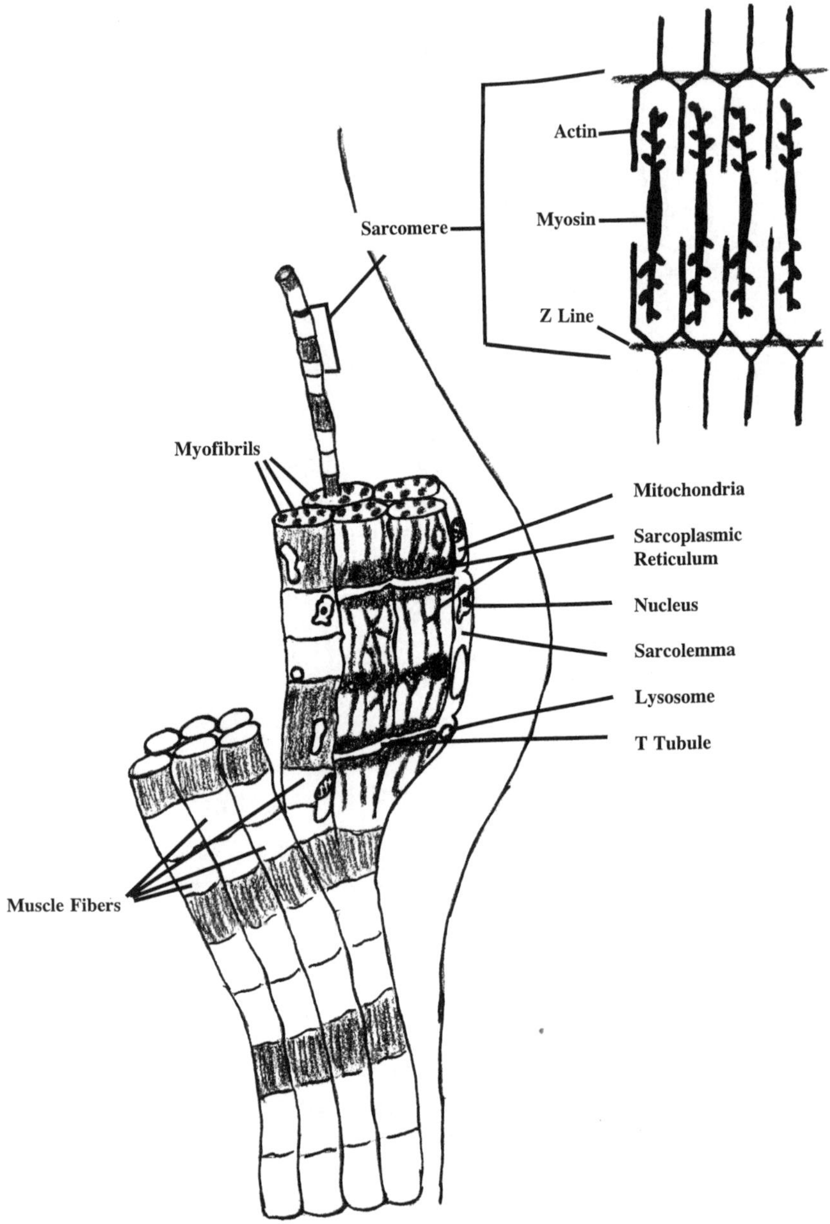

Figure 7.2. Diagrammatic representation of skeletal muscle cells and its components

Table 7.1. Enzymes and their locations in various cellular organelles

Cellular organelle	Function and enzyme
Membranes	Various ATPases involved in the transport of nutrients across the semipermeable membrane; dehydrogenases; and cytochrome oxidases.
Mitochondria	Enzymes required for the oxidation of carbohydrates, protein, and lipids as well as those involved in the electron transport chain; polyphenol oxidase in plants.
Nucleus	Main function is protein synthesis including DNA; the enzymes involved are polymerases and synthetases.
Microbodies or peroxisomes	Oxidation of toxic wastes including breakdown of hydrogen peroxide to water and oxygen; peroxidase; catalase; urate oxidase; D-amino oxidase; and polyphenol oxidase.
Lysosomes or spherosomes	Primary function is phagocytosis or pinocytosis; rich in hydrolytic enzymes including ribonuclease, phosphatases, lipases, cathepsins, collagenase, lysozyme, and lipoxygenase.
Golgi bodies	Secretion of proteins including plasma and lysosomal membranes, so enzymes involved in protein synthesis are present.
Endoplasmic reticulum	Aiding ribosomes in protein synthesis and channeling them through the cytoplasm; rich in phosphatases and diphosphatases.
Plastids	Enzymes required for biosynthesis of ATP, glucose and other biomolecules; lipoxygenase; and polyphenol oxidase.
Vacuole	Storage and secretion of hydrolytic enzymes, proteases, nucleases, glycosidases, and phosphatases

Compiled from Whitaker, 1996; Lehninger, 1975; and Crane et al., 1985.

metabolic processes such as fatty acid transport, oxidation, and resynthesis of glycerides and phospholipids. If not appropriately controlled during storage, however, the hydrolysis of lipids can lead to undesirable flavor and textural changes.

Lipases

Lipases are found in plants, animals, and microorganisms but often vary in specificity according to their preference for low– or high–molecular weight triacylglycerols, preference for position of fatty acid, preference for length of fatty acid, or stereospecificity preference. While the temperature optimum of most of the extensively studied lipases lies between 30 and 40°C, certain lipases have been shown to be active in frozen-food systems held as low as $-29°C$ (Alford and Pierce, 1961). Lipase activity has also been evident by the accumulation of free fatty acids in frozen fish samples (Bosund and Ganrot, 1969; Erickson, 1993).

Despite the reductions in activity that occur in response to lowering of temperature, freezing may accentuate lipolysis by disrupting the lysosomal membrane that releases hydrolytic enzymes. For example, Geromel and Montgomery (1980) demonstrated the release of an acid lipase from lysosomes in rainbow trout fillets subjected to low-temperature treatments. While fast and intermediate rates of freezing did not cause the release of acid lipase from the lysosomes, slow freezing and fluctuation of temperatures (-12 to $-35°C$ for 96 h) of the frozen fillets resulted in appreciable release of the acid lipase from lysosomes during the first month of storage.

Conditions before or during processing may also impact the release of lipases and/or modify the activity of the enzymes during frozen storage. Geromel and Montgomery (1980) found that fasting fish for 24 and 72 h before sacrifice resulted in higher acid lipase activities. Because fasting is typically applied to channel catfish before sacrifice and processing, FFA concentrations may be magnified in these fish over those that would have occurred in nonfasting fish. Similarly, processing treatments that involve disruption of the tissue would accelerate lysosomal release of lipases (Galliard et al., 1976).

Modification in lipolytic activities during freezing and storage may not always occur via changes in enzyme concentration. Various factors, including concentration of Ca^{2+} and NaCl, are known to affect the activity. As the amount of unfrozen water changes during freezing, these salt concentrations will also vary, thereby modifying the lipase activities.

In stored products, lipolytic activity has detrimental consequences. Release of sufficient quantities of short-chain free fatty acids may cause hydrolytic rancidity. Free fatty acids may also interact with proteins to form insoluble lipid-protein complexes that modify texture (Kolodziejska et al., 1994). Finally, the release of free fatty acids from a triacylglycerol matrix may accelerate their interaction with oxidative catalysts and hence accelerate the rate of lipid oxidation and generation of off flavors. The H^+ ions generated by this sequence of reactions have also been postulated to be responsible for the browning or yellowing of unblanched frozen green vegetables via substitution of the Mg^+ in chlorophyll by H^+ (Schwimmer, 1981b).

Phospholipases

Several types of phospholipases within foods differ in the position of attack on the phospholipid. Phospholipases A_1 and A_2 hydrolyze the acyl groups at the sn-1 and sn-2 positions of phospholipids, respectively, while phospholipase C, prevalent in molds and bacteria, hydrolyzes the ester linkage between the glycerol backbone and the phosphoryl group. Phospholipase D, usually found in plants, hydrolyzes the ester linkage on the other side of the phosphoryl group yielding phosphatidic acid (Richardson and Hyslop, 1985).

The liberation of free fatty acids from phospholipids, in contrast to the liberation from triacylglycerols, has been shown to inhibit lipid oxidation (Shewfelt and Hultin, 1983). An explanation offered for the difference is that release of free fatty acids leads to decreased interaction between oxidized and oxidizable fatty acids within the membrane matrix, thus inhibiting free-radical propagation reactions. An enzyme system in plant tissue, however, consisting of phospholipase D, diglyceride lipase, and lipoxygenase, has been described to accelerate lipid oxidation in membranes (Paliyath and Thompson, 1987), suggesting a potential role for lipoxygenase in the response of membranes to lipid hydrolysis.

While membrane lipid hydrolysis modifies the degree of oxidation, the extent of hydrolytic activity, in turn, may be dependent on the extent of membrane lipid oxidation. Oxidized fatty acids attached to phospholipids have been found to be more susceptible to hydrolysis by PLA_2 than those not oxidized (Sevanian and Kim, 1985). These oxidized free fatty acids have then subsequently been converted to stable nonpropagating species by the action of glutathione peroxidases or other repair enzymes (Sevanian et al., 1988; Shewfelt and Purvis, 1995). Consequently, the oxidative repercussions in frozen foods to phospholipase activity are complex and dependent on the presence or absence of oxidative or repair enzymes.

7.3.2 Proteolytic Enzymes

Hydrolysis of proteins to short-chain peptides and amino acids is catalyzed by proteases. Depending on their site of attack, proteases can be classified into four types: serine, sulfhydryl, aspartic, and metalloproteinases (Whitaker, 1994).

In meat, endogenous proteinases (cathepsins, calcium-dependent calpains, and collagenases) are considered beneficial, providing tenderization to muscle in rigor mortis. Degradation of Z lines and rupturing of the sarcomeres have been observed during frozen storage of chicken breast muscle (Smolinska and Abdul-Halim, 1992), but, in general, activity of these enzymes during frozen storage is not considered detrimental to the quality of meat. High levels of proteolytic enzymes, on the other hand, may harm a product's texture during frozen storage. Yamamoto et al. (1977) observed degradation of the proteins troponin T and myosin by cathepsins released from ruptured lysosomes. In the case of Pacific whiting, cathepsins are released from lysosomes as a defense against infection by the myxosporidian parasite (Erickson et al., 1983; Masaki et al., 1993). The released enzymes have subsequently led to the liquefaction or mushiness of the muscle during frozen storage (Erickson et al., 1983). In diseased and injured tissues of plants, levels of proteases have also been found to be higher than in normal tissue (Porter, 1966; Hislop et al., 1982), and expression of these added enzymes during frozen storage could similarly lead to quality deterioration of the product.

7.3.3 Pectin Methyl Esterase

Pectins, present in vegetable and fruit tissues, are polymers consisting of methylated D-galacturonic acid units linked by the alpha-D (1,4) glycosidic linkages (Haard and Chism, 1996). Endogenous pectin methyl esterase catalyzes the removal of the methoxyl group in pectins, converting the substrate to galacturonic acid residues (Schwimmer, 1981a). In frozen fruit juices that have not been pasteurized, the demethylation of pectin by this enzyme allows interlinking of the pectin molecules via divalent ions, followed by precipitation and loss of "cloud" (Haard, 1985). These same reactions in some varieties of strawberries have caused gelation during frozen storage (Haard, 1985).

7.3.4 Trimethylamine Oxide Demethylase

Another hydrolytic enzyme unique to gadoid fish is the breakdown of trimethylamine oxide (TMAO) to dimethylamine and formaldehyde by TMAO demethylase. Formaldehyde is highly reactive and may cross-link with muscle proteins, rendering them insoluble such that a spongy or cottony texture results (Santos-Yap, 1995). Oxygen has an inhibitory effect on TMAO demethylase (Lundstrom et al., 1982); hence, application of vacuum or modified atmosphere packaging to products containing this enzyme may magnify the enzyme's activity and accelerate toughening.

7.3.5 Pigment-Degrading Enzymes

Both chlorophyllases and anthyocyanases exist in plants. If not inactivated by blanching, these hydrolytic enzymes may catalyze destruction of the pigments chlorophyll and anthocyanin, respectively, during frozen storage (Chichester and McFeeters, 1971; Simpson et al., 1976), which ultimately affects color quality.

7.4 Summary

Hydrolytic activities in frozen foods can result in quality degradation. Such degradation is more often observed in meat products than in plant products because of the application of blanching or pasteurization treatments to vegetable and fruit products, respectively. While hydrolytic rancidity, textural softening, and color loss are direct consequences of hydrolytic enzyme activity, textural toughening and acceleration in lipid oxidation may be secondary consequences. To control excess proteolytic activity, food-grade inhibitors are available (Reppond and Babbitt, 1993; Piyachomkwan and Penner, 1995); however, no food-grade inhibitors exist to control lipolytic activity. As such, the only way to avoid the consequences of lipolysis in frozen foods is to minimize the release of the enzymes from the cellular organelles that house them through processes involving fast freezing and minimization of ice recrystallization in the product.

References

Alford, J.A. and Pierce, D.A. 1961. Lipolytic activity of microorganisms at low and intermediate temperatures. III. Activity of microbial lipases at temperatures below 0°C. *J. Food Sci.* **26**:518.

Boller, T. 1986. Roles of proteolytic enzymes in interaction of plants with other organisms. In *Plant Proteolytic Enzymes,* vol. 1. (Dalling, M.J., ed.) pp. 67–96, CRC Press, Boca Raton, Fla.

Bosund, I. and Ganrot, B. 1969. Lipid hydrolysis in frozen Baltic herring. *J. Food Sci.* **34**:13.

Chichester, C.O. and McFeeters, R. 1971. Pigment degeneration during processing and storage. In *Biochemistry of Fruits and Their Products*, part II, vol. 2. (Hulme, A.C., ed.) pp. 707–719, Academic Press, New York.

Crane, F.L., Low, H., and Clark, M.G. 1985. Plasma membrane redox. In *Enzymes of Biological Membranes*, vol. 4. (Martonosi, A., ed.) pp. 465–510, Plenum Press, New York.

Erickson, M.C. 1993. Compositional parameters and their relationship to oxidative stability of channel catfish. *J. Agric. Food Chem.* **41**:1213–1218.

Erickson, M.C., Gordon, D.T., and Anglemier, A.F. 1983. Proteolytic activity in the sarcoplasmic fluids of parasitized Pacific whiting (*Merluccius productus*) and unparasitized true cod (*Gadus macrocephalus*). *J. Food Sci.* **48**:1315–1319.

Foegeding, E.A., Lanier, T.C., and Hultin, H.O. 1996. Characteristics of edible muscle tissue. In *Food Chemistry.* (Fennema, O.R., ed.) pp. 879–942, Marcel Dekker, New York.

Galliard, T., Matthew, J.A., Fishwick, M.J., and Wright, A.J. 1976. The enzymic degradation of lipids resulting from physical disruption of cucumber (*Cucumis sativus*) fruit. *Phytochem.* **15**:1647–1650.

Geromel, E.J. and Montgomery, M.W. 1980. Lipase release from lysosomes of rainbow trout (*Salmo gairdneri*) muscle subjected to low temperatures. *J. Food Sci.* **45**:412–415, 419.

Haard, N.F. 1985. Characteristics of edible plant tissue. In *Food Chemistry*, 2d ed. (Fennema, O.R., ed.) pp. 857–911, Marcel Dekker, New York.

Haard, N.F. and Chism, G.W. 1996. Characteristics of edible plant tissues. In *Food Chemistry*, 3d ed. (Fennema, O.R., ed.) pp. 943–1011, Marcel Dekker, New York.

Hislop, E.C., Paver, J.L., and Keon, J.P.R. 1982. An acid protease produced by *Monilinia fructigena in vitro* and in infected apple fruits and its possible role in pathogenesis. *J. Gen. Microbiol.* **128**:799–807.

Kolodziejska, I., Niecikowska, C., and Sikorski, Z.E. 1994. Dimethylamine and formaldehyde in cooked squid (*Illex argentinus*) muscle extract and mantle. *Food Chem.* **50**:281–283.

Lehninger, A.L. 1975. *Biochemistry*. Worth Publishers, New York.

Lundstrom, R.C., Correia, F.F., and Wilhelm, K.A. 1982. Dimethyl amine production in fresh red hake (*Urophycis chuss*): The effect of packaging material, oxygen permeability, and cellular damage. *J. Food Biochem.* **6**:229–241.

Masaki, T., Shimomukai, M., Miyauchi, Y., Ono, S., Tuchiya, T., Mastuda, T., Akazawa, H., and Soejima, M. 1993. Isolation and characterization of the protease responsible for jellification of Pacific hake muscle. *Nippon Suisan Gakkaishi* **59**:683–690.

Nobel, P.S. 1974. *Introduction to Biophysical Plant Physiology*. W.H. Freeman, San Francisco.

Paliyath, G. and Thompson, J.E. 1987. Calcium- and calmodulin-regulated breakdown of phospholipid by microsomal membranes from bean cotyledons. *Plant Physiol.* **83**:63–68.

Piyachomkwan, K. and Penner, M.H. 1995. Inhibition of Pacific whiting surimi-associated protease by whey protein concentrate. *J. Food Biochem.* **18**:341–353.

Porter, F.M. 1966. Protease activity in diseased fruits. *Phytopathology* **56**:1424–1425.

Reppond, K.D. and Babbitt, J.K. 1993. Protease inhibitors affect physical properties of arrowtooth flounder and walleye pollock surimi. *J. Food Sci.* **58**:96–98.

Richardson, T. and Hyslop, D.B. 1985. Enzymes. In *Food Chemistry*, 2d ed. (Fennema, O.R., ed.) pp. 371–476, Marcel Dekker, New York.

Santos-Yap, E.M. 1995. Fish and seafood. In *Freezing Effects on Food Quality*. (Jeremiah, L.E., ed.) pp. 109–133, Marcel Dekker, New York.

Schwimmer, S. 1981a. Enzyme action and plant food texture. In *Source Book of Food Enzymology*. (Schwimmer, S., ed) pp. 511–531, AVI, Westport, Conn.

Schwimmer, S. 1981b. Color changes induced by enzyme action on nonphenolic substrates in plant foods. In *Source Book of Food Enzymology.* (Schwimmer, S., ed.) pp. 298–314, AVI, Westport, Conn.

Sevanian, A. and Kim, E. 1985. Phospholipase A_2 dependent release of fatty acids from peroxidised membranes. *Free Radical Biol. Med.* **1**:263–271.

Sevanian, A., Wratten, M.A., McLeod, L.L., and Kim, E. 1988. Lipid peroxidation and phospholipase A_2 activity in liposomes composed of unsaturated phospholipids: A structural basis for enzyme activation. *Biochim. Biophys. Acta.* **961**:316–327.

Shewfelt, R.L. and Hultin, H.O. 1983. Inhibition of enzymic and nonenzymic lipid peroxidation of flounder muscle sarcoplasmic reticulum by pretreatment with phospholipase A_2. *Biochim. Biophys. Acta.* **751**:432–438.

Shewfelt, R.L. and Purvis, A.C. 1995. Toward a comprehensive model for lipid peroxidation in plant tissue disorders. *Hort. Sci.* **30**:213–218.

Simpson, K.L., Lee, T.-C., Rodriguez, D.B., and Chichester, C.O. 1976. Metabolism in senescent and stored tissues. In *Chemistry and Biochemistry of Plant Pigments*, 2d ed., vol. 1. (Goodwin, T.W., ed.) pp. 779–842, Academic Press, New York.

Smolinska, T. and Abdul-Halim, F. 1992. The effects of freezing systems and frozen storage on the quality of broiler tissue muscle. *Arch. Geflugelkunde.* **2**:80–85.

Whitaker, J.R. 1994. *Principles of Enzymology for the Food Sciences*, 2d ed. Marcel Dekker, New York.

Yamamoto, K., Samejima, K., and Yasui, T. 1977. A comparative study of the changes in hen pectoral muscle during storage at 4 and −20°C. *J. Food Sci.* **42**(6):1642–1645.

8

Protein Denaturation and Functionality Losses

Youling L. Xiong

8.1 Introduction

Frozen storage is one of the most important preservation methods for maintaining microbiological and chemical stability and extending the shelf life of food products. Deteriorations in texture, flavor, and color, resulting from biochemical, enzymatic, and functional changes in proteins, however, are problems associated with freezing and subsequent storage at subfreezing temperatures for many fresh and processed foods. Freeze-induced protein denaturation, enzyme inactivation, and related functionality losses are commonly observed in frozen fish, meat, poultry, egg products, and doughs. Muscle proteins are particularly susceptible to freeze denaturation compared to plant-derived proteins, and this is especially true for fish species. Denaturation of proteins during freezing and frozen storage can be monitored by measuring alterations in protein surface hydrophobicity, amino acid composition, conformational stability, solubility, aggregation, and enzyme activity. Losses in functional properties of proteins are commonly assessed by comparing water-holding ability, viscosity, gelation, emulsification, foaming, and whipping properties.

Freeze-induced protein denaturation has been attributed to the formation of ice crystals, dehydration, and concentration of solutes in the tissue or protein solution. For some marine fish, formaldehyde derived from trimethylamine oxide in frozen tissue has been shown to modify myofibrillar proteins and collagen by cross-linking adjacent polypeptides, thereby forming insoluble aggregates. Recent studies have also demonstrated that oxidative reactions, such as lipid peroxidation, are involved in denaturation and deterioration in functional attributes of muscle proteins during frozen storage. Thus, various cryoprotectants (e.g., sugars, sorbitol, and polyols), and antioxidant additives are incorporated into food before freezing to minimize physicochemical changes in proteins and to prevent functionality losses. Moreover, the discovery of antifreeze proteins in fish has piqued considerable research interest in identifying alternate means for the preservation of protein stability and functionality during freezing and subsequent storage.

8.2 Mechanisms of Freeze-Induced Protein Changes

Freezing is a physical process involving the transformation of water molecules from an amorphous state to highly structured ice crystals. The phase change can lead to protein denaturation caused by alterations in the chemical and physical environment of the protein. Freeze-induced damages to proteins are both mechanical and chemical. A conventionally accepted theory is that the damages involve three major alterations in the protein microenvironment: (1) changes in moisture; (2) changes in lipids; or (3) changes in certain cellular metabolites. According to Shenouda (1980), changes in moisture can be subdivided into the following: (1) formation and accretion of ice crystals; (2) dehydration; and (3) increases in solute concentration. Table 8.1 lists the major causes of freeze-induced protein denaturation and some possible techniques to prevent the denaturation.

8.2.1 Mechanical Damage

The formation of ice crystals from either intracellular or extracellular water can result in mechanical damage caused by irregular ice crystals protruding through and disrupting the cell walls. The size and location of ice crystals formed during freezing is greatly influenced by the freezing rate, storage time, and temperature fluctuations. At a slow freezing rate, the exterior fluid of cells cools more rapidly than the interior fluid, and as the supercooled extracellular fluid reaches a critical temperature, water separates from the solute and forms ice crystals. As crystallization proceeds, extracellular salt becomes more concentrated, creating an osmotic pressure gradient across the cell membrane. In an attempt to balance the chemical potential, intracellular (hypoosmotic) moisture flows outward, leading to dehydration and an increase in the ionic strength of the cell. Water drawn from the interior of the cell will freeze onto the existing extracellular ice crystals, causing them to grow, thus distorting and damaging the membrane and proteins. In contrast, at a fast freezing rate, the interior moisture of the cell is cooled so rapidly that small ice crystals, usually spearlike and separated by proteins, form inside the cell. Therefore, less dehydration and mechanical damage to proteins occurs (Love, 1968).

The free energy of a crystalline phase is minimized when its structure is perfect (i.e., each structural motif making up the lattice is repeated regularly) and its size is infinite (i.e., no boundaries). Thus, thermodynamically, a small ice crystal is less stable than a large ice crystal. During frozen storage, especially when temperature fluctuates, ice crystals can undergo metamorphic changes referred to as recrystallization. In general, small ice crystals have a tendency to melt, recrystallize, or aggregate onto existing larger crystals (nuclei) because of differences in

Table 8.1. Causes of freeze-induced protein denaturation and possible prevention methods

Cause	Mode of action	Method to minimize denaturation
Ice formation	Mechanical damage Cell-wall disruption	Fast freezing rate Low thawing temperature
Recrystallization	Mechanical damage	Low temperature storage Prevent temperature fluctuation
Dehydration	Increased exposure of hydrophobic groups; unfolding	Cryoprotectants
Salt concentration	Modified electrostatic interaction; unfolding	Cryoprotectants
Oxidation	Modified amino acid side chains Peptide bond cleavage	Antioxidants
Lipids	Protein-lipid complexes	?
Metabolites	Formaldehyde-methylene bridge between polypeptides	Inactivate trimethylamine demethylase

the surface energy. This phenomenon is often observed for food stored at close to the critical freezing zone (-0.8 to $-5°C$). Because recrystallization involves enlargement in size, changes in shape and orientation, and movement of ice crystals, it usually causes tissue damage and accelerates protein denaturation.

8.2.2 Dehydration

Dehydration caused by ice formation can lead to protein denaturation, which can be explained in thermodynamic terms. In biological tissues, where moisture is the most abundant cellular constituent, proteins exposed to the aqueous environment have a hydrophobic interior and a polar surface (i.e., most hydrophobic amino acid residues are occluded inside and most charged or polar side chains tend to stay outside). This structural assemblage is entropy driven, meaning it facilitates hydration of proteins by interacting with water molecules via hydrogen bonds and lowers the total free energy of the system (Privalov and Makhatadze, 1993), as can be expressed by the equation

$$\Delta G = \Delta H - T\Delta S$$

where ΔG, ΔH, and ΔS are, respectively, changes in free energy, enthalpy, and entropy. The loss of entropy involved in the formation of ordered (folded) protein structure in its hydrated state is compensated for by a large entropy gain of the surrounding water molecules ($\Delta S > 0$) owing to minimization of interactions between apolar groups of proteins and water. This results in a net reduction in global free energy ($\Delta G < 0$) and stabilization of proteins. However, in a dehydrated state, protein-solvent interactions are disrupted and protein molecules are exposed to an essentially "organic" environment (less polar than water). This will result in increased exposure of hydrophobic side chains and segments and, hence, changes in protein conformation. To maintain the lowest possible free energy status ($\Delta G < 0$) upon migration of water to form ice crystals, increased protein-protein interactions via hydrophobic and ionic forces occur, resulting in further denaturation of proteins and formation of protein aggregates. Hydrophobic interactions are generally weakened as the temperature decreases. Hence, lowering the temperature to subfreezing temperatures will destabilize proteins whose native structures are critically maintained by hydrophobic forces; this has been the basis for some recent theories put forth to explain cold-induced denaturation of proteins (Privalov et al., 1986; Franks, 1995).

8.2.3 Concentration of Salts

Water in biological tissues, cellular suspensions, or protein aqueous solutions can be divided loosely into free and bound types. During freezing to −10°C or below, as much as 10% water in muscle tissue could stay unfrozen and is more or less "bound" to protein or other nonaqueous cellular components (Lawrie, 1991). However, the amount of unfreezable water depends on the concentration and type of solutes, including various ionic compounds or salts. Thus, as freezing progresses, proteins are exposed to increased concentrations of salts, and this process can theoretically continue until the final eutectic temperature (at which all solutes are also crystallized out) is reached. Because electrostatic interactions are one of the major forces maintaining protein tertiary and quaternary structures, an abrupt increase in ionic strength or salt concentration in the nonfrozen phase can cause competition with existing electrostatic bonds, which in turn leads to extensive modification of the protein native structure. Disruption of energetically and entropically favorable native protein conformation usually results in denaturation, dissociation of subunits, and, ultimately, aggregation of denatured molecules. The extent of protein damage is a function of the type of salts, the freezing rate, the temperature and time of storage, and the characteristics of the protein involved in the interaction. For instance, freeze denaturation of carp myosin and myofibrillar proteins has been shown to increase with KCl and NaCl concentration, reaching maxima at about −11 to −13°C for KCl, and −8 to −18°C for NaCl (Inoue et al., 1992; Takahashi et al., 1993). Since these temperatures correspond to the eutectic points of the salts, the concentration effects by both salts are implicated.

8.2.4 Oxidation

Oxidative processes also contribute to protein denaturation in many biological materials during frozen storage. This is particularly true in muscle or plant tissues and cellular suspensions that contain lipids and prooxidants. Proteins exposed to oxidizing environments are very susceptible to chemical modifications, such as amino acid destruction, peptide scission, and formation of protein-lipid complexes. Among the various oxidizing agents are enzymes, heme, and transition metals that can be released from confined cellular organelles or derived from inactive compounds during freezing and frozen storage. These oxidizing agents can react with proteins either directly or indirectly through lipid and nonlipid radicals, leading to physical and chemical alterations in proteins. Many lipid degradation products, including malondialdehyde, are also capable of cross-linking polypeptides and thus are responsible for the generation of insoluble protein aggregates (Buttkus, 1970).

8.2.5 Cellular Metabolites

Increased meat toughness and sponginess are common textural problems associated with freezing and frozen storage of fresh meat. The texture deterioration is a result of protein denaturation and loss of protein solubility. Some muscle-cellular metabolites, such as free fatty acids (FFA) and formaldehyde, are directly responsible for such changes. The production and accumulation of FFA in fish tissue during prolonged frozen storage has long been recognized (Dyer and Dingle, 1961). Free fatty acids in frozen muscle are mostly derived from enzymatic hydrolysis of lipids, particularly the phospholipids, since many lipases remain active in frozen muscle tissue (Olley et al., 1962). Through both hydrophobic and hydrophilic interactions, FFA are able to bind to proteins and cause an increase in the hydrophobicity of proteins. Given the composition of FFA, it can be reasonably assumed that unless oxidation occurs, no covalent bonds are involved in the binding between FFA and proteins. Some of the nonpolar amino acids or hydrophobic patches occluded in the native protein may in fact be exposed because of the hydrophobic attraction by FFA. It is conceivable that a reduced polarity of the protein microenvironment brought about by dehydration (as discussed earlier) would promote the formation of lipoprotein complexes. Subsequent conformational changes would then lead to insolubilization of the proteins through protein polymerization and aggregation. Myofibrillar proteins in frozen muscle are more susceptible than sarcoplasmic proteins to FFA-induced denaturation or loss in solubility. Among the myofibrillar proteins, myosin or actomyosin seems to be the major target. The rate of FFA-myofibrillar protein interaction and subsequent protein solubility loss are related directly to the degree of unsaturation of FFA and storage time and inversely related to the chain length of FFA (Shenouda, 1980). A recent study showed that both oleic (C18:1) and myristic (C14:0) acids could bind to the hydrophobic regions of fish actomyosin during frozen storage, resulting in major losses of ATPase activity, solubility, and viscosity of the protein (Careche and Tejada, 1994).

In gadoid fish, trimethylamine demethylase, an enzyme that remains active during both frozen storage and thawing, can catalyze the decomposition of trimethylamine oxide to form dimethylamine and formaldehyde (Sotelo et al., 1995). Formaldehyde is a reactive compound that can covalently cross-link adjacent polypeptides via methylene bridges, thereby causing protein-structure deformation, loss in functionality, and hardening of fish tissue. Protein side-chain groups that can react with formaldehyde include amine, amido, guanido, thiol, phenolic, imidazole, and indolyl residues (Shenouda, 1980).

8.3 Major Freeze-Induced Protein Changes

8.3.1 Methods for Assessing Freeze-Induced Protein Denaturation

The most commonly employed method for assessment of freeze denaturation of muscle proteins is solubility determination. Since solubility is considered a functional attribute, it will be discussed later in this chapter. Many enzymes are susceptible to denaturation during frozen storage, and hence, loss of activity. Therefore, measurement of enzyme activity is a sensitive method to monitor freeze-induced protein denaturation. An example is myosin (or actomyosin) whose Ca-ATPase activity in the head region decreases during freezing/thawing and prolonged frozen storage and therefore is a sensitive measure of the extent of protein denaturation (Takahashi et al., 1993; Careche and Tejada, 1994). Several other physical and chemical techniques, specifically designed to measure energy input and surface properties associated with protein unfolding, have also been adopted to evaluate freezing denaturation of muscle proteins. Examples are hydrophobicity (Li-Chan et al., 1985) and differential scanning calorimetry (DSC) analyses (Wagner and Añón, 1985; Davies et al., 1994). In DSC analysis, disruption of the native protein structure upon heating proceeds with heat absorption and is accompanied by an increase in enthalpy and entropy. Thus, compared to its native state, when a protein is destabilized by freezing treatments, it will either absorb a smaller amount of heat, unfold at lower temperature, or both. Other techniques routinely used to measure protein conformation changes, including circular dichroism (CD), optical rotatory dispersion (ORD), infrared (IR), ultraviolet (UV), and nuclear magnetic resonance (NMR) spectroscopies, can also be used to assess freeze-induced protein denaturation.

8.3.2 Enzyme Inactivation

Enzymes are proteins in nature, and their structures and conformational stability are dictated by balanced electrostatic, hydrophobic, and van der Waal's interactions; hydrogen bonds; and disulfide linkages. Thus, enzyme activities can be influenced by factors that affect physical and chemical properties of proteins. Since it is difficult to measure enzyme activity in frozen tissue directly, enzyme activities are commonly evaluated following thawing. Freezing effect on enzymes, as reported in the literature, should be interpreted with caution, because during thawing, some enzymes may be subjected to additional damage.

Lowering temperatures to below the freezing point can either activate or inactivate enzymes, and in both cases changes in enzyme activity occur because of the perturbation of the environment surrounding the enzymes (i.e., pH, ionic strength, and solute and substrate concentrations). Some oxidases and lipases, particularly those with globular structures, are known to tolerate freeze-thaw treatments and are partially responsible for lipid oxidation and development of rancidity in frozen foods. The behavior of enzymes during freezing and frozen storage varies, depending on whether the enzyme is located in a cellular or noncellular environment. As reviewed by Fennema (1982), many enzymes in frozen cellular systems exhibit increased activity as in the case of enzyme-catalyzed degradation of glycogen and high-energy phosphates; the accumulation of lactic acid in frog, fish, beef, and poultry muscle; hydrolysis of phospholipids in cod muscle; decomposition of peroxides in potatoes and peas; and oxidation of L-ascorbic acid in rose hips, strawberries, and Brussels sprouts. In noncellular systems, this behavior is uncommon and has been observed only when the sample is extremely dilute before freezing. Rate enhancements in enzyme-catalyzed reactions have been attributed to the increase in substrate concentration and the disruption of cell membranes, thereby facilitating enzyme release.

However, decreased enzyme activity in frozen products is more commonly observed whether the product is a cellular or noncellular system. Freeze-induced rate decreases have been demonstrated for reactions catalyzed by lipase, invertase, and lipoxygenase for nonmuscle foods (Fennema, 1982), as well as reactions catalyzed by myosin ATPase and lactate dehydrogenase in muscle tissue (Nambudiri and Gopakumar, 1992; Careche and Tejada, 1994). Sensitivity of muscle enzymes to freezing and frozen storage varies depending on specific species, muscle tissues, and the environmental conditions. For example, myosin ATPase from warm-water fish is less prone to freeze denaturation than myosin from cold-water fish. Kinetic data have indicated that the denaturation of myosin ATPase in frozen whole beef muscle takes place in two consecutive first-order reactions during storage at temperatures below −20°C, that is, an initial rapid reaction followed by a slower one (Wagner and Añón, 1986). Myosin ATPase activity in intact fish muscle is also lost during prolonged frozen preservation, but most of the ATPase activity lost in isolated solutions occurs during the freezing and/or thawing processes, that is water-to-ice phase transitions, with very small further loss taking place during storage (fig. 8.1).

Rate decreases in enzyme-catalyzed reactions probably stem from two major freeze-induced changes. First, the combination of lowering temperature and changes in the aqueous environment (pH, ionic strength, dehydration, or ice crystal formation) would induce irreversible conformational changes at the catalytic center of the enzyme so that the enzyme can no longer bind efficiently with substrates. Depending on the freezing rate and storage time, these changes may be

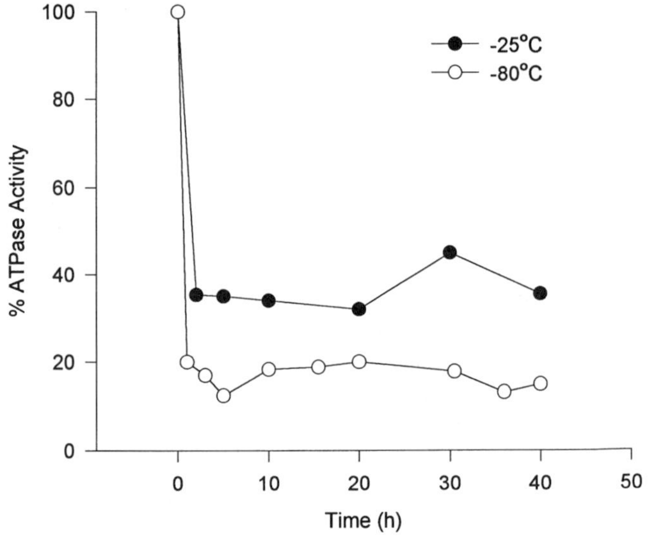

Figure 8.1. Percentage ATPase activity of soluble fraction of 0.2% cod myosin solution at various times of frozen storage at −25°C and −80°C (adapted from Ang and Hultin, 1989)

reversible or irreversible. Second, as the liquid-to-solid (ice crystal) phase transformation proceeds and the solute concentration increases, diffusivity of substrate and reaction products would become restrained, as would the mobility and flexibility of the enzyme. These physical changes apparently occur in tissues, ostensibly creating a condition unfavorable to enzymic reactions. There is also evidence that formaldehyde, derived from trimethylamine oxide present in many gadoid fish, is directly responsible for inactivation of myosin ATPase during freezing and storage (Ang and Hultin, 1989).

8.3.3 Denaturation of Muscle Proteins

Freezing preservation of muscle foods against microbial spoilage and biochemical deterioration is offset by protein denaturation. Structural damages in frozen muscle include disruption of sarcolemma and myofibrils and denaturation of proteins (Yamamoto et al., 1977; Kang et al., 1983). Deleterious protein changes, in turn, are associated with fiber shrinkage, exudation of fluid ("drip"), and texture toughening.

Proteins from marine species are particularly labile at subfreezing temperatures compared with land animal muscle proteins. When meat is frozen prerigor or subjected to extended storage, the detrimental effect of freezing and thawing is aggravated. In addition, the freezing and thawing rate markedly affects the extent of dripping when frozen muscle is thawed. Fast freezing (e.g., >4°C/min) facilitates the formation of small, intracellular ice crystals and therefore minimizes protein denaturation, fiber shrinkage, and exudation. However, the feasible rates of freezing commercially are much too slow to allow sufficient intracellular ice formation. Therefore, exudation of thawed meat is a common side effect of freezing preservation of muscle foods, especially for fish whose proteins are remarkably sensitive to subfreezing temperatures. It is worthy of mention that changes in muscle ultrastructure and proteins are a result of both freezing and thawing processes, and it is usually difficult to separate their individual effects.

It is generally agreed that the susceptibility of muscle proteins to freeze denaturation follows a specific order: myofibrillar proteins to sarcoplasmic proteins to stromal proteins. Increased surface hydrophobicity of rockfish protein salt extracts (mixed myofibrillar and sarcoplasmic proteins) during freezing (−10°C, 19 d) was observed by Li-Chan et al. (1985). Both heated and nonheated extracts from frozen protein samples exhibited greater hydrophobicity than those from nonfrozen extracts, suggesting that protein unfolding occurred during freezing and storage. The freezing effect was quite remarkable for proteins extracted at pH 5.5 and much less pronounced at pH 6.8. LeBlanc and LeBlanc (1992) noted marked increases in hydrophobicity of both sarcoplasmic and myofibrillar proteins from cod stored at −12 to −30°C for 300 d. The exposed hydrophobicity increased with storage temperature, coinciding with the increase in the disulfide and ester linkages, aldehydes, and free amino groups (table 8.2). Myosin is particularly susceptible to freeze-induced changes. In vitro cod myosin denaturation appears to be a function of freezing temperature and rate but not storage time. According to Ang and Hultin (1989), surface hydrophobicity of cod myosin suspension increased about 10- and 20-fold within the initial hour of freezing treatment at −80 and −25°C, respectively. At either freezing temperature, no further increase in hydrophobicity occurred during storage up to 40 h. The abrupt hydrophobicity increase was correlated with the reduction in ATPase activity (located at the myosin head), and the rate and extent of both changes were augmented by the addition of formaldehyde. In the absence of formaldehyde, no freeze-induced loss in free amino groups was detected. However, in 1–10 mM formaldehyde, decreases in 2,4,6-trinitrobenzenesulfonic acid (TNBS)–reactive amino groups occurred, and the binding of formaldehyde with free amino groups either insolubilized myosin or led to the formation of myosin polymers. Castrillón et al. (1996) noted a major loss of the sulfur-containing amino acids (cysteinyl and methionyl residues) in frozen (−20°C) sardine muscle after 4 mo storage. The losses of histidinyl, tyrosynyl, leusinyl, lysinyl, and phenylalanyl groups were also significant. The changes in these amino acids were accompanied by a decrease in protein solubility.

Using DSC analysis, Park et al. (1993) have demonstrated decreased stability of myofibrillar proteins during frozen storage (−28°C) of prerigor beef semimembranosus muscle. Maximum thermal transition temperatures for myosin and

Table 8.2. Chemical changes in myosin-rich fraction of cod fillet frozen stored for 1 yr

	Treatment				
Chemical trait	Fresh	−30°C	−22°C	−15°C	−12°C
Soluble protein (mg/ml)	27.9	22.8	21.6	16.6	7.6
Surface hydrophobicity	10.2	11.5	16.1	26.2	69.9
Energy transfer coefficient	84.5	84.1	83.3	79.2	70.2
Total SH (mmol/g protein)	0.070	0.070	0.070	0.074	0.14
S-S (mmol/g protein)	1.4	1.3	2.2	2.2	2.8
Aldehydes (μmol/g protein)	0.40	0.58	1.4	3.1	5.3
Free NH_2 (mmol/g protein)	0.43	0.43	0.49	0.65	1.0
Ester links (mmol/g protein)	0.23	0.28	0.37	0.88	3.0

Source: Adapted from LeBlanc and LeBlanc (1992)

actin from 5-month stored muscle in the freezer are 54 and 70.7°C, respectively, compared to 58.3 and 73.7°C for unfrozen myosin and actin. Frozen storage also tends to destabilize myofibrillar proteins of postrigor muscle; however, differences between frozen and unfrozen samples (transition temperatures as well as total enthalpy of denaturation) are very small. The extent of protein denaturation is ostensibly inversely related to the freezing rate that dictates ice crystallization. Freezing postrigor beef semitendinosus muscle at <0.1, 0.24–0.3, and >1.2°C/min does not alter the denaturation temperatures for myofibrillar proteins but decreases enthalpy of denaturation for myosin in intact muscle by 37.5, 31.2, and 18.7%, respectively (fig. 8.2, table 8.3) and in sarcoplasm-depleted tissue by 31.8, 27.7, and 19.5%, respectively (Wagner and Añón, 1985). More-rapid changes in the rate of enthalpy losses during storage are also noted at lower temperatures (Wagner and Añón, 1986). While actin is rather resistant to freeze denaturation, myosin typically loses its conformational stability rapidly in the initial frozen storage (e.g., <2 months) and more slowly thereafter. Since a linear relationship exists between the enthalpy reduction and the loss in ATPase activity (Wagner and Añón, 1986), it can be assumed that denaturation of the myosin head is a primary event during the early stage of frozen storage. The thin filament, however, would undergo minimal freeze changes. Water-soluble proteins are protective of myofibrillar proteins against freeze denaturation because the removal of sarcoplasmic components results in a decrease in the transition temperature for myosin and actin and facilitates contraction of myofibrils dring frozen storage (Wagner and Añón, 1985; Yoon et al., 1991).

Cold-water fish species generally show an increase in total enthalpy of denaturation on frozen storage, but species from warm habitats exhibit small decreases. In a recent study involving 20 fish species, Davies et al. (1994) observed a clear relationship between habitat temperature and freeze-induced denaturation of fish myofibrillar proteins. The onset and transition temperatures for myosin denatura-

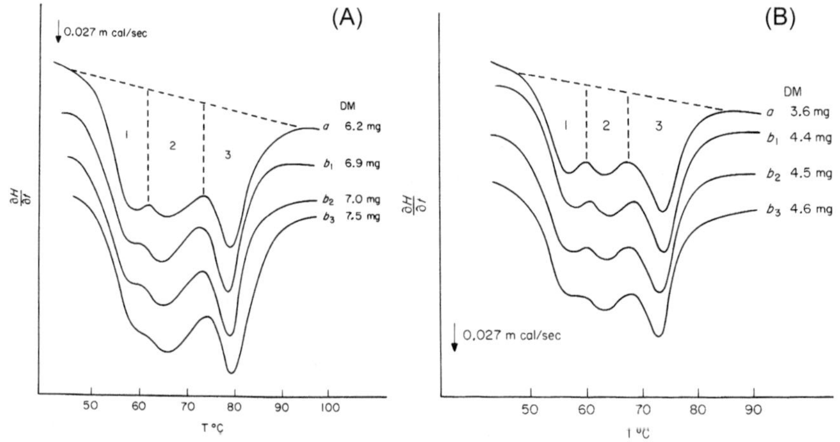

Figure 8.2. DSC thermal curves of (A) bovine whole muscle and (B) muscle depleted of sarcoplasmic proteins and connective tissue at pH 5.6. Letter *a* denotes fresh, unfrozen sample; letters b_1, b_2, and b_3 indicate, respectively, the following freezing rates: >1.2°C/min (fast), 0.24–0.3°C/min (intermediate), and <0.1°C/min (slow). DM: dry matter weight. (Adapted from Wagner and Añón, 1985.)

tion measured by DSC decreased (mean = −5.2°C) after storage at −20°C for up to 1 yr. The decrease was apparently greater for cold-water species, such as cod, than for warm-water species, such as red snapper (fig. 8.3). The observation that myosin transitions were markedly changed in red snapper whole muscle but not in isolated myosin and that the transitions were similar both in cod whole muscle and in isolated myosin led Davies et al. (1994) to conclude that some sarcoplasmic components in warm-water muscle were of a protective nature on myosin stability. It is not clear, however, whether the stability differences between red snapper and cod myosins were in fact due also to different distributions of muscle fibers, for example, red (slow twitch) versus white (fast twitch). In contrast, freezing and frozen storage have little effect on actin stability, irrespective of species.

Some detrimental effects of freeze-thaw cycles, mimicking temperature abuse conditions, have been demonstrated. Heat stability of both trout and Alaska pollock surimi was unaffected by freeze-thaw cycles (between −30 and 4°C, cycled 6 and 15 times) in the sense that the DSC peaks did not shift to different temperatures (Kim et al., 1986). The considerable enthalpy loss caused by temperature abuses, however, was a clear manifestation of less-ordered protein conformations.

Compared to myofibrillar or sarcoplasmic proteins, stromal proteins are relatively resistant to freeze denaturation. The water (~70%) in cod myocommata, the

Table 8.3. Effect of freezing rate on enthalpy of denaturation (ΔH) of myofibrillar proteins from bovine semitendinosus muscle as analyzed by differential scanning calorimetry

	Peak I (myosin)			Peak II			Peak III (actin)			Total		
Sample[a]	<0.1	0.24–0.3	>1.2°C/min[b]	<0.1	0.24–0.3	>1.2°C/min	<0.1	0.24–0.3	>1.2°C/min	<0.1	0.24–0.3	>1.2°C/min
Whole muscle	37.5	31.2	18.7[c]	5.0	5.0	4.6	9.1	9.1	7.4	14.6	13.2	9.6
Washed muscle	31.8	27.7	19.5	(+2.8)	3.5	1.4	11.6	14.4	9.3	15.7	16.8	11.2

[a]Whole muscle: pH 5.6; washed muscle: depleted of sarcoplasmic proteins and connective tissue by treating with Ringer solution (0.15 M NaCl, 3 mM KCl, 3 mM $CaCl_2$, pH 5.6)

[b]Freezing rate

[c]Percentage loss in enthalpy compared with the fresh, unfrozen sample

Source: Calculated from Wagner and Añón (1985)

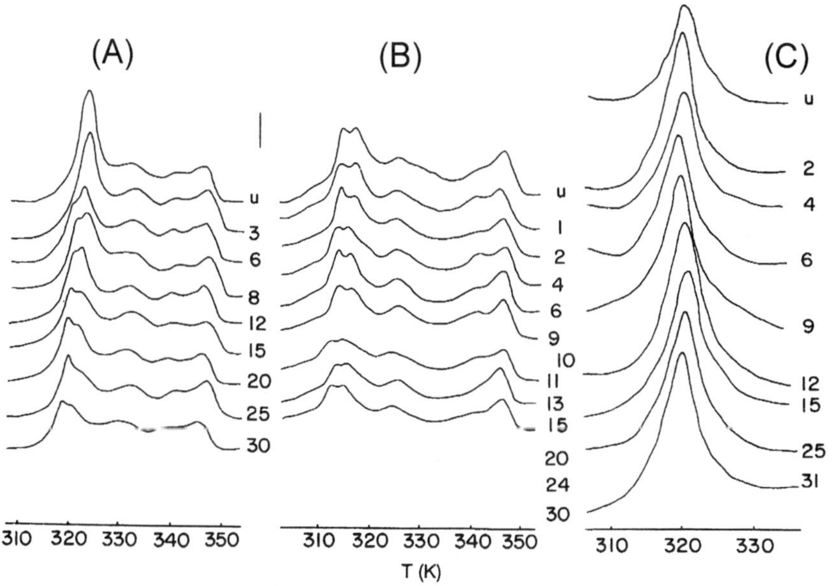

Figure 8.3. DSC thermal curves showing changes occurring during 30-wk frozen storage of (A) red snapper muscle (pH 6.3); (B) cod muscle (pH 6.6); and (C) isolated myosin from red snapper (pH 6.3). The numbers on the right of curves are weeks of frozen storage, and *u* represents the unfrozen sample. (Adapted from Davies et al., 1994.)

collagenous connective tissue in the muscle tissue, is nonfreezable even at −80°C, so long as the fish is frozen immediately after death (Love and Lavéty, 1972). Thus, no collagen changes caused by ice-crystal formation would occur. However, denaturation and formation of insoluble aggregates of collagen in hake muscle occur during frozen storage, and these changes are clearly correlated with the production of dimethylamide (Montero and Borderías, 1990).

8.3.4 Denaturation of Egg Proteins

Egg products are widely used as ingredients in a variety of food preparations. Currently, about one-third of the total liquid egg is frozen and utilized by hotel-restaurant-institution settings. During frozen storage, protein denaturation can occur resulting from physical and chemical reactions. Freeze-induced changes in liquid egg products have been reviewed by Cotterill (1990). Not all egg proteins are susceptible to freeze denaturation; generally, freezing causes only minor changes in raw egg white compared to raw egg yolk. Thinning of thick white, which sometimes occurs in frozen stored liquid white, is associated with denaturation of albumen. The loss of enthalpy of denaturation is accelerated by slow freezing rates, high thawing temperatures, high storage temperature, and long storage time (Wootton et al., 1981). Individually, conalbumin is more susceptible, and ovalbumin is less labile to freeze-induced changes than the whole egg albumen.

Commonly observed changes in frozen egg yolk at temperatures less than −6°C are increases in viscosity and spontaneous gelation caused by protein modifications (Herald et al., 1989). Fast freezing and thawing results in less gelation than slow freezing and thawing, probably because of decreased damage to protein structures. Conspicuously, the detrimental effect of freezing on yolk proteins results from the formation of ice crystals, dehydration, and increases in ionic strength and salt concentrations. There is evidence that some lipoprotein complexes are altered by freezing: when the protein moiety is hydrolyzed, gelation is inhibited (Cotterill, 1990). The behavior of raw whole egg is similar to egg yolk in freezing and frozen storage, but the magnitude of protein changes and resulting functionality alterations are less pronounced for the whole egg.

8.3.5 Denaturation of Proteins in Frozen Doughs

Baking quality of frozen doughs deteriorates gradually during frozen storage or freeze-thaw cycles. Among various possible reasons (e.g., decreased yeast activity and hence less gas production or release of disulfide-reducing substances from dead yeast cells), deleterious changes in gluten proteins seem to be a major factor (Inoue et al., 1994). Indeed, scanning electron microscopy has revealed destruction of gluten networks in frozen dough, presumably because of ice-crystal formation (Berglund et al., 1991). Electrophoresis performed under nonreducing conditions clearly showed the disappearance of glutenin subunits and oligomers in the dough subjected to repeated freezing and thawing, suggesting that structure of glutenin was altered by freeze-thaw cycles (Inoue et al., 1994). Storage at −20°C for 70 d slightly increased the percentage of ethanol-soluble proteins (gliadin), while three freeze-thaw cycles increased the proportion of water-soluble (albumins) and acid-soluble (glutenin) proteins.

8.4 Protein Functionality Changes

8.4.1 Functionality and Its Relation to Denaturation

Protein functionality is a general term referring to "any physicochemical property which affects the processing and behavior of protein in food systems as judged by the quality attributes of the final product" (Kinsella, 1976). Of the various functional properties, there are three major types: hydration (e.g., water binding and solubility), structure and texture (e.g., gelation, cohesion, adhesion, and viscoelasticity), and surface activity (e.g., emulsification and foaming) (table 8.4). These functional attributes are determined by the structure and conformation of proteins.

Table 8.4. Functional properties of food proteins

General property	Specific property	Description	Example
Hydration	Water binding/ uptake	Water entrapment	Dough formation
	Swelling	Water immobilization in protein matrix	Salted meat
	Solubility	Solvation, extraction	Whey
	Dispersion/ suspension	Solvation	Protein drinks
Structure/texture	Gelation	Protein matrix formation	Gelatin, egg custard
	Elasticity	Protein matrix deformation	Kamoboko
	Viscosity	Shear stress due to protein-protein and protein-water interaction	Baby formula
	Adhesion/cohesion	Protein-protein interaction	Boneless ham
Surface	Emulsification	Protein adsorbed at lipid-water interface to reduce surface tension	Creams
	Foaming	Protein film at water-air interface to reduce surface tension	Meringue
	Film Formation	Protein aggregation	Packaging materials

Consequently, any physical or chemical factors that can disturb protein native structure would lead to alterations in protein functionality. Frequently, denaturation is necessary for conferring desirable functional properties as long as the denaturation process is specifically controlled to allow ordered protein-protein interactions or interactions with other food components in a synergistic manner. For instance, protein gelation, which is responsible for adhesion of meat particles and binding water in comminuted and restructured products, requires that myofibrillar proteins be denatured so that the unfolded molecules can cross-link with one another to form three-dimensional viscoelastic gel networks. Similarly, to generate a foam product, the soluble proteins must diffuse to the air-water interface where the protein molecules can unfold and interact to form an air entrapment. Functional properties are imparted by highly cooperative sequential reactions or physicochemical processes. Therefore, if the protein is predenatured, as can be caused by freezing and frozen storage, or the food processing condition is not regulated to allow ordered denaturation aggregation, poor product quality can be expected.

8.4.2 Solubility

Solubility is a sensitive indicator and widely used index for protein denaturation. When denaturation occurs, the hydrophobic amino acid side chains occluded in the protein molecules (except membrane proteins) will be exposed to the surface. To maintain the lowest free energy, the apolar groups tend to interact with one another to form protein aggregates, thus decreasing water binding and solubility. In gadoid fish, loss of protein solubility and water binding in frozen muscle have been attributed to cross-linkages between adjacent polypeptides by formaldehyde derived from trimethylamine oxide, a natural constituent of muscle (Sotelo et al., 1995). The process of freezing and thawing produces some insoluble myosin as well as a small amount of protein in the high–molecular weight fraction (>200 kDa), and these changes are facilitated by the presence of formaldehyde (Ang and Hultin, 1989). Both myofibrillar and sarcoplasmic proteins are found to disappear with increasing formaldehyde concentrations. The myosin heavy chains and light chains and troponins are particularly susceptible. For whole fish muscle or muscle homogenates, protein solubility changes are linked to free fatty acids released from phospholipids, as well as degradation products of oxidized lipids. Results obtained from frozen poultry meat studies have confirmed the deleterious effect of oxidizing lipids on protein solubility (Smith, 1987). Decreased protein solubility by neutral, unoxidized lipids is probably due to the formation of protein-lipid complexes via hydrophobic interactions, but when oxidizing lipids are involved, protein free radicals may be generated. Solubility loss, as a result of aggregation of lipoprotein during frozen storage of liquid egg, is also well recognized. In frozen liquid whole egg, lipovitellin and γ-livetin are the most susceptible egg components, responsible for 31 and 56% solubility losses in frozen egg (−24°C, 80 d), respectively (Herald et al., 1989).

Losses of water-holding capacity of muscle proteins and the failure of the fibers to reabsorb moisture during meat processing are phenomena commonly observed in frozen and thawed meat. Increased damage is caused to both myofibrillar and sarcoplasmic proteins as the time and temperature of frozen storage increase. Awad et al. (1968) have shown that the number of protein components separated by gel electrophoresis of urea-treated bovine actomyosin decrease from six to three over 8-wk storage at −4°C. Decreased solubility of specific sarcoplasmic proteins has also been demonstrated. It is of particular interest to note that at −22°C, at which solubility of myofibrillar proteins is very low, ice crystals form within the I-band but not within the A-band (Rahelic et al., 1985). This may be because the water-holding capacity of actin filaments is less than that of myosin filaments.

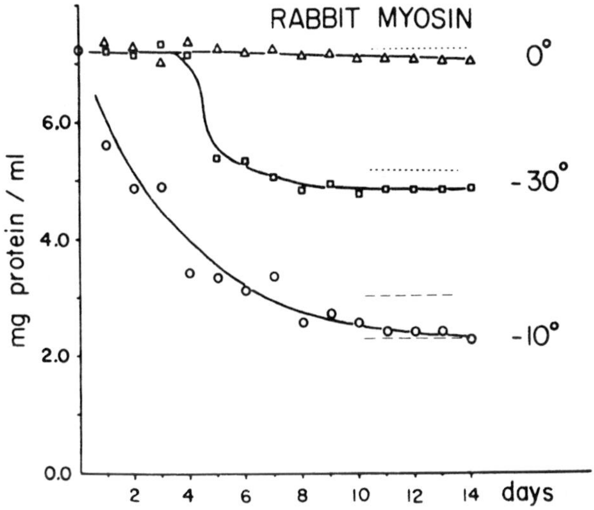

Figure 8.4. Relative rates of decreasing protein concentration in rabbit myosin solutions stored at 0, −10, and −30°C for up to 14 d. The myosin solutions contained 0.72% protein, 0.45 M KCl, 0.026 M potassium phosphate, and pH 6.9. After freezing, the protein solutions were thawed in 4 to 5°C water. (Adapted from Buttkus, 1970.)

Many studies have shown that protein-solubility loss during frozen storage is not directly related to the storage temperature. Buttkus (1970) has found that the rate of formation of insoluble, high–molecular weight protein aggregates in both rabbit and trout myosin solutions increases as the temperature decreases to below the freezing point and reaches a maximum near the eutectic point of the myosin-salt-water solution because of concentration effects. Thus, when the freezing temperature deviates from the eutectic point (−11°C) of myosin-KCl ($\Gamma = 0.5$) solution, aggregation and consequent insolubilization of myosin is decreased and approaches the rate observed at 0°C (fig. 8.4). During the initial stage of freezing, monomeric myosin aggregates, forming dimers and trimers in a side-to-side manner. The aggregation process, as well as the critical freezing point, can be decreased by ATP, pyrophosphate, glycerol, and magnesium chloride. Jiang et al. (1991) have also shown that although solubility (extractability) of actomyosin in frozen grass prawn decreases with time for all storage temperatures studied (−10, −20, −30, and −40°C), only the −10°C storage results in a marked loss in actomyosin solubility. The ionic environment of proteins is an important factor affecting protein solubility changes in frozen storage. Loss of carp myofibrillar protein solubility was found to be maximal at −13°C with KCl and −8 to −18°C with NaCl (Takahashi et al., 1993).

8.4.3 Viscosity

Proteins are hydrocolloids, and in aqueous solution they exhibit pseudoplastic flow characteristics. Since true viscosity of a protein in food systems is difficult to determine, the term *apparent viscosity*, defined as shear stress divided by shear rate, is commonly used to describe the flow behavior. Viscosity of protein suspensions is important because it influences consistency and fluidity of liquid and semiliquid foods (e.g., soups, spreads, muscle minces, and meat batters). When a protein is denatured, its conformation and surface properties (e.g., hydrophilicity and hydrophobicity) are altered, resulting in reduced protein-water interactions and solubility. With some exceptions, when freeze denaturation occurs, protein viscosity decreases.

Freezing induces changes in viscosity of muscle mince from many fish species, such as hake, sardine, megrim, and perch (Jiménez-Colmenero et al., 1988; Reddy et al., 1992). The decrease in viscosity during frozen storage is correlated with losses in protein solubility attributed to denaturation. Addition of oxidized lipids augments the viscosity changes, indicating oxidative deterioration. However, the removal of oxygen from the package before freezing does not prevent the viscosity loss. Egg-white proteins are also susceptible to freezing processes. Freezing causes decreases in the viscosity of egg white, and the viscosity loss is particularly severe at slow freezing rates (Wootton et al., 1981). For liquid whole egg or egg yolk, increases in viscosity are commonly observed during frozen storage presumably because of the formation of lipoprotein complexes and hydrocolloidal structures. Shear stress of liquid egg increases with frozen storage time, and the liquid exhibits Newtonian flow characteristics (fig. 8.5). The viscosity increase is more pronounced when the egg is pasteurized before freezing, apparently because of increased susceptibility caused by heating. Repetitive freeze-thaw processes generally disrupt protein structure, which frequently leads to significant alterations in the consistency of protein suspensions. However, viscosity of fish proteins is not always affected by freeze-thaw abuse, and the species of fish seems to be important. For instance, sand trout myofibrillar protein sols show the same flow behavior regardless of the number (up to 15) of cyclic freezing and thawing, but shear stress of sols made from pollock myofibrils significantly increases with the freeze-thaw cycles (Kim et al., 1986). It is possible that mild changes in protein conformation may actually promote protein-protein interactions yet not reduce the hydrophilicity of the protein aggregates necessary to keep the protein in a fully hydrated state. It must be pointed out that some nonsophisticated viscometers do not have the capability of separating shear stress caused by elastic deformation and shear stress caused by true viscous flow. Many protein suspensions do exhibit viscoelastic deformation characteristics when agitated. Hence, caution should be taken when interpreting some rheological data reported in the literature, for example, concerning freezing effect on protein viscosity changes.

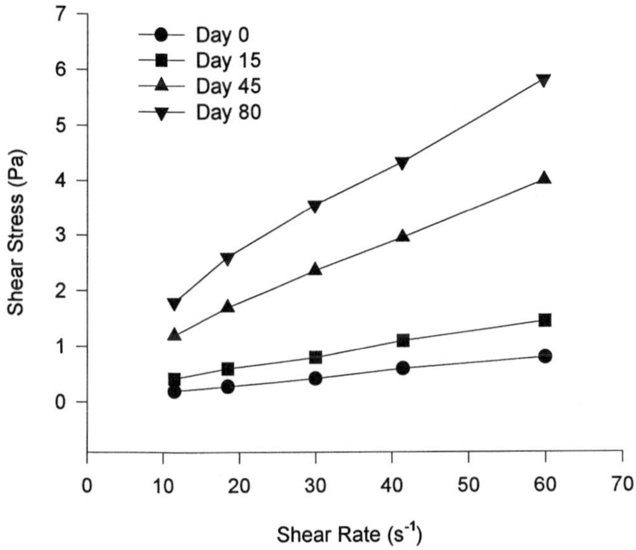

Figure 8.5. Effect of length of frozen storage at −24°C on the shear stress–shear rate relationship of liquid whole egg. (Adapted from Herald et al., 1989.)

8.4.4 Gelation

The gel-forming ability of proteins is an important quality index of muscle foods. The importance of muscle protein gelation is due to the gel imparting cohesion and adhesion in comminuted and restructured meats, as well as in surimi-based products, necessary for producing desirable texture, retaining moisture, and stabilizing fat particles in cooked products. Worldwide consumption of fish meat pastes and surimi analogs (e.g., imitation crab, lobster, and scallop) is increasing. The ability of proteins to form a gel is reduced by freezing and storage conditions that lead to physical and chemical deteriorations in proteins, including freezing rate, storage time and temperature, and temperature fluctuation during frozen storage. Sych et al. (1991) have observed a steady decline in cohesiveness but a rapid increase in firmness of cooked cod surimi gels during frozen storage of the surimi at −20°C. Increased gel firmness is apparently due to a loss in the water-holding capacity of myofibrillar proteins. Freezing and thawing abuse impairs gelling functionality of surimi. Using a torsion test, Kim et al. (1986) found a drastic reduction in shear modulus of cooked sand trout surimi gels after the surimi was repeatedly frozen and thawed. However, freeze-thaw abuse enhanced the gelling ability of Alaska pollock surimi (fig. 8.6). Since most sarcoplasmic constituents (e.g., salt-soluble proteins and iron) and lipids are removed during surimi preparation, losses in gelation properties of frozen surimi are probably due mainly to ice formation and dehydration, which

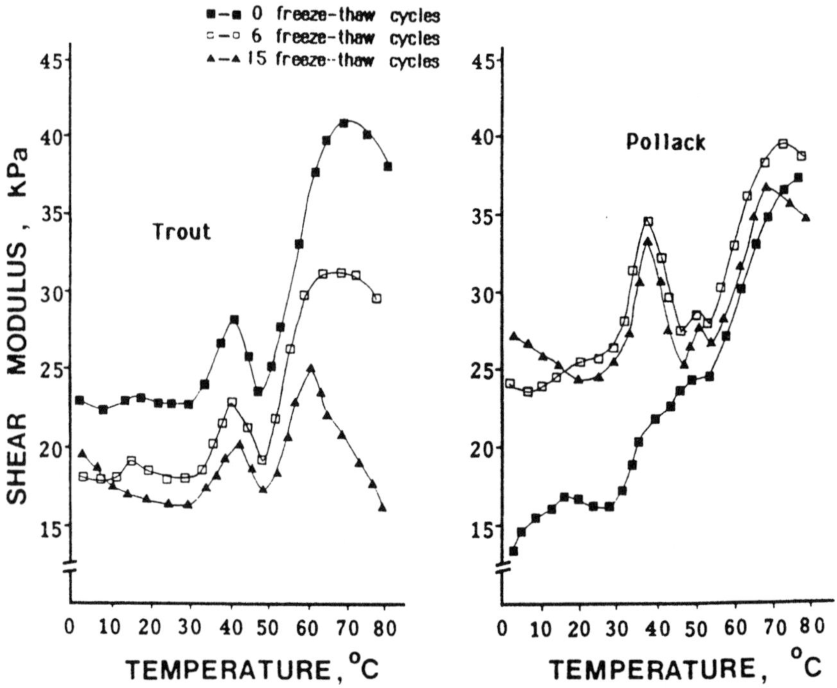

Figure 8.6. Rigidity profiles of freeze-thaw (−30 to 4°C) abused sand trout and Alaska pollock surimi sols/gels during heating. (Adapted from Kim et al., 1986.)

can cause irreversible damage to myofibril structure and deterioration in physicochemical attributes.

Losses in gelation properties of mammalian muscle proteins during storage have also been observed. Park et al. (1993) demonstrated a progressive decrease in shear stress and shear strain of cooked gels prepared from beef batters frozen stored at −28°C for up to 5 mo. The decreases in the gel rheological properties were apparently related to the destabilization of myofibrillar proteins during frozen storage. When mechanically deboned chicken meat is subjected to freezing and storage at subfreezing temperatures, losses in gel-forming ability of myofibrillar proteins also occur, and this functionality deterioration is aggravated by freeze-thaw abuse (Smith, 1987). Mechanically deboned chicken meat is a rather heterogeneous system containing high concentrations of prooxidants and lipids. Therefore, it is believed that losses in functional properties, including gelation and solubility, of myofibrillar proteins during frozen storage of the meat is due, in part, to oxidative reactions.

Spontaneous gelation of liquid whole egg upon freezing is an undesirable process. Gelation is attributed to the yolk fraction; dilution of liquid yolk by egg white reduces this incidence. In general, upon freezing and storage below −6°C, the viscosity of raw egg yolk increases and gelation occurs. Eventually, all fluidity is lost, making the gelled product difficult to process. The mechanism of egg yolk gelation is still not completely understood. Until recently, the gelation was thought to be totally irreversible. Low-density lipoproteins and lecithin-protein complexes may be involved, and treatments of the yolk with proteolytic enzymes can effectively prevent gelation (Cotterill, 1990). The gelation may result from aggregation of lipoproteins whose conformation can be altered when the concentrations of nonprotein constituents increase and the pH changes because of dehydration and ice formation. Addition of some small–molecular weight additives, such as salts, sugars, and glycerine, can reduce gelation. Commercially, a variety of cryoprotective agents (e.g., sodium chloride, sucrose, syrups, and gums) are added to liquid whole egg or yolk to prevent gelation during frozen storage.

8.4.5 Emulsification

Finely comminuted or homogenized meat products are emulsion systems in which muscle proteins act as emulsifiers to lower the surface tension at the water-oil interface. Emulsifying capacity is a function of protein hydrophobicity and solubility, while emulsion stability is related to the total concentration of proteins. Although sarcoplasmic proteins have considerable emulsifying ability, the salt-soluble myofibrillar proteins are the main contributor to emulsion characteristics and stability of finely chopped meat. The ability of myofibrillar proteins to stabilize fat particles in a complex meat system is decreased by freezing and prolonged storage below the freezing point. In frozen fish muscle homogenates, emulsifying capacity, viscosity, and protein solubility are often related to one another (Jiménez-Colmenero et al., 1988; Reddy et al., 1992). Decreased emulsifying ability of fish mince during extended frozen storage is commonly attributed to the loss of protein solubility, making the protein molecules unable to diffuse to the oil-water interface where protein-lipid association via hydrophobic interactions occurs. This simple attribution seems valid for purified proteins or washed muscle tissue but is not always justifiable for whole muscle homogenates that contain many protein-reactive substances. For those formaldehyde-forming species, emulsifying capacity declines rapidly in the course of frozen storage. Ostensibly, this is related to marked losses in protein solubility. However, some studies have also shown that in fish devoid of formaldehyde, emulsifying capacity of frozen stored muscle homogenates is independent of protein solubility (Huidobro and Tejada, 1993). In such cases, a balanced hydrophobicity-solubility may be a more suitable predictor of emulsifying properties. According to Li-Chan et al. (1985), when protein solubility is high (over 50% of total protein), surface hydrophobicity is the chief factor responsible for emulsifying capacity, whereas at lower levels of solubility, the main factor is the amount of soluble protein. Thus, if the formation of formaldehyde can be prevented (e.g., by adding a proper oxidizing agent to destroy the precursor trimethylamine oxide), emulsifying capacity would be less dependent on protein solubility. Conceivably, an inhibition of the enzyme responsible for formaldehyde formation would also prevent protein functionality losses in certain fish species during frozen storage.

8.4.6 Foaming

Many prepared foods, such as meringues and ice creams, are foam-type products in which proteins play a crucial role in air entrapment. While not popular in the U.S. market, some meat-based foam products are widely available in other parts of the world, such as France. The behavior of proteins at the gas-liquid interface is extremely important because formation of a protein-based flexible, cohesive film around the air bubbles is essential for foaming capacity and foam stability. Foaming capacity of proteins is influenced by the nature of the protein and the prevailing environmental conditions. There is a relationship between the molecular flexibility of proteins, film properties, and foam stability. Flexible, disordered proteins are more surface active than extensively cross-linked, stable, and compact globular proteins. Thus, α-casein can rapidly form films, whereas lysozyme has limited film-forming properties.

Freezing is detrimental to foaming properties of egg-white proteins. Foam stability of whole egg white is markedly affected by freezing rate and much less so by storage time and temperature of the freezer. Decreased foam stability by freezing is ascribed to denaturation of conalbumin and ovalbumin, which are very sensitive to freezing rate (table 8.5).

Compared to milk and egg proteins, relatively few studies have been published on foaming properties of muscle proteins, and reports on the freezing effect on muscle protein foams are even scarcer. The following description highlights results from a study by Huidobro and Tejada (1992) on foaming capacity of fish minces during frozen storage at $-18°C$ for up to a year. Foaming capacity of frozen stored minces initially was inhibited by the presence of the species' own lipids. Thus, foam volume of the minces at the beginning of frozen storage followed the order of blue whiting (low-fat species) to horse mackerel (intermediate fat) to mackerel (high fat) (fig. 8.7). Conspicuously, lipids acted to displace proteins from the air-water interface in a competitive manner by undergoing strong preferential adsorption without forming a resilient foam-stabilizing film. Protein solubility (in 0.86 M NaCl, pH 7.0) decreased steadily during frozen storage, most noticeably in blue whiting, coinciding with the production of dimethylamine and formaldehyde (Huidobro and Tejada, 1992). A more recent report by the same authors (Huidobro and Tejada, 1995) indicated that myosin heavy chain was the most susceptible protein, and the insoluble aggregates formed from myosin largely contributed to the protein-solubility loss during frozen storage. However, during prolonged storage, foaming capacity increased, but when the end of the storage period was approached, it declined again. These observations led Huidobro and Tejada (1992) to conclude that protein solubility was not an in-

Table 8.5. Effect of freezing rate on denaturation and functional properties of egg-white proteins

Freezing rate (°C/min)	ΔH(J/g) Egg white	Conalbumin	Ovalbumin	Foam instability (ml drained/g egg-white foam)	Egg-white viscosity (mPa s)
Unfrozen	15.2	5.4	9.6	0.40	4.42
3.10	14.4	5.1	9.2	0.39	4.34
0.40	14.0	4.9	8.7	0.36	4.21
0.20	12.8	4.4	8.0	0.32	4.21
0.07	12.0	4.1	7.8	0.29	3.92
0.06	11.9	3.9	7.8	0.26	3.93
0.03	11.1	3.4	8.4	0.23	3.71

Source: Adapted from Wootton et al. (1981)

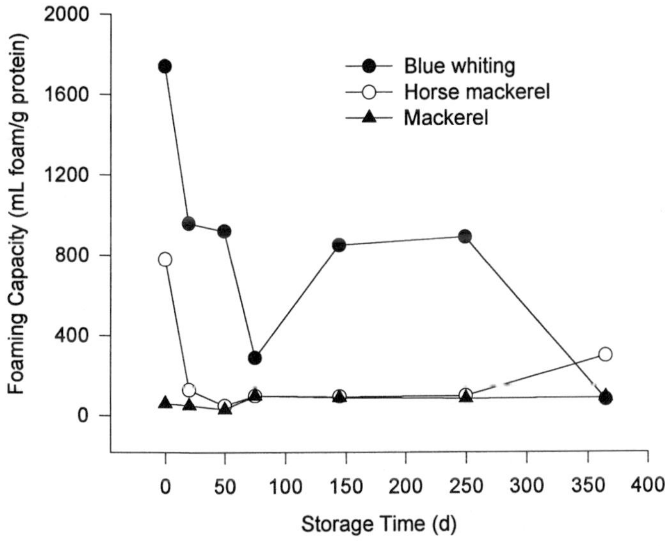

Figure 8.7. Foaming capacity of muscle homogenates of blue whiting (low fat), horse mackerel (intermediate fat), and mackerel (high fat) caught in winter and stored at −18°C. Homogenates were prepared by blending muscle mince with 0.86 M NaCl (pH 7) at a 1:20 ratio. (Adapted from Huidobro and Tejada, 1992.)

dispensable requisite for foaming capacity, and the mere dispersion of proteins or their aggregates would be sufficient. Nonetheless, the ultimate consequence of extended frozen storage was a decrease in foaming properties resulting from formation of large, indispersible protein aggregates due to excessive formaldehyde production.

8.5 Preservation of Protein Functionality through Preventing Protein Denaturation

Various cryoprotectants, that is, substances that can protect proteins from chemical deterioration, have been used to prevent protein denaturation and losses in functional properties. Also, antioxidants can be incorporated into muscle foods before freezing to minimize oxidative changes in proteins during storage. The subjects of using cryoprotectants and antioxidants to preserve frozen-food quality are covered more specifically and in greater detail in chapters 11 and 12, respectively. Hence, only a brief account of these useful food additives for protein stabilization will be given here. Some general properties of antifreeze proteins are also described herein because such proteins are of potential significance for commercial applications.

8.5.1 Cryoprotectants

Inhibition of protein denaturation in frozen fish surimi and minces is achieved in the fish industry by incorporation of sucrose, sorbitol (typically 4% each), and polyphosphate. Reducing sugars is normally not used to avoid unnecessary browning reactions. Since the discovery of cryoprotectants some three decades ago, numerous studies have been conducted to determine the specific effects of various cryoprotectants on fish surimi functionality. We have found that functional properties of isolated chicken myofibrillar proteins (similar to fish surimi) could also be protected by cryoprotectants. Without cryoprotectants, the proteins lost their solubility and gel-forming ability rapidly during frozen storage. However, when 4% sucrose and 4% sorbitol were incorporated in the protein pellet before freezing, the losses were inhibited (fig. 8.8).

The rather high level of sucrose and sorbitol (typically 4% for each) imparts a sweet taste to surimi that may be undesirable to the consumer. Therefore, numerous other cryoprotective materials have also been investigated for use as possible alternatives. Examples are low–molecular weight compounds, such as carboxylic acids and sodium glutamate, and large–molecular weight substances and mixtures such as Polydextrose, glucose syrups, protein hydrolysates from fish and casein, egg white, and a number of polysaccharide hydrocolloids. The effectiveness of these stabilizers is dependent on species and proteins (Sych et al., 1990, 1991).

Mechanisms of protein stabilization by the various cryoprotectants are not very clear. It is postulated that cryoprotectants afford their stabilizing effect by influencing ice-crystal formation, protein dehydration, and increases in salt concentration. Sugars and polyols stabilize proteins presumably through their effect on water structure and modification of hydrophobic interactions that may be essential for protein structures. Formation of hydrogen bonding between carbohydrates or polyols with protein side chains to increase protein hydration and decrease

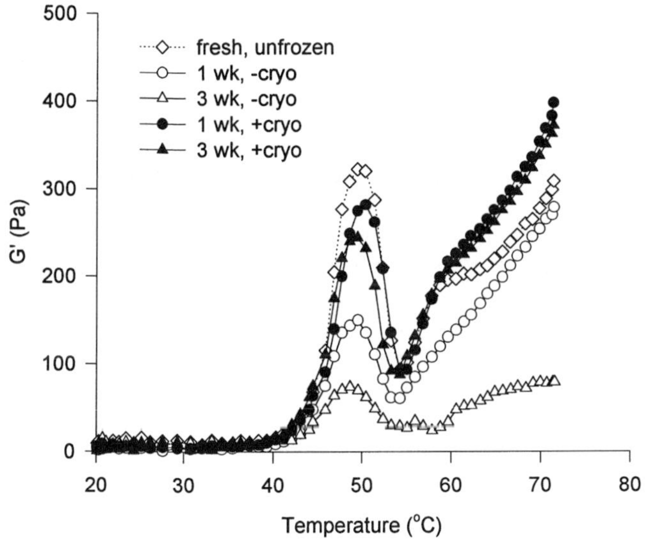

Figure 8.8. Gelation of chicken pectoralis myofibrillar protein before and after frozen storage at −20°C in the presence or absence of combined 4% sucrose and 4% sorbitol for different periods. Myofibrils were suspended in 0.6 M NaCl, 50 mM sodium phosphate buffer (pH 6.0) before gelation. G': storage modulus, an elastic component of gels. (Xiong and Blanchard, unpublished data.)

protein-protein interactions is also a possible mechanism. Increased hydration may reduce the quantity of frozen water in the protein matrix and minimize the accumulation of solutes and ions surrounding the proteins.

8.5.2 Antioxidants

One of the most important causes of quality deterioration in frozen foods is oxidation, which promotes physicochemical changes in addition to development of rancidity. Various prooxidants and oxidizing lipids naturally present in muscle can initiate protein oxidation; among them are linoleic acid peroxides, which are reportedly 10 times more potent than unoxidized linoleic acid in decreasing solubility of cod myofibrillar proteins during frozen storage (Jarenbäck and Liljemark, 1975). To inhibit oxidative changes in foods, many antioxidative measures, including vacuum packaging and uses of antioxidants, have been proposed. Decreased susceptibility of proteins to freezing changes by antioxidative treatment is due to the antioxidants' capability of scavenging free radicals, chelating metal ions, and donating electrons, therefore minimizing alterations in amino acid residues and conformation of the protein. The effectiveness of antioxidants, however, varies with animal species and muscle types.

Maintenance of protein functionality by antioxidant preservation during chill storage has been demonstrated (Xiong et al., 1993). However, information is limited concerning the effect of antioxidants on proteins during frozen storage. Smith (1987) showed that incorporation of the antioxidants BHA, propyl gallate, and citric acid into mechanically and hand-deboned turkey meat before freezing markedly inhibited lipid oxidation, but it was ineffective in protecting functional deteriorations in myofibrillar proteins during short-term storage (i.e., <16 wk at −20°C). However, during extended storage (>16 wk), antioxidant-treated meat exhibited better functionality (i.e., smaller decreases in protein solubility, myosin ATPase activity, cooking loss from meat batters, and firmness and cohesiveness of myofibrillar protein gels) compared to the control meat.

8.5.3 Antifreeze Proteins and Genetic Engineering

Antifreeze proteins were discovered some 20 years ago following the initial observations of the unusual effects of freezing on fish. Marine waters normally have salinities that cause their freezing points to be below −1.8°C, but fish in the Antarctic ocean can live under the supercool condition. It was recognized later that a group of glycoproteins found in the serum, named antifreeze proteins, were responsible for the antifreeze effect in muscle tissue. As many as eight glycosylated antifreeze proteins, based on their electrophoretic mobility ranging from 2.6 to 30 kDa in molecular weights, have been isolated from polar fish (Feeney and Yeh, 1993). The longer antifreeze proteins (10–30 kDa) have higher antifreeze activities in lowering the freezing temperature than the short proteins (2.6–5 kDa). Most active antifreeze proteins have a repeated Ala-Ala-Thr tripeptide unit, and the molecules appear to be flexible rods and have a locally defined helical structure. So far, only the shorter antifreeze proteins have been found in extracts of muscle tissue.

The mechanism of antifreeze by the proteins is thought to be caused by adsorption at the ice-solution interface. The proteins have electric dipole moments along their helical axes. The dipole could align itself in a way conducive to protein-water interactions on the prismatic facets of ice via dipole-dipole interaction, leading to adsorption. At the water-ice interface, protein hydrophilic groups face the ice surface, while the hydrophobic groups surround a specific ice facet, thus preventing water from effectively binding to the surface of ice. Consequently, not only do antifreeze proteins lower the freezing temperature, but they also retard recrystallization and the formation of large ice crystals during frozen storage (Feeney and Yeh, 1993).

According to Feeney and Yeh (1993), as little as 0.5 μg/mL antifreeze protein gives extensive inhibition of ice-crystal growth when the temperature is raised from −80 to −3.5°C. Therefore, antifreeze proteins are potentially useful in food products as antifreezing materials to prevent physical and chemical damage and deteriorations in proteins and other functional cellular constituents. Recent research has shown that it is also possible to incorporate antifreeze protein genes in warmer-water fish species, such as Atlantic salmon, through genetic manipulations. Furthermore, antifreeze proteins may be synthesized chemically and incorporated into food systems to improve storage stability. Antemortem injection of Antarctic cod antifreeze glycoproteins (0.01 μg/kg muscle) to lambs, followed by postmortem storage of meat at −20°C for 8 wk, showed a reduced ice-crystal size in frozen muscle tissue and less drip loss upon thawing the meat when compared with noninjected meat (Payne and Young, 1995). Obviously, more exploratory research is warranted before any commercial application of the antifreeze proteins becomes feasible.

References

Ang, J.F. and Hultin, H.O. 1989. Denaturation of cod myosin during freezing after modification with formaldehyde. *J. Food Sci.* **54**:814–818.

Awad, A., Powrie, W.D., and Fennema, O. 1968. Chemical deterioration of frozen bovine muscle at −4°C. *J. Food Sci.* **33**:227–235.

Berglund, P.T., Shelton, D.R., and Freeman, T.P. 1991. Frozen bread dough ultrastructure as affected by duration of frozen storage and freeze-thaw cycles. *Cereal Chem.* **68**:105–107.

Buttkus, H. 1970. Accelerated denaturation of myosin in frozen solution. *J. Food Sci.* **35**:558–562.

Careche, M. and Tejada, M. 1994. Hake natural actomyosin interaction with free fatty acids during frozen storage. *J. Sci. Food Agric.* **64**:501–507.

Castrillón, A.M., Alvarez-Pontes, E., Arias, M.T.G., and Navarro, P. 1996. Influence of frozen storage and defrosting on the chemical and nutritional quality of sardine (*Clupea pilchardus*). *J. Sci. Food Agric.* **70**:29–34.

Cotterill, O.J. 1990. Freezing egg products. In *Egg Science and Technology,* 3d ed. (Stadelman, W.J. and Cotterill, O.J., eds.) pp. 217–242, Haworth Press, Binghamton, New York.

Davies, J.R., Ledward, D.A., Bardsley, R.G., and Poulter, R.G. 1994. Species dependence of fish myosin stability to heat and frozen storage. *Int. J. Food Sci. Technol.* **29**:287–301.

Dyer, W.J. and Dingle, J.R. 1961. Fish proteins with special reference to freezing. In *Fish as Food. I. Biochemistry and Microbiology.* (Borgstrom, G., ed.) pp. 275–327, Academic Press, New York.

Feeney, R.E. and Yeh, Y. 1993. Antifreeze proteins: Properties, mechanism of action, and possible applications. *Food Technol.* **47**(1):82, 84–88, 90.

Fennema, O. 1982. Behavior of proteins at low temperatures. In *Food Protein Deterioration Mechanisms and Functionality.* (Cherry, J.P., ed.) pp. 109–133, ACS Symposium Series 206, American Chemical Society, Washington, D.C.

Franks, F. 1995. Protein destabilization at low temperatures. *Adv. Protein Chem.* **46**:105–139.

Herald, T.J., Osorio, F.A., and Smith, D.M. 1989. Rheological properties of pasteurized liquid whole egg during frozen storage. *J. Food Sci.* **54**:35–38, 44.

Huidobro, A. and Tejada, M. 1992. Foaming capacity of fish minces during frozen storage. *J. Sci. Food Agric.* **60**:263–270.

Huidobro, A. and Tejada, M. 1993. Emulsifying capacity of fish mince from several species during frozen storage. *J. Sci. Food Agric.* **61**:333–338.

Huidobro, A. and Tejada, M. 1995. Alteration of the electrophoretic pattern of myofibrillar proteins in fish mince during frozen storage. *Z. Lebensm. Unters. Forsch.* **200**:247–251.

Inoue, N., Takatori, K., Motoshige, T., and Shinano, H. 1992. Effect of storage temperature on the freeze denaturation of fish myosin B. *Nippon Suisan Gakkaishi* **58**:2357–2360.

Inoue, Y., Sapirstein, H.D., Takayanagi, S., and Bushuk, W. 1994. Studies on frozen doughs. III. Some factors involved in dough weakening during frozen storage and thaw-freeze cycles. *Cereal Chem.* **71**:118–121.

Jarenbäck, L. and Liljemark, A. 1975. Ultrastructural changes during frozen storage of cod. III. Effects of linoleic acid and linolenic acid hydroperoxides on myofibrillar proteins. *J. Food Technol.* **10**:437–452.

Jiang, S.-T., Hwang, B.-S., Moody, M.W., and Chen, H.-C. 1991. Thermostability and freeze denaturation of grass prawn (*Penaeus monodon*) muscle proteins. *J. Agric. Food Chem.* **39**:1998–2001.

Jiménez-Colmenero, F., Tejada, M., and Borderías, A.J. 1988. Effect of seasonal variations on protein functional properties of fish during frozen storage. *J. Food Biochem.* **12**:159–170.

Kang, J.O., Ito, T., and Fukazawa, T. 1983. Effect of frozen storage on the structure and enzymatic activities of myofibrillar proteins of rabbit skeletal muscle. *Meat Sci.* **9**:131–144.

Kim, B.Y., Hamann, D.D., Lanier, T.C., and Wu, M.C. 1986. Effects of freeze-thaw abuse on the viscosity and gel-forming properties of surimi from two species. *J. Food Sci.* **51**:951–956, 1004.

Kinsella, J.E. 1976. Functional properties of proteins in foods: a survey. *CRC Crit. Rev. Food Sci. Nutr.* **7**:219–280.

Lawrie, R.A. 1991. *Meat Science*, 5th ed. Pergamon Press, New York.

LeBlanc, E.L. and LeBlanc, R.J. 1992. Determination of hydrophobicity and reactive groups in proteins of cod (*Gadus morhua*) muscle during frozen storage. *Food Chem.* **43**:3–11.

Li-Chan, E., Nakai, S., and Wood, D.F. 1985. Relationship between functional (fat binding, emulsifying) and physicochemical properties of muscle proteins: effects of heating, freezing, pH, and species. *J. Food Sci.* **50**:1034–1040.

Love, R.M. 1968. Ice formation in frozen muscle. In *Low Temperature Biology of Foodstuffs*. (Hawthorn, J. and Rolfe, E.J., eds.) pp. 105–124, Pergamon Press, Oxford.

Love, R.M. and Lavéty, J. 1972. The connective tissues of fish. VII. Postmortem hydration and ice crystal formation in myocommata and their influence on gaping. *J. Food Technol.* **7**:431–441.

Montero, P. and Borderías, J. 1990. Behavior of myofibrillar proteins and collagen in hake (*Merluccius merluccius* L.) muscle during frozen storage and its effect on texture. *Z. Lebensm. Unters. Forsch.* **190**:112–117.

Nambudiri, D.D. and Gopakumar, K. 1992. ATPase and lactate dehydrogenase activity in frozen stored fish muscle as indices of cold storage deterioration. *J. Food Sci.* **57**:72–76.

Olley, J., Pirie, R., and Watson, H. 1962. Lipase and phospholipase activity in fish skeletal muscle and its relationship to protein denaturation. *J. Sci. Food Agric.* **13**:501–516.

Park, J.W., Lanier, T.C., and Pilkington, D.H. 1993. Cryostabilization of functional properties of prerigor and postrigor beef by dextrose polymer and/or phosphates. *J. Food Sci.* **58**:467–472.

Payne, S.R. and Young, O.A. 1995. Effects of pre-slaughter administration of antifreeze proteins on frozen meat quality. *Meat Sci.* **41**:147–155.

Privalov, P.L., Griko, Y.V., Venyaminov, Y.S., and Kutyshenko, V.P. 1986. Cold denaturation of myoglobin. *J. Mol. Biol.* **190**:487–497.

Privalov, P.L. and Makhatadze, G.I. 1993. Contribution of hydration to protein folding thermodynamics. II. The entropy and Gibbs energy of hydration. *J. Mol. Biol.* **232**:660–679.

Rahelic, S., Gawwad, A.H., and Puac, S. 1985. Structure of beef longissimus dorsi muscle frozen at various temperatures. Part 2: Ultrastructure of muscle frozen at −10, −22, −33, −78, and −115°C. *Meat Sci.* **14**:73–81.

Reddy, G.V.S., Srikar, L.N., and Sudhara, N.S. 1992. Deteriorative changes in pink perch mince during frozen storage. *Int. J. Food Sci. Technol.* **27**:271–276.

Shenouda, S.Y.K. 1980. Theories of protein denaturation during frozen storage of fish flesh. *Adv. Food Res.* **26**:275–311.

Smith, D.M. 1987. Functional and biochemical changes in deboned turkey due to frozen storage and lipid oxidation. *J. Food Sci.* **52**:22–27.

Sotelo, C.G., Piñeir, C., and Pérez-Martín, R.I. 1995. Denaturation of fish proteins during frozen storage: role of formaldehyde. *Z. Lebensm. Unters. Forsch.* **200**:14–23.

Sych, J., Lacroix, C., Adambounou, L.T., and Castaigne, F. 1990. Cryoprotective effects of lactitol, palatinit, and polydextrose on cod surimi proteins during frozen storage. *J. Food Sci.* **55**:356–360.

Sych, J., Lacroix, C., Adambounou, L.T., and Castaigne, F. 1991. The effect of low- and non-sweet additives on the stability of protein functional properties of frozen cod surimi. *Int. J. Food Sci. Technol.* **26**:185–197.

Takahashi, K., Inoue, N., and Shinano, H. 1993. Effect of storage temperature on freeze denaturation of carp myofibrils with KCl or NaCl. *Nippon Suisan Gakkaishi* **59**:519–527.

Wagner, J.R. and Añón, M.C. 1985. Effect of freezing rate on the denaturation of myofibrillar proteins. *J. Food Technol.* **20**:735–744.

Wagner, J.R. and Añón, M.C. 1986. Effect of frozen storage on protein denaturation in bovine muscle. 1. Myofibrillar ATPase activity and differential scanning calorimetric studies. *J. Food Technol.* **21**:9–18.

Wootton, M., Hong, N.T., and Thi, H.A.P. 1981. A study of the denaturation of egg white proteins during freezing using differential scanning calorimetry. *J. Food Sci.* **46**:1336–1338.

Xiong, Y.L., Decker, E.A., Robe, G.H., and Moody, W.G. 1993. Gelation of crude myofibrillar protein isolated from beef heart under antioxidative conditions. *J. Food Sci.* **58**:1241–1244.

Yamamoto, K., Samejima, K., and Yasui, T. 1977. A comparative study of the changes in hen pectoralis muscle during storage at 4°C and −20°C. *J. Food Sci.* **42**:1642–1645.

Yoon, K.S., Lee, C.M., and Hufnagel, L.A. 1991. Effect of washing on the texture and microstructure of frozen fish mince. *J. Food Sci.* **56**:294–298.

9

Lipid Oxidation: Flavor and Nutritional Quality Deterioration in Frozen Foods

Marilyn C. Erickson

When food products are subjected to decreasing temperatures, the ability of microorganisms on or in these foods to grow decreases. As a consequence, quality deterioration in many food products shifts from microbiological causes to chemical/enzymatic causes. Chief among these chemical causes is lipid oxidation. In this chapter this degradative pathway will be discussed by first addressing the basic mechanisms of lipid oxidation in frozen foods. The chapter will then focus on those factors, both physical and chemical, that influence the rate of lipid oxidation, followed by a discussion of the consequences of lipid oxidation in frozen foods.

9.1 Basic Mechanisms of Lipid Oxidative Processes

A free radical process is the basic mechanism upon which lipid oxidation proceeds. The following three stages characterize this process:

Initiation:	$In\bullet + RH \rightarrow InH + R\bullet$
Propagation:	$R\bullet + O_2 \rightarrow ROO\bullet$
	$ROO\bullet + RH \rightarrow R\bullet + ROOH$
Termination:	$2RO_2\bullet \rightarrow O_2 + RO_2R$
	$RO_2\bullet + R\bullet \rightarrow RO_2R$
	$2R\bullet \rightarrow R\text{–}R$

During initiation a hydrogen is abstracted from a fatty acid, leaving the fatty acid alkyl radical. Combining quickly with oxygen, this fatty acid alkyl radical is converted to the fatty acid peroxyl radical as one step of the propagation stage. As another step, the peroxyl radical abstracts a hydrogen from an adjacent fatty acid, forming a hydroperoxide molecule and a new fatty acid alkyl radical. Breakdown of the hydroperoxide, in turn, is responsible for further propagation of the free radical process. Given the bond dissociation energies of LOO–H (about 90 kcal/mol) and LO–OH (about 44 kcal/mol), spontaneous decomposition is unlikely at freezing temperatures (Terao, 1990). Hydroperoxide breakdown during frozen storage instead would be dominated by one-electron transfer from metal ions

$$Fe^{2+} + LOOH \longrightarrow Fe^{3+} + HO\bullet + OH^-$$

In food and biological systems, the major contributors to decomposition of lipid hydroperoxides would be heme and nonheme iron; reactions involving the ferrous ion are much faster than those involving ferric ion. Finally, when the free radical content becomes large enough, two free radicals may combine to terminate the free radical process. It is important to emphasize that, despite the simplified description just offered, lipid oxidation is a very complex process. In particular, initiation, propagation, and termination reactions are not separated in time. Initiation reactions exist throughout the entire storage life of a product. That these chemical and enzymatic reactions exist in living tissue is evidenced by the occurrence in aerobic organisms of enzymes that can eliminate or detoxify these compounds (e.g., superoxide dismutase, catalase, and glutathione peroxidase). Therefore, activated oxygen species are likely to be present in the food item even before it is harvested and not just produced during processing and storage.

9.2 Influence of Physical Properties on Lipid Oxidation

Lipid oxidation, whether it occurs through enzymic or nonenzymic pathways is influenced by a number of physical factors. These factors include temperature, water activity, pH, light, and oxygen pressure or solubility. Of these factors, changes in temperature, water activity, and oxygen solubility would be expected to have the greatest impact in frozen foods.

9.2.1 Temperature

In general, as temperatures are decreased, the rate of chemical or enzymic reactions also decreases. Reactions associated with lipid oxidation are no exception although temperature-dependent conformational changes in catalytic heme proteins may also play an important role (Valpuesta et al., 1986). For menhaden mince, storage at −20°C strongly retarded lipid oxidation compared to storage of sample at −7°C (Hwang and Regenstein, 1988). Similarly, uncooked ground beef patties at −12.2°C had higher thiobarbituric acid (TBA) values than those stored at −23.3°C or −34.4°C (Bhattacharya et al., 1988). However, decreased oxidative stabilities are noted in the range of 0 to −10°C. In this range, temperature induced deceleration is more than offset by the accelerating effect of increased concentration of reactants as water is removed in the form of pure ice. In addition, changes in membrane lipid interactions induced by ice formation have been noted (Sanderson et al., 1993). Illustrating this accelerating effect of ice crystal formation on lipid oxidation, microsomal membrane fractions were incubated in the presence and absence of miscible solvents (alcohols) that prevent freezing at temperatures below 0°C (Apgar and Hultin, 1982). By comparing the ratio of the reaction rate at 6°C to that at −12°C, the rate of lipid oxidation was found to decrease less in the presence of ice than in the presence of alcohols.

Ice crystal size and distribution in the intra- or extracellular space will vary with freezing rate. For example, high freezing rates induce the formation of many small, intracellular ice crystals, but at slow freezing rates the ice crystals form only extracellularly and disruption of membrane and other tissue structures occurs (Lepock et al., 1978; Jul, 1984; Nilsson and Ekstrand, 1994). Enhanced interaction of catalysts with lipid substrates would be expected upon cellular disruption, yet apart from very slow rates, freezing rate has little influence on the quality of the product. For example, in the study of Tomas and Anon (1990), chicken breast and salmon muscle were frozen either slowly or rapidly at −25°C and then stored at −5°C for up to 83 and 47 days, respectively. While there was a small but significant increase in TBA values during storage, no significant differences occurred between fast- and slow-frozen samples at any one time point. Such results demonstrate that storage temperature rather than freezing rate has a greater effect on oxidative stabilities. This is particularly true when temperature fluctuations occur during storage. Temperature fluctuations from −28°C to −10°C produced a significant increase in peroxide value for frozen Pacific herring (Bilinski et al., 1981), and when storage temperature alternated between −7° and −20°C for menhaden mince, lipid oxidation proceeded at rates approximating those at −7°C rather than midway between −7° and −20°C (Hwang and Regenstein, 1988). Order of time-temperature holding treatments, however, will markedly influence development of rancidity. When lamb was held at temperatures of −5 or −10°C before storage at −35°C, rancidity development was more than if stored at −35°C first, followed by storage at −5° or −10°C (Hagyard et al., 1993).

9.2.2 Water Activity

Water may exert both pro- and antioxidant effects on lipid oxidative reactions. While water can act as an antioxidant through hydration of metallic catalysts at very low water contents, the range of water activities found in frozen foods is still at levels that promote oxidation through solubilization of metal catalysts. Water activities found in the liquid portion of frozen tuna ranged from 0.810 to 0.968 (Ohta and Tanaka, 1978). Consequently, decreases in liquid moisture serve only to concentrate catalysts. Eventually, however, the food system reaches a glassy state in which diffusion of solubilized catalysts becomes limited (Nelson and Labuza, 1992).

9.2.3 Surface Area and Oxygen Solubility

Oxygen is an essential component for the occurrence of lipid oxidation. As such, frozen food systems are particularly susceptible to lipid oxidation because competing glycolytic and other oxygen-using reactions are inhibited (Gray, 1978). It stands to reason that oxygen removal from a food system would eliminate oxidative reactions. Numerous studies have demonstrated that vacuum packaging substantially retards lipid oxidation (Spanier et al., 1992; Bjerking and Johnsen, 1995). Ice glazes and coatings have also been applied to act as barriers (Ijichi, 1978; Khayat and Schwall, 1983; Stuchell and Krochta, 1995) that potentially reduce oxygen uptake below critical levels. Studies on model systems, however, suggest that these critical levels are quite low, with significant reduction in rates occurring only at oxygen tensions less than 100 mM Hg (Salaris and Babbs, 1989). Such low oxygen concentrations may exist in the interior portions of nonporous food materials and account for the longer shelf life of fish samples stored in block form than samples stored as fillets (Licciardello et al., 1980; Ravindranathan et al., 1987). Hence, the potential for oxidation is considered to be less in products with smaller surface areas than in products with large surface areas.

Oxygen permeability into a food product is dependent not only on the surface area but on the density of the product and on the amount of ice accumulation. Flink and Goodhart (1978) evaluated the oxygen permeability of ice and frozen minced fish flesh. Slowly frozen minced fish was somewhat more permeable to oxygen than an equal thickness of slowly frozen ice. Transport of oxygen through fish was proposed to occur through cracks and holes in the product, but in thicker samples, areas of greater resistance in series with more porous areas caused transport to became "plugged."

The concentration of oxygen to which a product is exposed appears to affect the reaction mechanism. Under conditions of decreasing oxygen concentration, increased formation of ethane and *n*-pentane were measured by Kostrucha and Kappus (1986), whereas malondialdehyde increased with increasing oxygen concentrations. These results are consistent with alkanes originating from unsaturated fatty acid monohydroperoxides and requiring only one molecule of oxygen and malondialdehyde formed by the breakdown of endoperoxy unsaturated fatty acid hydroperoxides following the incorporation of two molecules of oxygen. These results demonstrate that multiple oxidation measurements should be employed when evaluating treatments to modify oxygen concentrations.

9.3 Influence of Chemical Composition on Lipid Oxidation

Chemical composition of frozen foods has a direct bearing on the rate and extent of lipid oxidation that occurs in the product. The most notable component is the lipid substrate but others also play a significant role, including initiators, catalysts, and inhibitors. Within each of these categories, a wide range of components having varying impacts on the oxidative process exist. Consequently, while qualitative information is available on the effects of these components, the complex interactions that may occur among components have resulted in a paucity of information on the quantitative relationship between composition and shelf life. The following sections detail the state of our knowledge in the area of composition and lipid oxidation.

9.3.1 Sources of Initiation

Both enzymic and nonenzymic mechanisms have been identified as potential sources of initiation of lipid oxidation in foods. The relative importance of these sources in a food product will be dependent not only on their initial concentrations but may be impacted by exposure of the product to processing treatments such as heat prior to storage. Initiation reactions, however, do not start when the food product is processed. These same reactions take place in live plants and animals, so that even by the time the product is labeled as food, breakdown of preexisting peroxides may be one of the major reactions occurring.

Enzymic

One of the enyzmes that is most often classified as an initiator of lipid oxidation is lipoxygenase. Specifically, this class of enzymes catalyzes the addition of molecular oxygen to a *cis-cis*-4-pentadiene–containing unsaturated fatty acid, releasing a stereospecific conjugated diene hydroperoxy fatty acid product. Strictly speaking, these enzymes are not initiators since preformed hydroperoxides are necessary for their activation (Kanner et al., 1987). This feature, however, does not discount their importance in food since hydroperoxides, as pointed out previously, may be remnants of the living organism.

Lipoxygenases are ubiquitous in many plants and animals, including poultry and fish tissues (Hsieh, 1994). Although products of lipoxygenase have been associated with "fresh flavors" (German et al., 1992), lipoxygenase inactivation by blanching is required for the majority of frozen vegetables. In this regard, blanching is most commonly carried out by passing the vegetables through steam or hot water (Arthey, 1993). Failure to inactivate this enzyme can result in the generation of offensive flavors and decolorization of pigments. Williams and coworkers, for example, have shown that the loss of green color of blanched, frozen English green peas correlates with the amount of lipoxygenase activity in them (Williams et al., 1986). As a class, though, lipoxygenases are noted for their lack of stability. Within 3 h at 0°C, 50% of the original 12-lipoxygenase activity in trout or carp was lost, and a single freeze-thaw cycle inactivated the enzyme (German et al., 1992). Such observations would suggest that under frozen conditions lipoxygenase would not play a significant role in oxidative reactions of fish tissue.

Membrane systems that reduce iron constitute another enzymic system of importance in the process of lipid oxidation because they not only generate active catalysts but do so in an environment consisting of highly unsaturated membrane lipids. In muscle systems, both sarcoplasmic reticulum (McDonald and Hultin, 1987) and mitochondrial membranes (Luo, 1987) have been demonstrated to generate active oxygen species in the presence of NAD(P)H and Fe^{3+}. Distinguishing fish from other muscle systems was its significant activity at low refrigeration temperatures and at frozen temperatures (Apgar and Hultin, 1982). In fact, stimulation by this enzymic system would be expected to increase with time postmortem as NADH concentrations drop to levels that are maximal to the enzymic system (McDonald and Hultin, 1987).

Myeloperoxidase is another enzymic system that is capable of generating a strong oxidant and, hence, able to initiate lipid oxidation in muscle foods. In lipoproteins and phospholipid liposomes, hypochlorous acid, the product of the myeloperoxidase + H_2O_2 + Cl^- system, has been found to initiate lipid peroxidation (Panasenko et al., 1995). Myeloperoxidase, which is released from neutrophils after stimulation of the cells, may exert its activity in muscle tissue following contamination of muscle with blood. Even when the blood is washed off, residual concentrations of myeloperoxidase may be sufficient to accelerate oxidation in stored foods.

Nonenzymic

Redox-active transition metals are major factors catalyzing lipid oxidation in biological systems. Iron in particular is a well-known catalyst of lipid oxidation. In the nonheme form, it participates in the production of the reactive oxygen species, the hydroxyl radical (OH•), via the classical Fenton reaction

$$Fe^{2+} + H_2O_2 \longrightarrow Fe^{3+} + OH\bullet + OH^-$$

and is particularly effective when Fe^{3+} can be recycled to Fe^{2+} by various reducing agents.

At concentrations found in the low-molecular-weight (LMW) fraction of fish muscle, both iron (2.2 μM) and copper (1.4 μM) have been found to catalyze lipid oxidation in fish muscle sarcoplasmic reticulum (Decker et al., 1989). Surprisingly, the proportion of oxidation contributed by copper was greater at in situ levels of copper and iron than at concentrations typically used for in vitro systems. While total soluble and LMW iron concentrations in both the ordinary and dark muscles of mackerel were not affected by freezing, a 1.7-fold increase in the <10 kDa fraction of ordinary muscle and a 1.4-fold increase in the same fraction of dark muscle occurred for copper (Decker and Hultin, 1990a). In addition, storing frozen-thawed ordinary muscle resulted in a 1.9-fold increase and a 2.6-fold increase in the <10 kDa fraction for iron and copper, respectively (Decker and Hultin, 1990a). Iron and copper distributions, therefore, may vary in frozen product or previously frozen product and ultimately affect oxidative stabilities.

The ability of heme proteins to promote lipid peroxidation has been demonstrated by many researchers (Kendrick and Watts, 1969; Kaschnitz and Hatefi, 1975; Kanner et al., 1977). One mechanism of this activation involves decomposition of preformed fatty acid hydroperoxides to peroxyl radicals which in the presence of oxygen propagate lipid peroxidation (Tappel, 1953). A second mechanism by which heme proteins activate the oxidative process is through generation of a hydrogen peroxide–activated heme intermediate species capable of initiating the oxidation of lipids (Harel and Kanner, 1985; Kanner and Harel, 1985). Following more extensive oxidation, however, release of nonheme iron from myoglobin and hemoglobin has been demonstrated upon exposure to hydrogen peroxide (Schricker and Miller, 1983; Rhee et al., 1987; Puppo and Halliwell, 1988). These conditions may accentuate the participation of iron via generation of active oxygen species such as the hydroxyl radical. The ability of heme proteins to serve as biological reductants of LMW iron and copper complexes (Eguchi and Saltman, 1984; Hegetschweiler et al., 1987) would also lead to greater generation of soluble active oxygen species.

9.3.2 Lipids

Two types of studies have been conducted to understand the relationship between lipids and oxidative stability. In one type, the emphasis has been on measuring the generation of oxidative products in systems of varying but defined lipid compositions. In the second type, losses of the lipid substrate during storage are monitored and attributed to oxidation.

Fatty Acid Unsaturation

The lowest carbon-hydrogen bond dissociation energies of a fatty acid are found at bis-allylic methylene positions, that is, between adjacent double bonds (Gardner, 1989; Koppenol, 1990). Thus, these positions are the thermodynamically favored sites for attack by lipid peroxyl radicals in polyunsaturated fatty acids (PUFAs). Fatty acid chain length has no apparent effect on the rate or extent of radical formation (Wagner et al., 1994), but *n*-3 fatty acids autoxidized faster than *n*-6 fatty acids (Adachi et al., 1995).

Despite the occurrence of lipid oxidation in frozen-food products, it has not always been possible to detect losses of unsaturated fatty acids (Beltrán and Moral, 1990; Jeong et al., 1990; Polvi et al., 1991). No measurable losses in PUFAs were observed during frozen storage of carp and mackerel (Mai and Kinsella, 1979; Xing et al., 1993). Losses in these cases may be overshadowed by considerable fish-to-fish variation in quantities of lipid. Even when a paired fillet technique was used to minimize the variability in triacylglycerol composition (the muscle tissue from one side of the fish is compared to a control consisting of the muscle tissue from the other side of the fish), it was insufficient to detect losses in total lipids or the triacylglycerol fraction (Xing et al., 1993; Erickson, 1997). Focusing on specific lipid fractions rather than total lipid, however, has led to greater sensitivity in detection of losses of unsaturated fatty acids, with losses occurring primarily in the phospholipid fraction (Whang and Peng, 1987; Jeong et al., 1990; Erickson, 1997).

Class of Lipids

Lipids can be considered to perform two broadly separate physiological functions in food tissues: as a source of energy and as structural components of all cells. As a result of these separate functions, the location of lipid performing these functions may vary. Triacylglycerols are present primarily in adipose cells serving as an energy source, and phospholipids and cholesterol are found in membranes serving the structural role. In addition to the differences in structural organization of the lipids, storage lipids and membrane lipids are distributed in tissue differentially. While membrane lipids may be distributed throughout the tissue, such is not the case for storage lipids. In red meat muscle, adipose cells were found both between and within muscle fibers; in chicken and trout muscle, adipose cells were found primarily in peripheral subcutaneous fat (German, 1990); and in salmon muscle, adipose cells were distributed mainly in the myosepta and to a lesser extent in the connective tissue surrounding bundles of white muscle fibers (perimysium) (Zhou et al., 1995). Finely dispersed intracellular lipid droplets were also present in dark fish-muscle cells but not in white fish-muscle cells (Zhou et al., 1995). These differences in location of triacylglycerols and phospholipids will have an impact on the oxidative stability of the fractions since differences in accessibility to catalysts and to inhibitors would be expected. Exclusion of phospholipid as a significant source of oxidative products, however, is difficult since it will be present throughout tissue. For example, although triacylglycerols are the major lipid component in skin-side fish fillet sites, the adipose cells are bounded by membrane phospholipids that are also capable of being oxidized. It is inappropriate, therefore, to utilize quantity as the sole basis for expecting that one component is the major component being oxidized. Rather, the relative reactivity and accessibility to catalysts and inhibitors should be the major determinants for identification of the critical site of oxidation. Once oxidized, this critical lipid site may then spread to other lipid sites. Such was the case for a peroxidizing fish microsomal fraction system that led to oxidation of emulsified lipids added to the system (Slabyj and Hultin, 1984). Similarly, in model meat systems during frozen storage, phospholipids were the first to oxidize, while triacylglcyerols played a greater role in the later stages of storage (Igene et al., 1980). Carbon-centered lipid radicals which have been found in the extracellular medium of oxidizing cells may be the vehicle by which oxidation of adjacent lipid structures occurs (North et al., 1994).

Membranes in cells are made up of several different classes of phospholipids as well as cholesterol and small quantities of sphinigolipids, glycolipids, cardiolipin, and so forth (Hawthorne and Ansell, 1982). Within the phospholipid classes, variation in oxidative susceptibility exists that is dependent on the nature of the polar head group (choline, ethanolamine, serine, and inositol) and the degree of fatty acid unsaturation of the individual class (Fukuzawa et al., 1988). Phosphatidylethanolamine consistently changed more in terms of unsaturation during frozen storage than phosphatidylcholine. This supports the concept that phosphatidylethanolamine was more susceptible to autoxidation (Igene et al., 1980). In contrast, phosphatidylserine has been shown to inhibit lipid peroxidation induced by a ferrous-ascorbate system in the presence of phosphatidylcholine hydroperoxides (Yoshida et al., 1991). Effectiveness of inhibition was attributed to the iron-trapping ability of the negatively charged phosphatidylserine. Similarly, plasmalogen phospholipids (glycerophospholipids that contain a vinyl ethyl moiety at the *sn*-1 position) completely bound iron to liposomal lipids. This may account for its role in inhibition of lipid peroxidation (Zommara et al., 1995). Oxidation of cholesterol which has been observed in frozen foods (Pie et al., 1990, 1991) also slows down oxidation of adjacent phospholipids but does so by decreasing the fluidity in the hydrophobic domain of the membrane (Sunamoto et al., 1985).

9.3.3 Lipolysis

Lipolysis of both triacylglycerol and phospholipid fractions has been shown to occur during frozen storage of nonheated food systems (Ohshima et al., 1985; Jeong et al., 1990). In the absence of a detailed lipid profile analysis, therefore, loss of esterified fatty acids may not always be attributed to lipid oxidation. Confusing the picture even more is that when free fatty acids are released from esterified groups, they bind to proteins and other nonextractable components.

The source of the free fatty acids is critical to whether lipid hydrolysis has an accelerating or inhibiting effect on subsequent rates of lipid oxidation. If hydrolyzed from triacylglycerols, the increased surface activity of free fatty acids accelerates oxidation (Labuza, 1971; Shewfelt, 1981). In contrast, free fatty acids that have been hydrolyzed from phospholipids will inhibit oxidation (Mazeaud and Bilinski, 1976; Shewfelt et al., 1981), possibly through disruption of the fatty acid alignment in membranes that facilitated free radical chain propagation (Shewfelt and Hultin, 1983).

Temperature during frozen storage will impact on the relative effects of lipolytic enzymes on lipid oxidation. When oyster was stored at −35°C, the activity of lipase was suppressed much more than that of phospholipase (Jeong et al., 1990).

At lower temperatures, therefore, there is a greater tendency for lipolytic enzymes to inhibit oxidative processes.

Lipid oxidation has also been reported to modulate the activity of phospholipase A_2 in model systems by presenting the enzyme with substrates in more accessible physical arrangements (Sevanian et al., 1988). Studies conducted by Ham and Liston (1987, 1988) supported a linkage between phospholipid hydrolysis and lipid peroxidation in frozen fish muscle.

9.3.4 Prooxidants

For food materials that are subjected to frozen storage, soluble fractions from food systems have been tested for their effect on lipid peroxidation (Slabyj and Hultin, 1983; Decker and Hultin, 1990a,b; Erickson et al., 1990; Kanner, Salan, et al., 1991). Stimulation of peroxidation by these fractions may be attributed to either the presence of salts or reducing agents. They exert their effect through modification in the availability of reduced transition metals.

Reducing Agents

Both ascorbic acid and cysteine are capable of reducing LMW iron and thus stimulating the iron-redox cycle of lipid oxidation (Wills, 1966; Searle and Willson, 1983). Enhancement of oxidation may also be attributed to mobilization of iron from ferritin and other bound sites (Decker and Hultin, 1990b; Laudicina and Marnett, 1990). Kanner, Salan, et al. (1991) reported that turkey muscle cytosol contained reducing compounds at a concentration approximately equivalent to 150 μM ascorbic acid; in previously frozen mackerel, ordinary muscle ascorbate concentrations of 78 μM were measured. Removal of ascorbate from the soluble fraction of ordinary muscle decreased $FeCl_3$-stimulated lipid oxidation in model systems, suggesting that the ascorbate was capable of reducing metals in mackerel muscle (Decker and Hultin, 1990b).

During frozen storage of fish, ascorbic acid concentrations have been observed to decrease (Erickson, 1993a,b, 1994; Thed and Erickson, 1994; Brannan and Erickson, 1996b). Loss of ascorbic acid has been described by a first-order regression equation for both mince ($r = -0.89$) and intact fillets ($r = -0.92$), with loss of ascorbic acid in the mince occurring at a faster rate over the 6-month storage period than for the intact fillets (Brannan and Erickson, 1996a). Changes in concentration during storage could modify ascorbic acid's role in oxidation. Deng et al. (1978) found that ascorbic acid tended to function as a prooxidant at low concentrations and as an antioxidant at high concentrations, with shift concentrations varying with the muscle type. Also influencing ascorbic acid's activity, though, is compartmentation. For instance, intercellularly distributed ascorbic acid obtained through vacuum tumbling, dipping, or injection, while capable of regenerating tocopherol and protecting membrane phospholipids, simultaneously accelerated oxidation of triacylglycerols. On the other hand, application of ascorbic acid via metabolic absorption, which leads to intracellular distribution, did not lead to triacylglycerol oxidation (Erickson, 1997). Similarly, mincing, which disrupts localization of ascorbic acid, led to an acceleration in oxidation of triacylglycerols (Erickson and Sista, 1997). Accessibility to triacylglycerol deposits, therefore, impacts whether ascorbic acid exhibits prooxidative behavior.

Salts and Minerals

Salt (sodium chloride) is added to foods for a variety of purposes including flavor and the inhibition of microorganisms. Despite these beneficial aspects, sodium chloride has been shown to have an accelerating effect on lipid oxidation in frozen foods (Ellis et al., 1968; Matlock et al., 1984; King and Earl, 1988). Kanner, Harel, and Jaffe (1991) suggested that NaCl acts to displace iron ions from binding sites; Osinchak et al. (1992) have attributed the active component of the salt to the chloride ion. In the latter study, incubation of the soluble fraction of mackerel muscle with NaCl led to an approximate doubling of LMW iron in the press juice, but addition of ceruloplasmin (an iron chelator) inhibited the NaCl stimulation.

9.3.5 Antioxidants

Antioxidants are generally classified into two distinct categories (Kanner et al., 1987; Gordon, 1990; Shahidi and Wanasundara, 1992) as follows:

1. Primary antioxidants that react directly with lipid radicals, thus inhibiting propagation
2. Secondary antioxidants that inhibit initiation and branching reactions

Discussion in this section will focus on the fate during frozen storage of endogenous antioxidants in the food. Responses to exogenous antioxidant treatments will be covered in chapter 12.

Primary antioxidants function predominantly by donating hydrogen to the lipid free radical to become a free radical (LOO• + AH → LOOH + A•). The antioxidant free radical can then either combine with another lipid peroxyl radical or another antioxidant radical to terminate the reaction. Examples of primary antioxidants found endogenously in food systems include the tocopherols, ascorbic acid, flavonoids, carnosine, and glutathione.

Effectiveness of primary antioxidants is dependent on the chemical structure of the antioxidant, including its electron-dontaing ability (Gordon, 1990), its stabilization of the antioxidant radical (Burton et al., 1985), and its charge (Jovanovic and Simic, 1988). Buettner (1993) proposed a hierarchy or pecking order for free

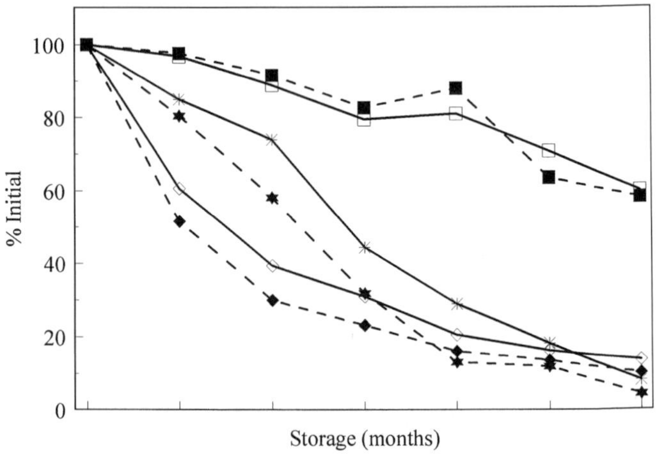

Figure 9.1. Loss of antioxidants in channel catfish muscle tissue during frozen storage. Solid lines = intact fillets; dashed lines = minced fish. ■ and □ = α-tocopherol; ◇ and ◆ = glutathione; ✶ and ✳ = ascorbic acid. (Adapted from Brannan and Erickson, 1996b.)

radical electron (hydrogen atom) transfer reactions based on one-electron reduction potentials. In agreement with this pecking order, the decline of glutathione occurred at a faster rate than that of ascorbic acid, which in turn was faster than the decline of α-tocopherol in frozen channel catfish fillets (fig. 9.1) (Brannan and Erickson, 1996b). Mincing, in turn, accelerated the loss of glutathione and ascorbic acid but had no significant effect on α-tocopherol in frozen channel catfish fillets (Brannan and Erickson, 1996b; Jia et al., 1996). However, degradation of antioxidants is greatly influenced by the food system. In examining frozen strawberries and reconstituted orange juice, systems that are not highly susceptible to oxidation during frozen storage, Jones et al. (1992) found glutathione contents comparable to the respective values for fresh strawberriers and oranges. Jia et al. (1996), on the other hand, found that mackerel exhibited more rapid losses of glutathione than bluefish, which was in line with the decline in sensory scores for both fish. Different rates of degradation were also observed for tocopherol in channel catfish, tilapia, striped bass, and hybrid striped bass (fig. 9.2a). Accelerated degradation after 6 months of storage corresponded to an increased production of oxidative products (fig. 9.2b) (Erickson, 1993a,b, 1994). The similar time period at which a shift in rate of degradation of tocopherol and generation of oxidative products occurred suggested that a critical concentration of tocopherol had been reached that was no longer able to compete effectively for peroxyl radicals and prevent the propagation step of oxidation. The effectiveness of tocopherol is therefore often dependent on the level of PUFAs (Leedle and Aust, 1986). Variabilities in PUFA levels would also account for the widely different degrees of rancidity development found in two different samples of frozen smoked salmon

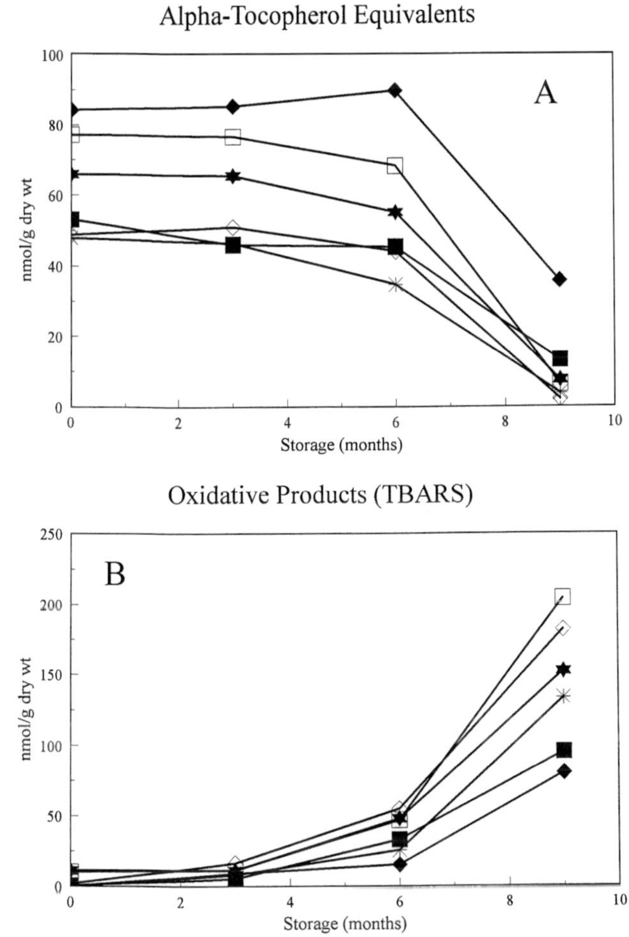

Figure 9.2. Oxidative response of minced fish muscle during frozen storage in terms of (a) α-tocopherol equivalents and (b) TBARS. ◆ = striped bass; □ = "blue" tilapia; ✶ = "red" tilapia; ◇ = "aqua" catfish; ■ = hybrid striped bass; and ✳ = "LSU" catfish. (Adapted from Erickson, 1993a,b, 1994.)

despite the samples having similar tocopherol concentrations initially (Ackman and Timmins, 1995).

An interesting by-product of tocopherol's H-donor role in oxidation is its ability to alter the profile of the generated oxidative products. When tocopherol donates a hydrogen and quenches the peroxy radicals, intramolecular cyclizations leading to dihydroperoxides are prevented. Thus, in the presence of sufficient concentrations of tocopherol, there is a prevalence of monohydroperoxides compared to dihydroperoxides. Breakdown of these peroxides, in turn, leads to characteristic

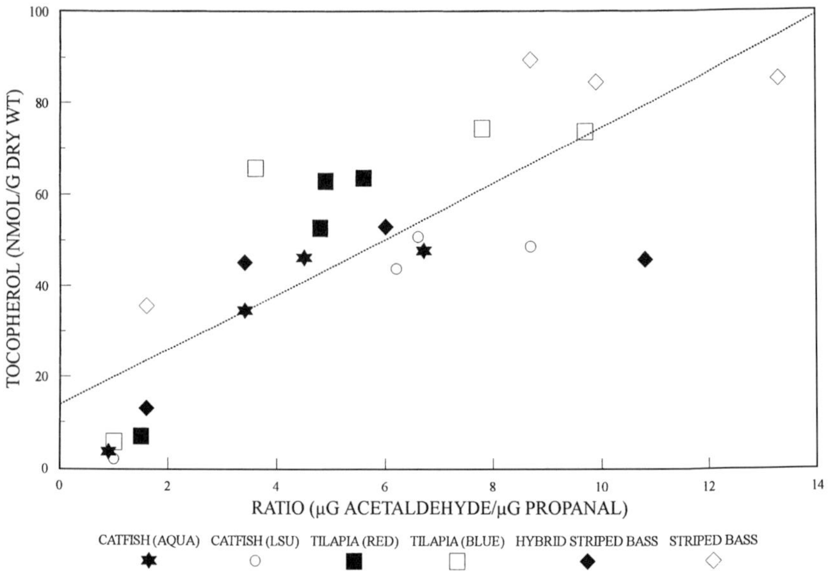

Figure 9.3. Generation of volatiles (acetaldehyde and propanal) in relation to tocopherol concentrations in fish tissue during frozen storage. (Adapted from Erickson, 1993a,b, 1994.)

volatiles. As shown in figure 9.3, acetaldehyde production tended to predominate when concentrations of tocopherol were high in frozen fish, whereas propanal generation became more prevalent when tocopherol concentrations were reduced.

Secondary antioxidants include both oxygen scavengers and chelators. Examples of oxygen scavengers that decrease the level of active oxygen species include superoxide dismutase, catalase, glutathione peroxidase, and carotenoids. Examples of chelators include citrate, phosphates, ceruloplasmin, and many of the free amino acids. As was the case for radical scavengers, concentrations or reactivities of several of the oxygen scavengers have been found to decrease during frozen storage (Bjerkeng and Johnsen, 1995; Jia et al., 1996). Since chelators only bind and minimize the reactivities of metals, on the other hand, their concentrations would not be expected to change during storage.

9.4 Consequences of Lipid Oxidation

The occurrence of lipid oxidation in frozen foods leads to loss of quality. While losses may not be evident in early stages of oxidation, in later stages losses may be seen in flavor, appearance, nutritional value, or protein functionality. The following sections characterize the losses in quality associated with the attributes of flavor, appearance (color), and nutritional value. Information on the types of protein functionality losses associated with lipid oxidation may be found in chapter 8.

9.4.1 Flavor Quality Deterioration

Hydroperoxides of fatty acids produced during propagation do not have a direct adverse effect upon the flavor and aroma of foods. Rather, it is the decomposition of hydroperoxides to aldehydes and ketones that is responsible for the characteristic flavors and aromas that collectively are often described by the term rancid. While decomposition proceeds slowly at frozen storage temperatures, heating the product accelerates breakdown of those hydroperoxides formed during storage. The number of double bonds in the hydroperoxide affects the breakdown: linolenate hydroperoxides are much less stable than oleate and linoleate hydroperoxides (Frankel, 1962).

Food chemists have investigated in detail the formation of volatiles during autooxidation of fats and oils. These volatiles, exhibiting a wide variety of aroma properties, include not only aldehydes but also ketones, alcohols, hydrocarbons, furans, and acids. The relative distribution of these volatiles is strongly dependent on the autoxidation conditions (Grosch et al., 1974), however, impact of volatile oxidation products on product flavor is dependent both on concentration and the threshold value of the compound in the product (Hamilton, 1989; Brewer and Vega, 1995). Consequently, since hydrocarbons, substituted furans, vinyl alcohols, and 1-alkenes have high threshold values, they are presumed to have little impact on flavor. In contrast, the more potent volatiles (2-alkenals, alkanals, *trans,trans*-2,4-dienals, isolated dienals, isolated *cis*-enals, *trans*-2,-*cis*-4-dienals, and vinyl ketones) vary in their impact as variations in concentration and threshold occur (Frankel, 1985). Furthermore, since solubility of many of the oxidative volatiles increases as fat content increases, one would expect decreased sensory perception for a given concentration of volatiles (Roozen et al., 1994). Sensory differences for the oxidized attribute, however, could not be found for fish samples varying by less than 2% fat (Brannan and Erickson, 1996a).

Several studies have attempted to relate one or more oxidative volatiles to sensory off-flavor scores in frozen products (Persson and von Sydow, 1972; McGill et al., 1977; Hwang et al., 1990; Brewer et al., 1992; Lai et al., 1995; Brannan and Erickson, 1996a). Whereas hexanal is often one of the major volatiles produced during oxidation, higher correlation coefficients to off-flavor scores have been found between a combination of volatiles (Persson and von Sydow, 1972; Wu and Sheldon, 1988) or total volatiles (Brannan and Erickson, 1996a) than to individual volatiles. Hence, these studies reemphasize that more than one volatile impacts on the sensory response.

In several cases, assessment of odor and taste revealed no clear relationship between chemical measurements of oxidation and taste panel results (Nakayama and Yamamoto, 1977; Zotos et al., 1995). In those cases, addition of other flavors such as smoke (Zotos et al., 1995) could have led to masking and nondetection of rancid flavors. Alternatively, perception of rancidity may be more difficult in the absence of degradation of desirable flavors such as desirable meaty flavors (St. Angelo et al., 1992). Numerous studies, on the other hand, have shown that specific oxidation products may be desirable flavor components (Josephson et al., 1983; Josephson et al., 1986; Karahadian and Lindsay, 1989a; Lindsay, 1990), particularly when formed in more precise (i.e., less random) reactions by the action of lipoxygenase enzymes (Josephson et al., 1984a,b; Hsieh and Kinsella, 1986; Hsieh and Kinsella, 1989; German et al., 1992; Hsieh, 1994) and/or by the modifying influence of tocopherol on autoxidation reactions (Karahadian and Lindsay, 1989b). That is, certain molecules which are part of the collection of volatile breakdown products of autoxidation when detected separately are not off-flavors per se but in fact contribute to the impact flavors associated with freshness itself. Frozen storage and associated oxidative reactions consequently lead more to quantitative changes than to qualitative changes.

Oxidative flavor deterioration can occur in both plant and animal products during frozen storage. It is more often identified with frozen muscle food products than frozen vegetable products because blanching is typically applied to the vegetables prior to freezing to inactivate peroxidase and lipoxygenase enzymes. In the absence of blanching, however, there is fairly strong evidence that lipoxygenase activity and associated off-flavor development is the limiting factor in frozen storage of corn, green beans, and green peas. Evidence to support this statement was reviewed by Barrett and Theerakulkait (1995). Adverse effects associated with blanching are permanent modification of cellular structure, solubilization and/or destruction of some nutrients and vitamins, and conversion of green chlorophylls to yellow green pheophytins (Katsaboxakis, 1984).

9.4.2 Color Quality Deterioration

Pigment degradation and associated color losses are another manifestation of lipid oxidation. This degradation is of major importance because the first impression consumers have about the acceptability of a food is visual. In red meat and fish products, the principal pigments of concern are the heme pigments, and the color is dictated by the relative concentrations of purple myoglobin (Mb), bright red oxymyoglobin (MbO_2), and brown metmyoglobin (MetMb). While discoloration will occur in refrigerated storage, freezing/thawing is known to accelerate the autoxidation of myoglobin (Brown and Mebine, 1969; Ledward and MacFarlane, 1971; Chow et al., 1988; Lanari et al., 1989; Lanari and Zaritzky, 1991). The apparent activation energy (E_a) calculated for the rate of metmyoglobin formation in skipjack tuna was 25.2 Kcal/mol above −2°C, 90.5 Kcal/mol between −2 and −5°C, and 18.3 Kcal/mol below −5°C (Miki and Nishimoto, 1984). Acceleration of reactions between −5 and −2°C is explained on the basis that there is a concentration of reactants that overbalances the temperature depressing effect. In terms of metmyoglobin generation in frozen beef, however, partial freezing and storage at −5°C was comparable to freezing at lower temperatures (−20°C) when storage exceeded 100 days (Lanari and Zaritzky, 1991).

While a close relationship undoubtedly exists between lipid oxidation and pigment oxidation, the initial culprit is debatable and may ultimately be dependent on the system of study. Both Akamittath et al. (1990) and Mikkelsen et al. (1991) were able to inhibit lipid oxidation in meat systems through additives without affecting the rate of pigment oxidation. In contrast, model system studies have demonstrated an acceleration in pigment oxidation by lipid oxidation reactions (Yin and Faustman, 1993). Since metmyoglobin can be reduced back to myoglobin by nonenzymic or enzymic reactions, the in situ levels of NADH and other possible reductants may be important determinants in myoglobin stability in meats.

Heme pigments are not the only pigments present in frozen products that are subject to oxidative degradation during storage. Carotenoid fading also adversely affected the appearance and value of salmonid flesh. Here again, loss of astaxanthin was faster in the frozen product (−20°C) than in the refrigerated (1–2°C) samples, with 47 and 90% pigment retention, respectively, after 14 days of storage (Chen et al., 1984). Lowering of water activity may be partially responsible for acceleration of carotenoid loss at temperatures below the freezing point. Pigments have been shown to deteriorate about 10 times faster when water activity was reduced from 0.4 to below 0.3 (Martinez and Labuza, 1968). Disruption of tissue compartmentation upon freezing may also be a contributing factor to carotenoid oxidation. Ground crayfish lost more than 65% of its astaxanthin in 3 months at −20°C, whereas whole crayfish lost only about 30% of the red pigment under the same conditions (Chen and Meyers, 1982). Both chemical and enzymatic pathways of lipid oxidation may contribute to oxidation of carotenoids. For example, free radicals propagated from fatty acids (e.g., ROO•) may combine with carotenoids (CH) in a coupled oxidation to form a lipid hydroperoxide (ROOH) and a carotenoid radical (C•) (Simpson, 1985). Alternatively, enzymes such as lipoxygenase (Tsukuda, 1970) or myeloperoxidases (Kanner and Kinsella, 1983) may be contributing factors in carotenoid decoloration.

Chlorophyll is another pigment present in frozen fruits and vegetables that is capable of serving as a secondary substrate in lipid oxidation. For example, the Abbott variety of kiwi fruit became yellowish green after 11 months of storage while the Hayward variety remained green (Fuster et al., 1994). Prefreezing treatments such as blanching or syruping minimize the loss of chlorophyll pigments from frozen products. The enzymatic oxidation of chlorophyll has been reviewed by Simpson et al. (1976).

9.4.3 Nutritional Quality Deterioration

While the process of lipid oxidation destroys lipids, proteins, and vitamins, freezing preservation, in general, is considered to deliver a product comparable in nutritional quality to that of the fresh product. When comparing the nutritive value of fresh and frozen foods, one should be careful to measure the nutritive value as it is eaten with the percentage of nutrient retained from the original unstored cooked product. Otherwise, variations in cooking yields may lead to misleading results.

As the major substrate for lipid oxidation, unsaturated fatty acids become modified and nutritionally unavailable. Only minor quantities of fatty acids need be destroyed to generate detectable levels of rancid volatiles. In fact, the variability in lipid composition in many muscle foods is often greater than the oxidative losses of unsaturated fatty acids. Xing et al. (1993), for instance, found no observable changes in the *n*-3 polyunsaturated fatty acids following storage at −20°C for 28 weeks. Erickson (1997), on the other hand, observed measurable changes for the polyunsaturated fatty acids of the phospholipid fraction in channel catfish following frozen storage and found that the losses represented only a small percentage of the total fatty acid pool. Consequently, while several of the unsaturated fatty acids are considered nutritionally essential or beneficial (i.e., 18:2 *n*-6, 18:3 *n*-3, 20:5 *n*-3, or 22:6 *n*-3), the losses do not become limiting for the majority of frozen products.

Two types of reactions, namely radical reactions and the reactions involving the secondary products, are possible in protein damage by oxidized lipids. Protein radicals (P•) formed in the reaction of protein (PH) with lipid hydroperoxides (LOOH) or lipid radicals (LOO• and LO•) eventually cause protein damage such as protein-to-protein cross-links, lipid-to-protein cross-links, protein scission, and amino acid damage. Involvement of radical reactions in the protein damage caused by oxidized lipids has been reviewed in detail (Gardner, 1979; Schaich, 1980). Of the secondary products of lipid oxidation, aldehydes are some of the most probable candidates for the molecules involved in the protein damage that forms fluorescent and cross-linked proteins.

Evaluation of protein nutritional quality has been studied by both in vitro and in vivo methods. In model systems, myofibrillar protein isolates have been reacted with either hexanal or hydroperoxides such that in vitro digestibility of the protein with pepsin/trypsin was reduced (Janitz, Pyrcz et al., 1990) and losses in available lysine and cysteine were found (Janitz, Grzeskowiak et al., 1990). Losses were greatest when myofibrillar proteins were thermally denatured, especially for those in which the lipid oxidation products were present before denaturation. In contrast, no nutritionally significant decrease in the lysine availability or in vitro digestibility was found for muscle proteins in samples of cod, whiting, herring, lemon sole, or skate when stored at −8°C for up to 20 months (Poulter and Lawrie, 1977). Similar findings were found when nutritional quality of proteins from frozen sardine pastes (6 months at −20°C) was studied in rats using the nitrogen balance technique (Navarro et al., 1987). Consequently, it would appear that while protein damage may very well have occurred in these frozen fish systems, the damage was minimal in comparison to the total available protein.

The degradation of vitamins during the freezing process, in contrast to lipid or protein degradation, generally has a more significant impact on the product's nutritional value. Whether vitamin losses are considered nutritionally important, though, is dependent on the abundance of the vitamin in the food compared to the recommended daily allowance (RDA) and the importance of the food in the average diet. Consequently, while ascorbic acid losses have been noted in frozen fish products (Thed and Erickson, 1994; Brannan and Erickson, 1996b), the loss is only a small proportion of the RDA. As a result, studies on ascorbic acid losses have generally emphasized fruits and vegetables which contribute a major proportion of the ascorbic acid requirement.

Ascorbic acid degradation during freezing and storage is expected to arise primarily from oxidative mechanisms due to it's role as an antioxidant. Lipid oxidation's contribution to degradation is not always evident when comparison of literature values from different studies is taken. In these cases, consideration should be given to the different types of packaging used in the studies, which may have had varying oxygen permeabilities. In addition, keep in mind that the freezing process in these studies often includes prefreezing treatments such as blanching as well as thawing. Therefore, when the entire freezing process was considered (blanching, freezing, storage for 6–12 months at −18°C, and thawing), average losses for L-ascorbic acid ranged from 12% for aspargus to 65% for spinach, but generally averaged about 50% for green beans, lima beans, broccoli, cauliflower, and green peas (Fennema, 1975). That blanching is a major contributor to degradation of ascorbic acid is apparent when one considers those products for which it is not required. For most popular fruits processed in a regular manner, less than 30% of their original vitamin C content is lost during the entire freezing process, while concentrated citrus juices lose less than 5% (Fennema, 1977). In both fruits and vegetables, however, vitamin C degradation is highly temperature dependent. A 10°C rise in storage temperature within the range of −18 to −7°C accelerated vitamin C degradation by a factor of 6–20 in green beans, green peas, cauliflower, and spinach, and by a factor of 30–70 for peaches, boysenberries, and strawberries (Fennema, 1975).

Although losses of other water soluble vitamins may be significant in frozen foods, there are conflicting reports in the literature. Variability in the types of analyses used, inherent variability between individual units of a food product, and variability that can be present within a food product (i.e., different cuts of meat) may all contribute to these conflicting results. In these studies, identification of degradative mechanisms is often lacking because emphasis has been placed only on the rate and/or extent of loss of the vitamins. Consequently, in these frozen storage studies, it is difficult to differentiate losses attributable to oxidative degradative mechanisms versus losses attributable to blaching or thaw drip. A review of the general stability of the water soluble vitamins to light or air provides some indication as to the prevalence of oxidation in destruction of vitamins during frozen storage (table 9.1). Variation in either physical or chemical parameters known to affect rates of lipid oxidation will further help to decipher its role in degradation of water soluble vitamins during freezing and storage.

Table 9.1. General stability of water-soluble vitamins to light or air[a]

Water-soluble vitamin	Air or Oxygen	Light
Ascorbic acid	U	U
Biotin	S	S
Choline	U	S
Vitamin B_{12}	U	U
Folic acid	U	U
Inositol	S	S
Niacin	S	S
Pantothenic acid	S	S
Vitamin B_6	S	U
Riboflavin	S	U
Thiamin	U	S

Adapted from Gregory (1985)

[a]S, stable (no important destruction); U, unstable (significant destruction)

Losses for both tocopherol and carotenes have been observed in frozen food products (Lusk et al., 1964; Chen et al., 1984; Pozo et al., 1988; Anderson et al., 1990; Ingemansson et al., 1993; Erickson, 1993a,b, 1994; Brannan and Erickson, 1996b; Erickson, 1997). The significance of these losses nutritionally, though, is dependent on an individual's total dietary intake of unsaturated fatty acids. For instance, Garrido et al. (1993) have shown that as the dietary level of polyunsaturated fatty acid increases, so also does the level of tocopherol required to protect tissues from oxidative stress. Dietary levels of carotenoids would also be expected to be critical in providing protection against diseases such as arteriosclerosis and cancer through their antioxidant activities. In that event, depletion of the antioxidant in the food could lead to suboptimal levels in the tissues.

9.4.4 Nutritional Consequences of Lipid Oxidation Products

Autoxidized lipids in foodstuffs have been shown to be a source of radicals and oxidation products in vivo (Miyazawa, Nagaoka, and Kaneda, 1983; Sato, and Kaneda, 1983; Miyazawa et al., 1985; Pryor, 1985; Staprans et al., 1996). This outcome is worthy of attention as there is increasing evidence of a pathogenic role of endogenous lipid oxidation in a number of chronic and acute disease states (Cross et al., 1987; Halliwell, 1987).

Whereas several animal studies associated with the feeding of oxidized fats have not revealed detrimental effects (Hung et al., 1983; Billek, 1985), other studies have demonstrated toxicological responses (Nolen et al., 1967; Desjardins et al., 1987; Oarada et al., 1988; Kanazawa, 1991). In many of the latter studies, the toxic dose levels of oxidation products that the animals have been exposed to are orders of magnitude higher than would be found in the average diet. In other cases, the response to oxidized fats is dependent on the concentration of pro- and antioxidants present in the diet. Broiler tissue thiobarbituric acid reactive substances (TBARS), for instance, decreased in oxidized oil–fed groups with increasing dietary α-tocopherol concentration (Galvin et al., 1994). In another case, when semipurified diets were supplemented with graded levels of iron (0 to 6,250 mg/kg diet) and fed to trout for 12–36 weeks, the concurrent effects of diet rancidity and iron overload (greater than 86 ng/kg) led to the development of unique histopathological signs, poor growth, and high mortalities in the trout (Desjardins et al., 1987). In contrast, when diet rancidity was low (less than 10 μg malonaldehyde/g diet), the toxic level of dietary iron was greater than 1,380 mg/kg diet.

The type of oxidation product present in a food is considered to have a direct bearing on the toxicity of the lipids. Secondary products are more toxic than hydroperoxides. This conclusion is based partly on a study by Yoshioka and Kaneda (1974) in which mice were fed methyl linoleate that had been autoxidized to different degrees. Minimal effects on toxicity were observed in mice fed the hydroperoxides, whereas the greatest toxicities occurred when carbonyl values were at their highest. Lower toxicity by hydroperoxides may result from conversion in the stomach prior to ordinary absorption. Terao and Fukino (1993) found that dietary hydroperoxides decomposed in an acidic environment with loss of conjugated diene structure as well as cleavage of the hydroperoxyl group. Conversion of the hydroperoxylacyl group to the hydroxy-epoxide and/or triol was likely under their conditions. However, the presence of a reductase enzyme in the gut could also contribute to detoxification of hydroperoxides. In this regard it has been suggested that glutathione peroxidase has a role. In a recent study using everted sacs of rat small intestine and peroxidized methyl linoleate, adding either glutathione or an inhibitor of the glutathione peroxidase system decreased or increased, respectively, the amounts of secondary oxidation products on the contraluminal side (Kowalski et al., 1990).

Secondary oxidation products have been shown definitively to be absorbed into the circulatory system following oral administration (Kanazawa et al., 1985; Oarada et al., 1986; Marquez-Ruiz et al., 1992), with consequent effects on lipid oxidation products present in body tissues and fecal matter. Hepatic metabolic disturbances have also been noted with orally administered secondary products, including decreased lipogenesis and increased lipolysis (Ashida et al., 1988). Since hydrolysis and absorption of polymers in the digestive tract is limited (Yoshida and Alexander, 1983), aldehydes were pinpointed as the major compounds responsible for the metabolic alterations.

While the verdict is currently out regarding oxidized dietary lipids and their role in atherosclerosis, cancer, and other major diseases, the frozen food industry should be prepared for the repercussions should additional evidence accumulate supporting its role. In the past 30 years, nutritional issues (salt, fiber, saturated fat, and total fat) have driven the major changes in food products produced. Such could be the case with products susceptible to oxidation. To ensure that frozen foods remains a viable entity in the marketplace, the industry should expand its research efforts in the following three areas:

1. Identification of chemical markers for shelf life prediction (i.e., fatty acid unsaturation, antioxidant status, and prooxidant status)
2. Development of shelf life indicators of oxidative quality that could be used by the consumer
3. Optimization of product stability through application of antioxidants

References

Ackman, R.G. and Timmins, A. 1995. Stability of α-tocopherol in frozen smoked fish. *J. Food Lipids* **2**:65–71.

Adachi, S., Ishiguro, T., and Matsuno, R. 1995. Autoxidation kinetics for fatty acids and their esters. *J. Am. Oil Chem. Soc.* **72**:547–551.

Akamittath, J.G., Brekke, C.J., and Schanus, E.G. 1990. Lipid oxidation and color stability in restructured meat systems during frozen storage. *J. Food Sci.* **55**:1513–1517.

Anderson, H.J., Bertelsen, G., Christophersen, A.G., Ohlen, A., and Skibsted, L.H. 1990. Development of rancidity in salmonid steaks during retail display. *Z. Lebensm. Unters. Forsch.* **191**:119–122.

Apgar, M.E. and Hultin, H.O. 1982. Lipid peroxidation in fish muscle microsomes in the frozen state. *Cryobiol.* **19**:154–162.

Arthey, D. 1993. Freezing of vegetables and fruits. 1993. In *Frozen Food Technology*. (Mallett, C.P., ed.) pp. 237–269, Chapman & Hall, New York.

Ashida, H., Kanazawa, K., and Natake, M. 1988. Comparison of the effects of orally administered linoleic acid, and its hydroperoxides and secondary autoxidation products on hepatic lipid metabolism in rats. *Agric. Biol. Chem.* **52**:2007–2014.

Barrett, D.M. and Theerakulkait, C. 1995. Quality indicators in blanched, frozen, stored vegetables. *Food Technol.* **49**(1):62, 64–65.

Beltrán, A. and Moral, A. 1990. Gas chromatographic estimation of oxidative deterioration in sardine during frozen storage. *Lebensm. Wiss. Technol.* **23**:499–504.

Bhattacharya, M., Hanna, M.A., and Mandigo, R.W. 1988. Lipid oxidation in ground beef patties as affected by time-temperature and product packaging parameters. *J. Food Sci.* **53**:714–717.

Bilinski, E., Jonas, R.E.E., and Peters, M.D. 1981. Treatments affecting the degradation of lipids in frozen Pacific herring, *Clupea harengus* pallasi. *Can. Inst. Food Sci. Technol. J.* **14**:123–127.

Billek, G. 1985. Heated fats in the diet. In *The Role of Fats in Human Nutrition*. (Padley, F.B. and Podmore, J., eds.) pp. 163–172, Ellis Horwood, Chichester.

Bjerkeng, B. and Johnsen, G. 1995. Frozen storage quality of rainbow trout (*Oncorhynchus mykiss*) as affected by oxygen, illumination, and fillet pigment. *J. Food Sci.* **60**:284–288.

Brannan, R.G. and Erickson, M.C. 1996a. Sensory assessment of frozen stored channel catfish in relation to lipid oxidation. *J. Aquat. Food Prod. Technol.* **5**:67–80.

Brannan, R.G. and Erickson, M.C. 1996b. Quantification of antioxidants in channel catfish during frozen storage. *J. Agric. Food Chem.* **44**:1361–1366.

Brewer, M.S. and Vega, J.D. 1995. Detectable odor thresholds of selected lipid oxidation compounds in a meat model system. *J. Food Sci.* **60**:592–595.

Brewer, M.S., Ikins, W.G., and Harbers, C.A.Z. 1992. TBA values, sensory characteristics, and volatiles in ground pork during long-term storage: effects of packaging. *J. Food Sci.* **57**:558–580.

Brown, W.D. and Mebine, L.B. 1969. Autoxidation of oxymyoglobin. *J. Biol. Chem.* **244**:6696–6701.

Buettner, G.R. 1993. The pecking order of free radicals and antioxidants: lipid peroxidation, α-tocopherol, and ascorbate. *Arch. Biochem. Biophys.* **300**:535–543.

Burton, G.W., Doba, T., Gabe, E.J., Hughes, L., Lee, F.L., Prasad, L., and Ingold, K.U. 1985. Autoxidation of biological molecules. 4. Maximizing the antioxidant activity of phenols. *J. Am. Chem. Soc.* **107**:7053–7065.

Chen, H.M. and Meyers, S.P. 1982. Extraction of astaxanthin pigment from crawfish waste using a soy oil process. *J. Food Sci.* **47**:892–896, 900.

Chen, H.-M., Meyers, S.P., Hardy, R.W., and Biede, S.L. 1984. Color stability of astaxanthin pigmented rainbow trout under various packaging conditions. *J. Food Sci.* **49**:1337–1340.

Chow, E.-J., Ochiai, Y., Watabe, S., and Hashimoto, K. 1988. Effect of freezing and thawing on the discoloration of tuna meat. *Nippon Suisan Gakkaishi* **54**:639–648.

Cross, C.E., Halliwell, B., Borish, E.T., Pryor, W.A., Ames, B.N., Saul, R.L., McCord, J.M., and Harman, D. 1987. Oxygen radicals and human disease. *Ann. Intern. Med.* **107**:526–545.

Decker, E.A. and Hultin, H.O. 1990a. Factors influencing catalysis of lipid oxidation by the soluble fraction of mackerel muscle. *J. Food Sci.* **55**:947–950, 953.

Decker, E.A. and Hultin, H.O. 1990b. Nonenzymic catalysts of lipid oxidation in mackerel ordinary muscle. *J. Food Sci.* **55**:951–953.

Decker, E.A., Huang, C.-H., Osinchak, J.E., and Hultin, H.O. 1989. Iron and copper: role in enzymic lipid oxidation of fish sarcoplasmic reticulum at in situ concentrations. *J. Food Biochem.* **13**:179–186.

Deng, J.C., Watson, M., Bates, R.P., and Schroeder, E. 1978. Ascorbic acid as an antioxidant in fish flesh and its degradation. *J. Food Sci.* **43**:457–460.

Desjardins, L.M., Hicks, B.D., and Hilton, J.W. 1987. Iron catalyzed oxidation of trout diets and its effect on the growth and physiological response of rainbow trout. *Fish Physiol. Biochem.* **3**:173–182.

Eguchi, L.A. and Saltman, P. 1984. The aerobic reduction of Fe(III) complexes by hemoglobin and myoglobin. *J. Biol. Chem.* **259**:14337–14338.

Ellis, R., Currie, G.T., Thornton, F.E., Bollinger, N.C., and Gaddis, A.M. 1968. Carbonyls in oxidizing fat. 11. The effect of the pro-oxidant activity of sodium chloride on pork tissue. *J. Food Sci.* **33**:555–561.

Erickson, M.C. 1993a. Ability of chemical measurements to differentiate oxidative stabilities of frozen minced muscle tissue from farm-raised striped bass and hybrid striped bass. *Food Chem.* **48**:381–385.

Erickson, M.C., 1993b. Compositional parameters and their relationship to oxidative stability of channel catfish. *J. Agric. Food Chem.* **41**:1213–1218.

Erickson, M.C. 1994. Comparison of chemical measurements to differentiate oxidative stability of frozen minced tilapia fish muscle. *Int. J. Food Sci. Technol.* **29**:585–591.

Erickson, M.C. 1997. Oxidative susceptibility of catfish lipids as affected by the microenvironment of ascorbic acid. *J. Food Lipids*. In press.

Erickson, M.C. and Sista, R.V. 1997. Influence of microenvironment on oxidative susceptibility of seafood lipids. In *Flavor and Lipid Chemistry in Seafood*. (Shahidi, F., ed.) American Chemical Society, Washington, D.C. In press.

Erickson, M.C., Hultin, H.O., and Borhan, M. 1990. Effect of cytosol on lipid peroxidation in flounder sarcoplasmic reticulum. *J. Food Biochem.* **14**:407–419.

Fennema, O. 1975. Effects of freeze-preservation on nutrients. In *Nutritional Evaluation of Food Processing*, 2nd ed. (Harris, R.S. and Karmas, E., eds.) p. 244, AVI Publishing, Westport, Conn.

Fennema, O. 1977. Loss of vitamins in fresh and frozen foods. *Food Technol.* **31**:32–38.

Flink, J.M. and Goodhart, M. 1978. Transport of oxygen through ice and frozen minced fish. *J. Food Process Preserv.* **2**:229–248.

Frankel, E.N. 1962. Hydroperoxides. In *Symposium on Foods: Lipids and Their Oxidation*. (Schultz, H., Day, E.A., and Sinnhuber, R.O., eds.) pp. 51–78, AVI Publishing, Westport, Conn.

Frankel, E.N. 1985. Chemistry of autoxidation: Mechanism, products and flavor significance. In *Flavor Chemistry of Fats and Oils*. (Min, D.B. and Smouse, T.H., eds.) pp. 1–37, American Oil Chemists' Society, Champaign, Ill.

Fukuzawa, K., Tadokoro, T., Kishikawa, K., Mukai, K., and Gebicki, J.M. 1988. Site-specific induction of lipid peroxidation by iron in charged micelles. *Arch. Biochem. Biophys.* **260**:146–152.

Fuster, C., Prestamo, G., and Cano, M.P. 1994. Drip loss, peroxidase and sensory changes in kiwi fruit slices during frozen storage. *J. Sci. Food Agric.* **64**:23–29.

Galvin, K., Morrissey, P.A., and Buckley, D.J. 1994. Effect of dietary oil quality and alpha-tocopherol supplementation on the oxidative stability of broiler tissues. *Proc. Nutr. Soc.* **53**:13A.

Gardner, H.W. 1979. Lipid hydroperoxide reactivity with proteins and amino acids: a review. *J. Agric. Food Chem.* **27**:220–229.

Gardner, H.W. 1989. Oxygen radical chemistry of polyunsaturated fatty acids. *Free Radical Biol. Med.* **7**:65–86.

Garrido, A., Garate, M., Campos, R., Villa, A., Nieto, S., and Valenzuela, A. 1993. Increased susceptibility of cellular membranes to the induction of oxidative stress after ingestion of high doses of fish oil: Effect of aging and protective action of DL-α-tocopherol supplementation. *J. Nutr. Biochem.* **4**:118–122.

German, J.B. 1990. Muscle lipids. *J. Muscle Foods* **1**:339–361.

German, J.B., Zhang, H., and Berger, R. 1992. Role of lipoxygenases in lipid oxidation in foods. In *Lipid Oxidation in Food.* (St. Angelo, A.J., ed.) pp. 74–92, American Chemical Society, Washington, D.C.

Gordon, M.H. 1990. The mechanism of antioxidant action in vitro. In *Food Antioxidants.* (Hudson, B.J.F., ed.) pp. 1–18, Elsevier Applied Science, New York.

Gray, J.I. 1978. Measurement of lipid oxidation: a review. *J. Am. Oil Chem. Soc.* **55**:697–704.

Gregory, J.F., III. 1985. Chemical changes of vitamins during food processing. In *Chemical Changes in Food During Processing.* (Richardson, T. and Finley, J.W., eds.) pp. 373–408, AVI Publishing, Westport, Conn.

Grosch, W., Laskawy, G., and Fischer, K.-H. 1974. Oxidation of linolenic acid in the presence of haemoglobin, lipoxygenase, or by singlet oxygen. Identification of the volatile carbonyl compounds. *Lebensm. Wiss. Technol.* **6**:335–338.

Hagyard, C.J., Keiller, A.H., Cummings, T.L., and Chrystall, B.B. 1993. Frozen storage conditions and rancid flavour development in lamb. *Meat Sci.* **35**:305–312.

Halliwell, B. 1987. Oxidants and human disease: some new concepts. *FASEB J.* **1**:358–364.

Hamilton, R.J. 1989. The chemistry of rancidity in foods. In *Rancidity in Foods.* (Allen, J.C. and Hamilton, R.J., eds.) pp. 1–21, Elsevier Applied Science, London, England.

Han, T.-J. and Liston, J. 1987. Lipid peroxidation and phospholipid hydrolysis in fish muscle microsomes and frozen fish. *J. Food Sci.* **52**:294–296, 299.

Han, T.-J. and Liston, J. 1988. Correlation between lipid peroxidation and phospholipid hydrolysis in frozen fish muscle. *J. Food Sci.* **53**:1917, 1919.

Harel, S. and Kanner, J. 1985. Muscle membranal lipid peroxidation initiated by H_2O_2-activated metmyoglobin. *J. Agric. Food Chem.* **33**:1188–1192.

Hawthorne, J.N. and Ansell, G.B. 1982. *Phospholipids.* Elsevier Biomedical Press, Amsterdam.

Hegetschweiler, K., Saltman, P., Dalvit, C., and Wright, P.E. 1987. Kinetics and mechanisms of the oxidation of myoglobin by Fe(III) and Cu(II) complexes. *Biochim. Biophys. Acta* **912**:384–397.

Hsieh, R.J. 1994. Contribution of lipoxygenase pathway to food flavors. In *Lipids in Food Flavors.* (Ho, C-T. and Hartman, T.G., eds) pp. 30–48, American Chemical Society, Washington, D.C.

Hsieh, R.J. and Kinsella, J.E. 1986. Lipoxygenase-catalyzed oxidation of *n*-6 and *n*-3 polyunsaturated fatty acids: relevance to and activity in fish tissue. *J. Food Sci.* **51**:940–945, 996.

Hsieh, R.J. and Kinsella, J.E. 1989. Lipoxygenase generation of specific volatile flavor carbonyl compounds in fish tissues. *J. Agric. Food Chem.* **37**:279–286.

Hung, S.S.O., Walker, B.L., and Slinger, S.J. 1983. Effect of oxidized oil on the liver fatty acids of rainbow trout (*Salmo gairdneri*). *Comp. Biochem. Physiol.* **76B**:349–353.

Hwang, K.T. and Regenstein, J.M. 1988. Protection of menhaden mince lipids from rancidity during frozen storage. *J. Food Sci.* **54**:1120–1124.

Hwang, S-Y., Bowers, J.A., and Kropf, D.H. 1990. Flavor, texture, color, and hexanal and TBA values of frozen cooked beef packaged in modified atmosphere. *J. Food Sci.* **55**:26–29.

Igene, J.O., Pearson, A.M., Dugan, L.R., Jr., and Price, J.F. 1980. Role of triglycerides and phospholipids on development of rancidity in model meat systems during frozen storage. *Food Chem.* **5**:263–276.

Ijichi, K.E. 1978. Evaluation of an alginate coating during frozen storage of red snapper and silver salmon. M.S. Thesis. Univ. of California, Davis.

Ingemansson, T., Pettersson, A., and Kaufmann, P. 1993. Lipid hydrolysis and oxidation related to astaxanthin content in light and dark muscle of frozen stored rainbow trout (*Oncorhynchus mykiss*). *J. Food Sci.* **58**:513–518.

Janitz, W., Grzeskowiak, B., Korczak, J., and Berghofer, E. 1990. Effects of technological measures on reaction of oxidized fats with meat proteins. III. Changes in nutritional value. *Fleischwirtschaft* **70**:600, 603–604.

Janitz, W., Pyrcz, J., Flaczyk, E., and Berghofer, E. 1990. Effects of technological measures on reaction of oxidized fats with meat proteins. II. Change in in vitro protein digestibility and the sensitivity of connective tissue to thermal hydrolysis. *Fleischwirtschaft* **70**:345–347.

Jeong, Y., Ohshima, T., Koizumi, C., and Kanou, Y. 1990. Lipid deterioration and its inhibition of Japanese oyster *Crassostrea gigas* during frozen storage. *Nippon Suisan Gakkaishi* **56**:2083–2091.

Jia, T.-D, Kelleher, S.D., Hultin, H.O., Petillo, D., Maney, R., and Krzynowek, J. 1996. Comparison of quality loss and changes in the glutathione antioxidant system in stored mackerel and bluefish muscle. *J. Agric. Food Chem.* **44**:1195–1201.

Jones, D.P., Coates, R.J., Flagg, E.W., Eley, J.W., Block, G., Greenberg, R.S., Gunter, E.W., and Jackson, B. 1992. Glutathione in foods listed in the National Cancer Institute's health habits and history food frequency questionnaire. *Nutr. Cancer* **17**:57–75.

Josephson, D.B., Lindsay, R.C., and Olafsdottir, G. 1986. Measurement of volatile aroma constituents as a means for following sensory deterioration of fresh fish and fishery products. In *Seafood Quality Determination*. (Kramer, D.E. and Liston, J., eds.) pp. 27–47, Elsevier Science, Amsterdam.

Josephson, D.B., Lindsay, R.C., and Stuiber, D.A. 1983. Identification of compounds characterizing the aroma of fresh whitefish (*Coregonus clupeaformis*). *J. Agric. Food Chem.* **31**:326–330.

Josephson, D.B., Lindsay, R.C., and Stuiber, D.A. 1984a. Variations in the occurrences of enzymically derived volatile aroma compounds in salt- and freshwater fish. *J. Agric. Food Chem.* **32**:1344–1347.

Josephson, D.B., Lindsay, R.C., and Stuiber, D.A. 1984b. Biogenesis of lipid-derived volatile aroma compounds in the emerald shiner (*Notropis atherinoides*). *J. Agric. Food Chem.* **32**:1347–1352.

Jovanovic, S.V. and Simic, M.G. 1988. Redox properties of oxy and antioxidant radicals. In *Oxygen Radicals in Biology and Medicine*. (Simic, M.G, Taylor, K.A., Ward, J.F., and von Sonntag, C., eds.) pp. 115–130, Plenum Press, New York.

Jul, M. 1984. *The Quality of Frozen Foods*. Academic Press, Orlando, Fla.

Kanazawa, K. 1991. Hepatic dysfunction caused by dietary products of lipid peroxidation. *Int. J. Tissue React.* **13**:67–78.

Kanazawa, K., Kanazawa, E., and Natake, M. 1985. Uptake of secondary autoxidation products of linoleic acid by the rat. *Lipids* **20**:412–419.

Kanner, J. and Harel, S. 1985. Initiation of membranal lipid peroxidation by activated metmyoglobin and methemoglobin. *Arch. Biochem. Biophys.* **237**:314–321.

Kanner, J. and Kinsella, J.E. 1983. Lipid deterioration initiated by phagocytic cells in muscle foods: β-Carotene destruction by a myeloperoxidase-hydrogen peroxide-halide system. *J. Agric. Food Chem.* **31**:370–376.

Kanner, J., German, J.B., and Kinsella, J.E. 1987. Initiation of lipid peroxidation in biological systems. *CRC Crit. Rev. Food Sci. Nutr.* **25:**317–364.

Kanner, J., Harel, S., and Jaffe, R. 1991. Lipid peroxidation of muscle food as affected by NaC1. *J. Agric. Food Chem.* **39**:1017–1021.

Kanner, J., Mendel, H., and Budowski, P. 1977. Carotene oxidizing factors in red pepper fruits (*Capsicum annum* L.): peroxidase activity. *J. Food Sci.* **42**:1549–1551.

Kanner, J., Salan, M.A., Harel, S., and Shegalovich, I. 1991. Lipid peroxidation of muscle food: the role of the cytosolic fraction. *J. Agric. Food Chem.* **39**:242–246.

Karahadian, C. and Lindsay, R.D. 1989a. Role of oxidative processes in the formation and stability of fish flavors. In *Flavor Chemistry. Trends and Developments*. (Teranishi, R., Buttery, R.G., and Shahidi, F., eds.) pp. 60–75, American Chemical Society, Washington, D.C.

Karahadian, C. and Lindsay, R.C. 1989b. Action of tocopherol-type compounds in directing reactions forming flavor compounds in autoxidizing fish oils. *J. Am. Oil Chem. Soc.* **66**:1302–1308.

Kaschnitz, R.M. and Hatefi, Y. 1975. Lipid oxidation in biological membranes. Electron transfer proteins as initiators of lipid autoxidation. *Arch. Biochem. Biophys.* **171**:292–304.

Katsaboxakis, K.Z. 1984. The influence of the degree of blanching on the quality of frozen vegetables. In *Thermal Processing and Quality of Foods*. (Zeuthen, P., Cheftel, J.C., Ericksson, C., Jul, M., Leniger, H., Linko, P., Varela, G., and Vos, G., eds.) pp. 559–566, Elsevier Applied Science Publishers, New York.

Kendrick, J. and Watts, B.M. 1969. Acceleration and inhibition of lipid oxidation by heme compounds. *Lipids* **4**:454–458.

Khayat, A. and Schwall, D. 1983. Lipid oxidation in seafood. *Food Technol.* **37**(7):130–140.

King, A.J. and Earl, L.A. 1988. Effect of selected sodium and potassium salts on TBA values of dark meat turkey patties. *J. Food Sci.* **53**:723–726.

Koppenol, W.H. 1990. Oxy radical reactions: from bond-dissociation energies to reduction potentials. *FEBS Lett.* **264**:165–167.

Kostrucha, J. and Kappus, H. 1986. Inverse relationship of ethane or *n*-pentane and malondialdehyde formed during lipid peroxidation in rat liver microsomes with different oxygen concentrations. *Biochim. Biophys. Acta* **879**:120–125.

Kowalski, D.P., Feeley, R.M., and Jones, D.P. 1990. Use of exogenous glutathione for metabolism of peroxidized methyl linoleate in rat small intestine. *J. Nutr.* **120**:1115–1121.

Labuza, T.P. 1971. Kinetics of lipid oxidation in foods. *CRC Crit. Rev. Food Sci. Technol.* **2**:55–405.

Lai, S.-M., Gray, J.I., Booren, A.M., Crackel, R.L., and Gill, J.L. 1995. Assessment of off-flavor development in restructured chicken nuggets using hexanal and TBARS measurements and sensory evaluation. *J. Sci. Food Agric.* **67**:447–452.

Lanari, M.C. and Zaritzky, N.E. 1991. Effect of packaging and frozen storage temperature on beef pigments. *Int. J. Food Sci. Technol.* **26**:629–640.

Lanari, M.C., Bevilacqua, A.E., and Zaritzky, N.E. 1989. Pigments modifications during freezing and frozen storage of packaged beef. *J. Food Process. Eng.* **12**:49–66.

Laudicina, D.C. and Marnett, L.J. 1990. Enhancement of hydroperoxide-dependent lipid peroxidation in rat liver microsomes by ascorbic acid. *Arch. Biochem. Biophys.* **278**:73–80.

Ledward, D.A. and MacFarlane, J.J. 1971. Some observations on myoglobin- and lipid oxidation in frozen beef. *J. Food Sci.* **36**:987–989.

Leedle, R.A. and Aust, S.D. 1986. Importance of the polyunsaturated fatty acid to vitamin E ratio in the resistance of rat lung microsomes to lipid peroxidation. *Free Radical Biol. Med.* **2**:397–403.

Lepock, J.R., Morse II, P.D., Keith, A.D., and Kruuv, J. 1978. Freeze-thaw damage in isolated lobster sarcoplasmic reticulum membranes: a model system for membrane damage. *Cryobiol.* **15**:643–653.

Licciardello, J.J., Ravesi, E.M., and Allsup, M.G. 1980. Extending the shelf life of frozen Argentine hake. *J. Food Sci.* **45**:1312–1317.

Lindsay, R.C. 1990. Fish flavors. *Food Rev. Int.* **6**:437–455.

Luo, S.-W. J. 1987. NAD(P)H-dependent lipid peroxidation in trout muscle mitochondria. Ph.D. Thesis. University of Massachusetts, Amherst.

Lusk, G., Karel, M., and Goldblith, S.A. 1964. Astacene pigment loss occurring in freeze-dried shrimp and salmon during storage. *Food Technol.* **18**:157–158.

Mai, J. and Kinsella, J.E. 1979. Changes in lipid composition of cooked minced carp (*Cyprinus carpio*) during frozen storage. *J. Food Sci.* **44**:1619–1624.

Marquez-Ruiz, G., Perez-Camino, M.C., and Dobarganes, M.C. 1992. Digestibility of fatty acid monomers, dimers, and polymers in the rat. *J. Am. Oil Chem. Soc.* **69**:930–934.

Martinez, F. and Labuza, T.P. 1968. Rate of deterioration of freeze-dried salmon as a function of relative humidity. *J. Food Sci.* **33**:241–247.

Matlock, R.G., Terrell, R.N., Savell, J.W., Rhee, K.S., and Dutson, T.R. 1984. Factors affecting properties of precooked frozen pork sausage patties made with various NaCl/phosphate combinations. *J. Food Sci.* **49**:1372–1375.

Mazeaud, F. and Bilinski, E. 1976. Free fatty acids and the onset of rancidity in rainbow trout (*Salmo gairdneri*) flesh. Effect of phospholipase A. *J. Fish. Res. Board Can.* **33**:1297–1302.

McDonald, R.E. and Hultin, H.O. 1987. Some characteristics of the enzymic lipid peroxidation system in the microsomal fraction of flounder skeletal muscle. *J. Food Sci.* **52**:15–21, 27.

McGill, A.S., Hardy, R., and Gunstone, F.D. 1977. Further analysis of the volatile components of frozen cold stored cod and the influence of these on flavor. *J. Sci. Food Agric.* **28**:200–205.

Miki, H. and Nishimoto, J. 1984. Kinetic parameters of freshness-lowering and discoloration based on temperature dependence in fish muscles. *Bull. Jpn. Soc. Sci. Fish.* **50**:281–285.

Mikkelsen, A., Bertelsen, G., and Skibsted, L.H. 1991. Polyphosphates as antioxidants in frozen beef patties. Lipid oxidation and colour quality during retail display. *Z. Lebensm. Unters. Forsch.* **192**:309–318.

Miyazawa, T., Chiba, T., and Kaneda, T. 1985. Spin trapping of oxygen-centered lipid radicals in liver of oxidized oil-dosed rats. *Agric. Biol. Chem.* **49**:3081–3083.

Miyazawa, T., Nagaoka, A., and Kaneda, T. 1983. Tissue lipid peroxidation and ultraweak chemiluminescence in rats dosed with methyl linoleate hydroperoxide. *Agric. Biol. Chem.* **47**:1333–1339.

Miyazawa, T., Sato, C., and Kaneda, T. 1983. Antioxidative effects of alpha-tocopherol and riboflavin-butyrate in rats dosed with methyl linoleate hydroperoxide. *Agric. Biol. Chem.* **47**:1577–1582.

Nakayama, T. and Yamamoto, M. 1977. Physical, chemical and sensory evaluations of frozen-stored deboned (minced) fish flesh. *J. Food Sci.* **42**:900–905.

Navarro, M.P., Castrillon, A.M., and Varela, G. 1987. Effect of frozen storage on protein quality of sardines. *Int. J. Food Sci. Technol.* **22**:77–80.

Nelson, K.A. and Labuza, T.P. 1992. Relationship between water and lipid oxidation rates. Water activity and glass transition theory. In *Lipid Oxidation in Food.* (St., Angelo, A.J., ed.) pp. 93–103, American Chemical Society, Washington, D.C.

Nilsson, K. and Ekstrand, B. 1994. Refreezing rate after glazing affects cod and rainbow trout muscle tissue. *J. Food Sci.* **59**:797–838.

Nolen, G.A., Alexander, J.C., and Artman, N.R. 1967. Long-term rat feeding study with used frying fats. *J. Nutr.* **93**:337–348.

North, J.A., Spector, A.A., and Buettner, G.R. 1994. Cell fatty acid composition affects free radical formation during lipid peroxidation. *Am. J. Physiol.* **267**:C177–C188.

Oarada, M., Ito, E., Terao, S., Miyazawa, T., Fujimoto, K., and Kaneda, T. 1988. The effect of dietary lipid hydroperoxide on lymphoid tissues in mice. *Biochim. Biophys. Acta* **960**:229–235.

Oarada, M., Miyazawa, T., and Kaneda, T. 1986. Distribution of ^{14}C after oral administration of (U-^{14}C)-labeled methyl linoleate hydroperoxides and their secondary oxidation products in rats. *Lipids* **21**:150–154.

Ohshima, T., Wada, S., and Koizumi, C. 1985. Accumulation of lyso-form phospholipids in several species of fish flesh during storage at −5°C. *Bull. Jpn. Soc. Sci. Fish.* **51**:965–971.

Ohta, F. and Tanaka, K. 1978. Some properties of the liquid portion in the frozen fish muscle fluid. *Bull. Jpn. Soc. Sci. Fish.* **44**:59–62.

Osinchak, J.E., Hultin, H.O., Zajicek, O.T., Kelleher, S.D., and Huang, C.-H. 1992. Effect of NaCl on catalysis of lipid oxidation by the soluble fraction of fish muscle. *Free Radical Biol. Med.* **12**:35–41.

Panasenko, O.M., Evgina, S.A., Driomina, E.S., Sharov, V.S., Sergienko, V.I., and Vladimirov, Y.A. 1995. Hypochlorite induces lipid peroxidation in blood lipoproteins and phospholipid liposomes. *Free Radical Biol. Med.* **19**:133–140.

Persson, T. and von Sydow, E. 1972. A quality comparison of frozen and refrigerated cooked sliced beef. 2. Relationships between gas chromatographic data and flavor scores. *J. Food Sci.* **37**:234–239.

Pie, J.E., Spahis, K., and Seillan, C. 1990. Evaluation of oxidative degradation of cholesterol in food and food ingredients: identification and quantification of cholesterol oxides. *J. Agric. Food Chem.* **38**:973–979.

Pie, J.E., Spahis, K., and Seillan, C. 1991. Cholesterol oxidation in meat products during cooking and frozen storage. *J. Agric. Food Chem.* **39**:250–254.

Polvi, S.M., Ackman, R.G., Lall, S.P., and Saunders, R.L. 1991. Stability of lipids and omega-3 fatty acids during frozen storage of Atlantic salmon. *J. Food Proc. Preserv.* **15**:167–181.

Poulter, R.G. and Lawrie, R.A. 1977. Studies on fish muscle protein. Nutritional consequences of the changes occurring during frozen storage. *J. Sci. Food Agric.* **28**:701–709.

Pozo, R., Lavety, J., and Love, R.M. 1988. The role of dietary alpha-tocopherol (vitamin E) in stabilising the canthaxanthan and lipids of rainbow trout muscle. *Aquaculture* **73**:165–175.

Pryor, W.A. 1985. Free radical involvement in chronic diseases and aging. The toxicity of lipid hydroperoxides and their decomposition products. In *Xenobiotic Metabolism: Nutritional Effects*. (Finley, J.W. and Schwass, D.E., eds) pp. 77–96, American Chemical Society, Washington, D.C.

Puppo, A. and Halliwell, B. 1988. Formation of hydroxyl radicals from H_2O_2 in the presence of Fe. (Is Hb a biological Fenton reagent?) *Biochem. J.* **249**:185–190.

Ravindranathan, N.P., Chinnamma, G., Thampuran, N., Perigreen, P.A., and Gopakumar, K. 1987. Studies on frozen storage characteristics of individually quick frozen and block frozen mackerel. *Fish. Technol.* **24**:103–108.

Rhee, K.S., Zibru, Y.A., and Ordonez, G. 1987. Catalysis of lipid oxidation in raw and cooked beef by metmyoglobin-hydrogen peroxide, nonheme iron and enzyme systems. *J. Agric. Food Chem.* **35**:1013–1017.

Roozen, J.P., Frankel, E.N., and Kinsella, J.E. 1994. Enzymic and autoxidation of lipids in low fat foods: model of linoleic acid in emulsified hexadecane. *Food Chem.* **50**:33–38.

Salaris, S.C. and Babbs, C.F. 1989. Effect of oxygen concentration on the formation of malondialdehyde-like material in a model of tissue ischemia and reoxygenation. *Free Radical Biol. Med.* **7**:603–609.

Sanderson, P.W., Williams, W.P., Cunningham, B.A., Wolfe, D.H., and Lis, L.J. 1993. The effect of ice on membrane lipid phase behavior. *Biochim. Biophys. Acta* **1148**:278–284.

Schaich, K.M. 1980. Free radical initiation in proteins and amino acids by ionizing and ultraviolet radiations and lipid oxidation. III. Free radical transfer from oxidizing lipids. *Crit. Rev. Food Sci. Nutr.* **13**:189–244.

Schricker, B.R. and Miller, D.D. 1983. Effects of cooking and chemical treatments on heme and nonheme iron in meat. *J. Food Sci.* **48**:1340–1342, 1349.

Searle, A.J.F. and Willson, R.L. 1983. Stimulation of microsomal lipid peroxidation by iron and cysteine. Characterization and the role of free radicals. *Biochem. J.* **212**:549–554.

Sevanian, A., Wratten, M.L., McLeod, L.L., and Kim, E. 1988. Lipid peroxidation and phospholipase A_2 activity in liposomes composed of unsaturated phospholipids: a structural basis for enzyme activation. *Biochim. Biophys. Acta* **961**:316–327.

Shahidi, F. and Wanasundara, P.K.J.P.D. 1992. Phenolic antioxidants. *Crit. Rev. Food Sci. Nutr.* **31**:67–103.

Shewfelt, R.L. 1981. Fish muscle lipolysis: a review. *J. Food Biochem.* **5**:79–100.

Shewfelt, R.L. and Hultin, H.O. 1983. Inhibition of enzymic and non-enzymic lipid peroxidation of flounder muscle sarcoplasmic reticulum by pretreatment with phospholipase A_2. *Biochim. Biophys. Acta* **751**:432–438.

Shewfelt, R.L., McDonald, R.E., and Hultin, H.O. 1981. Effect of phospholipid hydrolysis on lipid oxidation in flounder muscle microsomes. *J. Food Sci.* **46**:1297–1301.

Simpson, K.L. 1985. Chemical changes in natural food pigments. In *Chemical Changes in Food During Processing*. (Richardson, T. and Finley, T.W., eds.) pp. 409–441, AVI Publishing, Westport, Conn.

Simpson, K.L., Lee, T.-C., Rodriguez, D.B., and Chichester, C.O. 1976. Metabolism in senescent and stored tissues. In *Chemistry and Biochemistry of Plant Pigments*, 2d ed. vol. 1. (Goodwin, T.W., ed.) pp. 439–457, Academic Press, New York.

Slabyj, B.M. and Hultin, H.O. 1983. Microsomal lipid peroxidation system from herring light and dark muscle: effect of cytosolic factors. *J. Food Biochem.* **7**:107–114.

Slabyj, B.M. and Hultin, H.O. 1984. Oxidation of a lipid emulsion by a peroxidizing microsomal fraction from herring muscle. *J. Food Sci.* **49**:1392–1393.

Spanier, A.M., Vercellotti, J.R., and James, C., Jr., 1992. Correlation of sensory, instrumental, and chemical attributes of beef as influenced by meat structure and oxygen exclusion. *J. Food Sci.* **57**:10–15.

St. Angelo, A.J., Spanier, A.M., and Bett, K.L. 1992. Chemical and sensory evaluation of flavor in untreated and antioxidant-treated meat. In *Lipid Oxidation in Food*. (St. Angelo, A.J., ed.) pp. 140–160, American Chemical Society, Washington, D.C.

Staprans, I., Rapp, J.H., Pan. X.-M., and Feingold, K.R. 1996. Oxidized lipids in the diet are incorporated by the liver into very low density lipoprotein in rats. *J. Lipid Res.* **37**:420–430.

Stuchell, Y.M. and Krochta, J.M. 1995. Edible coatings on frozen King salmon: effects of whey protein isolate and acetylated monoglycerides on moisture loss and lipid oxidation. *J. Food Sci.* **60**:28–31.

Sunamoto, J., Baba, Y., Iwamoto, K., and Kondo, H. 1985. Liposomal membranes. XX. Autoxidation of unsaturated fatty acids in liposomal membranes. *Biochim. Biophys. Acta* **833**:144–150.

Tappel, A.L. 1953. The mechanism of the oxidation of unsaturated fatty acids catalyzed by hematin compounds. *Arch. Biochem. Biophys.* **44**:378–395.

Terao, J. 1990. Reactions of lipid hydroperoxides. In *Membrane Lipid Oxidation.* (Vigo-Pelfrey, Carmen, eds.) pp. 219–238, CRC Press, Boca Raton, Fla.

Terao, J. and Fukino, N. 1993. Decomposition of phospholipid hydroperoxide in simulated gastric juiçe. *J. Food Lipids* **1**:79–83.

Thed, S.T. and Erickson, M.C. 1994. Alteration in lipid oxidation of channel catfish during frozen storage in response to ascorbate applications. *J. Muscle Foods* **5**:37–48.

Tomas, M.C. and Anon, M.C. 1990. Study on the influence of freezing rate on lipid oxidation in fish (salmon) and chicken breast muscles. *Int. J. Food Sci. Technol.* **25**:718–721.

Tsukuda, N. 1970. Studies on the discoloration of red fishes. VI. Partial purification and specificity of the lipoxygenase-like enzyme responsible for carotenoid discoloration in fish skin during refrigerated storage. *Bull. Jpn. Soc. Sci. Fish.* **36**:725–733.

Valpuesta, J.M., Ondarroa, M., and Goni, F.M. 1986. Temperature dependence of cytochrome c–induced phospholipid oxidation in liposomes. *Biochim. Biophys. Acta* **878**:435–439.

Wagner, B.A., Buettner, G.R., and Burns, C.P. 1994. Free radical-mediated lipid peroxidation in cells: oxidizability is a function of cell lipid bis-allylic hydrogen content. *Biochem.* **33**:4449–4453.

Whang, K. and Peng, I.C. 1987. Lipid oxidation in ground turkey skin and muscle during storage. *Poultry Sci.* **66**:458–466.

Williams, D.C., Lim, M.H., Chen, A.O., Pangborn, R.M., and Whitaker, J.R. 1986. Blanching of vegetables for freezing—which indicator enzyme to choose. *Food Technol.* **40**:130–140.

Wills, E.D. 1966. Mechanisms of lipid peroxide formation in animal tissues. *Biochem. J.* **99**:667–676.

Wu, T.C. and Sheldon, B.W. 1988. Flavor components and factors associated with the development of off-flavors in cooked turkey rolls. *J. Food Sci.* **53**:49–54.

Xing, Y., Yoo, Y., Kelleher, S.D., Nawar, W.W., and Hultin, H.O. 1993. Lack of changes in fatty acid composition of mackerel and cod during iced and frozen storage. *J. Food Lipids* **1**:1–14.

Yin, M.-C. and Faustman, C. 1993. Influence of temperature, pH, and phospholipid composition upon the stability of myoglobin and phospholipid: a liposome model. *J. Agric. Food Chem.* **41**:853–857.

Yoshida, H. and Alexander, J.C. 1983. Enzymatic hydrolysis in vitro of thermally oxidized sunflower oil. *Lipids* **18**:611–616.

Yoshida, K., Terao, J., Suzuki, T., and Takama, K. 1991. Inhibitory effect of phosphatidylserine on iron-dependent lipid peroxidation. *Biochem. Biophys. Res. Commun.* **179**:1077–1081.

Yoshioka, M. and Kaneda, T. 1974. Toxicity of hydroperoxyalkenals in autoxidized oils. *Proc. Int. Congr. Food Sci. Technol.* **1**:276–282.

Zhou, S., Ackman, R.G., and Morrison, C. 1995. Storage of lipids in the myosepta of Atlantic salmon (*Salmo salar*). *Fish Physiol. Biochem.* **14**:171–178.

Zommara, M., Tachibana, N., Mitsui, K., Nakatani, N., Sakono, M., Ikeda, I., and Imaizumi, K. 1995. Inhibitory effect of ethanolamine plasmalogen on iron- and copper-dependent lipid peroxidation. *Free Radical Biol. Med.* **18**:599–602.

Zotos, A., Hole, M., and Smith, G. 1995. The effect of frozen storage of mackerel (*Scomber scombrus*) on its quality when hot-smoked. *J. Sci. Food Agric.* **67**:43–48.

10

Relationship of Frozen-Food Quality to Microbial Survival

David A. Golden and Lilliam Arroyo-Gallyoun

10.1 Introduction

The mechanical freezing of foods was first utilized in the late 1800s. However, it was not until 1929 that Clarence Birdseye developed and introduced frozen food technology to the commercial marketplace. Freezing is among the most successful techniques used to prolong the storage life of food, with benefits including retention of nutrients and a product that, once thawed, closely resembles the fresh form of the food. The preservation properties of freezing are based on the application of temperatures sufficiently low to minimize chemical and biochemical reactions and halt metabolic processes. Thus, it stands to reason that freezing can have a lethal or inhibitory effect on microbiological systems. Freezing of foods and inhibition of microbial growth in frozen foods may actually be considered as coincidental events. It is well known that freezing temperatures can prevent the growth of microorganisms. However, prolonged shelf life of frozen foods is limited more by adverse chemical (enzymatic and oxidative) and physical (freezer burn) changes than by microbiological concerns. This does not imply that freezing is not important from the standpoint of microbiological safety and quality. In fact, aside from irradiation, freezing is probably the method of preservation that is simultaneously least destructive to foods and among the most inhibitory to microbial growth.

Since the introduction of Birdseye's technology, the process of freezing foods has become an important means of preserving foods from microbiological spoilage and controlling proliferation of foodborne pathogens. However, freezing can also be utilized to preserve microorganisms that are important in industrial settings. In some foods, such as frozen bread doughs and frozen yogurt, microbial stability under frozen storage is a valuable asset to the functionality, quality, and overall success of the product. Thus, freezing may be considered as a method of preserving foods from microbial activity and preserving microbial activity in frozen foods.

10.2 Effects of Freezing on Microorganisms

When microorganisms are subjected to an environmental stress such as freezing, some cells may express no detrimental effects, some are killed, and some may undergo sublethal or metabolic injury. Calcott and MacLeod (1974) suggested that the detrimental effects of freezing on microbial cells are due to (1) thermal (cold) shock, (2) concentration of extracellular solutes, (3) toxicity of concentrated intracellular solutes, (4) cell dehydration, (5) internal ice formation, and (6) attainment of a minimum cell volume. Each of these effects could lead to lethality. Schmidt-Lorenz (1976) described two kinds of lethal effects to microorganisms when subjected to freezing: (1) the immediate and direct consequence of freezing and thawing, and (2) killing during frozen storage. Cold shock responses (a form of cellular injury and death) are common when microorganisms in the exponential phase of growth are suddenly exposed to low temperatures (Ingram and Mackey, 1976). Cold shock is more commonly associated with Gram-negative bacteria, and it has been suggested that yeasts are not susceptible to cold shock. Sublethally injured microorganisms are of particular importance to the food industry because, if the injury-inducing stress is removed or reduced, these organisms can resuscitate and continue to grow and behave normally. Furthermore, because of their debilitated physiological condition and an increased sensitivity to selective chemical agents, injured microorganisms may be difficult to detect in a frozen product, particularly when selective media are used for recovery and enumeration (Hurst, 1980; Golden et al., 1988). In the absence of conditions needed for injury repair to occur, injured microorganisms will eventually die, although the time required for complete cell death to occur may exceed the storage stability of the frozen-food product.

10.2.1 Effect of Freezing on Water Activity (a_w)

During the freezing process, aqueous solutions remain in their liquid phase until reaching their freezing point at some temperature below 0°C. When foods freeze, ice separates from the unfrozen solution, and the vapor pressure of the ice is at equilibrium with the vapor pressure of the unfrozen solution (Corry, 1987). As the ice separates, dissolved solutes in the food are concentrated in the remaining liquid. As solute concentration rises in the unfrozen water, the freezing point and a_w decrease. Concentration of solutes continues as temperature continues to decrease, until the eutectic point is reached and the solution solidifies (Davies and Obafemi, 1985). The a_w can be calculated under these conditions as the ratio of the vapor pressure of ice to the vapor pressure of water (P_{ice}/P_{water}). The relationship between freezing temperatures, vapor pressures, and a_w of a pure water/ice system is presented in table 10.1. At 0°C, the vapor pressures for liquid water and ice are equal (4.579 mm Hg), thus the a_w of water at 0°C is 1.0. As temperature is decreased, the vapor pressure of ice falls below that of water, resulting in a decrease in a_w.

Table 10.1. Effect of freezing on vapor pressure and a_w of pure water and ice

	Vapor pressure (mm Hg)		
Temperature (°C)	Liquid water	Ice	$a_w(P_{ice}/P_{water})$
0	4.579	4.579	1.00
−5	3.163	3.013	0.953
−10	2.149	1.950	0.907
−15	1.436	1.241	0.864
−20	0.943	0.776	0.823
−25	0.607	0.476	0.784
−30	0.383	0.286	0.750
−40	0.142	0.097	0.680
−50	0.048	0.030	0.620

Source: Scott (1962)

From table 10.1, it can be concluded that all biological and chemical reactions that occur in frozen food substrates take place at a_w below that which exists in the same food in the unfrozen state. It would be expected that microorganisms capable of growth at subzero temperatures must be capable of growth at reduced a_w. Although this is often the case, foods are complex systems containing water and solutes (such as sugars and salts) that contribute to a_w. Elevated solute concentrations in some foods tend to maintain a_w at levels higher than expected for pure water (Jay, 1996). As a result, some microorganisms not recognized for their ability to grow at low a_w are capable of growth at subzero temperatures. With a few exceptions, it is generally accepted that microorganisms do not grow at temperatures below −10°C, which corresponds to a_w 0.907 for pure water. However, growth of the filamentous fungus *Xeromyces bisporus* has been reported at a_w 0.61 (Pitt and Christian, 1968). This observation and others reported in the literature make it clear that microbial growth inhibition in frozen foods is not entirely dependent on a_w. If this were the case, some microorganisms would be capable of growth at temperatures as low as −50°C (i.e., a_w 0.62).

10.2.2 Ice Crystal Formation

Microbial cells suspended in aqueous solutions during freezing behave much like solute molecules and become concentrated in the unfrozen portion of the solution as ice crystals form (Brown, 1991). The size of the ice crystals formed is proportional to the rate of freezing. That is, the faster a solution is frozen, the smaller the size of the crystal. Slow freezing supports the formation of large extracellular ice crystals, while small intracellular crystals develop during faster rates of freezing (Jay, 1996). Extracellular crystals move water out of the cell, resulting in cell dehydration and concentration of intracellular solutes. Microbial cells may show distortion and loss of membrane integrity because intracellular crystals are small enough to disrupt the cell membrane (Frazier and Westhoff, 1978; Brown, 1991). Damage to cell membranes can result in leakage of intracellular constituents and loss of the ability of the cell to maintain its internal environment. Yeast cells (Higuchi and Nemura, 1959) and *Escherichia coli* (Lindenberg and Lode, 1963) release phosphates, nucleic acids, proteins, and peptides following freezing.

10.2.3 Effects of Freezing on Cellular Membranes

Souzu et al. (1989) suggested that freeze-induced injury in microbial cells is due primarily to dehydration of lipid-rich membranes. This is further supported by the observations that ice crystal formation and subsequent dehydration occur in the cell membrane. Changes in the lipid state of microbial cell membranes is brought about by freezing; the phospholipid fraction is the most involved in membrane structural changes (Souzu, 1989). As temperatures are reduced, increased unsaturation of fatty acids results in greater lipid fluidity. The lipid solidification theory (Gaughran, 1947; Allen, 1953) suggests that higher levels of unsaturated fatty acids in microbial cell membranes serve to maintain the lipid in a fluid and mobile state, thereby preserving the functionality of the cell membrane.

10.3 Effects of Freezing Rate and Storage Temperature

Typically foods are frozen either quickly or slowly. Quick or fast freezing is usually achieved by reducing the temperature of the food to about −20°C within 30 min. During slow freezing, the desired temperature is usually reached within 3 to 72 h (Jay, 1996). Temperature decreases of around 1 to 10°C/h typify slow freezing, while decreases of more than 50°C/h are common for quick freezing (Brown, 1991). Most home freezers utilize the slow freezing process. Quick freezing typically favors microbial survival. The advantages and disadvantages of quick versus slow freezing from the standpoint of microbial survival were described by Borgstrom (1961) and are listed as follows:

Quick freezing	Slow freezing
Advantages	*Disadvantages*
Small ice crystals are formed	Large ice crystals are formed
Metabolism is blocked or suppressed	Breakdown of metabolic rapport
Brief exposure of microorganisms to concentration of adverse constituents	Longer exposure of microorganisms to concentration of adverse constituents
Disadvantages	*Advantages*
No adaptation to low temperatures	Gradual adaptation to reduced temperature
Cold shock from rapid transition	No cold shock effect
No protective effect	Protective effect from accumulation of concentrated solutes

The rate of microbial death may be high during the initial states of freezing, but typically decreases with time. Some microorganisms are highly resistant to prolonged freezing and have been observed to survive during 10 years of frozen storage (Ayers et al., 1980). Below −60°C, microbial destruction is slow. Between −2 and −10°C, more organisms are killed or injured than at −15°C; at −30°C the effect of freezing is even less pronounced (Christophersen, 1968). Temperatures higher than −10°C are more lethal to microorganisms partly due to the reduction in protein denaturation that occurs at lower temperatures (Borgstrom, 1961). El-Kest and Marth (1992a) reported that *Listeria monocytogenes* was more susceptible to injury and death when frozen and stored at −18°C as compared with freezing and storage at −198°C. Similarly, survival of a multiple strain starter culture used in making buffalo Mozzarella cheese was more tolerant to freezing at −196°C as compared with freezing at −30 or −80°C; however, survival during storage was best at −80°C (Coppola and Coppola, 1993).

10.4 Effects of Thawing on Microorganisms

It has been suggested that microorganisms die during the thawing process, rather than during freezing or frozen storage. This has not been proven and may not be readily demonstrated. When frozen foods begin to thaw, ice crystals grow and exert the same physical stresses on microbial cells that are observed during the freezing process. In addition, as the frozen solution begins to melt, the medium surrounding microbial cells is diluted and the cells are exposed to osmotic shock (Calcott, 1978). Calcott (1978) suggested that the rate of thawing has little effect on the survival of microorganisms that have been previously frozen. However, others (Mazur, 1966; MacLeod and Calcott, 1976) have reported that microbial survival in rapidly thawed foods is greater than in foods thawed slowly. These observations do not take into account the use of a heating source to facilitate rapid thawing. Syed Ziauddin et al. (1993) demonstrated that higher microbial counts were obtained when frozen buffalo meat was thawed at room temperature as compared with meat thawed in a microwave. In general, there is poor agreement on whether the rate of thawing has a significant effect on the survival of microorganisms during thawing (Ingram and Mackey, 1976). However, repeated freezing and thawing cycles do lead to greater loss in microbial viability than does a single freezing-thawing treatment (Meyer et al., 1975; El-Kest and Marth, 1992a).

Regardless of thawing rate, the temperature during thawing is of tremendous importance. During thawing, the surfaces of large frozen products will reach higher temperatures sooner and will be exposed longer than interior portions of the product. This situation can result in microbial growth on the surface of partially thawed products, the consequences of which are obvious. Thawing foods using higher temperatures increases the risk of microbial spoilage and growth of pathogenic microorganisms. Although thawing at refrigeration temperatures would be more prudent, holding at refrigeration temperatures for too long could lead to growth of psychrotrophic pathogens and spoilage organisms.

10.5 Variations in Microbial Tolerance to Freezing

Microorganisms vary tremendously in their abilities to tolerate freezing. Environmental factors not withstanding, survival is affected by the type and age of the microorganism.

10.5.1 Growth Stage/Rate

Exponential growth phase cells are much more sensitive to freeze damage than stationary phase cells (Brown, 1991), and lag phase cells are typically not susceptible to rapid decreases in temperature (Hegarty and Weeks, 1940). Actively growing yeast cells have been demonstrated to be more sensitive to freezing than nongrowing cells (Meyer et al., 1975). Differences between growth phases in sensitivity to freezing are attributed to differences in physiological activities of the cells (Ray and Speck, 1973). For example, it has been demonstrated that conditions leading to slower growth rates of microorganisms can result in increases in unsaturated fatty acid content of cells (Dawson and Craig, 1966). Increases in the proportion of unsaturated fatty acids in microbial cells can be accompanied by increased tolerance to freezing and low temperatures in general (Olson and Nottingham, 1980; Golden and Beuchat, 1992; Golden et al., 1994).

10.5.2 Growth Temperature

Microorganisms can be classified in terms of optimum temperature for growth. Psychrotrophic organisms are those that grow well at or below 7°C and have an optimum growth temperature between 20 and 30°C (Jay, 1996). Psychrophiles, a term sometimes used interchangeably with psychrotrophs, include microorganisms that grow over the range of subzero to 20°C, with an optimum range of 10–15°C (Morita, 1975). Mesophilic microorganisms grow between 20 and 45°C, with growth optima between 30 and 40°C, while thermophiles grow above 45°C with optima between 55 and 65°C (Jay, 1996). In general, tolerance of microorganisms to freezing increases as their optimum temperatures for growth decrease. Additionally, freezing induces cold shock more readily in thermophiles and mesophiles than in psychrophiles and psychrotrophs. Many bacteria that grow at low temperatures synthesize larger amounts of enzymes to compensate for reduced enzymatic activity. Furthermore, increased production of energy-generating enzymes may also occur, thereby compensating for the slower metabolic rates of microorganisms consequent to suboptimal growth temperatures (Herbert and Bhakoo, 1979). Psychrotrophic bacteria and other organisms typically increase synthesis of unsaturated fatty acids at reduced incubation temperatures (Hunter and Rose, 1972; McMurrough and Rose, 1973; Jay, 1996).

10.5.3 Microbial Categories

Gram-positive bacteria are generally more resistant to freezing than are Gram-negative bacteria (Mazur, 1966; Ingram and Mackey, 1976; Olson and Nottingham, 1980), with cocci being more tolerant than bacilli. Yeasts and molds are typically more resistant than bacteria, partly because fungi are generally more tolerant to reduced a_w (Ingram and Mackey, 1976; Jay, 1996). It is interesting to note that in psychrotrophic bacteria, there is a predominance of Gram-negative species and relatively few Gram-positives (Herbert and Bhakoo, 1979; Jay, 1996). Partmann (1975) reported that the microflora of raw meat, before freezing at −30°C, consisted of 15% Gram-positive and 85% Gram-negative bacteria. After freezing, the proportions reversed to 70 and 30%, respectively, further indicating the sensitivity of Gram-negative bacteria to freezing conditions.

Gram-positive staphylococci, micrococci, and streptococci are relatively resistant to freezing, with survival exceeding 50% (Mazur, 1966), as are many Gram-positive bacilli including *Bacillus, Clostridium,* and *Lactobacillus* species. However, *Clostridium perfringens* is very sensitive to frozen storage (Olson and Nottingham, 1980; Trakulchang and Kraft, 1977). Gram-negative organisms such as *Escherichia, Salmonella, Serratia, Pseudomonas, Acinetobacter, Moraxella,* and *Vibrio* are much more sensitive to both freezing and frozen storage (Ingram and Mackey, 1976; Olson and Nottingham, 1980). Nevertheless, some Gram-negative enteric pathogens, if present in sufficiently high numbers before freezing, may survive during long periods of storage.

Various fungi are also able to survive freezing virtually unharmed. For example, *Aposphaera* and *Sphaeropsis* are fungi commonly present in frozen foods. *Arthrinium (Papularia), Phialophora, Botryotichum,* and *Cephalosporium* have all been isolated from frozen foods. *Aureobasidium (Pullularia)* grows at low temperatures and can be a problem in the frozen food industry (Beneke and Stevenson, 1987).

10.5.4 Microbial Products

Bacterial spores are extremely resistant to freezing and storage at subzero temperatures, with surviving levels exceeding 90%. Likewise, fungal spores are also highly resistant to freezing. Although *Clostridium botulinum* spores are resistant to freezing, this is not of particular significance if frozen foods are thawed properly to prevent toxin production. However, the toxins of *C. botulinum* and *Staphylococcus aureus* are highly resistant to inactivation by freezing, so it is possible for poisoning to occur upon ingestion of a frozen product that contains preformed toxins (Olson and Nottingham, 1980). Likewise, biogenic amines such as histamine and histidine may be formed by decarboxylating bacteria in scombroid fish prior to freezing (Taylor, 1988), and these toxins are not destroyed by freezing (Brown, 1991).

10.6 Detrimental Aspects of Microbial Tolerance to Freezing

Freezing can be used to destroy *Trichinella spiralis* (etiologic agent of trichinosis), although the effectiveness is dependent upon food thickness, freezing time, and temperature of storage. Generally, freezing becomes more destructive to *T. spiralis* as the temperature is lowered (Zimmerman et al., 1985; Kotula et al., 1990). Protozoa also are destroyed at temperatures below −5°C (Mazur, 1966). Despite the fact that freezing is one of the most effective methods available for controlling microbial growth in foods, it is not a reliable method for destruction of microorganisms. Although a large portion of microbial cells in a food substrate are killed or damaged by freezing, frozen storage, and thawing, spoilage and safety problems in the thawed product are not necessarily precluded. After frozen foods are thawed they should be handled with the same care as fresh foods. Since freezing treatment can result in disruption of the cellular structure and integrity of foods, it seems likely that thawed foods would be more susceptible to microbial growth than unfrozen foods. However, studies have revealed that frozen and thawed meat (Sulzbacher, 1952) and chicken (Elliot and Straka, 1964) spoil at about the same rates as the unfrozen products.

Microbial spoilage of thawed foods is commonly attributed to improper food handling practices before and/or after the foods are frozen, and foods being of initially poor microbiological quality (Mayes and Telling, 1993). The list of organisms commonly associated with spoilage of frozen and thawed foods is immense. The microbial flora on thawed foods is related to the microorganisms originally present on the unfrozen food, but will be modified by the freezing process which reduces the proportions of the more sensitive mesophilic and Gram-negative organisms. Table 10.2 describes the predominant spoilage bacteria commonly associated with thawed foods.

Pathogenic microorganisms are also commonly isolated from frozen and thawed foods. Golden et al. (1988) demonstrated that populations of various strains of *L. monocytogenes* were not appreciably reduced after storage for 14 days at −18°C, although up to 82% of the surviving population was injured during freeze treatment. Mackey and Derrick (1980) observed little loss of viability of nonpathogenic *E. coli* in meat during 220 days of −20°C storage. Although some consider that *Vibrio parahaemolyticus* is sensitive to freezing, Johnson and Liston (1973) reported that this organism survived for up to 130 days in oysters, fish fillets, and crabmeat stored at −15 and −30°C. *Yersinia enterocolitica* survived well on frozen meat stored for up to 28 days at −18 to −20°C (Hanna et al., 1977). Conner and Hall (1994) reported that > 50% of *E. coli* O157:H7 cells

Table 10.2. Predominant spoilage bacteria of frozen foods

Category	Obligate aerobes	Coryneform bacteria	Entero-bacteriaceae	*Lactobacillus*
Examples	*Pseudomonas, Alcaligenes, Moraxella, Alteromonas, Flavobacterium*	*Corynebacterium, Kurthia, Arthrobacter, Brochothrix*	*Escherichia, Citrobacter, Klebsiella, Erwinia, Proteus*	Lactic acid bacteria
Gram reaction	Negative	Positive	Negative	Positive
Catalase	Positive	Positive	Positive	Negative
Morphology	Rods	Rods	Rods	Rods and coccobacilli
Foods most affected	Protein foods (meat, fish, poultry, eggs)	Fresh vegetables, cured meats, sausages	Most proteinaceous foods	Milk, vegetables, pate, meats

Source: Mayes and Telling (1993); copyright © 1993, Blackie Academic & Professional, an imprint of Chapman & Hall

survived during 18 months of storage at −20°C in ground chicken breast meat. Similarly, Foster and Mead (1976) determined that > 60% of salmonellae survived for 126 days in ground chicken meat held at −20°C. The frequent isolation of *Salmonella* spp. from a wide range of frozen food products is evidence that freezing alone is insufficient for preventing the dissemination of foodborne salmonellosis (Olson and Nottingham, 1980).

It is worth mentioning that there is often the misconception that once meat, poultry, and other foods are thawed they cannot be refrozen for reasons of food safety. As long as frozen foods are thawed properly, particularly under refrigeration without prolonged storage, they can be refrozen. In fact, enhanced lethality of microorganisms resulting from freeze-thaw cycling is known to occur, as is described in detail earlier in this chapter. Textural changes in foods during freezing and thawing, and surface condensation of water on thawed products could feasibly facilitate the growth of microorganisms. Therefore, it cannot be overemphasized that safe and proper thawing procedures must be utilized to prevent the growth of foodborne pathogens, particularly if thawed foods are to be refrozen. Most processors of frozen foods advise against refreezing foods after they have been thawed. Loss of product quality through drip or leakage is probably of greater significance than safety when foods are thawed and refrozen.

Some microorganisms are actually capable of growth at freezing temperatures, although growth rates are typically markedly reduced. Growth of microorganisms below −10°C is rare but is relatively common at −5 to −7°C (Davies and Obafemi, 1985). *Micrococcus cryophilus* and *Vibrio* AF-1 have generation times (the time required for the cell population to double) of 28.33 and 23 h, respectively, at 0°C (Herbert and Bell, 1977). Relatively rapid growth can occur, as observed with *Bacillus psychrophilus* which has a generation time of 6.3 h at −5 to −7°C (Larkin and Stokes, 1968). In their review, Michener and Elliott (1964) reported the minimum growth temperatures of several microorganisms in various foods. Among those reported, a pink yeast grew on oysters during storage at −17.8°C, halophilic bacteria grew at −10°C on bacon, osmophilic yeasts grew in concentrated orange juice at −10°C, and *Pseudomonas fluorescens* and *Lactobacillus* spp. grew on peas at −4°C.

10.7 Beneficial Aspects of Microbial Tolerance to Freezing

Microbial resistance to inactivation by freezing can be of significant industrial importance, as is the case with frozen bread doughs and frozen yogurt. In the baking industry, yeast tolerance to freezing and thawing is a necessary trait when post/thaw leavening activity is required. Although functionality and dough strength are important considerations for frozen dough quality, the survival of yeast after freezing is perhaps the most important factor related to frozen dough stability (Hsu et al., 1979). Loss of yeast viability in frozen dough leads to inadequate leavening in thawed dough. Standard doughs typically contain about 3% viable yeasts as leavening agents, whereas 4–5% yeast is used in frozen doughs to compensate for loss of yeast viability during freezing and thawing. The requirement to use more yeast in frozen doughs represents a significant increase in cost (Lorenz, 1974). Similar concerns apply to bacteria in frozen yogurt which typically contains added *Lactobacillus bulgaricus* and *Streptococcus thermophilus* cultures. The addition of some species of lactic acid bacteria to frozen yogurts and other dairy products has been demonstrated to aid in lactose digestion (Gilliland, 1979; Klaenhammer, 1982; Shah and Jelen, 1990) and may provide various therapeutic benefits in the human intestinal tract (Wander and Reddy, 1974; Gilliland and Speck, 1977). However, it appears that added microbial cultures must be alive when consumed for such benefits to be realized (Gilliland, 1979; Klaenhammer, 1982).

Examination of frozen yogurts has revealed that these products typically contain greater than 1×10^6 lactic acid bacteria/g, suggesting that these organisms are capable of retaining viability during commercial freezing and storage of yogurt (Whitehead et al., 1993). However, decreases in bacterial numbers do occur in frozen yogurt (Gilliland and Lara, 1988), and numbers of viable bacteria differ substantially among yogurt brands (Kosikowski, 1981; Speck and Hansen, 1983). Ertugay et al. (1993) froze yeasted bread doughs, using a cryogenic process, at −20 and −40°C and by deep freezing at −20°C and determined that yeast performance in thawed doughs was best when doughs were frozen cryogenically at −20°C. Furthermore, these authors reported that yeast performance decreased as frozen storage time increased. Torner et al. (1992) concluded that frozen storage for up to 3 months has no adverse effect on the invertase and maltase activities, viability, and fermentative activity of pressed baker's yeast. In addition to time and temperature, yeast viability is adversely affected by fluctuations in freezing temperature, and prolonged thawing (Dunas, 1991).

10.8 Effects of Frozen Substrate Composition

Survival of microorganisms during freezing conditions is dependent on the composition of the substrate (MacLeod and Calcott, 1976). Some food constituents such as acids (Faville et al., 1951) reduce microbial tolerance to freezing. Conner and Hall (1994) observed that the rate of sublethal injury and death of *E. coli* O157:H7 was increased in frozen ground chicken meat supplemented with 4% (v/w) of a 60% solution of sodium lactate. Storage under oxygen atmospheres tends to be more deleterious to microorganisms than storage under nitrogen (Ray and Speck, 1973).

Cryoprotectants are compounds that protect cells (microbial and food substrate) from damage caused by freezing, thereby promoting cell viability or integrity during frozen storage (Fennema, 1973). Typical cryoprotectants are low molecular weight compounds such as sugars, glycerol, diethylene glycol, dimethyl sulfoxide (DMSO), and amino acids. Salts, specifically NaCl, may function as cryoprotectants (Golden and Beuchat, 1992; Hudson, 1992) or may increase microbial sensitivity to freezing (Conner and Hall, 1994). Complex molecules such as starch, milk and other proteins, and yeast extract may also provide cryoprotection (Woodborne and Strong, 1960; Moss and Speck, 1963; Janssen and Busta, 1973; Ray and Speck, 1973). Cryoprotectants used in foods are more thoroughly described in chapter 11.

Improvement of the cryotolerance of yeasts and bacteria can be achieved to some degree by altering the culturing conditions prior to freezing. Gélinas et al. (1991) reported that improved cryoresistance of *S. cerevisiae* grown under fed-batch conditions, as compared with batch-grown cells, was correlated with a higher ratio of free sterols to phospholipids. The addition of cryoprotective agents to doughs or yogurt improves microbial stability, although use may be restricted due to development of unsatisfactory physical and sensory qualities in the product. Kim and Park (1993) reported that 85 and 91% of *Lactobacillus acidophilus* survived after 90 days of storage in skim milk at −20 and −70°C, respectively, when β-glycerophosphate was added as a cryoprotective agent. Ethanol has been demonstrated to provide significant cryoprotection to *S. cerevisiae* under rapid freezing conditions, although protection under slow freezing is diminished (Lewis et al., 1993).

Glycerol is a commonly used cryoprotectant in foods and generally is quite effective at protecting microorganisms from freeze damage. El-Kest and Marth (1992a) reported that glycerol, milk fat, lactose, and casein in phosphate buffer protected *L. monocytogenes* from sublethal injury and death for up to 20 weeks at

−18°C, with glycerol exerting the greatest cryoprotective effect (table 10.3). Conversely, large losses in viability of osmotolerant *Zygosaccharomyces* spp. was reported in broth containing 20% glycerol during a 5-month storage period at −25°C (Jermini and Schmidt-Lorenz, 1987). Fat can offer protection, as demonstrated by Kraft et al. (1979) who reported that various spoilage bacteria survived better in beef patties containing 30% fat than in patties containing 20% fat. Low numbers of *L. monocytogenes* Scott A survived in ice cream for 5 months of frozen storage at −18°C, with no indication of sublethal injury (Lammerding and Doyle, 1990). Destruction of *Y. enterocolitica* at −20°C was more rapid in water than in milk (Toora et al., 1992), further suggesting that milk components can serve as cryoprotective agents. *Listeria monocytogenes* has been demonstrated to survive in solutions containing up to 16% NaCl for 33 days at −18°C, with no significant reduction in cell numbers (Hudson, 1992).

A positive relationship typically exists between the presence of solutes, and resistance to inactivation of yeasts by freezing. Murdock and Hatcher (1978) reported that yeasts suspended in 65° Brix orange concentrate were more tolerant to freezing temperatures than yeasts suspended in 45° Brix orange concentrate. Similar observations were made by Ingram (1951), who demonstrated that survival of yeasts in orange juice at −17.8°C was enhanced as the concentration of sucrose was increased. *Saccharomyces cerevisiae* was protected from freezing in the presence of 0.88% (0.15 M) NaCl (Mazur, 1966; MacLeod and Calcott, 1976). Golden and Beuchat (1992) reported that *Zygosaccharomyces rouxii* was

Table 10.3. Death and injury of *Listeria monocytogenes* suspended in phosphate buffer with and without added glycerol or milk components[a]

		Dead cells (% of initial population) after frozen storage			Injured cells (% of remaining viable population) after frozen storage		
Suspending fluid	Additive (%)	1 day	4 weeks	20 weeks	1 day	4 weeks	20 weeks
Phosphate buffer (PB)	—	55	88	95	65	78	90
PB + glycerol	2	10	14	4	31	20	29
	4	0	0	0	23	12	12
PB + milk fat	2	3	85	100	12	61	—
	4	17	83	100	0	34	—
PB + lactose	2	15	88	98	14	22	67
	4	25	36	85	5	42	54
PB + casein	2	10	29	72	15	44	73

[a] Preparations frozen and stored at −18°C and thawed at 35°C before testing

Source: El-Kest and Marth (1992a); copyright ©1992, IAMFES, Inc.

more tolerant to freezing at −18°C for up to 120 days in culture media supplemented with 10% NaCl, 33% glucose, or 45% sucrose than in media not supplemented with solutes.

The mechanism by which solutes increase resistance to freezing has not been fully defined. One theory proposes that cryoprotectants bind water at or near the cell surface, thereby resulting in protection from freezing (Morichi et al., 1963). Meryman et al. (1977) suggested that cryoprotectants reduce the amount of intracellular ice formation or increase the time required for water to leave the cell during the freezing process. Neutral solutes such as glycerol and DMSO may function as cryoprotective agents by preventing the buildup of toxic levels of electrolytes that normally occurs during freezing (Lovelock, 1953). Osmophilic yeasts accumulate compatible solutes (e.g., glycerol and arabitol) in response to reduced a_w (Brown, 1974; Hocking, 1988). These compatible solutes essentially replace intracellular water. Since freeze injury and lethality result, in part, from structural damage caused by ice crystal formation, it is possible that adaptive mechanisms that promote tolerance to reduced a_w also contribute to resistance to inactivation by freezing. Cell dehydration caused by high extracellular solute content has been proposed as a mechanism for increased resistance of microorganisms to heat inactivation (Corry, 1976) and may also play a role in increasing microbial tolerance to freezing.

10.9 Effects of Prefreezing Treatments of Food

Microbial resistance to freezing can be tremendously influenced by the conditions of growth in which microorganisms exist prior to exposure to freezing temperatures. As described earlier, exposure of microorganisms to temperatures well below their optima can result in cold shock. In response to cold shock, *E. coli* produces several specific proteins (cold shock proteins) that enable the organism to survive at reduced temperatures (Jones et al., 1987). This phenomenon has also been demonstrated in *S. cerevisiae* (Kondo and Inouye, 1991). It has been suggested that sensitivity of yeast to freeze-thaw damage in frozen bread dough is influenced mainly by fermentation prior to freezing, or delayed CO_2 production due to frozen storage (Dunas, 1991). Nakagawa and Ouchi (1994) reported that heat treatment of bread dough at 44°C for 10 min prior to freezing resulted in a reduction of freeze damage to the leavening yeast. It was proposed that increased freeze tolerance may be attributed to increased levels of intracellular trehalose, a cryoprotectant, or induction of heat shock proteins. El-Kest and Marth (1992b) demonstrated that susceptibility of *L. monocytogenes* to inactivation by freezing was increased when cells were treated with lipase or lysozyme prior to freezing at −18°C. Culturing *Z. rouxii* in media containing 300 ppm potassium sorbate prior to frozen storage at −18°C resulted in an increase in sensitivity to freezing (Golden and Beuchat, 1992).

Nitrogen-limited cells of *E. coli* are more resistant to freezing than cells grown under conditions of carbon limitation owing to the accumulation of higher carbohydrate contents (Calcott and MacLeod, 1974, 1975). These observations support the hypothesis that polyglucose and glycogen-like reserve materials serve as cryoprotective agents in *E. coli* by strengthening the outer membrane. Cryosurvival of streptococci is reportedly enhanced by protective polymers synthesized in response to altered growth conditions (Gilliland and Speck, 1974). Typically, reduced incubation temperatures result in enhanced tolerance of microorganisms to frozen storage. At low incubation temperatures, *Leuconostoc* and *Pediococcus* spp. increase production of extracellular dextrans (Jay, 1996) that could serve as cryoprotective agents.

Through genetic manipulation, genes encoding for the production of cold shock proteins may be cloned into organisms whose survival during frozen storage is desired. A gene that encodes for an "antifreeze" protein produced by the winter flounder has been introduced into baker's yeast, resulting in a two- to fourfold increase in the ability of this yeast to retain viability after freezing (Breakthrough, 1991). Attfield et al. (1992) developed strains of *S. cerevisiae* that were capable of accumulating elevated levels of trehalose, resulting in tolerance to freeze stresses greater than that of strains that normally produce low levels of trehalose. Sheu and Marshall (1993) reported that about 40% more lactobacilli survived freezing in ice milk when cells were entrapped in calcium alginate gels than when cells were not entrapped. Microentrapment of yogurt bacteria is a viable method of improving bacterial survival and reportedly does not affect overrun, texture, or body of frozen ice milk (Sheu et al., 1993). Development of freeze-resistant strains of yeasts and bacteria would be a tremendous asset for use in frozen doughs, yogurt, and other products whose success relies on the stability and subsequent activity of microorganisms after thawing.

Exposing microorganisms to various environmental stresses prior to freezing can augment the lethal effects of freezing. Various factors, intrinsic and extrinsic to foods, such as pH, a_w, preservatives, heat treatment (e.g., blanching), and cold storage can each exert an inhibitory effect on microorganisms. If two or more of these factors are applied in combination, microbial inhibition can be substantially improved, compared to inhibition obtained when one of these factors is used alone. This approach to controlling microbial growth in foods, often referred to as the hurdle effect, has tremendous application and can be used in conjunction with freezing to improve the microbial stability of frozen foods.

References

Allen, M.B. 1953. The thermophilic aerobic sporeforming bacteria. *Bacteriol. Rev.* **17**:125-173.

Attfield, P.V., Raman, A., and Northcott, C.J. 1992. Construction of *Saccharomyces cerevisiae* strains that accumulate low concentrations of trehalose, and their application in testing the contribution of the disaccharide to stress tolerance. *FEMS Microbiol. Lett.* **94**:271–276.

Ayers, J.C., Mundt, J.O., and W.E. Sandine. 1980. Prevention of food spoilage. In *Microbiology of Foods.* pp. 44–71, W.H. Freeman, San Francisco.

Beneke, E.S. and Stevenson, K.E. 1987. Classification of food and beverage fungi. In *Food and Beverage Mycology,* 2d ed. (Beuchat, L.R., ed.) pp. 30–50, AVI Publishing, Westport, Conn.

Borgstrom, G. 1961. Unsolved problems in frozen food microbiology. In *Proceedings: Low Temperature Microbiology Symposium.* pp. 197–251, Campbell Soup Company, Camden, N.J.

Breakthrough in freeze protection. 1991. *Bioeng. News* **12**:2.

Brown, A.D. 1974. Microbial water relations: features of the intracellular composition of sugar-tolerant yeasts. *J. Bacteriol.* **18**:769–777.

Brown, M.H. 1991. Microbiological aspects of frozen foods. In *Food Freezing: Today and Tomorrow.* (Bald, W.B., ed.) pp. 15–25, Springer-Verlag, New York.

Calcott, P.H. 1978. Manifestation of freeze-thaw injury. In *Cryoprotectants.* (Cook, J.G., ed.) pp. 40–48, Meadowfield Press, Shildon, England.

Calcott, P.H. and MacLeod, R.A. 1974. Survival of *Escherichia coli* from freeze-thaw damage: a theoretical and practical study. *Can. J. Microbiol.* **20**:671–681.

Calcott, P.H. and MacLeod, R.A. 1975. The survival of *Escherichia coli* from freeze-thaw damage: the relative importance of wall and membrane damage. *Can. J. Microbiol.* **21**:1960–1968.

Christophersen, J. 1968. Effect of freezing and thawing on the microbial population of foodstuffs. In *Low Temperature Biology of Foodstuffs.* (Hawthorn, J. and Rolfe, E.J., eds.) pp. 251–270, Pergamon Press, London.

Conner, D.E. and Hall, G.S. 1994. Efficacy of selected media for recovery of *Escherichia coli* O157:H7 from frozen chicken meat containing sodium chloride, sodium lactate or polyphosphate. *Food Microbiol.* **11**:337–344.

Coppola, R. and Coppola, S. 1993. Low-temperature storage of starters for buffalo milk Mozzarella cheesemaking. *Latte* **18**:666–673.

Corry, J.E.L. 1976. The effect of sugars and polyols on the heat resistance and morphology of osmophilic yeasts. *J. Appl. Bacteriol.* **40**:269–276.

Corry, J.E.L. 1987. Relationships of water activity to fungal growth. In *Food and Beverage Mycology,* 2d ed. (Beuchat, L.R., ed.) pp. 51–99, Van Nostrand Reinhold, New York.

Davies, R. and Obafemi, A. 1985. Response of microorganisms to freeze-thaw stress. In *Microbiology of Frozen Foods.* (Robinson, R.K., ed.) pp. 83–107, Elsevier Applied Science Publishers, New York.

Dawson, P.S.S. and Craig, B.M. 1966. Lipids of *Candida utilis:* changes with growth. *Can. J. Microbiol.* **12**:775–785.

Dunas, F.E.T. 1991. Yeast in frozen bread doughs (thesis abstract). *Dissertation Abstr. Int.* **C52**(4):535.

El-Kest, S.E. and Marth, E.H. 1992a. Freezing of *Listeria monocytogenes* and other microorganisms: a review. *J. Food Protect.* **55**:639–648.

El-Kest, S.E. and Marth, E.H. 1992b. Lysozyme and lipase alter unfrozen and frozen/thawed cells of *Listeria monocytogenes. J. Food Protect.* **55**:777–781.

Elliot, R.P. and Straka, R.P. 1964. Rate of microbial deterioration of chicken meat at 2°C after freezing and thawing. *Poultry Sci.* **43**:81–86.

Ertugay, Z., Koca, A.F., and Celik, I. 1993. Yeast performance and bread properties of white pan bread produced by the frozen dough process. *Doga Turkish J. Agric. Forest.* **17**:226–238.

Faville, L.W., Hill, E.C., and Parish, E.C. 1951. Survival of microorganisms in concentrated orange juice. *Food Technol.* **5**:33–36.

Fennema, O.R. 1973. Freezing injury and cryoprotectants. In *Low-Temperature Preservation of Foods and Living Matter.* (Fennema, O.R., Powrie, W.D., and Marth, E.H., eds.) pp. 476–503, Marcel Dekker, New York.

Foster, R.D. and Mead, G.C. 1976. Effect of temperature and added polyphosphate on the survival of salmonellae in poultry meat during cold storage. *J. Appl. Microbiol.* **41**:505–510.

Frazier, W.C. and Westhoff, D.C. 1978. Preservation by use of low temperatures. In *Food Microbiology,* 3rd ed. pp. 130–142, McGraw-Hill, New York.

Gaughran, E.R.I. 1947. The thermophilic microorganisms. *Bacteriol Rev.* **11**:189–225.

Gélinas, P., Fiset, G., Willemot, C., and Goulet, J. 1991. Lipid content and cryotolerance of baker's yeast in frozen doughs. *Appl. Environ. Microbiol.* **57**:463–468.

Gilliland, S.E. 1979. Beneficial interrelationships between certain microorganisms and humans: candidate microorganisms for use as dietary adjuncts. *J. Food Protect.* **42**:164–167.

Gilliland, S.E. and Lara, R.C. 1988. Influence of storage at freezing and subsequent refrigeration temperatures on β-galactosidase activity of *Lactobacillus acidophilus. Appl. Environ. Microbiol.* **54**:898–902.

Gilliland, S.E. and Speck, M.L. 1974. Relationship of cellular components to the stability of concentrated lactic streptococcus cultures at −17°C. *Appl. Microbiol.* **27**:793–796.

Gilliland, S.E. and Speck, M.L. 1977. Antagonistic action of *Lactobacillus acidophilus* toward intestinal and food-borne pathogens in associative cultures. *J. Food Protect.* **40**:820–823.

Golden, D.A. and Beuchat, L.R. 1992. Interactive effects of solutes, potassium sorbate and incubation temperature on growth, heat resistance, and tolerance to freezing of *Zygosaccharomyces rouxii. J. Appl. Bacteriol.* **73**:524–530.

Golden, D.A., Beuchat, L.R., and Brackett, R.E. 1988. Inactivation and injury of *Listeria monocytogenes* as affected by heating and freezing. *Food Microbiol.* **5**:17–23.

Golden, D.A., Beuchat, L.R., and Hitchcock, H.L. 1994. Changes in fatty acid composition of various lipid components of *Zygosaccharomyces rouxii* as influenced by solutes, potassium sorbate, and incubation temperature. *Int. J. Food Microbiol.* **21**:293–303.

Hanna, M.O., Stewart, J.C., Zink, D.L., Carpenter, Z.L., and Vanderzant, C. 1977. Development of *Yersinia enterocolitica* on raw and cooked beef and pork at different temperatures. *J. Food Sci.* **42**:1180–1184.

Hegarty, C.P. and Weeks, O.B. 1940. Sensitivity of *E. coli* to cold shock during the logarithmic growth phase. *J. Bacteriol.* **39**:475–484.

Herbert, R.A. and Bell, C.R. 1977. Growth characteristics of an obligately psychrophilic *Vibrio* sp. *Arch. Microbiol.* **113**:215–220.

Herbert, R.A. and Bhakoo, M. 1979. Microbial life at low temperatures. In *Cold Tolerant Microbes in Spoilage and the Environment.* (Russell, A.D. and Fuller, R., eds.) pp. 1–16, Academic Press, New York.

Higuchi, M. and Nemura, T. 1959. Release of nucleotides from yeast cells. *Nature* **141**:1381–1385.

Hocking, A.D. 1988. Strategies for microbial growth at reduced water activities. *Microbiol. Sci.* **5**:280–284.

Hsu, K.H., Hoseney, R.C., and Seib, P.A. 1979. Frozen dough. I. Factors affecting stability of yeasted doughs. *Cereal Chem.* **56**:419–424.

Hudson, J.A. 1992. Efficacy of high sodium chloride concentrations for the destruction of *Listeria monocytogenes. Lett. Appl. Microbiol.* **14**:178–180.

Hunter, K. and Rose, A.H. 1972. Lipid composition of *Saccharomyces cerevisiae* as influenced by growth temperature. *Biochim. Biophys. Acta* **260**:639–653.

Hurst, A. 1980. Injury and its effect on survival and recovery. In *Factors Affecting Life and Death of Microorganisms.* (*Microbial Ecology of Foods,* vol. 1) (Silliker, J.H., Elliott, R.P., Baird-Parker, A.C., Bryan, F.L., Christian, J.H.B., Clark, D.S., Olson, J.C., Jr., and Roberts, T.A., eds.) pp. 205–214, Academic Press, New York.

Ingram, M. 1951. The effect of cold on microorganisms. *Proc. Soc. Appl. Bacteriol.* **14**:243–260.

Ingram, M. and Mackey, B.M. 1976. Inactivation by cold. In *Inhibition and Inactivation of Vegetative Microbes.* (Skinner, F.A. and Hugo, W.B., eds.) pp. 111–151, Academic Press, New York.

Janssen, D.W. and Busta, F.F. 1973. Influence of milk components on the injury, repair of injury, and death of *Salmonella anatum* cells subjected to freezing and thawing. *Appl. Microbiol.* **26**:725–732.

Jay, J.M. 1996. Low-temperature food preservation and characteristics of psychrotrophic microorganisms. In *Modern Food Microbiology,* 5th ed. pp. 328–346, Chapman and Hall, New York.

Jermini, M.F.G. and Schmidt-Lorenz, W. 1987. Influence of frozen storage and thaw-freeze stresses on the viability of osmotolerant yeasts. *J. Food Protect.* **50**:414–417.

Johnson, H.C. and Liston, J. 1973. Sensitivity of *Vibrio parahaemolyticus* to cold in oysters, fish fillets, and crabmeat. *J Food Sci.* **38**:437–441.

Jones, P.G., VanBogelen, R.A., and Neidhardt, F.C. 1987. Induction of proteins in response to low temperature in *Escherichia coli. J. Bacteriol.* **169**:2092–2095.

Kim, D.D. and Park, S.K. 1993. Studies on the lyophilized acidophilus milk culture. *Korean J. Dairy Sci.* **15**:195–202.

Klaenhammer, T.R. 1982. Microbiological considerations in selection and preparation of *Lactobacillus* strains for use in dietary adjuncts. *J. Dairy Sci.* **65**:1339–1349.

Kondo, K. and Inouye, M. 1991. TIP1, a cold shock–inducible gene of *Saccharomyces cerevisiae. J. Biol. Chem.* **266**:17537–17544.

Kosikowski, F.V. 1981. Properties of commercial flavored frozen yogurts. *J. Food Protect.* **44**:853–856.

Kotula, A.W., Sharar, A.K., Paroczay, E., Gamble, H.R., Murrell, K.D., and Douglass, L. 1990. Infectivity of *Trichinella spiralis* from frozen pork. *J. Food Protect.* **53**:571–573.

Kraft, A.A., Reddy, K.V., Sebranek, J.G., Rust, R.E., and Hotchkiss, D.K. 1979. Effect of composition and method of freezing on microbial flora of ground beef patties. *J. Food Sci.* **44**:350–354.

Lammerding, A.M. and Doyle, M.P. 1990. Stability of *Listeria monocytogenes* to nonthermal processing conditions. In *Foodborne Listeriosis.* (Miller, A.J., Smith, J.L., and Somkuti, G.A., eds.) pp. 195–202, Elsevier Science, New York.

Larkin, J.M. and Stokes, J.L. 1968. Growth of psychrophilic microorganisms at subzero temperatures. *Can. J. Microbiol.* **14**:97–101.

Lewis, J.G., Learmonth, R.P., and Watson, K. 1993. Role of growth phase and ethanol in freeze-thaw stress resistance of *Saccharomyces cerevisiae. Appl. Environ. Microbiol.* **59**:1065–1071.

Lindenberg, G. and Lode, A. 1963. Release of ultraviolet-absorbing material from *Escherichia coli* at subzero temperatures. *Can J. Microbiol.* **9**:523–530.

Lorenz, K. 1974. Frozen dough—present trend and future outlook. *Bakers Digest* **48**:14–22, 30.

Lovelock, J.E. 1953. The mechanism of the protective action of glycerol against haemolysis by freezing and thawing. *Biochim. Biophys. Acta* **11**:28–36.

Mackey, B.M. and Derrick, C.M. 1980. The recovery of sublethally injured *Escherichia coli* from frozen meat. *J. Appl. Bacteriol.* **48**:315–324.

MacLeod, R.A. and Calcott, P.H. 1976. Cold shock and freezing damage to microbes. In *The Survival of Vegetative Microbes.* (Gray, T.R.G. and Postgate, J.R., eds.) pp. 81–109, Cambridge University Press, New York.

Mayes, T. and Telling, G. 1993. Product safety from factory to consumer. In *Frozen Food Technology.* (Mallett, C.P., ed.) pp. 96–102, Blackie Academic & Professional, Glasgow.

Mazur, P. 1966. Physical and chemical basis of injury in single-celled microorganisms subjected to freezing and thawing. In *Cryobiology.* (Meryman, H.T., ed.) pp. 214–317, Academic Press, New York.

McMurrough, I. and Rose, A.H. 1973. Effects of temperature variation on the fatty acid composition of a psychrophilic *Candida* species. *J. Bacteriol.* **114**:451–452.

Meryman, H.T., Williams, R.J., and Douglas, M.S.J. 1977. Freezing injury from "solution effects" and its prevention by natural and artificial cryoprotection. *Cryobiol.* **14**:287–302.

Meyer, E.D., Sinclair, N.A., and Nagy, B. 1975. Comparison of the survival and metabolic activities of psychrophilic and mesophilic yeasts subjected to freeze-thaw stress. *Appl. Microbiol.* **29**:739–744.

Michener, H.D. and Elliott, R.P. 1964. Minimum growth temperatures for food-poisoning, fecal-indicator, and psychrophilic microorganisms. *Adv. Food Res.* **13**:349–396.

Morichi, T., Irie, R., Yano, N., and Kembo, H. 1963. Protective effect of glutamic acid and related compounds on bacterial cells subjected to freeze-drying. *J. Gen. Appl. Microbiol.* **9**:149–161.

Morita, R.Y. 1975. Psychrophilic bacteria. *Bacteriol. Rev.* **39**:144–167.

Moss, C.W. and Speck, M.L. 1963. Injury and death of *Streptococcus lactis* due to freezing and frozen storage. *Appl. Microbiol.* **11**:326–329.

Murdock, D.I. and Hatcher, W.S. 1978. Effect of temperature on survival of yeast in 45° and 65° Brix orange concentrate. *J. Food Protect.* **41**:689–691.

Nakagawa, S. and Ouchi, K. 1994. Improvement of freeze tolerance of commercial baker's yeast in dough by heat treatment before freezing. *Biosci. Biotech. Biochem.* **58**:2077–2079.

Olson, J.C. and Nottingham, P.M. 1980. Temperature. In *Microbial Ecology of Foods,* vol. 1. (ICMSF) pp. 1–37, Academic Press, New York.

Partmann, W. 1975. The effects of freezing and thawing in food quality. In *Water Relationships of Foods.* (Duckworth, R.B., ed.) pp. 505–539, Academic Press, New York.

Pitt, J.I. and Christian, J.H.B. 1968. Water relations of xerophilic fungi isolated from prunes. Appl. Microbiol. **16**:1853–1858.

Ray, B. and Speck, M.L. 1973. Freeze-injury in bacteria. *Crit. Rev. Clin. Lab. Sci.* **4:**161–213.

Schmidt-Lorenz, W. 1976. Über die bedeutung der anwesenheit von mikroorganismen in gefrorenen lebensmitteln. *Lebensm. Wiss. Technol.* **9**:263–273.

Scott, W.J. 1962. Available water and microbial growth. In *Proceedings: Low Temperature Microbiology Symposium.* pp. 89–105, Campbell Soup Company, Camden, N.J.

Shah, N. and Jelen, P. 1990. Survival of lactic acid bacteria and their lactase under acidic conditions. *J. Food Sci.* **55**:506.

Sheu, T.-Y. and Marshall, R.T. 1993. Microentrapment of lactobacilli in calcium alginate gels. *J. Food Sci.* **54**:557–561.

Sheu, T.-Y., Marshall, R.T., and Heymann, H. 1993. Improving survival of culture bacteria in frozen desserts by microentrapment. *J. Dairy Sci.* **76**:1902–1907.

Souzu, H. 1989. Changes in chemical structure and function in *Escherichia coli* cell membranes caused by freeze-thawing. I. Change in lipid state in bilayer vesicles and in the original fragments depending on rate of freezing. *Biochim. Biophys. Acta* **978**:105–111.

Souzu, H, Sato, M., and Kojima, T. 1989. Changes in chemical structure and function in *Escherichia coli* cell membranes caused by freeze-thawing. II. Membrane lipid state and response of cells to dehydration. *Biochim. Biophys. Acta* **978**:112–118.

Speck, M. and Hansen, A. 1983. Properties of frozen non-fruit yogurt. *Cult. Dairy Prod. J.* **18**:6–9.

Sulzbacher, W.L. 1952. Effect of freezing and thawing on the growth rate of bacteria in ground meat. *Food Technol.* **6**:341–342.

Syed Ziauddin, K., Rao, D.N., Ramesh, B.S., and Amla, B.L. 1993. Effect of freezing, thawing, and frozen storage on microbial profiles of buffalo meat. *J. Food Sci. Technol. India* **30**:465–467.

Taylor, S.L. 1988. Marine toxins of microbial origin. *Food Technol.* **42**:94–98.

Toora, S., Budu-Amoako, E., Ablett, R.F., and Smith, J. 1992. Effect of high-temperature short-time pasteurization, freezing and thawing, and constant freezing on the survival of *Yersinia enterocolitica* in milk. *J. Food Protect.* **55**:803–805.

Torner, M.J., Pulido, E., and Martinez-Anaya, M.A. 1992. Effect of freezing on enzyme activity and breadmaking quality of commercial pressed yeast. *Rev. Espan. Cien. Tecnol. Aliment.* **32**:653–666.

Trakulchang, S.P. and Kraft, A.A. 1977. Survival of *Clostridium perfringens* in refrigerated and frozen meat and poultry items. *J. Food Sci.* **2**:518–521.

Wander, E. and Reddy, B.S. 1974. Metabolic epidemiology of colorectal cancer. *Cancer* **34**:801–806.

Whitehead, W.E., Bodyfelt, F.W., and Sandine, W.E. 1993. Evaluation of the microbial quality of frozen yoghurt products. *Cult. Dairy Prod. J.* **28**:21–22.

Woodborne, M.J. and Strong, D.H. 1960. Survival of *Salmonella typhimurium*, *Staphylococcus aureus*, and *Streptococcus faecalis* in simplified food substrates. *Appl. Microbiol.* **8**:109–113.

Zimmerman, W.J., Olson, D.G., Sandoval, A., and Rust, R.E. 1985. Efficacy of freezing in eliminating infectivity of *Trichinella spiralis* in boxed pork products. *J. Food Protect.* **48**:196–199.

SECTION III

Techniques to Minimize Quality Losses

11

Cryoprotectants for Improving Frozen-Food Quality

Grant A. MacDonald
and Tyre C. Lanier

11.1 Introduction

Cryoprotectants are compounds that improve the quality and extend the shelf life of frozen foods. The term *cryoprotectant* includes all compounds that help to prevent deleterious changes in foods caused by freezing and thawing processes or frozen storage. These substances may be added during processing and product formulation or produced naturally in the living organism from which the food is derived.

For this discussion, it is worthwhile to make a distinction between formulated, noncellular foods in which intimate mixing of cryoprotectants and food material is possible, and the more complex situation in intact, cellular foods. In formulated foods cryoprotectants may be needed to stabilize one important component in the food such as the myofibrillar proteins in surimi, a freeze-stabilized semi-processed fish mince. Conversely, for cellular foods such as fruits, vegetables, and meats, an increased shelf life may be concerned with stabilizing cellular structure and controlling water movement to prevent cellular collapse, as well as stabilizing proteins. These distinctions are perhaps analagous to the stabilization of proteins in the pharmaceutical industry and cryopreservation of cells and organs in the medical field.

A wide variety of compounds will cryoprotect labile proteins during freeze-thawing (reviewed in MacDonald and Lanier, 1991). These include sugars, amino acids, polyols, methyl amines, carbohydrate polymers, synthetic polymers (e.g., polyethylene glycol, PEG), other proteins (e.g., bovine serum albumin, BSA), and even inorganic salts (e.g., potassium phosphate and ammonium sulfate). For most proteins, the cryoprotectant must be at a high concentration to confer maximum protection. Exceptions are polymers such as PEG, BSA, and polyvinylpyrrolidone (PVP), which at concentrations of less than 1% (wt/vol) fully protect sensitive enzymes such as lactate dehydrogenase or phosphofructokinase (Carpenter et al., 1994). In addition, for freeze-tolerant organisms, specific antifreeze proteins, stress proteins (cold and heat "shock" proteins), and ice-nucleating agents may also be important. Antioxidants and metal chelators may also extend the shelf life of frozen foods.

Most preservative compounds that chemically stabilize food molecules at ambient temperatures can be equally effective at preventing these same changes during freezing, frozen storage, or thawing. Some compounds, such as many common antioxidants, may be useful additives to achieve both ambient and frozen storage stability but are seldom referred to as cryoprotectants. Many sugars and polyols that are effective in, but seldom added for, thermal stabilization of proteins are well-known as cryoprotectants of proteins in fish and meat, as well as their nonfood applications such as ensuring viability of frozen cell isolates.

This overlap between cryoprotection and other types of food stabilization is an important point. Many of the deteriorative processes that may be accelerated or are most apparent in frozen products in fact also proceed at ambient temperatures. Few cryoprotectants are unique to frozen systems. However, their importance to the stability of the food may grow in relative significance during frozen storage because of the cessation of microbial growth or because solutes become freeze concentrated and therefore more reactive.

Two physical changes—low temperature and ice-crystal formation—distinguish frozen systems from ambient. Low temperature slows most deteriorative reactions by decreasing molecular mobility. Additionally, intramolecular hydrophobic interactions that stabilize many macromolecules, particularly proteins, become weaker as the temperature decreases. More dramatic is the change in the state of water, which in most foods makes up the bulk of the volume. Water separates and suspends the remaining constituents. Ice-crystal formation per se may not be a threat to food quality when crystal size is small, but concentration of solutes as water crystallizes out of solution can effect serious imbalances in the colloidal chemistry of the food system. In intact animal and plant tissues, dehydration of cells may disrupt membranes or cell walls and distort tissue structure, adversely affecting thawed texture and water-holding properties. Large ice crystals will aggravate this tissue damage, particularly when they form intracellularly.

Leistner (1978) introduced the *hurdle* concept to stabilization of foods against microbial attack: individual additives or process steps that inhibit or destroy microorganisms can be viewed as "hurdles" to the growth of spoilage or harmful microorganisms in foods. Each such hurdle, while on its own is incapable of ensuring stability, contributes additively, or sometimes synergistically, to the overall microbial stability of the food. In the same way, chemical and physical stability of frozen foods are advanced by individual factors, one of which can be cryoprotectant additives. Because the class of additives known as cryoprotectants is diverse, each mechanistic type may be viewed as a separate hurdle to chemical or physical instability caused by freezing (Fennema, 1996).

We will address several mechanistic actions of cryoprotective compounds in this chapter. These actions do not necessarily separate compounds into chemically similar groups, because often a compound may act by more than one mechanism to stabilize the food system. We first address mechanisms that primarily affect protein stability, especially during freezing/thawing, and then mechanisms that affect cell integrity or ice-crystal formation more, especially during low-temperature storage.

11.2 Solute Exclusion from Protein Surfaces

Biochemists have known for many years that certain substances, such as sucrose and glycerol, when present at high concentration (>0.5 M), stabilize the activity of enzymes in solution and during freezing and thawing. Natural production of glucose by the liver is also thought to be partially responsible for freeze tolerance in certain subarctic frog species (Storey, 1990) and other species. Based on the results of freeze-thaw experiments on labile enzymes (Carpenter et al., 1988), myofibrillar proteins (MacDonald and Lanier, 1991, 1994; MacDonald, Lanier, and Griesbrecht, 1996; MacDonald, Lanier, Swaisgood, and Hamann, 1996), and a review of the literature on protein freezing, we have determined that cryopreservation can be explained by the same general mechanism that Timasheff, Arakawa, and others have defined for solute-induced protein stabilization in nonfrozen, aqueous solution (Gekko and Timasheff, 1981; Arakawa and Timasheff, 1982a,b, 1984a,b, 1985; Lee and Timasheff, 1981; Timasheff and Arakawa, 1989).

We have summarized the various aspects of this general mechanism for cosolvent stabilization of proteins in figure 11.1. Extensive studies on a variety of cosolvents that lower the solubility of globular proteins or stabilize them (e.g., glycerol, sugars, amino acids, and salts) have shown that these cosolvents are preferentially excluded from the domain of the protein, that is, in their presence the protein is preferentially hydrated (fig. 11.2). Such an exclusion is thermodynamically unfavorable since it increases the chemical potential (activity) of both the protein and the cosolvent. In the protein unfolding (denaturation) equilibrium, the effect of the solvent is defined by the relative affinities (exclusion or binding) of the solvent for the protein in its two end states, native (N) and denatured (D).

$$N \underset{\xleftarrow[\text{Stabilizer}]{}}{\overset{K(S)}{\Leftrightarrow}} D, \quad K = \frac{[D]}{[N]} \tag{11.1}$$

The effect of the stabilizer on the equilibrium constant of the reaction, K, is a function of the stabilizer (S) and can be expressed in terms of the equilibrium constant by the Wyman linkage (Wyman, 1964).

$$-\left(\frac{\delta \Delta G}{\delta \mu}\right) = \Delta \nu_{\mathrm{S}} = \nu^{\mathrm{D}} - \nu^{\mathrm{N}} \tag{11.2}$$

where $\Delta\, \upsilon_{\mathrm{S}}$ is the difference in the number of cosolvent molecules bound by the denatured and native molecules, μ is the chemical potential (partial molar free

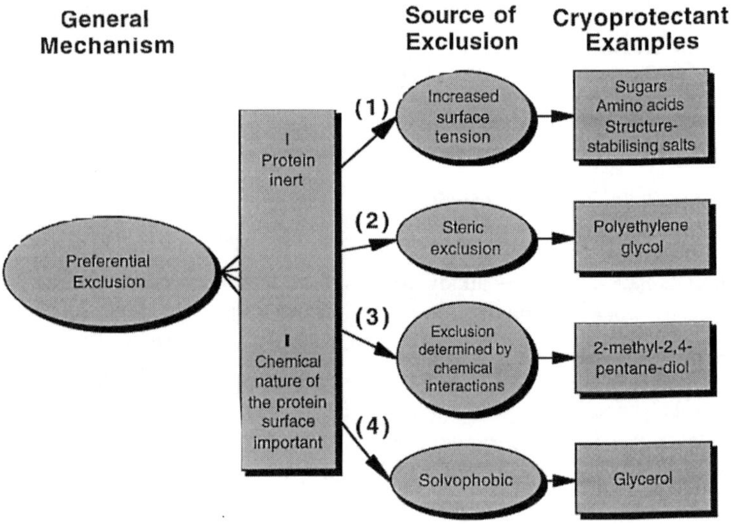

Figure 11.1. Overview of key mechanisms for protein stabilization in solution

energy), and ΔG is the Gibbs free energy. Since $\Delta G = -\mathrm{RT}\ \ln K$, K being the equilibrium constant for the reaction, and $\mu = \mathrm{RT}\ \ln a_s$, where a_s is the thermodynamic activity coefficient for additive S, then

$$\left(\frac{\delta \ln K}{\delta \ln a_S}\right) = \Delta \nu_S = \nu^D - \nu^N \tag{11.3}$$

Hence, if an additive is a stabilizer, the reaction is shifted to the left and from equation 11.3, $\Delta \nu_S$ would be negative. This means there must be less binding of the stabilizer to the denatured state.

Excluded cosolvents can be classified into two categories:

1. Those whose interaction with the protein is determined strictly by the properties of the solvent—the protein is chemically inert toward the cosolvent; and
2. Those in which the chemical nature of the protein surface determines the interactions with the cosolvent.

In the first category, there is always stabilization, since the protein-solvent interface increases on unfolding. In the second category, it becomes necessary to know the nature of the interactions in the two end states to ascertain whether a cosolvent excluded from the native protein will be a stabilizer. Obviously, there can also be intermediate cases where both mechanisms operate simultaneously.

In the first category two specific mechanisms can be identified:

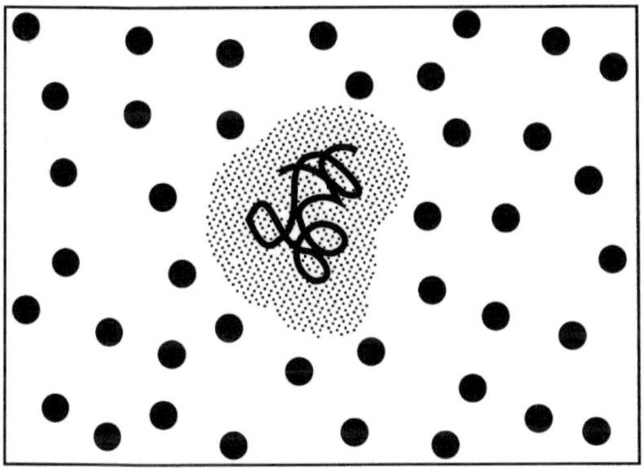

Figure 11.2. Solute exclusion from the cavity occupied by the protein and its hydration shell (shaded area), the black dots signifying solute molecules (Reprinted with permission from MacDonald and Lanier, 1991.)

1. Preferential exclusion may result from the *increase in surface tension of water* upon addition of the cosolvent (sugars, amino acids, or salting out salts). Since the surface of contact between protein and solvent constitutes an interface, there must be in this surface an interfacial (surface) tension. Additives perturb the cohesive force of water and hence its surface tension. The excess or deficiency of the additive in the surface layer can be calculated by the Gibbs adsorption isotherm (van Holde, 1985)

$$d\gamma = \Gamma d\mu \tag{11.4}$$

where γ is surface tension and Γ is the adsorption (excess concentration) of cosolvent at the surface.

Thus, if a substance increases the surface tension of water, its excess in the surface layer will be negative; that is preferentially excluded from the interface. Timasheff and Arakawa (1989) compared the ratio of the preferential interaction measured experimentally by dialysis equilibrium to that calculated from the surface tension effect for a range of proteins and stabilizers and concluded that this mechanism appears to give rise to the observed preferential exclusion of sugars, amino acids, and structure-stabilizing salts. Stabilization by increased surface tension appears to be the most widespread mechanism in protein-stabilizing solvent systems.

2. A second mechanism in the first category is *steric hindrance of high-molecular-weight (MW) macromolecules* where a zone enriched with

water is formed around the protein because the additive molecule cannot penetrate the protein structure. The smaller water molecule can penetrate within the shell, forming a zone enriched with water. This is the mechanism attributed to freeze-thaw stabilization of labile enzymes by PEG, PVP, and dextran (Carpenter and Crowe, 1988; Carpenter et al., 1994). In their studies they found PEG effectiveness increased with molecular weight; on a molar basis PEG MW 8,000 was 5,000 times more effective than PEG MW 400.

In the second category, the exclusion is determined by chemical interactions between the solvent components and protein structure.

3. This case occurs when preferential exclusion is due to *charges on the surface of the protein*. For example, 2-methyl-2-4-pentanediol (MPD) is strongly repelled from charges on the protein surface and migrates into the bulk solvent, leaving a layer enriched with water (Pittz and Timasheff, 1978).
4. Glycerol, on the other hand, stabilizes proteins but lowers surface tension of solutions (Gekko and Morikawa, 1981). In this case glycerol is preferentially excluded by the *solvophobic effect* (Gekko and Timasheff, 1981), a consequence of the net balance between the binding of glycerol, a hydrophilic compound, to polar regions and exclusion from nonpolar regions of proteins. Since denaturation, or unfolding, involves an increase in surface contact between protein and solvent, and in particular exposes additional hydrophobic residues to contact with solvent, the presence of glycerol favors the native (more compact) state.

A primary factor contributing to the instability of proteins at subambient temperatures is the decreasing strength of intramolecular hydrophobic interactions, which stabilize the protein native structure, with decreasing temperature (Privalov, 1990). Since the hydrophobic interaction is a thermodynamic response to the structuring of water by apolar groups on the protein surface, agents that affect water structure can thus affect protein stability. While the protein intramolecular hydrophobic interactions are weakened by lowering temperature, there is a counteracting effect of increasing solute concentration and thus increasing solute surface tension effected by freezing pure water into ice.

11.3 Ligand Binding

An exception to the general destabilization of proteins by cosolvents that bind to the protein is the case when some oligomeric proteins are stabilized by binding with specific ligands. For example, the nucleotides ATP, ADP, and IMP exerted a protective effect on fish actomyosin stored at −20°C while the nucleotide catabolites inosine and hypoxanthine destabilized these proteins (Jiang et al., 1987). Myosin is a hexamer that consists of two identical heavy chains and two pairs of essential and regulatory light chains with a specific nucleotide binding site. Binding ATP induces a conformational change in the myosin structure, which leads to increased stability of the protein (Mornet et al., 1977). Similarly, recent studies have shown that the cold labile hexameric ATPase, chloroplast coupling factor 1 (CF1), is stabilized by binding ATP and ADP (Hightower and McCarty, 1996). Interestingly, in this study ADP stabilization of CF1 was much enhanced by the presence of Mg^{2+} ions. Possibly the Mg^{2+} provided additional contact between the α and β subunits and the ADP binding sites that caused an increased stability of the complex.

The ligand-binding stabilization mechanism may be more general because many multimeric proteins are cold labile. The dissociation of these complexes in the cold is thought to be a consequence of hydrophobic interactions between the subunits. As the temperature decreases, the strength of these interactions decreases and the complex dissociates (Privalov, 1990). It appears that binding nucleotides and other ligands induces conformational changes in the quarternary structure that confers increased stability, possibly by increasing intersubunit contact.

The importance of the specificity of the ligand-protein interaction is perhaps shown by the contrary findings of Carpenter and colleagues (Carpenter et al., 1986; Carpenter, Crowe, and Crowe, 1987; Carpenter, Martin, Crowe, and Crowe, 1987; Hazen et al., 1988) who showed that low concentrations of Zn^{2+} ions acted synergistically with sugars to stabilize the tetrameric ATPase, phosphofructosekinase, during freeze-thawing and freeze-drying, whereas MacDonald, Lanier, and Giesbrecht (1996) demonstrated that even at very low levels Zn^{2+} destabilized fish actomyosin during freeze-thawing.

11.4 Antioxidants

Polyphosphates at 0.2–0.3% commonly are added as synergists to the cryoprotective effect of carbohydrate additives in the manufacture of surimi (Park and Lanier, 1987; Park et al., 1988; Kumazawa et al., 1990). Phosphates may serve to chelate calcium ions that can induce protein aggregation (Saeki, 1996). As antioxidants of lipids they may also protect proteins from denaturation induced by hydrolysis or autooxidation of phospholipids (Wessels et al., 1981). Other antioxidants may also serve a useful function in protecting unsaturated lipids and proteins during frozen storage; these are reviewed in chapter 12.

11.5 Freezing-Point Depression

Ice formation concentrates reactants, particularly salts and hydrogen ions (pH depression) that thermodynamically destabilize proteins (Shenouda, 1980; Volkin and Klibanov, 1989). The volume of ice formed increases with decreasing temperature below the freezing point. Thus, by depressing the freezing point with low-molecular-weight solutes, the deleterious effects of freeze concentration on proteins at any subfreezing storage temperature (above the operative glass transition temperature; see later discussion) should be lessened.

The presence of any solute will lower the freezing point of water in a colligative fashion. Ideally, its performance is related to its concentration according to Raoult's law in which the mole fraction of solute is equal to the proportional reduction of solution vapor pressure compared to the pure solvent.

The freezing point of water is defined as the temperature at which the vapor pressure of the solid phase (ice) is equal to the vapor pressure of the liquid phase (Pauling, 1953; DeVries, 1988). Thus, on an equal mass basis, lower-molecular-weight solutes will have the greater molarity and therefore the greater effect on freezing-point depression. Interactions between solute and water molecules that affect the vapor pressure of the water may cause deviations from ideal behavior (Meryman, 1971; van Holde, 1985). The most general cause for nonideality in solutions of macromolecules arises from the size and shape as expressed by their molecular excluded volume (van Holde, 1985). This excluded volume, which accounts for the volume occupied by the solute and is therefore not available to other molecules in the solution, can be unexpectedly large for assymetric rods or random coils.

11.5.1 Antifreeze Proteins

A special situation of nonideality arises in the case of certain peptides and glycopeptides found to be produced by animals, insects, and plants in especially cold climates (Knight and Duman, 1986; Lee, 1991; Schmitt et al., 1994). These so-called *antifreeze proteins* are remarkable in their ability to noncolligatively depress the freezing point of their solutions with the number of molecules not affecting the depression (DeVries, 1988). Antifreeze glycopeptides (AFGP) and peptides (AFP) adsorb to ice and impede crystal growth by inhibiting the normal propagation of ice-growth layers across the crystal surface (Eastman and DeVries, 1986). Exactly how the AF(G)P molecules bind to the ice is not well understood, however. New ice layers that meet glycopeptide bound to the basal plane of an underlying layer, however, divide into many small fronts that become highly curved. Such curved fronts have a high ratio of surface area to volume, a condition that halts their growth if the temperature of the surrounding liquid does not fall. This difference between the nonequilibrium freezing point and the equilibrium freezing point is termed *thermal hysteresis,* which can be detected upon melting. AFP are thus sometimes referred to as thermal hysteresis proteins (THP).

Since some antifreeze proteins are now synthesized chemically or by genetic engineering, the use of antifreeze proteins for improving the quality of frozen foods is being actively researched (see Feeney and Yeh, 1993, for a review). Arai et al. (1984) prepared amphiphilic gelatin by an enzymatic modification process that displayed antifreeze (thermal hysteresis) properties similar to the naturally occurring peptides and glycopeptides. They suggested that this product might have future application as a protein cryoprotectant in foods.

11.5.2 Other Considerations

Low-MW sugars and polyols that—among carbohydrates—have the greatest effect on depression of freezing point are concentrated on the freezing out of ice. This also increases the surface tension of the unfrozen solvent, thereby enhancing the stabilization of proteins.

Large-MW polymers (PEG, starch hydrolysis products [SHPs]) have lesser effects on freezing point at additive concentrations feasible for food use. However, the addition of small amounts of polymers to sucrose solutions has a much greater effect on freezing-point depression than if it were added to water (Muhr, 1983). Ice-crystal growth in sucrose solutions of common nongelling ice-cream stabilizers is reduced one- to two-thirds, more than anticipated from considerations of their mobility and effect on freezing point. This may be because concentration of these polymers upon freezing will dramatically increase viscosity of the unfrozen milieu because of the strong intermolecular solute interactions and/or polymer entanglements.

11.6 Decreased Molecular Mobility and Enhanced Glass Formation

Carpenter and Crowe (1988) theorized that certain high-MW polymers, such as PVP, PEG, and dextran, are good cryoprotectants because they are sterically excluded from the protein surface by their size (Carpenter and Crowe, 1988; Carpenter et al., 1994). However, an additional mechanism has been postulated by other workers to explain the cryoprotective effects of many high-MW polyols (Lanier and Akahane, 1986) and glucose polymers (starch hydrolysis products; Levine and Slade, 1986) that takes into account the dramatic increase in viscosity that occurs as these polymers are freeze concentrated and form glasses. Because glass transitions in frozen foods have already been covered in depth in chapter 3, we summarize briefly in this section some of the implications of understanding and manipulating glass transitions for improving the stability and quality of frozen foods.

With continued lowering of the temperature below the freezing point, the rates of diffusion-limited deteriorative reactions decrease steadily as ice freezes out and solution viscosity increases, lowering diffusivity of the dissolved reactants (i.e., molecular mobility) (Parker and Ring, 1995; Fennema, 1996). Solutes that most interact with water to affect solution viscosity and water mobility would have a greater effect in reducing reaction rates at any subfreezing temperature (Meryman, 1971). Solution viscosity reaches a maximum at the glass transition temperature (T_g) of a solution (Levine and Slade, 1988a,b).

Figure 11.3 illustrates the glass transition temperature of a simple solution as a function of solute concentration. At higher concentrations of solute, T_g occurs at temperatures above freezing. Thus, the mixture cools to form a "candy" glass directly from the liquid state. At solute concentrations below the point C_g', when the temperature falls below the freezing curve, the solution will exist either as a viscous supersaturated solution in the liquid state or more commonly as a mixture of ice crystals and supersaturated solution. Under these conditions, the system is termed a *rubber,* exhibiting a high viscosity because of the presence of ice crystals and/or strong intermolecular solute interactions/entanglements. The rubber state is characterized by a departure from Arrhenius kinetics of chemical and enzymatic reactions, following instead Williams-Landel-Ferry kinetics. The rate of freezing and the constancy of storage temperature will influence ice-crystal sizes and numbers and likewise the solute concentration (Franks, 1985a,b). Thus, T_g may be altered by time factors not shown in this two-dimensional phase diagram.

As the glass transition temperature is approached, reactions become diffusion limited because of immobilization of the water within a solute structure (Treloar, 1970; Parker and Ring, 1995). The glassy state is attained when solute-solute interactions supersede solute-water attractions and occurs at the T_g for the given solute and concentration. Viscosities near 10^{14} Pa s are characteristic (Franks, 1985a). The structure is amorphous, like that of the liquid, with no crystallinity other than that of the enmeshed ice crystals. If complete vitrification (glass formation) is achieved, no further ice formation will accompany decreases in temperature.

According to the cryostabilization theory of Levine and Slade, deteriorative processes cease in the practical time frame once the glass state is achieved. Mobility or diffusion in a glassy system is claimed to be virtually nonexistent (Levine and Slade, 1988a,b; Slade et al., 1993; Slade and Levine, 1995). However, Fennema (1996) argues that no food system is so simple that a single T_g can be ascribed; microenvironments within the food will have successively lower T_g values. For this reason, Fennema asserts the concept of a glass transition zone of temperature, below which ultimate stability is achieved because of water immobilization.

The requirement in cryostabilization for use of polymers exhibiting relatively large MW arises from their ability to form glasses at higher temperatures. This is due to their propensity to entangle and to form hydrogen and other bonds, imparting a greater viscosity at any given concentration. The glass transition of a carbohydrate is weakly dependent on structure with a much stronger dependence on molecular weight. Levine and Slade (1986, 1989) published tables and graphs that indicate a generally direct relationship between molecular weight (or, inversely, dextrose equivalent [DE] for starch hydrolysis products) and T_g', the T_g

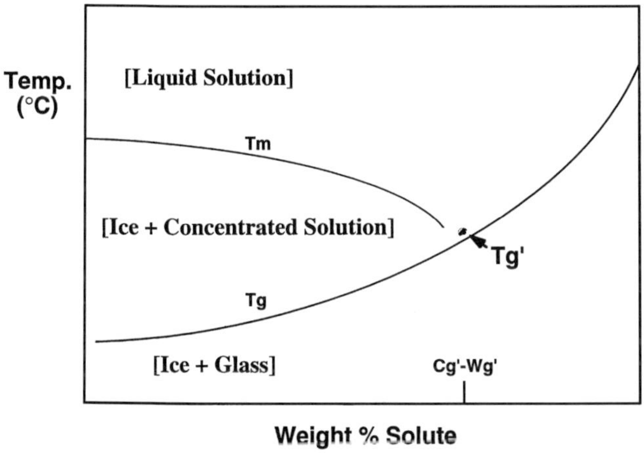

Figure 11.3. Schematic state diagram for a binary system in which the solute component does not crystallize

of a maximally frozen aqueous system (fig. 11.3). Roos and Karel (1991) and Roos (1995) also found a relationship between increasing T_g' and increasing molecular weight for a series of glucose polymers, although they found that the effective molecular weight was a better predictor than the molecular weight average of a polymer's ability to change T_g'. Branching of the molecule will affect the solution properties of polymers such that a strictly direct relationship between MW and T_g' will hold only for a homologous series.

It is possible to increase the overall glass transition temperature (or zone) of a food and therefore theoretically improve the food's stability by adding polymers that form glasses at high temperatures. The simplest equation that has been used to estimate the T_g' is the Gordon and Taylor equation (Gordon and Taylor, 1952)

$$T_g = \frac{w_1 T_{g_1} + kw_2 T_{g_2}}{w_1 + kw_2} \tag{11.5}$$

where T_{g_1} and T_{g_2} are the glass transition values (in K) of components 1 (water) and 2 (solute) of the sample, respectively; w_1 and w_2 are the weight fractions of components 1 and 2 of the sample; and k is an empirical constant.

Although Roos and Karel (1991) found the Gordon and Taylor equation (which was originally developed for binary polymer systems) useful for predicting the T_g' in amorphous carbohydrates, care should be taken when using it to estimate the T_g' of complex foods. Further and more complex models are discussed in Roos (1995).

At present the Holy Grail of freeze preservationists seems to be the attainment of a vitrified biological system—one in which no (or practially no) ice is present and all the fluid is immobilized in a glass. Aspects of this state apparently contribute to the ability of certain insects to survive freezing (Wasylyk et al., 1988) and to freeze resistance in plants (Hirsh, 1987). Research on this topic is seriously being pursued in cryobiology with the purpose of eventually being able to successfully freeze, without loss of viability, entire human organs (Fahy et al., 1984, 1987; Hirsh, 1987; MacFarlane, 1987). Attaining such a physical state in a biological system will likely require a combined approach of adding glass-forming cryoprotectants to raise T_g, high pressure during freezing and/or thawing to lower freezing/melting point (see later discussion) and raise T_g, and extremely fast freezing rates. Successful vitrification of small cells and embryos has been obtained in the laboratory (Ishimori, 1996), but there are equal problems of thawing the specimen without massive recrystallization and damage to the organism. For example, Kresin et al. (1994) determined that even if cooling rates high enough to avoid substantial crystallization are achieved, the critical warming rates needed to avoid devitrification and ice recrystallization are two to three orders of magnitude higher.

For practical reasons, it is unlikely that total vitrification (i.e., without ice formation) will be a feasible approach to food preservation. Yet the studies on cryoprotective additives and their interaction with pressure and freezing conditions should be followed closely to determine whether spin-off applications in food science might appear.

11.7 Enhanced Heterogeneous Ice Nucleation

Heterogeneous ice nucleation is enhanced by certain phytopathogenic strains of *Pseudomonas, Erwinia,* and *Xanthomonas,* causing frost injury in plants. So-called ice nucleation active (INA) proteins are the responsible nucleation agents. They vary somewhat in molecular size but have distinguishing domains that presumably result in a secondary structure (as yet undetermined), which serves as a template for orienting water into ice crystals. Pressure-pasteurized preparations of these organisms are being considered for food-additive applications.

The primary potential benefit of such additives to the quality of frozen foods may arise from the more rapid and complete freezing that can be attained, particularly at higher freezing temperatures, when INA preparations are applied to the food surface. Supercooling is greatly reduced because of enhanced nucleation at higher subzero temperatures such that freezing rates up to 33% faster at −5°C have been measured with fish fillets (Li and Lee, 1995).

Additionally, it may be possible to induce a more even distribution of small ice crystals at conventional freezing rates in the extracellular region of intact fruits, vegetables, and meats, a desirable state of ice in such products for quality retention. Ice-nucleating proteins have been isolated from insects and in marine mollusks (Aunaas, 1982; Zachariassen, 1985) and more recently in the plasma of freeze-tolerant frogs (Storey et al., 1992), which are thought to protect these species by inducing controlled freezing of extracellular water at multiple nucleation sites. Similarly, there is some evidence that microcrystalline cellulose facilitates the formation of small ice crystals in ice cream by serving as an evenly distributed ice nucleator (Barringer, 1997). Although addition of ice-nucleating bacteria (*Pseudomonas syringae*) to ice-cream mix resulted in no difference in ice-crystal size between seeded and unseeded mixes (Hartel, 1996), this may indicate that sufficient seeding was already accomplished by conventional additives to ice-cream mix.

Further applications of microcrystalline cellulose as an ice nucleator should be explored, because it is a safe and widely accepted food additive. Because of safety concerns, there has been no sensory evaluation of foods frozen with INA suspensions to confirm suspected benefits to the quality of conventional frozen foods (Li and Lee, 1995).

11.8 Inhibition of Ice Crystallization and Recrystallization

In cellular foods the growth of ice crystals disrupts structure by both physical fracture and the osmotic pressures exerted by extracellular concentration of solutes as ice freezes out. Except for freeze-concentration effects, ice-crystal growth and especially ice-crystal size may be of little consequence in formulated foods unless the product has a gel structure, like tofu, or is consumed in the frozen state, like ice cream. Because ice-crystal size is so crucial to the "mouth feel" of ice cream and similar frozen confectionary, there has been considerable research to determine how various additives moderate the growth of ice crystals in these products. Conceivably some of this understanding might be applicable to other food systems in which ice-crystal growth is undesirable.

Ice crystal size (and in cellular foods, ice location; Love, 1962) can be regulated to some extent by the freezing rate. A rapid freezing rate favors nucleation at the expense of ice-crystal growth such that many small crystals are formed (George, 1993). While this is normally the most desirable state of ice in food products, too rapid a freezing rate can lead to intracellular ice nucleation in cellular foods such as meat and fish (Dyer, 1971; Love, 1988; Calvelo, 1981; Wagner and Añón, 1985). At first this presents no problems, but with the inevitable cycling of temperatures experienced in most commercial cold store facilities there will be ice-crystal growth, termed recrystallization, or in frozen confectionary, "heat shock."

For this reason, pressure-assisted freezing is being evaluated for future food-processing applications (Kalichevsky et al., 1995). The formation of ice crystals is associated with increased volume that, according to Le Chatelier's principle, would be restricted under increased pressure. Consequently, the freezing point of pressurized water is reduced, allowing products to be supercooled to very low temperatures without ice-crystal formation. For example, water pressurized at 200 MPa can be supercooled to about −20°C (Kalichevsky et al., 1995). The first step in pressure-assisted freezing is thus pressurization of an unfrozen sample and cooling the sample under pressure to a desired subzero temperature. Upon subsequent release of the pressure there is rapid and uniform ice-crystal formation because of an increased ice-nucleation rate resulting from supercooling (about 10-fold for each kelvin of supercooling). Ice crystals formed under these conditions are considerably smaller than those formed by freezing under normal atmospheric pressures.

However, the pressures that may be required to supercool foods sufficiently may also cause denaturation of proteins, a special problem in the freezing of meats and seafoods. Solute-exclusion theory would predict that common cryoprotective compounds may also baroprotect proteins. Such baroprotection has been demonstrated recently by Ashie and Lanier (1996) for several carbohydrate cryoprotectants on meat and fish myofibrillar proteins, illustrating that the solute-exclusion principle is more widely applicable to stabilization of proteins under a variety of conditions. A side benefit of using carbohydrates to baroprotect during pressure-assisted freezing is their ability to cryoprotect the product upon subsequent storage. Even maltodextrins showed mild baroprotective ability, in inverse relation to their mean molecular weight (Ashie and Lanier, 1996), as would be predicted by solute-exclusion theory. These compounds become more effective as cryoprotectants (both in protection against protein denaturation and in inhibiting ice recrystallization) as the storage temperature decreases because of their greater ability to enhance the viscosity of the unfrozen milieu (Parker and Ring, 1995; Carvajal, Lanier, and MacDonald, 1997), perhaps even allowing the glass state to be reached (Levine and Slade, 1988a,b).

Many other polymeric food additives besides maltodextrins have been found useful for controlling ice-crystal size in frozen confections. Investigations of these effects have led to debate over whether such substances primarily affect ice-crystal nucleation, growth, recrystallization, or a combination.

11.8.1 Modification of Ice Crystallization by Hydrocolloids

Muhr et al. (1986) concluded that conventional ice-cream stabilizers have an insignificant effect on the nucleation of ice crystals in frozen confectionary and that there is insufficient evidence that they retard ice-crystal growth in ice cream. A later publication by these authors (Muhr and Blanshard, 1986) indicated that gums that do not form a gel network slightly retard the rate of ice-crystal growth but do not affect ice morphology. They suggested that because all stabilizers affect ice growth similarly, these substances affect transport kinetics and not interface kinetics (i.e., adsorption to the ice interface, as had been suggested previously).

Budiaman and Fennema (1987a) found that guar and locust bean gum had the greatest effects in retarding ice crystallization compared to carrageenan and alginate. They noted that the former materials have numerous galactose side chains that facilitate hydration and probably hinder water crystallization, whereas the latter are linear polymers without side groups of substantial size. However, all of the gums they tested retarded ice crystallization less than sucrose, suggesting that this was not the mechanism whereby stabilizers effectively protect against heat shock. In a subsequent paper (Budiaman and Fennema, 1987b), these authors explored the relationship between viscosity and effect on ice-crystal growth. Linear crystallization rates induced by the various hydrocolloids were not constant at constant initial viscosity. Budiaman and Fennema also suggested that resistance to crystallization may depend on molecular associations—polymer-polymer and polymer-water—that affect transport kinetics (Parker and Ring, 1995). All of the hydrocolloids at 1-M concentration were highly ineffective retardants of crystal growth compared to sucrose.

Muhr and Blanshard (1986) found that hydrocolloids that develop even weak gel structures change the ice morphology and severely retard ice-crystal growth. The presence of gel networks does not greatly affect diffusion of micromolecules such as sucrose (Muhr and Blanshard, 1982) so the effect must be one of interaction between network and ice. The presence of a network mechanically impedes the growth of ice crystals, thus affecting their rate of propagation and their morphology. As ice propagates, gel fibers are subjected to a tension that may result in rupture, especially in coarse networks.

Nongelling polymeric carbohydrates that effectively inhibit large ice-crystal formation in frozen confectionary probably function at lower temperatures in abusive storage to prevent growth in the size of ice crystals or recrystallization. Such stabilizers do not influence the glass transition temperature, T_g, significantly (Levine and Slade, 1989). Goff et al. (1993) measured T_g' values below −30°C on ice creams containing these stabilizers, but such values are too low to attain the glass state in commercial storage. However, these authors believed that the demonstrated retardation of ice-crystal growth by these stabilizers resulted from their effects on diffusion kinetics. Such polymers could still be expected to slow all diffusion-related phenomena, such as ice recrystallization, at temperatures well above the glass transition temperature (Parker and Ring, 1995; Hartel, 1996). In support of this, Barringer (1997) found from DSC studies that guar gum decreased the amount of freezable water in model ice-cream solutions and that CMC suppressed the diffusion of D_2O molecules in sucrose solutions.

11.8.2 Inhibition of Moving Ice-Grain Boundary by Peptides

While the most celebrated property of the so-called antifreeze proteins and peptides is their noncolligative depression of freezing point, both antifreeze glycopeptides (AFGP) and peptides (AFP) also inhibit recrystallization on frozen storage. The degree of inhibition is concentration dependent, but amounts as small as 0.5 μg/mL give extensive inhibition (Feeney and Yeh, 1993). For example, recent research in New Zealand has shown that preslaughter administration of AFGP from Antarctic cod reduces ice-crystal size and drip loss in frozen lamb (Payne and Young, 1995). A patent by Warren et al. (1992) describes several other applications of AFP to foods, such as in controlling ice-crystal size in ice pops during abusive storage (Feeney and Yeh, 1993).

Knight et al. (1995) showed that small peptides (37 AA long) that were analogs of the antifreeze peptides but did not induce thermal hysteresis in solution more or less equally inhibited ice recrystallization in the absence of salts. Yet when salt levels were increased, even to levels below physiological saline (10 mM), only those peptides with antifreeze (freezing-point-lowering) activity were effective in inhibiting ice recrystallization. Such salts would induce liquid inclusions in the ice at −6°C. The ability of the nonantifreeze peptides to prevent recrystallization was explained as arising from the ability of particles to inhibit boundary movement of ice grains necessary for ice-grain coarsening. Yet in systems containing salts, typical of most food systems, peptides present would be concentrated in the liquid droplets, which would tend to move away from a moving ice-grain boundary. Antifreeze proteins in this liquid could migrate to the surface of the ice grain and attach, thereby inhibiting ice-crystal growth just as they do during supercooling. It is assumed from this work that ice recrystallization upon warming of vitrified (i.e., glass-state) samples would not be especially inhibited by antifreeze proteins because there would be no liquid inclusions at the ice-grain boundaries during the glass-to-ice transformation.

Rubinsky et al. (1992) noted that AFGP increase the viability of vitrified live cells upon thawing, but AFP do not. They assumed that the mechanism of enhancing viability has nothing to do with the antifreeze activity of AFGP but rather is some protective effect exerted by their binding to cell membranes, since the membranes were always more intact when samples were vitrified with AFGP than with AFP. They noted that ice-crystal growth that does occur develops primarily on the basal plane to which the antifreezes do not adsorb, resulting in very small, needlelike crystals.

Carpenter and Hansen (1992) found that low levels of AFP from winter flounder enhanced survival of red blood cells cryopreserved in hydroxyethyl starch solutions. The effect was most apparent when the samples were warmed at suboptimal rates where ice recrystallization would be exaggerated. Much higher concentrations of AFP were even more effective in inhibiting extracellular recrystallization, but extensive growth of ice around the cells and concomitant cell damage occurred. They proposed that a delicate balance exists between AFP inhibition of ice recrystallization and preferential growth of ice around the cells.

11.8.3 Alteration of Membrane Permeability

Schmitt et al. (1994) found that AFP isolated from acclimated cabbage leaves decreased solute permeability and resultant osmotic shock of spinach thylakoid membranes, leading to increased hardiness of the plants. The food scientist may see in this work a potential for cryoprotection of frozen vegetables to achieve better texture and appearance, as well as opportunities for AFP applications in cryoprotection and the production of AFPs (and other natural cryoprotectants) by acclimation of plants and animals.

Note that decreased membrane permeability is not always desired in cryoprotection, however. Alcohols enhanced the viability of frozen yeast (*S. cerevisiae*) cultures when very fast freezing rates were used, ostensibly because the alcohols induced increased membrane permeability, allowing rapid water equilibration during extracellular freezing and avoidance of intracellular ice-crystal formation (Lewis et al., 1994). At slower freezing rates alcohols acted as cryosensitizers.

The heat/cold-shock response is a universal phenomenon in biological systems (Schlesinger et al., 1982). It is a homeostatic response to protect cells against environmental temperature insult (Maresca et al., 1988). The response is only one of a class of general biological responses to stresses capable of producing changes in cellular metabolism, such as uncoupling oxidative phosphorylation, inhibition of electron transport, inhibition of select enzymes, recovery from anoxia, and so forth (Ashburner and Bonner, 1979).

Obuchi et al. (1993) showed that the amount of nonfrozen water in *Sacchraromyces cerevisiae* (yeast) cells almost doubled as a heat-shock response after treatment at 43°C for 30–60 min, associated with the production of heat-shock proteins. Kalichevsky et al. (1995) suggested that this might indicate the potential for producing such heat-shock proteins for use in modifying the ice-crystallization properties of foods. However, a recent review (Berry, 1996) notes that while production of both heat- and cold-shock proteins by organisms can confer resistance to loss of viability at low temperature, there is no research to support any role of these proteins in reduced freezing point and other factors. Rather, the heat/cold-shock proteins seem to alter membrane properties that confer greater resistance to loss of viability upon freezing.

This raises a point that has not been addressed well to date: freezing conditions and adding cryoprotectants that may confer greater quality to a food can also protect the viability of microorganisms. Certainly freezing is not commonly relied on as a pasteurization method, yet significant changes to the thawed shelf life of foods could result from cryoprotection of spoilage microorganisms in sufficient loadings. Conversely, cryoprotectants can protect the viability of useful fermentation starters (De Valdez and Diekmann, 1993).

11.9 Surimi: A Model for Food Cryoprotectant Study

Myosin, the primary myofibrillar constituent of animal muscle, forms a protein gel network when solubilized in saline and heated. This gel imparts texture by binding fat, water, and muscle particles and is the basis of many processed muscle foods such as frankfurters, bologna, and surimi-based foods including shellfish analog products and Asian kamaboko and fishballs (Acton and Dick, 1989).

Surimi is a semirefined concentrate of the myofibrillar proteins of fish muscle, which are more unstable than the muscle proteins of homeotherms (Connell, 1961). Thus, the manufacture of frozen surimi requires the addition of cryopreservative compounds to stabilize its functional properties (Shimizu and Fujita, 1985).

Many compounds have been evaluated for this role. Sucrose and sorbitol are primarily used commercially at present (MacDonald and Lanier, 1991). Noguchi (1974) surveyed a wide variety of chemical compounds for their ability to maintain the solubility of carp actomyosin in dilute solutions over brief periods of frozen storage. This model system was demonstrated to predict the ability of compounds to cryoprotect the functionality of surimi during extended frozen storage. Besides numerous carbohydrate compounds, including most of the mono- and disaccharides evaluated and several low-molecular-weight polyols, numerous amino acids and carboxylic acids were also found to be cryoprotective. Other workers have also reported the cryoprotective action of a number of amino acids, quaternary amines, and other compounds with regard to the stability of various proteins and enzymes (Jiang et al., 1987; Loomis et al., 1988, 1989).

As already noted, the nucleotides ATP, ADP, and IMP exert a protective effect on fish actomyosin stored at −20°C, while the nucleotide catabolites inosine and hypoxanthine destabilized these proteins (Jiang et al., 1987). This finding may help explain why fresher fish, with consequent higher concentrations of ATP, ADP, and IMP, are more stable during frozen storage than less-fresh fish (Fukuda et al., 1984; Dyer and Peters, 1969).

Watanabe et al. (1988) demonstrated the cryoprotective ability of certain surfactants, particularly certain polyoxyethylene sorbitan esters and sucrose esters, in preventing loss of gel-forming ability in surimi. A cryoprotective effect has even been attributed to triglycerides, since free fatty acids that may be released through hydrolysis of phospholipids and react to denature proteins are thought instead to preferentially react with triglyceride, thus indirectly protecting the proteins (Wessels et al., 1981). Careche and Tejada (1990a,b) later found that lipids were mainly cryoprotective because they prevented the formation of formaldehyde, a protein denaturant, in fish species known to experience enzymatic breakdown of trimethylamine (Haard, 1992). MacDonald et al. (1990) had found that added sugar not only directly stabilized fish proteins but also inhibited TMAO breakdown during frozen storage of minces.

Mechanistic studies on the cryopreservative properties of these and other compounds for stabilizing fish muscle proteins have paralleled those in pharmaceutical and other nonfood applications and could lead to the development of more efficacious mixtures of compounds for the cryopreservation of food proteins and proteinaceous cellular foods. Our studies using sodium lactate (SL) to stabilize and cryoprotect fish actomyosin in model systems illustrated well the mechanism whereby low-molecular-weight additives stabilize proteins (MacDonald and Lanier 1994; MacDonald, Lanier, Swaisgood, and Hamann, 1996). In these studies we compared the effectiveness of SL to sucrose at various concentrations. Interestingly, SL had an optimum concentration for both freeze-thaw and heat stabilization, and the explanation for this anomalous behavior reinforces the key role of surface tension in stabilizing proteins in solution.

11.9.1 Freeze-Thaw Studies with Sodium Lactate in Fish Actomyosin

Myosin molecules consist of two regions: two globular heads and a fibrous rod. The heads are the site of the ATPase activity important in muscle contraction, while the rod, which has no enzyme activity, is involved in packing the myosin molecules into a filament (Rawn, 1989). While both sections of the myosin molecule can be involved in gel formation (Samejima et al., 1988), conformational changes usually occur in the head region at a temperature somewhat lower than that at which unfolding takes place in the rod (Acton and Dick, 1989; Sharp and Offer, 1992; Yamamoto, 1990). Thus, myosin ATPase activity can be used as a sensitive indicator of myosin denaturation related to unfolding of the tertiary structure in the head region. It follows that ATPase activity is also related directly to the functional property of gelation (Fukuda, 1989).

Figure 11.4 shows the results of freeze-thaw experiments on denaturation of tilapia (*Tilapia nilotica* × *Tilapia aurea*) actomyosin in the presence and absence

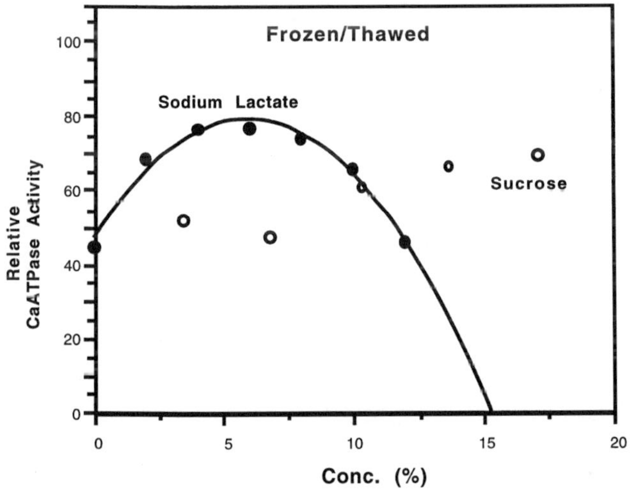

Figure 11.4. Cryoprotection of tilapia actomyosin by sodium lactate and sucrose. Actomyosin (2.5 mg/mL, pH 7.0, with 25-mM Tris-maleate in 0.6-M KCl) was frozen in liquid nitrogen for 3 min then held in a bath at −5°C for 60 min and thawed for 5 min at 25°C. Unfrozen control was taken as 100% Ca^{2+}ATPase activity. Results are the average of replicate experiments (MacDonald and Lanier, 1994).

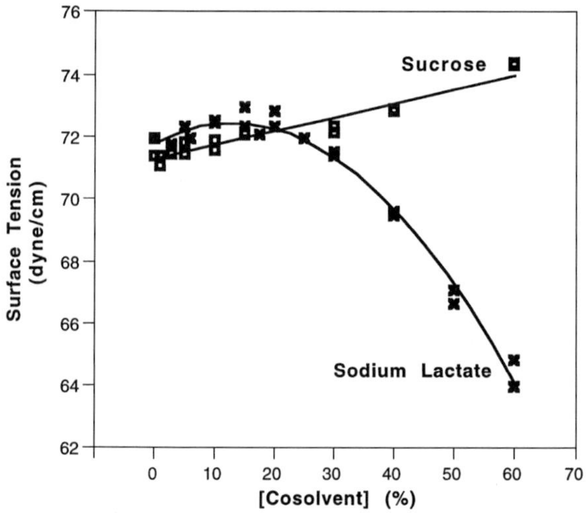

Figure 11.5. Surface tension of sodium lactate and sucrose solutions at 25°C as a function of concentration (Data from MacDonald, Lanier, Swaisgood, and Hamann, 1996.)

of two cosolvents, sucrose and sodium lactate. A loss of Ca^{2+}ATPase activity shows the extent of denaturation during freeze-thawing. The cryoprotective effect increased to a maximum at about 6% (w/v) SL concentration with 80% of the activity recovered. An optimum concentration for sodium lactate has also been observed in heat denaturation of actomyosin; however, maximal stabilization occurred at concentrations of 15 and 25%, variations in optimum being dependent on the method used to measure denaturation (MacDonald and Lanier, 1994; MacDonald, Lanier, Swaisgood, and Hamann, 1996). The level of cryoprotection observed in the tilapia freeze-thaw experiments for sucrose, on the other hand, increased monotonically with sugar concentration (fig. 11.5). On a percentage basis, SL appeared to be about four times more effective than sucrose since 25% sucrose would be needed to give an equivalent degree of cryoprotection to the optimum for SL. Note that this is true despite the lower molecular weight of lactate compared to sucrose, highlighting the nonideality of lactate in its interaction with water and the role it plays in cryoprotection. Surface-tension measurements showed that the concentration of SL that corresponded to stabilization and destabilization of heated actomyosin closely correlated with concentrations where surface tension either increased or decreased (fig. 11.5). In comparison, surface tension of sucrose solutions increased with increasing concentration, which also correlated with previous freezing-thawing and heat denaturation studies showing a similar relationship between stability and concentration (fig. 11.5). Surface tensions of sucrose solutions were also lower than the corresponding SL solutions up to 20% SL.

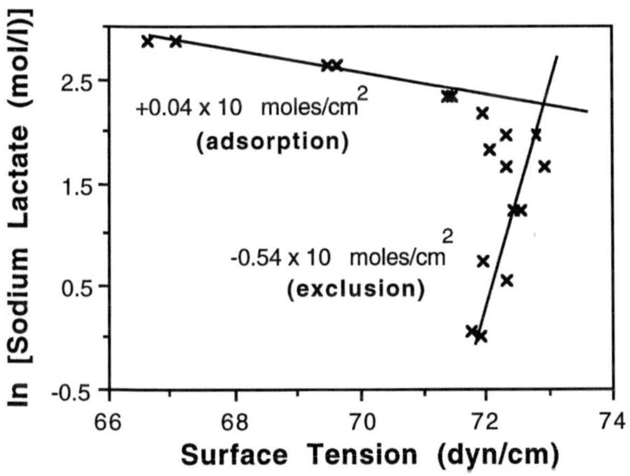

Figure 11.6. Gibbs adsorption isotherm relating sodium lactate concentration and surface tension of solutions (Data from MacDonald, Lanier, Swaisgood, and Hamann, 1996.)

It is possible to estimate the concentration of solute present at the interface compared to the bulk concentration by approximating the increasing and decreasing portions of the surface-tension curve with linear models and using the expanded form of equation 11.4, Gibbs adsorption isotherm (Tinoco et al., 1985), where chemical potential is approximated by RT ln *c*.

$$\Gamma = -\frac{1}{RT}\frac{d\gamma}{d\ln c} \qquad \textbf{(11.6)}$$

where Γ = adsorption (excess concentration) of solute at surface, mol m^{-2}
γ = surface tension, N m^{-1}
R = gas constant = 8.314 J deg^{-1} mol^{-1}
c = concentration of solute in bulk solution

Thus, using equation 11.6 and the linear models from figure 11.6, it was estimated that from 0 to 17.5% the concentration of SL at the interface was -0.54×10^{-10} mol/cm^2. The negative value shows that solute was excluded from the interface compared to the bulk concentration. Using the same procedure, adsorption of solute at the interface for the decreasing surface-tension portion of the graph (25–60% SL) was estimated to be 0.04×10^{-10} mol/cm^2, which is low compared to values of $0.2–4.0 \times 10^{-10}$ mol/cm^2 for typical detergents (Rosen, 1978).

To verify the hypothesis that SL stabilizes mainly by increasing the surface tension of the solvent, we measured the surface tension of solutions of different SL concentrations at the corresponding thermal transition (T_r) for tilapia actomyosin. The resultant plot of surface tension versus concentration and temperature

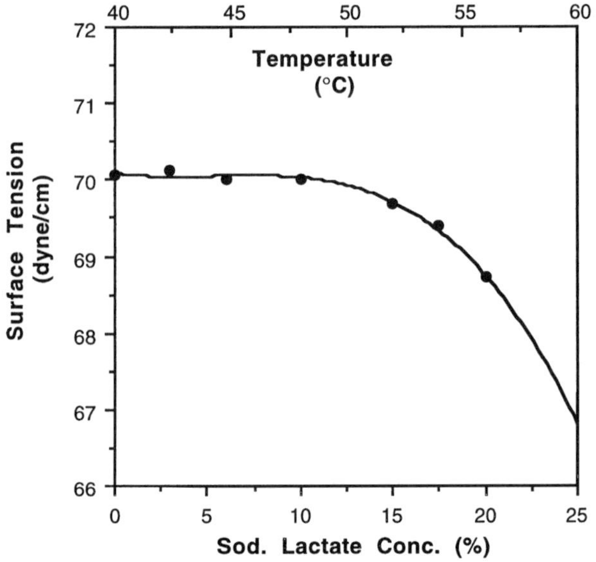

Figure 11.7. Surface tension as a function of temperature and sodium lactate concentration. The combination of sodium lactate concentration and temperature was derived from the thermal transition temperature (T_r) for aggregation of tilapia actomyosin at various concentrations of sodium lactate (MacDonald, Lanier, Swaisgood, and Hamann, 1996.)

showed that up to about 15%, the surface tension was—within experimental error—close to constant (fig. 11.7). At higher SL concentrations and temperatures, in the region where previously actomyosin proteins were destabilized, surface tension was lower ($p < 0.05$). These results verify, experimentally, the results of Lee and Timasheff (1981), who calculated a similar relationship between the surface tension at the transition temperature and sucrose concentration for denaturation of three proteins. The increase in actomyosin T_r in the presence of SL therefore appears to be related closely to the need for lowering the surface tension at the protein-solvent interface to a level at which the free energy change provided by the protein expansion is sufficient to overcome the pressure of the solvent that counteracts the process.

At higher concentrations of sodium lactate (25–60%), surface tension again played a role in the observed effects. At these concentrations, however, polymerization of lactate to the dimer lactoyllactate (Holten et al., 1971) had occurred. Inspection of the chemical structure of lactoyllactate (fig. 11.8) shows that it is amphiphilic, exhibiting one side that is more hydrophobic and the other more hydrophilic. Hence lactoyllactate would be expected to preferentially absorb at an air-water interface and lower surface tension. This would then explain the rela-

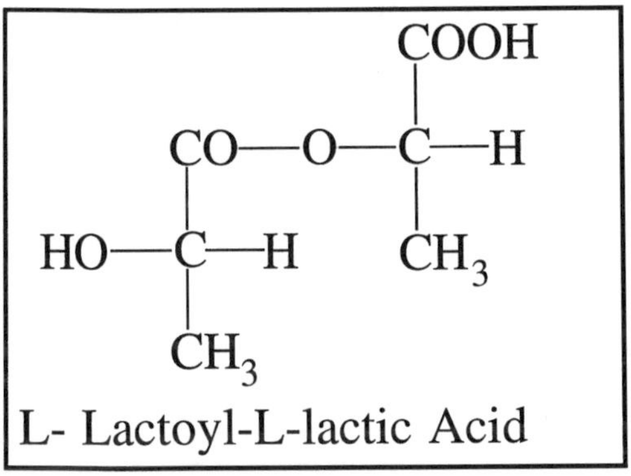

Figure 11.8. Structure of lactoyllactate ester

tionship between surface tension and SL concentration, since as the concentration of lactoyllactate increased with SL concentration, surface tension would be expected to decrease.

Lactate, thus, is an example of compounds that cryoprotect and otherwise stabilize proteins by inducing conditions that thermodynamically favor the maintenance of the native protein state. Sodium lactate is a particularly useful example of an additive that stabilizes by the preferential exclusion mechanism because higher concentrations of SL destabilized actomyosin and the transition from stabilization to destabilization was associated with either an increase or decrease in solution surface tension with SL concentration.

Alternately, certain high-molecular-weight polymers may cryostabilize proteins by impeding molecular mobility in the system at conventional freezer temperatures. The surimi or fish actomyosin system has also served as a model for the testing of this mechanism, originally by Lim (1989) and coworkers and more recently by us.

11.9.2 Validity of the Glass-Formation Mechanism in Cryoprotection

Carvajal, Lanier, and MacDonald (1997) recently evaluated the cryoprotectant properties of a range of maltodextrins (DE 5–25) versus sucrose or a sucrose-sorbitol mixture (1:1) (table 11.1) when added (8% level) to freshly washed Alaska pollock mince, including a control with no additives. Minces were either stored for 3 mo at −8, −14, and −20°C or subjected to six repeat freeze-thaw (F/T) cycles, where blocks of mince were tempered for 8 hr at 4°C and refrozen (16 hr at the appropriate temperature). After each storage or freeze-thaw treatment the Ca^{2+}ATPase of actomyosin extracts was measured as an index of protein denaturation.

Table 11.1. Relationship between maltodextrin DE, average molecular weight, and glass transition temperature (Tg′) (from Grain Processors Corporation, Muscatine, Iowa.)

Maltodextrin	DE	Mean MW	T_g'
M050	5	3,600	−5
M100	10	1,800	−11
M150	15	1,200	−13
M180	18	1,000	−15
M200	20	900	−16
M250	25	720	−19
Sucrose	N/A	350	−40

ATPase activities of treatments at the initiation of testing (following 6-wks storage at −20°C) were slightly higher for the sugar/sorbitol treatments compared with the maltodextrins, with only slight effect of MW on maltodextrin effectiveness (fig. 11.9). When comparing the activities of these treatments with that of the control, it is clear that during the freezing process and initial isothermal storage, maltodextrins had a significant cryoprotective effect.

Figure 11.10 plots the residual ATPase activity, as a percentage of the initial activity remaining after 3-mo additional frozen storage or six F/T cycles, as affected by average molecular weight of the carbohydrate and storage temperature. The 3-mo storage data varied according to storage temperature while, as expected, freezing temperature had little effect on F/T cycling effects. The 3-mo storage data exhibited a slight decline in the remaining enzyme activity as MW of the cryoprotectant increased, an effect that was more pronounced at the higher storage temperature. In contrast, for samples subjected to F/T cycling, the residual activity was clearly reduced as the molecular weight of the cryoprotectant increased.

Note that some residual ATPase values exceed 100%. This increase in ATPase activity likely reflects an initial unfolding that results in subsequent aggregation. Mornet et al. (1989) discussed the need for flexibility in the myosin head for both binding ATP and ATP hydrolysis. Cheung (1969) showed that maximum activity of myosin ATPase was gained with low urea concentrations (less than 2 M urea), whereas increasing the concentration above 3 M urea resulted in rapid inactivation. MacDonald, Lanier, and Giesbrecht (1996) also showed that the addition of low concentrations of zinc ions to actomyosin enhanced ATPase activity whereas with zinc concentrations above 0.1 M ATPase activity was reduced and denaturation and aggregation was induced.

Cryoprotection of the maltodextrins relative to that of sucrose/sorbitol treatments increased as the storage temperature was decreased. Yet sucrose/sorbitol treatments demonstrated their capacity to cryoprotect nearly equally well during

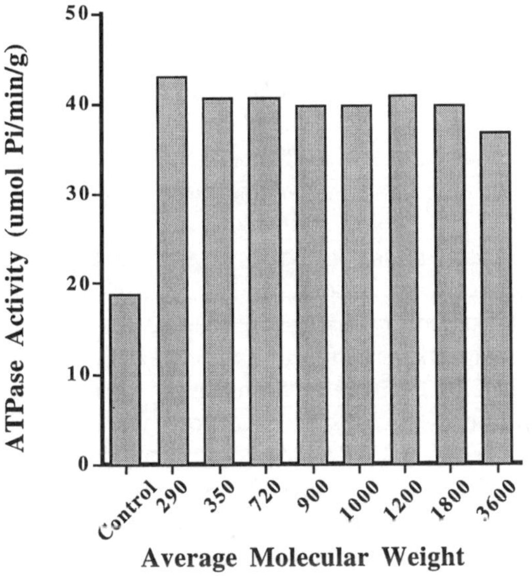

Figure 11.9. ATPase activity of surimi at initial evaluation, 6 wk following freezing (From Carvajal, Lanier, and MacDonald, 1997.)

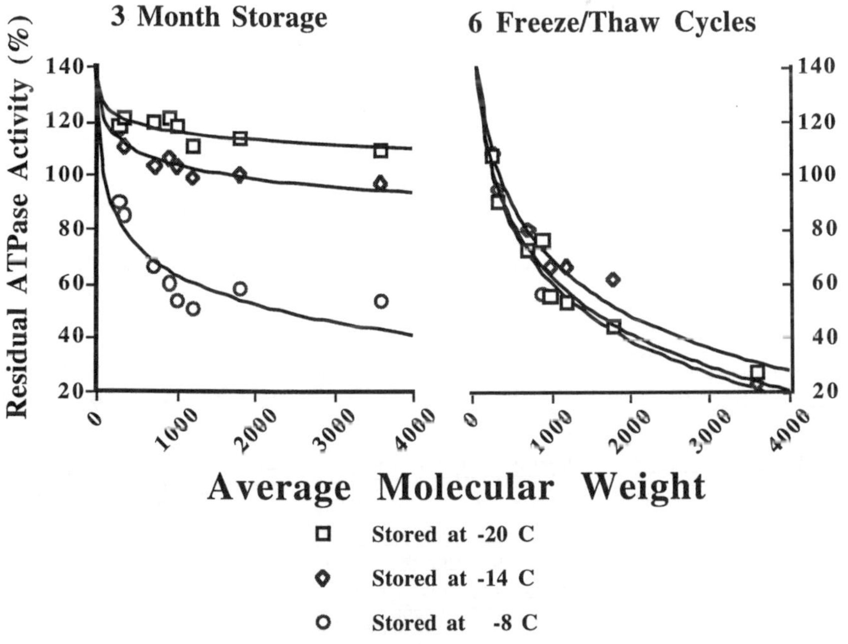

Figure 11.10. Residual ATPase activity following additional storage or freeze/thaw cycling, relative to values in figure 11.9 (From Carvajal et al., 1997.)

both isothermal storage and F/T cycling. The cryoprotectancy of maltodextrins was nearly equal to that of sucrose/sorbitol at −20 and −14°C, slightly decreasing with increasing mean MW. At −8°C isothermal storage, and especially under F/T cycling at any temperature, cryoprotectancy diminished markedly with increasing mean MW of the cryoprotectant.

This evidence supports dual mechanisms for cryoprotection by higher-MW carbohydrates, which vary depending on their relative MW. The lower-MW polymers enhance surface tension of the solution in conjunction with the preferential exclusion mechanism for protein stabilization. MacDonald (in Carvajal et al., 1997) recently measured the effects of increasing concentration on surface tension of starch hydrolysis products versus sucrose (fig. 11.11). Surface tension increased with concentration for 15- and 20-DE maltodextrins similar to sucrose. This lends credence to the supposition that the lower-mean-MW starch hydrolysis products may cryoprotect (at least in part) by the same mechanism (preferential exclusion) as sucrose. Evidence for this theory is found in their cryoprotectant effect under F/T cycling (fig. 11.10). The decreasing effectiveness of maltodextrins of increasing mean MW to stabilize proteins under F/T cycling is an indication that restriction of molecular mobility contributes little to protein stabilization under those conditions.

Yet the higher MW maltodextrins were effective cryoprotectants at storage temperatures near or below their respective T_g' values where their effect on molecular mobility of the system should be dramatic. It is unlikely that the 4- and 10-DE maltodextrins, which cryoprotected well during isothermal storage at −14 and −20°C (fig. 11.10), acted via a preferential exclusion mechanism since these were ineffective protectants in F/T cycling and 10-DE maltodextrin does not increase the surface tension of solutions (fig. 11.11).

Actually the addition of maltodextrins, especially the low-DE types, at effective levels can detrimentally affect gelation of surimi (MacDonald and Lanier, 1991; Carvajal et al., 1997) presumably due to competition for water, thus posing a practical limitation on the use of maltodextrins in products where gelation is an important functional property. The infusion of intact meats with cryoprotectants shows much promise as a means of stabilizing their proteins to the denaturing effects of frozen storage (Carvajal and Lanier, 1997). In such applications maltodextrins would have advantages in low cost and lack of sweetness. Infusion of high-MW maltodextrins within the cells of intact muscle might not be required if lower-molecular-weight cryoprotectants that penetrate the membranes to protect the proteins were also added (Carvajal and Lanier, 1997). The maltodextrins could situate extracellularly and thereby control the ice-crystal growth that occurs primarily in this location under the slow freezing and fluctuating storage conditions employed industrially (Love, 1962). Added sweetness of the low-molecular-weight components could be minimized by the use of less-sweet polyols such as lactitol or by dextrose if browning is not a problem.

Apart from the deleterious changes, such as enzymatic processes, associated with phase changes as muscle or minces are frozen and thawed, it is possible that

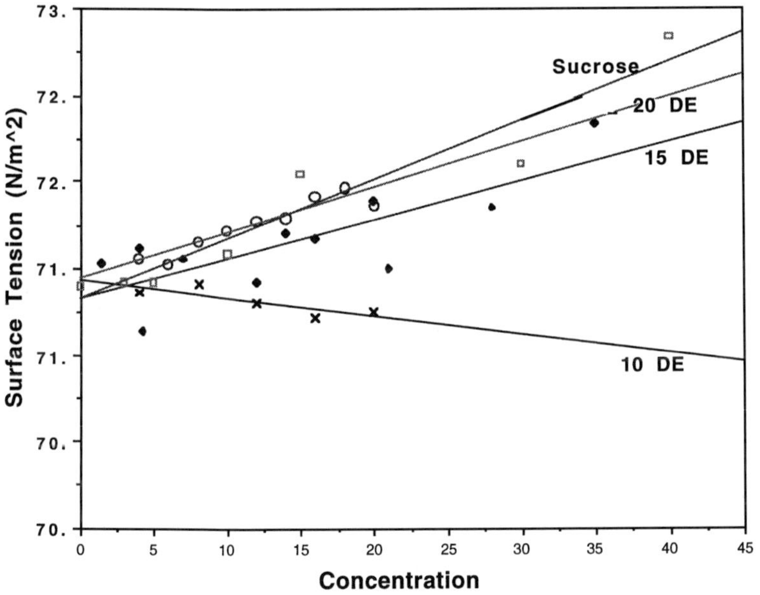

Figure 11.11. Surface tension of sucrose or maltodextrins (10–20 DE) as a function of concentration (From Carvajal et al., 1997.)

cold destabilization of myofibrillar proteins (Privalov, 1990) is a major factor in the instability of fish proteins. Therefore, some of the principal mechanisms for myosin denaturation, aggregation, and the subsequent loss of functionality or quality may not be diffusion controlled and hence may not be greatly affected by glass formation. Likewise, very small molecules such as oxygen may yet be able to diffuse within the glass formed at or below the T_g of the system such that oxidation of lipids and proteins may go unhindered (Fennema, 1996). Thus, total reliance on the restriction of molecular mobility by high-MW components to achieve cryoprotection may be unadvisable. For these and other practical reasons we again suggest developing cryoprotectant "cocktails" that are designed specifically for each food and its processing/storage environment (MacDonald and Lanier, 1991).

11.10 Conclusions

Cryoprotection of foods is a developing field with many potential new applications. For example, the great success of cryoprotected fish mince, or surimi, in the fishing industry has stimulated research efforts in the red-meat industries to apply similar technology to gain quality or other functional benefits from frozen red-meat raw materials (Park, Lanier, Hamann, and Keeton, 1987). The U.S. beef industry imports about 3% of its meat from New Zealand and about 6% from Australia. This meat is frozen, and recent research from this laboratory has shown that it could potentially be cryoprotected to improve functionality (Park, Lanier, Hamann, and Keeton, 1987; Park, Lanier, Swaisgood, Hamann, and Keeton, 1987; Park et al., 1993).

Although examples of the use of cryoprotectants to improve the stability of foods (e.g., surimi, ice cream, sugared eggs or yolks; for the latter see Ijichi et al., 1970; Cotterill, 1986; Stadelman et al., 1988) are familiar to many, the concept of cryoprotection has not yet gained wide acceptance in the food industry. Cost of additives and additional processing are a factor, as is the perception that cryoprotectants are generally sweet compounds that may adversely affect flavor and in some cases, color during heating due to Maillard browning. There is also a considerable trend away from the use of additives or preservatives in foods that might discourage investigations of cryoprotection. But it is likely that many potential food applications for cryoprotectants have not yet been thoroughly investigated. Cryoprotectant additives must be viewed as adjuncts to handling procedures and material selection that optimize the inherent stability of the food system.

In the pharmaceutical industry cryoprotectant use is very common. Vitrification and cryoprotectant use is also being explored for the storage of human organs and tissues. It is particularly notable that many organisms, some as complex as frogs, survive cold climates by the natural production of endogenous cryoprotectants (Costanzo et al., 1991). Since these situations represent both isolated proteins in solution and cellular systems it is worthwhile noting the concepts or practices that may relate to foods.

As illustrated by the suggestion that a combination of high- and low-molecular-weight compounds might be efficacious in the cryoprotection of intact muscle, in the future we may be able to combine various compounds to create cryoprotectant cocktails, which may be more effective than single compounds alone. We should especially explore interactions between high- and low-molecular-weight carbohydrates and certain ions, organic acids, and/or amino acids to uncover possible synergies. For example, Carpenter et al. (1988) found a remarkable synergy between sugars and zinc ions in the stabilization of phosphofructokinase.

Delivery of cryoprotectants, particularly in cellular foods, could pose a significant challenge. Consequently, there may be potential to either genetically manipulate the raw-material organism to enhance frozen-storage stability or induce cryoprotectant production before harvest by stimulating the organism's own metabolism in response to various stresses (e.g., cold, heat, and anoxia). This, along with the hurdles of cryoprotectant cocktail addition and optimized freezing/thawing conditions, such as pressure-assisted freezing and thawing, could significantly increase our ability to deliver fresh quality in foods previously frozen.

References

Acton, J.C. and Dick, R.L. 1989. Functional roles of heat induced protein gelation in processed meat. In *Food Proteins*. (Kinsella, J.E. and Soucie, W.G., eds.) pp. 195–209, AOCS: Protein and Coproducts Div., Champaign, Ill.

Arai, S., Watanabe, M., and Tsuji, R.F. 1984. Enzymatically modified gelatin as an antifreeze protein. *Agric. Biol. Chem.* **48:**2173–2175.

Arakawa, T. and Timasheff, S.N. 1982a. Preferential interactions of proteins with salts in concentrated solutions. *Biochem.* **21**(25):6545–6552.

Arakawa, T. and Timasheff, S.N. 1982b. Stabilization of protein structure by sugars. *Biochem.* **21**(25):6536–6544.

Arakawa, T. and Timasheff, S.N. 1984a. Mechanism of protein salting in and salting out by divalent cation salts: balance between hydration and salt binding. *Biochem.* **23**(25): 5912–5923.

Arakawa, T. and Timasheff, S.N. 1984b. Protein stabilization and destabilization by guanidinium salts. *Biochem.* **23**(25):5924–5929.

Arakawa, T. and Timasheff, S.N. 1985. The stabilization of proteins by osmolytes. *Biophys. J.* **47:**411–414.

Ashburner, M. and Bonner, J.J. 1979. The induction of gene activity in Drosophila by heat shock. *Cell.* **17**:241–254.

Ashie, I.N.A. and Lanier, T.C. 1996. Baro- and cryo-protection of proteins in pressure-assisted freezing of seafoods. *Proceedings of the Joint Atlantic Fisheries Technologists and Seafood Science Americas Conference,* University of Florida, Gainesville, FL. Sea Grant. In press.

Aunaas, T. 1982. Nucleating agents in the haemolymph of an intertidal mollusc tolerant to freezing. *Experientia.* **38:**1456–1457.

Barringer, S.A. 1997. Personal communication.

Berry, E.D. 1996. Characterization of the cold shock response of *Bacillus cereus.* Ph.D. Thesis. North Carolina State University, Raleigh.

Budiaman, E.R. and Fennema, O. 1987a. Linear rate of water crystallization as influenced by temperature of hydrocolloid suspensions. *J. Dairy Sci.* **70:**534–546.

Budiaman, E.R. and Fennema, O. 1987b. Linear rate of water crystallization as influenced by viscosity of hydrocolloid suspensions. *J. Dairy Sci.* **70:**547–554.

Calvelo, A. 1981. Recent studies on meat freezing. In *Developments in Meat Science,* vol. 2. (Lawrie, R., ed.) pp. 125–158, Applied Science Publishers, Barking, UK.

Careche, M. and Tejada, M. 1990a. The effect of neutral and oxidized lipids on functionality in hake (*Merluccius merluccius L.*): a dimethylamine- and formaldehyde-forming species during frozen storage. *Food Chem.* **36:**113–128.

Careche, M. and Tejada, M. 1990b. Effect of neutral and oxidized lipids on protein functionality in megrim (*Lepidorhombus whiffiagonis W.*) and sardine (*Sardina pilchardus R.*) during frozen storage. *Food Chem.* **37:**275–287.

Carpenter, J.F. and Crowe, J.H. 1988. The mechanism of cryoprotection of proteins by solutes. *Cryobiol.* **25:**244–255.

Carpenter, J.F., Crowe, L.M., and Crowe, J.H. 1987. Stabilization of phosphofructokinase with sugars during freeze-drying: characterization of enhanced protection in the presence of divalent cations. *Biochim. Biophys. Acta* **923:**109–115.

Carpenter, J.F., Hand, S.C., Crowe, L.M., and Crowe, J.H. 1986. Cryoprotection of phosphofructokinase with organic solutes: characterization of enhanced protection in the presence of divalent cations. *Arch. Biochem. Biophys.* **250:**505–512.

Carpenter, J.F. and Hansen, T.N. 1992. Antifreeze protein modulates cell survival during cryopreservation: mediation through influence on ice crystal growth. *Proc. Natl. Acad. Sci. USA.* **89:**8953–8957.

Carpenter, J.F., Martin, B., Crowe, L.M., and Crowe, J.H. 1987. Stabilization of phosphofructokinase during air-drying with sugars and sugar/transition metal mixtures. *Cryobiol.* **24:**455–464.

Carpenter, J.F., Martin, B., Loomis, S.H., and Crowe, J.H. 1988. Long-term preservation of dried phosphofructokinase by sugars and sugar/zinc mixtures. *Cryobiol.* **25:**372–376.

Carpenter, J.F., Petrelski, S.J., Anchordoguy, T.J., and Arakawa, T. 1994. Interactions of stabilizers with proteins during freezing and drying. In *Formulation and Delivery of Proteins and Peptides,* vol. 567. (Cleland, J.L. and Langer, R., eds.) pp. 134–147, American Chemical Society, Washington, D.C.

Carvajal, P.A. and Lanier, T.C. 1997. Cryoprotection of intact muscle by carbohydrate infusion. *J. Food Sci.* In press.

Carvajal, P.A., Lanier, T.C., and MacDonald, G.A. 1997. Cryoprotection of Alaska pollock surimi by maltodextrins. *Cryobiol.* In press.

Cheung, H.C. 1969. Conformation of myosin: effects of substrates and modifiers. *Biochem. Biophys. Acta* **194:**478– 484.

Connell, J.J. 1961. The relative stabilities of the skeletal muscle myosins of some animals. *Biochem. J.* **80:**503–509.

Costanzo, J.P., Lee, R.E., and Wright, M.F. 1991. Glucose loading prevents freezing injury in rapidly cooled wood frogs. *Am. J. Physiol.* **261:**R1549–R1553.

Cotterill, O.J. 1986. Freezing egg products. In *Egg Science and Technology,* 3d ed. (Stadelman, W.J. and Cotterill, O.J., eds.) pp. 217–242, AVI Publishing, Westport, Conn.

DeValdez, G.F. and Diekmann, H. 1993. Freeze-drying conditions of starter cultures for sourdoughs. *Cryobiol.* **30:**185–190.

DeVries, A.L. 1988. The role of antifreeze glycopeptides and peptides in the freezing avoidance of Antarctic fishes. *Comp. Biochem. Physiol.* **90B:**611–621.

Dyer, W.J. 1971. Speed of freezing and quality of frozen fish. In *Fish Inspection and Quality Control.* (Kreuzer, R., ed.) pp. 75–81, Fishing News Books, London.

Dyer, W.J. and Peters, J. 1969. Factors influencing quality changes during frozen storage and distribution of frozen products, including glazing, coating, and packaging. In *Freezing and Irradiation of Fish.* (Kreuzer, R., ed.) Fishing News Books, London.

Eastman, J.T. and DeVries, A.L. 1986. Antarctic fishes. *Sci. Am.* **255:**106–117.

Fahy, G.M., MacFarlane, D.R., Angell, C.A., and Meryman, H.T. 1984. Vitrification as an approach to cryopreservation. *Cryobiol.* **21:**407–426.

Fahy, G.M., Levy, D.I., and Ali, S.E. 1987. Some emerging principles underlying the physical propertries, biological actions, and utility of vitrification solutions. *Cryobiol.* **24:**196–213.

Feeney, R.E. and Yeh, Y. 1993. Anti-freeze proteins: properties, mechanisms of action, and possible applications. *Food Technol.* **47:**82–90.

Fennema, O. 1996. Water and ice. In *Food Chemistry,* 3d ed. (Fennema, O., ed.) pp. 1–94, Marcel Dekker, New York.

Franks, F. 1985a. *Biophysics and Biochemistry at Low Temperatures.* Cambridge University Press, UK.

Franks, F. 1985b. Complex aqueous systems at subzero temperatures. In *Properties of Water in Foods.* (Simantos, D. and Multon, J.L., eds.) pp. 497–509, Martinus Nijhoff Publishers, Dordrecht, The Netherlands.

Fukuda, Y. 1989. ATPase as an indicator of surimi quality. Paper presented at Pacifichem 1989; subsymposium: Use of fish minces and surimi products by the food industry: fish meat as raw material, Hawaii.

Fukuda, Y., Tarakita, Z., and Arai, K. 1984. Effect of freshness of chub mackerel on the freeze denaturation of myofibrillar protein. *Bull. Jpn. Soc. Sci. Fish.* **50**(5):845–852.

Gekko, K. and Morikawa, T. 1981. Thermodynamics of polyol-induced thermal stabilization of chymotrypsinogen. *J. Biochem.* **90**(1):51–60.

Gekko, K. and Timasheff, S.N. 1981. Mechanism of protein stabilization by glycerol: preferential hydration in glycerol-water mixtures. *Biochem.* **20:**4667–4676.

George, R.M. 1993. Freezing processes used in the food industry. *Trends Food Sci.Technol.* **4:**134–138.

Goff, H.D., Caldwell, K.B., and Stanley, D.W. 1993. The influence of polysaccharides on the glass transition in frozen sucrose solutions and ice cream. *J. Dairy Sci.* **76:**1268 1277.

Gordon, M. and Taylor, J.S. 1952. Ideal copolymers and the second-order transitions of synthetic rubbers. I. Non-crystalline copolymers. *J. Appl. Chem.* **2:**493–501.

Haard, N.F. 1992. Biochemical reactions in fish muscle during frozen storage. In *Seafood Science and Technology.* (Bligh, E.G., ed.) pp. 176–209, Fishing News Books, Oxford.

Hartel, R.W. 1996. Ice crystallization during the manufacture of ice cream. *Trends Food Sci. Technol.* **7:**315–321.

Hazen, K.C., Bourgeois, L.D., and Carpenter, J.F. 1988. Cryoprotection of antibody by organic solutes and organic solute/divalent cation mixtures. *Arch. Biochem. Biophys.* **267:**363–371.

Hightower, K.E. and McCarty, R.E. 1996. Influence of nucleotides on the cold stability of chloroplast coupling factor 1. *Biochem.* **35:**10051–10057.

Hirsh, A.G. 1987. Vitrification in plants as a natural form of cryoprotection. *Cryobiol.* **24:**214–228.

Holten, C.H., Muller, A., and Rehbinder, D. 1971. *Lactic Acid: Properties and Chemistry of Lactic Acid and Its Derivatives.* Verlag Chemie GmbH, Weinheim.

Ijichi, K., Palmer, H.H., and Lineweaver, H. 1970. Frozen whole eggs for scrambling. *J. Food Sci.* **35:**695–698.

Ishimori, H. 1996. Vitrification of mouse and bovine embryos using a mixture of ethylene glycol and dimethyl sulfoxide. *Snow Brand R&D Rep.* **106:**115–152.

Jiang, S.T., Hwang, B.O., and Tsao, C.T. 1987. Protein denaturation and changes in nucleotides of fish muscle during frozen storage. *J. Agric. Food Chem.* **35:**22–27.

Kalichevsky, M.T., Knorr, D., and Lillford, P.J. 1995. Potential food applications of high-pressure effects on ice-water transitions. *Trends Food Sci. Technol.* **6:**253–259.

Knight, C.A. and Duman, J.G. 1986. Inhibition of recrystallization of ice by insect thermal hysteresis proteins: a possible cryoprotective role. *Cryobiol.* **23:**256–262.

Knight, C.A., Wen, D., and Laursen, R.A. 1995. Nonequilibrium antifreeze peptides and the recrystallization of ice. *Cryobiol.* **32:**23–34.

Kresin, M., Heschel, I., and Rau, G. 1994. Expectations and limitations of vitrification as a method for cryopreservation. *Cryobiol.* **30:**573.

Kumazawa, Y., Oozaki, Y., Iwami, S., Matsumoto, I., and Arai, K. 1990. Combined protective effect of inorganic pyrophosphate and sugar on freeze-denaturation of carp myofibrillar protein. *Nippon Suisan Gak.* **56**(1):105–113.

Lanier, T.C. and Akahane, T. 1986. Method of retarding denaturation of meat products. U.S. patent no. 4,572,838.

Lee, R. 1991. Principles of insect low temperature tolerance. In *Insects at Low Temperature.* (Lee, R.E. and Denlinger, D.L., eds.) pp. 17–41, Chapman & Hall, NewYork.

Lee, J.C. and Timasheff, S.N. 1981. The stabilization of proteins by sucrose. *J. Biol. Chem.* **256**(14):7193–7201.

Leistner, L. 1978. Hurdle effect and energy saving. In *Food Quality and Nutrition.* (Downey, W.K., ed.) pp. 553–557, Elsevier, Essex, UK.

Levine, H. and Slade, L. 1986. A polymer physico-chemical approach to the study of commercial starch hydrolysis products (SHPs). *Carbohydr. Polym.* **6:**213–244.

Levine, H. and Slade, L. 1988a. A food polymer science approach to the practice of cryostabilization technology. *Comments Agric. Food Chem.* **1:**315–396.

Levine, H. and Slade, L. 1988b. Principles of "cryostabilization" technology from structure/property relationships of carbohydrate/water systems: a review. *Cryo-Lett.* **9:**21–63.

Levine, H. and Slade, L. 1989. Response to the letter by Simatos, Blond, and Le Meste on the relation between glass transition and stability of a frozen product. *Cryo-Lett.* **10:**347–370.

Lewis, J.G., Learmonth, R.P., and Watson, K. 1994. Cryoprotection of yeast by alcohols during rapid freezing. *Cryobiol.* **31:**193–198.

Li, J. and Lee, T.-C. 1995. Bacterial ice nucleation and its potential application in the food industry. *Trends Food Sci. Technol.* **6:**259–265.

Lim, M.H. 1989. Studies of reaction kinetics in relation to thermal behavior of solutes in frozen systems. Ph.D. Thesis. University of California–Davis.

Loomis, S.H., Carpenter, T.J., Anchordoguy, T.J., Crowe, J.H., and Branchini, B.R. 1989. Cryoprotective capacity of end products of anaerobic metabolism. *J. Exp. Zoo.* **252:**9–15.

Loomis, S.H., Carpenter, T.J., and Crowe, J.H. 1988. Identification of strombine and taurine as cryoprotectants in the intertidal bivalve *Mytilus edulis. Biochim. Biophys. Acta* **94:**113–118.

Love, R.M. 1962. New factors involved in the denaturation of frozen cod muscle protein. *J. Food Sci.* **27:**544–550.

Love, R.M. 1988. *The Food Fishes: Their Intrinsic Variation and Practical Implications.* pp. 130–140, Van Nostrand Reinhold, New York.

MacDonald, G.A. and Lanier, T.C. 1991. The role of carbohydrates as cryoprotectants in meats and surimi. *Food Technol.* (March):150–159.

MacDonald, G.A. and Lanier, T.C. 1994. Actomyosin stabilization to freeze-thaw and heat denaturation by lactate salts. *J. Food Sci.* **59**(1):101–105.

MacDonald, G.A., Lanier, T.C., and Giesbrecht, F.G. 1996. Interaction of sucrose and zinc for cryoprotection of surimi. *J. Agric. Food Chem.* **44**(1):113–118.

MacDonald, G.A., Lanier, T.C., Swaisgood, H.E., and Hamann, D.D. 1996. Mechanism for stabilization of fish actomyosin by sodium lactate. *J. Agric. Food Chem.* **44**(1):106–112.

MacDonald, G.A., Wilson, N.D.C., and Lanier, T.C. 1990. Stabilised mince: an alternative to the traditional surimi process. In *Chilling and Freezing of New Fish Products.* pp. 69–76, International Institute of Refrigeration, Aberdeen, Scotland.

MacFarlane, D.R. 1987. Physical aspects of vitrification in aqueous solutions. *Cryobiol.* **24:**181–195.

Maresca, B., Patriarca, E., Goldenberg, C., and Sacco, M. 1988. Heat shock and cold adaptation in Antarctic fishes: a molecular approach. *Comp. Biochem. Physiol.* **90B:**623–629.

Meryman, H.T. 1971. Cryoprotective agents. *Cryobiol.* **8:**173–183.

Mornet, D., Bonet, A., Audemard, E., and Bonicel, J. 1989. Functional sequences of the myosin head. *J. Musc. Res. Cell Motility* **10:**10–24.

Mornet, D., DerTerrossian, E., Pradel, L.A., Kassab, R., and Barman, T.E. 1977. The reaction of myosin with bromoalkyl analog of adenosine triphosphate. *FEBS Lett.* **84:**362–366

Muhr, A.H. 1983. The influence of polysaccharides on ice formation in sucrose solutions. Ph.D. Thesis. Nottingham University.

Muhr, A.H. and Blanshard, J.M.V. 1982. Diffusion in gels. *Polymer.* **23:**1012–1026.

Muhr, A.H. and Blanshard, J.M.V. 1986. Effect of polysaccharide stabilizers on the rate of growth of ice. *J. Food Technol.* **21:**683–710.

Muhr, A.H., Blanshard, J.M.V., and Sheard, S. 1986. Effects of polysaccharide stabilizers on the nucleation of ice. *J. Food Technol.* **21:**587–604.

Noguchi, S. 1974. The control of denaturation of fish muscle proteins during frozen storage. D.Sc. Thesis. Sophia University, Tokyo, Japan.

Obuchi, K., Iwahashi, H., Kaul, S.C., Ishimura, M., Fujii, S., Satoh, A., and Komatsu, Y. 1993. In *High Pressure Bioscience and Food Science.* (Hayashi, R., ed.) pp. 147–156, San-E Publishing, Kyoto, Japan.

Park, J.W. and Lanier, T.C. 1987. Combined effects of phosphates and sugar or polyol on protein stabilization of fish myofibrils. *J. Food Sci.* **52:**1509–1513.

Park, J.W., Lanier, T.C., and Green, D.P. 1988. Cryoprotective effects of sugar, polyols, and/or phosphates on Alaska pollock surimi. *J. Food Sci.* **53:**1–3.

Park, J.W., Lanier, T.C., Hamann, D.D., and Keeton, J.T. 1987. Use of cryoprotectants to stabilize functional properties of prerigor salted beef during frozen storage. *J. Food Sci.* **52:**537–542.

Park, J.W., Lanier, T.C., and Pilkington, D.H. 1993. Cryostabilization of functional properties of pre-rigor and post-rigor beef by dextrose polymer and/or phosphates. *J. Food Sci.* **58:**467–472.

Park, J.W., Lanier, T.C., Swaisgood, H.E., Hamann, D.D., and Keeton, J.T. 1987. Effects of cryoprotectants in minimizing physicochemical changes of bovine natural actomyosin during frozen storage. *J. Food Biochem.* **11:**143–161.

Parker, R. and Ring, S.G. 1995. A theoretical analysis of diffusion-controlled reactions in frozen solutions. *Cryo-Lett.* **16:**197–208.

Pauling, L. 1953. *General Chemistry,* 2d ed. pp. 344–347, W.H. Freeman, San Francisco.

Payne, S.R. and Young, O.A. 1995. Effects of pre-slaughter administration of antifreeze proteins on frozen meat quality. *Meat Sci.* **41:**147–155.

Pittz, E.P. and Timasheff, S.N. 1978. Interaction of ribonuclease A with aqueous 2-methyl-2,4-pentanediol at pH 5.8. *Biochem.* **17:**615–623.

Privalov, P.L. 1990. Cold denaturation of proteins. *Crit. Rev. Biochem. Mol. Biol.* **25:**281–305.

Rawn, J.D. 1989. *Biochemistry.* Neil Patterson Publishers, Burlington, N.C.

Roos, Y.H. 1995. *Phase Transitions in Foods.* Academic Press, San Diego.

Roos, Y.H. and Karel, M. 1991. Water and molecular weight effects on glass transitions in amorphous carbohydrates and carbohydrate solutions. *J. Food Sci.* **56:** 1676–1681.

Rosen, M.J. 1978. *Surfactants and interfacial phenomena.* John Wiley & Sons, New York.

Rubinsky, B., Arav, A., and DeVries, A.L. 1992. The cryoprotective effect of antifreeze glycopeptides from Antarctic fishes. *Cryobiol.* **29:**69–79.

Saeki, H. 1996. Gel-forming ability and cryostability of frozen surimi processed with $CaCl_2$-washing. *Fish. Sci.* **62:**252–256.

Samejima, K., Oka, Y., Yamamoto, K., Asghar, A., and Yasui, T. 1988. Effects of SH groups, ϵ-NH_2 groups, ATP, and myosin subfragments on heat-induced gelling of cardiac myosin and comparison with skeletal myosin and actomyosin gelling capacity. *Agric. Biol. Chem.* **52:**63.

Schlesinger, M.J., Ashburner, M., and Tisseries, A. (eds.) 1982. *Heat Shock from Bacteria to Man.* CSH, New York.

Schmitt, J.M., Sieg, F., and Hincha, D.K. 1994. Noncolligative cryoprotection by proteins. *Cryobiol.* **31:**560.

Sharp, A. and Offer, G. 1992. The mechanism of formation of gels from myosin molecules. *J. Sci. Food Agric.* **58:**63–73.

Shenouda, S.Y. 1980. Theories of protein denaturation during frozen storage of fish flesh. *Adv. Food Res.* **26:**275–311.

Shimizu, Y. and Fujita, T. 1985. Stability of unwashed and washed fish mince during frozen storage. *Bull. Jpn. Soc. Sci. Fish.* **51:**1187–1194.

Slade, L. and Levine, H. 1995. Glass transitions and water-food structure interactions. *Adv. Food Nutr. Res.* **38:**103–269.

Slade, L., Levine, H., Ievolella, J., and Wang, M. 1993. The glassy state phenomenon in applications for the food industry: application of the food polymer science approach to structure-function relationships of sucrose in cookie and cracker systems. *J. Sci. Food Agric.* **63:**133–176.

Stadelman, W.J., Olson, V.M., Shemwell, G.A., and Pasch, S. 1988. Nutritional value of eggs. In *Egg and Poultry-Meat.* (Morton, I.D., Scott, R., Watson, D.H., and Lewis, M., eds.) pp. 20–39, Ellis Horwood, Chichester, England.

Storey, K.B. 1990. Life in a frozen state: adaptive strategies for natural freeze tolerance in amphibians and reptiles. *Am. J. Physiol.* **258:**R559–R568.

Storey, K.B., Baust, J.G., and Wolanczyk, J.P. 1992. Biochemical modification of plasma ice nucleating activity in a freeze-tolerant frog. *Cryobiol.* **29:**374–384.

Timasheff, S.N. and Arakawa, T. 1989. Stabilization of protein structure by solvents. In *Protein Structure: A Practical Approach.* (Creighton, T.E., ed.) pp. 331–346, IRL Press, Oxford.

Tinoco, I., Sauer, K., and Wang, J.C. 1985. *Physical Chemistry: Principles and Applications in Biological Sciences,* 2d ed. Prentice-Hall, Englewood Cliffs, N.J.

Treloar, L.R.G. 1970. *Introduction to Polymer Science.* Springer-Verlag, New York.

van Holde, K.E. 1985. *Physical Biochemistry,* 2d ed. Prentice-Hall, Englewood Cliffs, N.J.

Volkin, D.B. and Klibanov, A.M. 1989. Minimizing protein inactivation. In *Protein Function: A Practical Approach.* (Creighton, T.E., ed.) pp. 1–23, IRL Press, New York.

Wagner, J.R. and Añón, M.C. 1985. Effect of freezing rate on the denaturation of myofibrillar proteins. *J. Food Technol.* **20:**735–744.

Warren, C.J., Mueller, G.M., and McKown, R.L. 1992. Ice crystal growth suppression polypeptides and methods of preparation. U.S. patent no. 5,118,792.

Wasylyk, J.M., Tice, A.R., and Baust, J.G. 1988. Partial glass formation: a novel mechanism of insect cryoprotection. *Cryobiol.* **25:**451–458.

Watanabe, T., Kitabatake, N., and Doi, E. 1988. Protective effects of non-ionic surfactants against denaturation of rabbit skeletal myosin by freezing and thawing. *Agric. Biol. Chem.,* **25**(10):2517–2523.

Wessels, J.P.H., Simmonds, C.K., Seaman, P.D., and Avery, L.W.J. 1981. Annual Report, Fishing Industry Research Institute, Capetown, South Africa.

Wyman, J. 1964. Linked functions and reciprocal effects in hemoglobin: a second look. *Adv. Protein Chem.* **19:**223–286.

Yamamoto, K. 1990. Electron microscopy of thermal aggregation of myosin. *J. Biochem.* **108**(6):896–898.

Zachariassen, K.E. 1985. Physiology of cold tolerance in insects. *Physiol. Rev.* **65:**799–832.

12

Antioxidants and Their Application to Frozen Foods

Marilyn C. Erickson

12.1 Introduction

As urbanization continues and the time between harvest and consumption of food increases, problems concerning a limited storage life become more critical. Most frozen foods contain some proportion of lipid and thus are often susceptible to oxidative degradation and production of off flavors during frozen storage. To stem the generation of off flavors, antioxidants—substances capable of delaying, retarding, or preventing the chemical and enzymatic reactions associated with lipid oxidation—are often utilized. This chapter will describe the mechanisms and factors affecting the stabilization of foods by antioxidants, the sources for these antioxidants, and the methods by which antioxidants may be incorporated into the food. The chapter will then focus on characterizing the response of a variety of frozen foods to antioxidant applications.

12.2 Antioxidant Mechanisms

The basic mechanisms of lipid oxidative reactions have been reviewed in chapter 9 but briefly can be characterized by three distinctive steps: initiation, propagation, and termination reactions. In the initiation step, the abstraction of a bis-allylic hydrogen from an unsaturated fatty acid is facilitated by light, irradiation, heat, metal, or radicals; in the propagation step, the carbon-centered lipid radical reacts rapidly with oxygen to form a peroxyl radical that in turn is capable of abstracting a hydrogen atom from an adjacent unsaturated lipid molecule.

The role of antioxidants in lipid oxidation is very simple: to prevent the oxidation of lipids. This can be achieved by either diminishing the formation of initiating lipid radicals or scavenging chain-carrying peroxyl radicals. The former is performed by a preventive antioxidant and the latter by a chain-breaking antioxidant.

12.2.1 Preventive Antioxidants

Chelators

Although not truly antioxidants, chelators bind to transition metal ions and reduce their catalytic activity in lipid oxidation. Examples of chelators include ethylenediamine tetraacetic acid (EDTA), phosphates, citrate, flavonoids such as quercetin and rutin, amino acids, carnosine, phytic acid, xanthan, phosvitin, and chitin. Iron requires at least one free coordination site at which binding of the peroxide occurs. Occupation of all iron coordination sites by a chelator, such as phytate, would prevent binding of hydrogen peroxide and subsequent formation of the initiating species, the hydroxyl radical (Graf et al., 1984). Similarly, binding of hydroperoxides would be inhibited and decomposition to hydroxyl and lipid alkoxyl radicals would be prevented. Other chelators that do not block the coordination site, such as EDTA and citrate, require an excess of the chelator to block the coordination site of the metal ion; otherwise, at lower concentrations, these chelators may increase the solubility and the oxidation-reduction potential of iron (Mahoney and Graf, 1986). Phosphates, in contrast, accelerate the precipitation of iron, thus restricting the accessibility of these ions to lipid sites. Bear in mind that little free iron, is normally present in a food system. It will already be bound to other constituents such as proteins and lipids. Thus, a chelator added to the food system must reach the bound site to compete for the metal ion. There, both thermodynamic and kinetic factors will affect the binding; however, pH of the food medium will affect chelation strength.

Scavenging of Oxygen and Oxygen Active Species

Because of its high affinity for oxygen, ascorbic acid may scavenge molecular oxygen from aqueous solutions (Schuler, 1990). In doing so, ascorbic acid is oxidized to dehydroascorbic acid. Because metal ions are simultaneously reduced in the reaction, this mechanism would be beneficial only under conditions of limited oxygen concentrations. Otherwise, the reduced metal ions could interact with unscavenged oxygen to form active oxygen species capable of initiating lipid oxidation. A more important function of ascorbic acid and derivatives such as ascorbyl palmitate, when in sufficient concentration, is to scavenge active oxygen species such as the superoxide radical, perhydroxyl radical, hydroxyl radical, hydrogen peroxide, and singlet oxygen (Fessenden and Verma, 1978; Nanni et al., 1980; Cabelli and Bielski, 1983; Dalton et al., 1986; Ueda et al., 1996). Such reactions occur nonenzymatically in animal tissues, but in plant tissues enzyme catalysis may account for the reaction of hydrogen peroxide with ascorbic acid (Dalton et al., 1986).

Compounds that inhibit photo-induced autooxidation can also be classified as preventive antioxidants. Singlet oxygen, derived from photo-induced autooxidation, reacts 1,000 times faster than triplet oxygen with unsaturated lipids (Dawson and Gartner, 1983); therefore, quenching singlet oxygen is desirable. Both carotenoid pigments and tocopherols possess the ability to quench singlet oxygen, with the quenching mechanism consisting of both a physical component and a chemical reaction to dissipate the excited energy in singlet oxygen (Krinsky, 1989). The relationship between structure and quenching ability of the carotenoids has been studied, with quenching increasing as the number of conjugated double bonds increases, acyclic carotenoids quenching more than cyclic compounds and conjugated keto groups or a cyclopentane ring on the carotenoid stimulating quenching (Hirayama et al., 1994). For the tocopherol homologs, the relative physical quenching efficiencies were found to decrease in the order $\alpha \geq \beta > \gamma > \delta$-tocopherol, while chemical reactivity followed the sequence $\alpha > \gamma > \delta > \beta$ (Kaiser et al., 1990). From these results, it was concluded that methylation in position 5 of the chromane ring enhanced physical quenching of singlet oxygen, whereas chemical reactivity was favored by a methylated position 7.

Superoxide dismutase and catalase are two enzymes in biological tissues that work cooperatively and provide a defense against the destructive action of active oxygen species. Superoxide dismutase, which exists in three different metalloforms, converts superoxide anion to hydrogen peroxide while catalase converts the hydrogen peroxide to water and oxygen. The effectiveness in addition of either one of these enzymes to a frozen-food system would be dependent on the active endogenous concentrations present and whether either enzyme is limiting. Although superoxide dismutase has been patented as an antioxidant in foods (Michelson and Monod, 1975), it is perceived to have limited usefulness since it becomes inactivated by singlet oxygen, peroxyl radicals, and its own product, hydrogen peroxide (Pigeolet et al., 1990; Escobar et al., 1996).

Another group of chemicals capable of scavenging active oxygen species is the flavonoids. These chemicals are phenol derivatives, being hydroxylated, methoxylated, and/or glycosylated derivatives of the 2-phenyl-benzopyrane ring. They are synthesized in substantial amounts (0.5–1.5%) and widely distributed in plants. More than 4,000 individual flavonoids have been identified in higher and lower plants (Harborne, 1988), with six subclasses being recognized: flavones, flavanones, isoflavones, flavonols, flavanols, and anthocyanins. Both flavonoid molecular structure and type of active oxygen species being scavenged affect the activity of individual flavonoids. Hu et al. (1995) found that the presence of a free hydroxyl group in position 3, hydroxyl groups in ring B, a saturated C2-C3 bond, or the absence of a carbonyl group at the C4 position enhanced the superoxide scavenging activity of flavonoids. In terms of hydroxyl radical scavenging capabilities, flavonoids were found to fall into three groups: those with no apparent scavenging capabilities (3-hydroxy flavone, 7-hydroxy flavone, chrysin, apigenin, galangin, kaempferol, and fisetin); those with scavenging abilities (7,8-dihydroxy flavone, rutin, (+)-catechin, and (−)-epicatechin); and those that seemed to enhance hydroxyl radical production (baicalein, quercetin, myricetin, and morin) (Hanasaki et al., 1994).

Inhibition of Initiating Enzymes

Flavonoids have been reported to inhibit both 1-, 5-, and 12-lipoxygenases (Yoshimoto et al., 1983; King and Klein, 1987). In the case of xanthine oxidase, competitive inhibition could be ascribed to the antioxidant properties of 7-hydroxyflavones with the C-7 OH of flavones taking the place of C-2 or C-6 OH of xanthine in the active site of the enzyme (Cotelle et al., 1996).

Conversion of Lipid Hydroperoxides to Alcohols

Another enzyme capable of acting as a preventive antioxidant is glutathione peroxidase. The conversion of hydroperoxides to alcohols by glutathione peroxidase eliminates the contribution of initiating lipid alkoxyl radicals formed upon the decomposition of hydroperoxides by Fe^{2+}. As was the case for superoxide dismutase and catalase, the usefulness of glutathione peroxidase added to frozen foods may be limited since inactivation may occur with hydrogen peroxide and hydroxyl radicals (Pigeolet et al., 1990; Tabatabaie and Floyd, 1994).

12.2.2 Chain-Breaking Antioxidants

At normal oxygen pressures, oxygenation of alkyl radicals to peroxyl radicals is very fast. Under reduced oxygen conditions, therefore, scavenging of alkyl radicals may occur. Alkyl radicals are in general reducing agents and scavenged by electron acceptors (Gordon, 1990). Examples of electron acceptors that have exhibited antioxidant properties in biological tissues include vitamin K and tocopherolquinone (Stryer, 1975; Harman, 1982; Burton and Ingold, 1984).

Scavenging of Peroxyl Radicals

Antioxidants acting as peroxyl scavengers are generally phenols and to a lesser extent compounds with an amino group. Among the monophenols are the tocopherols, tocotrienols, butylhydroxytolutene (BHT), butylhydroxyanisole (BHA), vanillin, and eugenol. Examples of polyphenols are carnosic, carnosol, hydroxytyrosol, and the catechins of tea. Examples of antioxidants having an amino group are uric acid, amino acids, carnosine, and Maillard reaction products.

Two conditions are necessary for a compound to act as a peroxyl scavenging antioxidant: the ability to donate a hydrogen atom rapidly to a lipid radical and stabilization of the radical derived from the antioxidant. The phenol molecule itself is inactive as an antioxidant, but substitution of alkyl groups into the 2, 4, or 6 positions increases the electron density on the hydroxyl group by an inductive effect and thus increases the reactivity with lipid radicals. Substitution at the 4 position with an ethyl or *n*-butyl group rather than a methyl group improves the activity of a phenolic antioxidant, but longer chain or branched alkyl groups in this position decrease the activity. The radical formed by reaction with a peroxyl radical is stabilized by delocalization of the unpaired electron around the aromatic ring, reducing the rate of propagation of the autooxidation chain reaction. The stability of the phenoxyl radical is further increased by bulky groups in the 2 and 6 positions as in BHT, since these substituents increase the steric hindrance in the region of the radical and thereby further reduce the rate of propagation reactions involving the antioxidant radical. The presence of bulky constituents in the 2 and 6 positions, however, also reduces the rate of reaction of the phenol with lipid radicals, a steric effect that opposes the increased stabilization of the radical. Both effects must be considered in assessing the overall activity of an antioxidant.

Tocopherols are composed of eight different compounds belonging to the families, tocols and tocotrienols, with the prefix, α, β, γ, and δ, depending on the number and position of methyl groups attached to the chromane rings. In tocols, the side chain is saturated, whereas in tocotrienols it is unsaturated. This phytyl side chain contributes little to the antioxidant activity, but it facilitates the incorporation, retainment, and mobility of tocopherol in membranes. There, the reaction of tocopherol with a peroxyl radical is favored by a great degree over the reaction of a polyunsaturated fatty acid (PUFA) with a peroxyl radical so that only a small amount of tocopherol is needed to protect a large amount of PUFA (Burton et al., 1993). The higher antioxidant potency in membranes observed for tocotrienols compared to tocopherols has been attributed to three properties of tocotrienols: (1) their higher recycling efficiency from chromanoxyl radicals; (2) their more uniform distribution in membrane bilayers; and (3) their stronger disordering of membrane lipids, which makes interaction of chromanols with lipid radicals more efficient (Serbinova et al., 1991).

While the main mechanism for tocopherol's antioxidant action is believed to be scavenging of peroxyl radicals and conversion to a hydroperoxide, an additional mechanism has been proposed by Dmietriev et al. (1994). These investigators suggested that tocopherol is capable of reducing the peroxyl radical back to the lipid molecule with the concurrent formation of oxidized tocopherol and either molecular oxygen or the superoxide anion radical. Alternately, Therond et al. (1996) contended that α-tocopherol exerted a stabilizing effect toward hydroperoxides, limiting their further degradation into peroxyl radicals.

The effectiveness of flavonoids in retarding lipid oxidation in foods appears to be related not only to their chelating capacity but also to their ability to act as free-radical acceptors. While metal complexing capacity of flavonoids is centered around the C-3′ and C-4′ atoms (van Acker et al., 1996), the number of hydroxyl groups of flavonoids, as well as their location on ring A or B, was found to be of negligible importance for radical scavenging (Foti et al., 1996). The antioxidant activity of flavonoids, instead, was dictated more by the existence than the location of two orthophenolic functions. The greater reactivity of the ortho-dihydroxy system was postulated to be due to the smaller dissociation energy of the O-H bond in comparison with the 1,3-system, owing to the greater stability of the transient radical involved. In addition, the presence in the central ring of the carbonyl and the 2,3-double bond led to increased antioxidant activity by participating in radical stabilization. The 3-OH group in flavonoids also seemed to have a more favorable antioxidant effect than the 5-OH group (Roginsky et al., 1996).

Carotenoids are also capable of scavenging peroxyl radicals but do so in a manner exclusive of hydrogen abstraction. Exertion of their antioxidative effect is postulated to occur through addition of the peroxyl radical to the conjugated system such that the resulting carbon-centered radical is stabilized by resonance. Under low-oxygen conditions, addition of a second peroxyl radical to the carbon-centered radical produces a nonradical polar product. At high oxygen pressures, however, carotenoids act as prooxidants because the carbon-centered radical may add oxygen in a reversible reaction resulting in an unstable chain-carrying peroxyl radical, which can further degrade to radical and nonradical polar products with no net inhibition of oxidation (Jørgensen and Skibsted, 1993). The stability of carotenoids in an oxidizing system is dictated by their structure, which in turn affects the degree of inhibition, the order of suppression being astaxanthin > canthaxanthin > β-carotene > zeaxanthin (Jørgensen and Skibsted, 1993).

Amino acids, dipeptides, and Maillard reaction products (MRP) have all been reported to exhibit antioxidative effects. While one of their antioxidative mechanisms involves chelation, peroxyl radicals may also be scavenged through reaction with amino and imidazole groups of amino acids and dipeptides, such as carnosine (Kohen et al., 1988). For MRP, antioxidative effects were found to be a function of their reducing power, which were independent of their degree of brownness (Eichner, 1980; Tanaka et al., 1988).

Regeneration of Antioxidants

Substances capable of regenerating primary antioxidant radicals or their degradation products are often referred to as *synergists*. Ascorbic acid is the most notable synergist, with the ability to donate a hydrogen to the tocopheryl radical and other phenol-based antioxidant radicals in line with their one-electron reduction potentials (Packer et al., 1979; Kagan et al., 1990; Buettner, 1993). Similarly, regeneration of tocopherol by flavonoids is thermodynamically feasible based on their reduction potentials (Jovanovic et al., 1994).

12.3 Environmental Impact on Antioxidant Effectiveness

Physical and chemical factors within a food product will influence the relative effectiveness of an antioxidant applied to that food product. Tissue compartmentation and type of lipid substrate requiring protection are examples of those physical and chemical factors, respectively.

Accessibility to substrate and other antioxidants is critical for an antioxidant or synergist to exhibit an inhibiting effect. For example, Erickson (1997) showed that ascorbic acid absorbed and metabolically distributed within live fish exhibited an antioxidant effect during frozen storage when the fish were subsequently sacrificed. In contrast, ascorbic acid displayed prooxidative effects if administered through vacuum tumbling. Different responses were attributed to differences in site of distribution: an intracellular location for metabolically absorbed ascorbic acid and an intercellular location for vacuum-tumbled ascorbic acid. Intercellularly distributed ascorbic acid, while capable of regenerating tocopherol and protecting membrane phospholipids, simultaneously accelerated oxidation of triglycerides. Stimulation of phospholipid oxidation by intracellularly distributed ascorbic acid, however, was offset by regeneration of tocopherol.

The effect of different lipid substrates has a significant impact on the activity of different antioxidants according to their hydrophilic or lipophilic character. This phenomenon, termed the *polar paradox* effect, primarily has been demonstrated in model systems (Porter et al., 1989; Frankel et al., 1994, 1996) but briefly describes the greater effectiveness of polar antioxidants toward nonpolar lipids and nonpolar antioxidants toward polar lipids. Differences observed in the efficiency of antioxidants with polar and nonpolar lipids have been explained by their affinities toward the air-oil interfaces in bulk oil and oil-water interfaces in emulsions or membranes.

12.4 Source of Antioxidants

While synthetic antioxidants such as BHA and BHT traditionally have been added to foods, there has been a growing interest in replacing them with natural ingredients from oilseeds, spices, and other plant materials (Nakatani, 1992; Pratt, 1992). Currently, the use of natural ingredients is limited to tocopherol extracts and rosemary extracts. To remedy this limited selection, recent studies continue to explore the active principles in many different source materials (table 12.1). Commercialization of additional natural antioxidants will be dependent on sufficient concentrations of the active components in the source material such that extraction may be accomplished cost-effectively.

12.5 Methods for Antioxidant Application

A number of different methods for applying antioxidants have arisen over the years. These methods are described in table 12.2, along with their advantages and disadvantages. In applications involving mixing, antioxidants may be applied directly to the product before mixing or dissolved in an appropriate medium before application and mixing (i.e., water, oil, or alcohol) to aid in the dispersion of the

Table 12.1. Recent studies exploring the antioxidant constituents in a variety of source materials

Source material	Reference(s)
Ginger	Jitoe et al., 1992; Kikuzaki and Nakatani, 1993; Masuda and Jitoe, 1994
Mustard flour	Shahidi et al., 1994
Oregano	Vekiari et al., 1993
Oriental herbs	Kim et al., 1994
Peanut hulls	Yen and Duh, 1993, 1994, 1995
Rice hulls	Asamarai et al., 1996
Rosemary	Cuvelier et al., 1994
Sage	Cuvelier et al., 1996
Teas	Cao et al., 1996
Thyme	Schwarz et al., 1996
Vegetables	Cao et al., 1996

antioxidant throughout the product. The latter process is particularly desirable in the application of water-insoluble antioxidants (i.e., BHA, BHT, *tert*-butylhydroquinone [TBHQ], and tocopherols) to food products having moisture as the major component (i.e., muscle tissue).

Relatively few studies have been conducted to compare the effectiveness of the various methods of application. Brotsky (1976) compared injection and overnight marination and found no significant differences among these treatment methods for broiler parts subsequently stored for 9 months at −21°C.

The time at which antioxidants are applied to the food can impact their ultimate effectiveness. Kelleher et al. (1992) demonstrated this effect with washed minced mackerel in which antioxidants were added at all processing steps or only added before the storage step. In this study, the initial sensory quality of product where antioxidant had been incorporated during processing was significantly better than product when the antioxidant had only been added just before frozen storage. More importantly, this difference in rating was maintained over 24-wk storage at −20°C. In concurrence, thiobarbituric acid reactive substance (TBARS) values were consistently higher in samples not treated with antioxidants during processing.

The idea of minimizing oxidative reactions in early stages of processing is also the basis for adding antioxidants into the diet of animals. Dietary application, however, is also distinguished from the other methods of application in the site of distribution for the antioxidant (i.e., within the membrane for α-tocopherol and intracellularly rather than intercellularly for ascorbic acid). Environmental impact on antioxidant effectiveness has been described in section 12.3. An improved understanding of the mechanism by which antioxidants are used to protect tissues against oxidative stress no doubt will lead to the development of more effective methods for application of antioxidants.

Table 12.2. Advantages and disadvantages to methods for application of antioxidants

Application method	Advantages	Disadvantages
Mixing *The antioxidant is applied to the food and then blended throughout the product.*	1. Requires inexpensive equipment 2. Simple process 3. Can use visual spices and insoluble particles	1. Disruption of tissue required for distribution of antioxidant 2. No assurance of even distribution in product
Dipping/marination *The food is held in an aqueous solution containing the antioxidant(s) for a period of 1 min to several hours under refrigeration.*	1. Requires no process equipment 2. Very simple process 3. Can use visual spices and insoluble particles 4. Least-expensive process 5. Wide range of product shapes and sizes can be accomodated	1. Time-consuming for dips/marinades extending several hours 2. Excess marinade regarded as loss 3. May yield inconsistent results in response to varying uptake during dipping 4. Surface appearance may change because of exposure to high concentrations of chemicals in soaking solution
Injection *An aqueous solution of the antioxidant is mechanically injected into the food through the utilization of needles and pressure. The method relies on diffusion of antioxidant from the site of injection to other areas in the food product.*	1. Extremely fast 2. Can be incorporated into a process line 3. Accurate and consistent application 4. Recirculation of solution decreases waste	1. Requires expensive equipment 2. Needles clog if incorporating visual spices or insoluble particles 3. Cleaning techniques may be cumbersome 4. Localization of antioxidants around site of injection 5. Product must be of uniform size
Vacuum tumbling *An aqueous solution of the antioxidant is absorbed into the food by applying a vacuum while rotating the application container.*	1. Quicker than dipping/marination 2. Less-expensive process than injection 3. Can use visual spices and insoluble particles in application solution	1. Requires specific equipment 2. No assurance that antioxidant is distributed to all product 3. Addition of water and tumbling action may alter texture and appearance of product

(*continued*)

Table 12.2. (*continued*)

Application method	Advantages	Disadvantages
Spraying/glazing *The antioxidant is dissolved in an appropriate medium and distributed with pressure through a nozzle onto the product as a fine mist of particles. In the case of glazing, an aqueous solution is sprayed onto a frozen product to form an ice layer.*	1. Can be applied to nonuniform intact products 2. Layer of ice acts as a barrier to diffusion of oxygen	1. Antioxidant effective only at surface of product
Dietary *Antioxidants are incorporated into the diet of an animal at a fortified level.*	1. Antioxidants distributed to cellular sites where maximum effectiveness may be observed 2. Initiation reactions are prevented during harvest and processing	1. Heat labilization of many of the antioxidants requires higher doses in fortification of extruded feeds 2. Dietary supplementation generally occurs over a period of several weeks and metabolic concentration of water-soluble antioxidants would be limited because of the animal's acceleration in degradation of these compounds 3. Monetary compensation is not always made to producer applying fortified feeds 4. Fortification of fish is expected to be inconsistent because of variable feed uptake and inability to monitor and control uptake

12.6 Response of Frozen Foods to Antioxidant Applications

A wealth of studies have been conducted on frozen foods to monitor the response to application of antioxidants. Tables 12.3, 12.4, 12.5, and 12.6 summarize a large portion of these antioxidant responses in fish/shellfish, poultry, pork, and beef/lamb, respectively. In recent years, an emphasis has been on investigation of natural antioxidants, but treatments with synthetic antioxidants are quite often simultaneously conducted to serve as a comparison. This shift in emphasis of investigation is in response to concerns about the toxicity of synthetic antioxidants and the assumption that natural antioxidants would naturally be safer. Additional information about the toxicological aspect of antioxidants may be found in the review by Barlow (1990). Consult your governmental regulatory agency as to which antioxidants and concentrations may be applied legally to your frozen-food product.

A wide variation in response, from activation to severe inhibition, has been noted with many of the antioxidants (tables 12.3, 12.4, 12.5, and 12.6). These variations may be due to the type of product being stored, the temperature or length of storage of the product, the oxidative measurement used to judge the effectiveness, the concentration of antioxidants used with the product, or the processing/packaging applied to the product. For instance, McNeill et al. (1988) suggested that the ineffectiveness of tripolyphosphates to inhibit lipid oxidation in raw chicken meat may have been due to their hydrolysis by active muscle phosphatases within the product. In other cases, ineffectiveness may be due to the lack of significant oxidation occurring during storage in either the treated or untreated samples. Such was the case for uncooked pork that had been treated with either 100 or 200 ppm tocopherol (Whang et al., 1986). Similarly, rancidity was not detected in frozen minced trout or frozen fish flakes when sensory methods were used to judge quality (Moral et al., 1986; Boyd et al., 1993). Chemical measurements, however, were able to differentiate antioxidant-treated samples from nonantioxidant-treated samples in these same studies. These discrepancies lead us to question which measurement is more accurate. Since ultimately it is consumers who evaluate the final stored product, sensory measurements when conducted properly should be the truest measure (see chapter 18 for additional information on sensory measurements). Chemical measurements should not be abandoned, however, because while they are more sensitive in differentiating treatments, they can provide important information on the antioxidant's mechanism.

Another reason variations in response to an antioxidant may be seen is that different concentrations of the antioxidant had been applied to the product. Addition of NaCl to a sodium tripolyphosphate (STPP) tumbling treatment led to increased absorption and retention of the phosphate that could account for the

Table 12.3. Studies evaluating the inhibition of lipid oxidation by antioxidants and chelators during frozen storage of fish and shellfish

Antioxidant	Concentrations[a]	Method of application	Product	Storage conditions	Oxidative measure	Effectiveness	Reference
EDTA	0.5%	Dipping	Mullet fillets	−18°C, 9 mo	TBARS	8% inhibition	Deng et al., 1977
Ascorbic acid	2%					55% inhibition	
TBHQ	0.025%					35% inhibition	
Ascorbic acid/TBHQ	2%/0.025%					55% inhibition	
Ascorbic acid/EDTA	2%/0.5%					55% inhibition	
TBHQ/EDTA	0.025%/0.5%					50% inhibition	
Na erythorbate	1%	Dipping, 1–5 min	Pacific herring fillets	−28°C, 8 mo	peroxide value	84–96% inhibition	Bilinski et al., 1979
Na erythorbate	1.0%	Dipping	Whole herring	−28°C, 4 mo	peroxide value	85% inhibition	Bilinski et al., 1981
Na tripolyphosphate	1.0%					No effect	
Arginine/xylose Maillard reaction products (MRP)	crude (0.1%) or purified (0.001–0.1%)	Mixing	Minced herring	−20°C, up to 28 wk	peroxide value	No effect	Beckel et al., 1985
Histidine/glucose MRP	crude (0.1%)					No effect	
Ascorbic acid	1.5%	Dipping	Cooked mussels	−12°C, up to 20 wk	TBARS	Ascorbic acid, alone or in presence of chelating agents, arrested progression of	Ablett et al., 1986
Ascorbic acid/citrate/EDTA	1.5%/0.5%/0.5%						

(continued)

[a]Unless otherwise indicated, water-insoluble antioxidants based on fat content of food product

Table 12.3. (*continued*)

Antioxidant	Concentrations[a]	Method of application	Product	Storage conditions	Oxidative measure	Effectiveness	Reference
						oxidation from 10 to 20 w, compared to control	
Sorbitol/MSG/BHA/BHT	(3%/0.5%/0.01%/0.01%)	Mixing	Minced trout	−12, −18, or −24°C, up to 365 d	TBARS	In general, significantly lower for the antioxidant treatments than control at storage temperatures	Moral et al., 1986
Tenox 20 (TBHQ/citric acid)	0.02%/10% of fat	Mixing	Minced menhaden	−7°C, up to 120 d	peroxide value	No antioxidant activity	Hwang and Regenstein, 1988
Ascorbic acid (AA)	0.5% of fat					91–94% inhibition	
Erythorbic acid	0.5% of fat					90–91% inhibition	
Citric acid	0.5% of fat					29–64% inhibition	
EDTA	0.1% of fat					8–18% inhibition	
Herbalox-W (rosemary)	0.1% of fat					No effect	
Herbalox-O (rosemary)	0.1% of fat					41–61% inhibition	
Covi-ox T30P (tocopherol mixture)	0.1% of fat					17–35% inhibition	
Covi-ox T30P/AA	0.1%/0.5%					83–85% inhibition	

(*continued*)

Table 12.3. (*continued*)

Antioxidant	Concentrations[a]	Method of application	Product	Storage conditions	Oxidative measure	Effectiveness	Reference
Tocopherol	20 × basal diet (2.6 mg/100 g)	Dietary	Whole rainbow trout	−12°C, up to 210 d	Rancid flavor	No effect	Pozo et al., 1988
Erythorbic acid	0.5%	Dipping, 1 min	Mackerel	−7°C, up to 9 mo	Flavor	No extension in shelf life	Santos and Regenstein, 1990
			Hake			1–1.5-wk extension in shelf life	
Ascorbic acid	0.5%	Dipping, 10 min	Squid	−20°C, 9 mo	Acceptability	3-mo extension in acceptability, from 6 to 9 mo	Selvaraj et al., 1991
BHA	0.01%	Mixing	Oyster homogenate	−25°C, up to 90 d	TBARS	68–83% inhibition	Hatate and Kochi, 1992
Tocopherol	0.1%					45–57% inhibition	
Na ascorbate	0.2%					32–46% inhibition ≤ 60 d; prooxidative effect at 90 d	
TBHQ	0.02%	Mixing	Minced mackerel	−20°C, up to 24 wk	Peroxide value	45–84% inhibition	Kelleher et al., 1992
Propyl gallate	0.02%					88–97% inhibition	
BHT	0.02%					2–63% inhibition	
BHA	0.02%					43–54% inhibition	
Tenox GT-2 (tocopherol mixture)	0.03%					56–86% inhibition	
Ascorbate/EDTA	0.2%/1.5 mM					27–58% inhibition	

(*continued*)

Table 12.3. (continued)

Antioxidant	Concentrations[a]	Method of application	Product	Storage conditions	Oxidative measure	Effectiveness	Reference
BHT	200/30 ppm for all antioxidants tested	Mixing	Ground fish	−20°, 1 wk	TBARS	47/35% inhibition	Ramanathan and Das, 1992
Rutin						44/17% inhibition	
Quercetin						84/62% inhibition	
Morin						77/40% inhibition	
Myricetin						89/70% inhibition	
Kaempferol						52/19% inhibition	
Tannic acid						87/84% inhibition	
Ellagic acid						85/84% inhibition	
L-ascorbic acid						223/373% activation	
α-tocopherol						50/25% inhibition	
α-tocopherol	0.05%	Mixing	Previously frozen fish meat	−30°C, up to 6 d	TBARS	19–37% inhibition	Wada and Fang, 1992
Tocopherol/rosemary	0.05%/0.02%					23–48% inhibition	
Rosemary	0.02%					12–35% inhibition ≤ 4 d activation at 6 d	
BHA	0.02%					29–31% inhibition ≤ 4 d activation at 6 d	

(continued)

Table 12.3. (continued)

Antioxidant	Concentrations[a]	Method of application	Product	Storage conditions	Oxidative measure	Effectiveness	Reference
Herbalox-W (rosemary)	0.25%	Dipping, 1 min	Cooked fish flakes	−20°C, 90 d	TBARS	54% inhibition	Boyd et al., 1993
TBHQ/ascorbic acid	0.05%/2%					69% inhibition	
Combination of above antioxidants						72% inhibition	
Tocopherol	300-mg α-tocopherol acetate/kg diet	Dietary	Salmon fillets	−18°C, 5 wk	Rancid flavor	80–95% inhibition depending on PUFA content of diet	Waagbø et al., 1993
Xanthan	0.2 or 0.4%	Mixing	Ground cod	−25°C, 3 mo	TBARS	Inhibition in direct relation to xanthan concentration used; addition of citric acid led to further inhibition	Stodolnik, 1994
Xanthan/citric acid	0.2 or 0.4%/0.02%						
Lem-O-Fos, Formula 101 (85% Na tripolyphosphate and 15% lemon juice conc.)	6.7 mL/100 g	Injection	Channel catfish fillets	−20°C, up to 6 mo	TBARS	33–50% inhibition of skin-side fillet sites up to 4 mo; no effect at 6 mo	Freeman and Shannon, 1994

(continued)

Table 12.3. (continued)

Antioxidant	Concentrations[a]	Method of application	Product	Storage conditions	Oxidative measure	Effectiveness	Reference
Na ascorbate	0.3% soln, pH 5	Metabolic absorption, 24 h	Channel catfish fillets	−6°C, 6 mo	TBARS	46% inhibition	Thed and Erickson, 1994
	50 mg/40 fillets	Vacuum tumbling				34% inhibition	

Table 12.4. Studies evaluating the inhibition of lipid oxidation by antioxidants and chelators during frozen storage of poultry products

Antioxidant	Concentrations[a]	Method of application	Product	Storage conditions	Oxidative measure	Effectiveness	Reference
Kena (Na tripolyphosphate/Na hexametaphosphate)	3%	Marination	Coated fried chicken parts	−21°C, 9 mo	Rancid flavor	70% inhibition	Brotsky, 1976
Na pyrophosphate	1, 3, or 6%	Marination	Cooked broiler halves	−34°C, 2 mo	TBARS	6% > 3% > 1% 16–66% inhibition	Ang and Young, 1987
Na tripolyphosphate	0.1, 0.3, or 0.5%	Mixing	Mechanically separated chicken meat	−10°C, up to 28 wk	TBARS	No effect	McNeill et al., 1988
Na caseinate	0.5, 1.0, or 2.0%					No effect	
Rosemary	0.01, 0.02, or 0.03%					33–43% inhibition	
Dadex (20% BHA/ 6 % PG/4% CA)	0.02, 0.05, or 0.08%					Effective only at higher concentration (60% inhibition)	
Rosemary	200 ppm	Mixing	Turkey sausage	−18°C, 20 d	TBARS	46% inhibition	Barbut et al., 1988
BHA & BHT/ citric acid	200 ppm/ 20 ppm					66% inhibition	
Na acid pyrophosphate	0.4%	Mixing	Ground chicken	−18°C, 6 wk	TBARS	17% inhibition in breast 44% inhibition in mechanically deboned chicken meat	Barbut et al., 1989

(*continued*)

[a]Unless otherwise indicated, water-insoluble antioxidants based on fat content of food product.

Table 12.4. (continued)

Antioxidant	Concentrations[a]	Method of application	Product	Storage conditions	Oxidative measure	Effectiveness	Reference
Tenox 6 (10% BHA, 10% BHT, 6% PG, 6% CA)	2%	Antioxidant coated on pure salt and mixed with product	Ground turkey with and without added cupric and ferric ions	−25°C, up to 6 mo	TBARS	0–34% inhibition	Salih et al., 1989
VE (0.0215% Tocopherols/0.0005% citric acid/0.045% ascorbyl palmitate)	2%					3–11% inhibition	
Na tripolyphosphate	0.3 or 0.5%	Mixing	Ground turkey meat	−14°C, 5 mo in vacuum package followed by 6 wk in non-vaccuum package	Hexanal	53–97% inhibition	Craig et al., 1991
Na ascorbate monophosphate	0.3 or 0.5%					85–94% inhibition	
Na tripolyphosphate (STPP)	0.3%	Mixing	Ground chicken	−20°C, 6 mo	TBARS	45–65% inhibition	Lai et al., 1991
Rosemary	0.05 or 0.1%					11–39% inhibition	
STPP/rosemary	0.3%/0.05 or 0.1%					63–73% inhibition	
STPP/TBHQ	0.3%/0.02%					84–87% inhibition	

Table 12.5. Studies evaluating the inhibition of lipid oxidation by antioxidants and chelators during frozen storage of pork products

Antioxidant	Concentrations[a]	Method of application	Product	Storage conditions	Oxidative measure	Effectiveness	Reference
Na tripoly-phosphate	0.375%	Mixing	Precooked sausage patties	−9°C, 8 wk	TBARS	50% inhibition	Matlock et al., 1984
Lem-O-Fos	0.441%					61% inhibition	
Ascorbic acid BHA BHT Propyl gallate Combinations of above	0.05% AA 0.1% BHA 0.1% BHT 0.03% PG	Rubbed on surface	Bacon slices	−30°C, 124 d	Flavor desirability	Combination of BHA, BHT, and PG only antioxidant treatment effective for extending frozen storability	Jeremiah, 1985
Tocopherol	100 or 200 ppm	Mixing	Ground pork, raw or cooked	−20°C, 14 d if cooked; 60 d if raw	TBARS and rancid flavor	25–45% inhibition for cooked samples; no effect on raw samples	Whang et al., 1986
Porcine plasma, spray-dried	2.5 or 5%	Mixing	Ground pork	−15°C, up to 120 d	TBARS	2.5% plasma, 38–68% inhibition; 5% plasma, 74–86% inhibition	Faraji et al., 1991
BHT	0.02%					BHT, 19–27% inhibition	
BHA/BHT (1:1) Rosemary extract Tea extract	200 ppm 1000 ppm 500 ppm	Mixing	Ground pork	−20°C, 4 mo	Pentane	94% inhibition 87% inhibition 71% inhibition	Loliger, 1991

(*continued*)

[a]Unless otherwise indicated, water-insoluble antioxidants based on fat content of food product

Table 12.5. (*continued*)

Antioxidant	Concentrations[a]	Method of application	Product	Storage conditions	Oxidative measure	Effectiveness	Reference
Na tripolyphosphate (STPP)	0.5%	Mixing	Restructured pork steaks	−30°, 8 mo	TBARS	69% inhibition	Liu et al., 1992
STPP/rosemary oil-soluble	0.5%/0.05%					74% inhibition	
STPP/rosemary water-soluble	0.5%/0.1%					67% inhibition	
STPP/TBHQ	0.5%/0.02%					80% inhibition	
STPP/sage	0.5%/0.05%					67% inhibition	
Tocopherol	10, 100, or 200 mg α-tocopherol acetate	Dietary	Pork chops	Vacuum packaged at −20°C for 4 mo, then refrigerated at 4°C for 2, 4, 6, and 8 d	TBARS	200 = 100 > 10 75–90% inhibition after 2-d storage	Monahan et al., 1994
BHT/PG/citric acid (CA)	0.01%/0.01%/0.01%	Mixing	Pork sausage	−20°C, up to 16 wk	TBARS and surface color	Rosemary extract was as effective as BHT/PG/CA but showed more surface discoloration	Ho et al., 1995
Rosemary	500 ppm						
Na lactate	3%	Mixing	Pork sausage	−20°C, up to 252 d	TBARS	Acted as a prooxidant	Krahl et al., 1995

Table 12.6. Studies evaluating the inhibition of lipid oxidation by antioxidants and chelators during frozen storage of beef and lamb products

Antioxidant	Concentrations[a]	Method of application	Product	Storage conditions	Oxidative measure	Effectiveness	Reference
BHT	0.2%	Spraying	Beef muscle pieces	−10°C	TBARS	21% inhibition	Caldironi and Bazan, 1982
Ascorbic acid/ citric acid/ EDTA	0.01%					No effect	
BHA	0.2% based on fat content	Mixing	Ground meat beef:pork (1:1)	−10°C, 4–20 wk	TBARS	TBHQ = BHA = BHA and TBHQ	Chastain et al., 1982
BHA and TBHQ							
TBHQ					Flavor	88–91% avg. inhibition	
Onion juice	20%	Mixing	Ground lamb	−20°C, up to 30 d	TBARS	40% inhibition	Jurdi-Haldeman et al., 1987
Garlic juice	4.8%					15% inhibition	
TBHQ	0.02% of fat content	Mixing in fat section before vacuum blending	Raw restructured beef steaks	−20°C, up to 12 mo	TBARS	51–63% inhibition	Crackel et al., 1988
Ascorbyl palmitate: mixed tocopherols/ citric acid	5:1/0.3% or 1:1/0.3% CA					24–39% inhibition 10–28% inhibition	
Tocopherol	370 IU/animal/d	Dietary	Beef patties	−18°C, 1.5 and 3 mo	TBARS	84% and 46% inhibition at 1.5 and 3 mo, respectively	Faustman et al., 1989

(*continued*)

[a] Unless otherwise indicated, water-insoluble antioxidants based on fat content of food product

Table 12.6. (*continued*)

Antioxidant	Concentrations[a]	Method of application	Product	Storage conditions	Oxidative measure	Effectiveness	Reference
Diphosphate (P_2O_7) Triphosphate (P_3O_{10}) Trimetaphosphate (P_3O_9)	0.2 or 0.5%	Mixing	Ground beef patties	Retail-type freezer cabinet	TBARS	$P_2O_7 = P_3O_{10} > P_3O_9$	Mikkelsen et al., 1991
Na tripolyphosphate Rosemary Na tripolyphosphate/ rosemary Na tripolyphosphate/ TBHQ	0.3% 0.05 or 0.10% 0.3%/0.05 or 0.10% 0.3%/0.02%	Mixing	Restructured beef steaks	−20°C, 6 mo	TBARS	42% inhibition 39–54% inhibition 55–66% inhibition 64% inhibition	Stoick et al., 1991
Tocopherol	2,100 IU/kg head/d	Dietary	Longissimus beef	−20°C, dark or illuminated	Saturation index	Extension of shelf life by 214 and 101 d under dark and illuminated conditions, respectively	Lanari et al., 1993
Tocopherol	2,100 IU α-tocopherol acetate/animal/d	Dietary	Beef loin	−20°, 80 d	TBARS	85% inhibition	Lanari et al., 1994

greater effectiveness of the treatment in lipid stabilization when both NaCl and STPP were use than when STPP was used alone (Ang and Young, 1987).

Vacuum packaging can also modify the response to antioxidant treatments. Vacuum packaging alone was sufficient to control lipid oxidation during frozen storage of low-fat pork sausage, whereas a noticeable improvement in control of lipid oxidation during frozen storage was obtained upon antioxidant addition to high-fat vacuum-packaged pork sausage (Ho et al., 1995). Santos and Regenstein (1990) have also observed synergistic effects between vacuum packaging and antioxidants.

Addition of antioxidants may not always lead to a superior product. Cooked fish flakes were characterized with green aroma and flavor notes upon application of a rosemary extract (Boyd et al., 1993). With chicken marinated in a phosphate solution, a chemical flavor was noted (Brotsky, 1976). Caution should therefore be exercised in using these and other antioxidants that may taint a product's flavor.

12.7 Summary

To apply antioxidants intelligently to frozen foods, some knowledge of the mechanisms by which they function is necessary. A summary of our current understanding has been presented in this chapter. However, additional research in this area is needed, particularly with regard to the impact of microenvironment on an antioxidant's effectiveness. Consideration should be given to the method of application, product being stored, and concentrations that may be applied legally before selecting and using an antioxidant. In all likelihood, more than one antioxidative intervention may be required because of lipid oxidation's complexity.

References

Ablett, R.F., Gould, S.P., and Sherwood, D.A. 1986. Frozen storage performance of cooked cultivated mussels (*Mytilus edulis. L.*): influence of ascorbic acid and chelating agents. *J. Food Sci.* **51**:1118–1121.

Ang, C.Y.W. and Young, L.L. 1987. Effect of marination with sodium pyrophosphate solution on oxidative stability of frozen cooked broiler leg meat. *Poultr. Sci.* **66**:676–678.

Asamarai, A.M., Addis, P.B., Epley, R.J., and Krick, T.P. 1996. Wild rice hull antioxidants. *J. Agric. Food Chem.* **44**:126–130.

Barbut, S., Draper, H.H., and Hadley, M. 1988. Effects of freezing method and antioxidants on lipid oxidation in turkey sausage. *J. Food Protect.* **51**:878–882.

Barbut, S., Draper, H.H., and Hadley, M. 1989. Lipid oxidation in chicken nuggets as affected by meat type, phosphate, and packaging. *J. Food Protect.* **52**:55–58.

Barlow, S.M. 1990. Toxicological aspects of antioxidants used as food additives. In *Food Antioxidants*. (Hudson, B.J.F., ed.) pp. 253–307, Elsevier Applied Science, New York.

Beckel, R., Lingnert, H., Lundgren, B., Hall, G., and Waller, G.R. 1985. Effect of Maillard reaction products on the stability of minced herring in frozen storage. *J. Food Sci.* **50**:501–502, 530.

Bilinski, E., Jonas, R.E.E., and Lau, Y.C. 1979. Control of rancidity in frozen Pacific herring, *Clupea harengus pallensi*: use of sodium erythorbate. *J. Fish Res. Board Can.* **36**:219–222.

Bilinski, E., Jonas, R.E.E., and Peters, M.D. 1981. Treatments affecting the degradation of lipids in frozen Pacific herring, *Clupea harengus pallasi*. *Can. Inst. Food Sci. Technol. J.* **14**:123–127.

Boyd, L.C., Green, D.P., Giesbrecht, F.B., and King, M.F. 1993. Inhibition of oxidative rancidity in frozen cooked fish flakes by *tert*-butylhydroquinone and rosemary extract. *J. Sci. Food Agric.* **61**:87–93.

Brotsky, E. 1976. Automatic injection of chicken parts with polyphosphate. *Poultr. Sci.* **55**:653–660.

Buettner, G.R. 1993. The pecking order of free radicals and antioxidants: lipid peroxidation, α-tocopherol, and ascorbate. *Arch. Biochem. Biophys.* **300**:535–543.

Burton, G.W., Hughes, L., Foster, D.O., and Pietrzak, E. 1993. Antioxidant mechanisms of vitamin E and β-carotene. In *Free Radicals: From Basic Science to Medicine.* (Poli, G., Albano, E., and Dianzani, M.U., eds.) pp. 388–399, Birkhauser Verlag, Basel.

Burton, G.W. and Ingold, K.U. 1984. β-carotene: an unusual type of lipid antioxidant. *Science* **224**:569–573.

Cabelli, D.E. and Bielski, B.H.J. 1983. Kinetics and mechanism for the oxidation of ascorbic acid/ascorbate by HO_2/O_2^- radicals: a pulse radiolysis and stopped-flow photolysis study. *J. Phys. Chem.* **87**:1809–1812.

Caldironi, H.A. and Bazan, N. 1982. Effect of antioxidants on malonaldehyde production and fatty acid composition in pieces of bovine muscle and adipose tissue stored fresh and frozen. *J. Food Sci.* **47**:1329–1332, 1337.

Cao, G., Sofic, E., and Prior, R.L. 1996. Antioxidant capacity of tea and common vegetables. *J. Agric. Food Chem.* **44**:3426–3431.

Chastain, M.F., Huffman, D.L., Hsieh, W.H., and Cordray, J.C. 1982. Antioxidants in restructured beef/pork steaks. *J. Food Sci.* **47**:1779–1782.

Cotelle, N., Bernier, J.-L., Catteau, J.-P., Pommery, J., Wallet, J.-C., and Gaydou, E.M. 1996. Antioxidant properties of hydroxy-flavones. *Free Radical Biol. Med.* **20**:35–43.

Crackel, R.L., Gray, J.I., Booren, A.M., Pearson, A.M., and Buckley, D.J. 1988. Effect of antioxidants on lipid stability in restructured beef steaks. *J. Food Sci.* **53**:656–657.

Craig, J., Bowers, J.A., and Seib, P. 1991. Sodium tripolyphosphate and sodium ascorbate monophosphate as inhibitors of off-flavor development in cooked, vacuum-packaged, frozen turkey. *J. Food Sci.* **56**:1529–1531, 1561.

Cuvelier, M.-E., Berset, C., and Richard, H. 1994. Antioxidant constituents in sage (*Salvia officinalis*). *J. Agric. Food Chem.* **42**:665–669.

Cuvelier, M.-E., Richard, H., and Berset, C. 1996. Antioxidative activity and phenolic composition of pilot-plant and commercial extracts of sage and rosemary. *J. Am. Oil Chem. Soc.* **73**:645–652.

Dalton, D.A., Russell, S.A., Hanus, F.J., Pascoe, G.A., and Evans, H.J. 1986. Enzymatic reactions of ascorbate and glutathione that prevent peroxide damage in soybean root nodules. *Proc. Natl. Acad. Sci. USA* **83**:3811–3815.

Dawson, L.E. and Gartner, R. 1983. Lipid oxidation in mechanically deboned poultry. *Food Technol.* **37**:112–116.

Deng, J.C., Matthews, R.F., and Watson, C.M. 1977. Effect of chemical and physical treatments on rancidity development of frozen mullet (*Mugil cephalus*) fillets. *J. Food Sci.* **42**:344–347.

Dmitriev, L.F., Ivanova, M.V., and Lankin, V.Z. 1994. Interaction of tocopherol with peroxyl radicals does not lead to the formation of lipid hydroperoxides in liposomes. *Chem. Phys. Lipids* **69**:35–39.

Eichner, K. 1980. Antioxidative effect of Maillard reaction intermediates. In *Autooxidation in Food and Biological Systems*. (Simic, M.G. and Karel, M., eds.) pp. 367–385, Plenum Press, New York.

Erickson, M.C. 1997. Oxidative susceptibility of catfish lipids as affected by the microenvironment of ascorbic acid. *J. Food Lipids.* In press.

Escobar, J.A., Rubio, M.A., and Lissi, E.A. 1996. SOD and catalase inactivation by singlet oxygen and peroxyl radicals. *Free Radical Biol. Med.* **20**:285–290.

Faraji, H., Decker, E.A., and Aaron, D.K. 1991. Suppression of lipid oxidation in phosphatidylcholine liposomes and ground pork by spray-dried porcine plasma. *J. Agric. Food Chem.* **39**:1288–1290.

Faustman, C., Cassens, R.G., Schaefer, D.M., Buege, D.R., Williams, S.N., and Scheller, K.K. 1989. Improvement of pigment and lipid stability in Holstein steer beef by dietary supplementation with vitamin E. *J. Food Sci.* **54**:858–862.

Fessenden, R.W. and Verma, N.C. 1978. A time-resolved electron spin resonance study of the oxidation of ascorbic acid by hydroxyl radical. *Biophys. J.* **24**:93–101.

Foti, M., Piattelli, M., Baratta, M.T., and Ruberto, G. 1996. Flavonoids, coumarins, and cinnamic acids as antioxidants in a micellar system: structure-activity relationship. *J. Agric. Food Chem.* **44**:497–501.

Frankel, E.N., Huang, S.-W., Kanner, J., and German, J.B. 1994. Interfacial phenomena in the evaluation of antioxidants: bulk oils vs. emulsions. *J. Agric. Food Chem.* **42**:1054–1059.

Frankel, E.N., Huang, S.-W., Prior, E., and Aeschbach, R. 1996. Evaluation of antioxidant activity of rosemary extracts, carnosol, and carnosic acid in bulk vegetable oils and fish oil and their emulsions. *J. Sci. Food Agric.* **72**:201–208.

Freeman, D.W. and Shannon, C.W. 1994. Rancidity in frozen catfish fillets as influenced by antioxidant injection and storage. *J. Aquat. Food Prod. Technol.* **3**:65–76.

Gordon, M.H. 1990. The mechanism of antioxidant action in vitro. In *Food Antioxidants*. (Hudson, B.J.F., ed.) pp. 1–18, Elsevier Applied Science, New York.

Graf, E., Mahoney, J.R., Bryant, R.G., and Eaton, J.W. 1984. Iron-catalyzed hydroxyl radical formation: stringent requirement for free iron coordination site. *J. Biol. Chem.* **259**:3620–3624.

Hanasaki, Y., Ogawa, S., and Fukui, S. 1994. The correlation between active oxygens scavenging antioxidative effects of flavonoids. *Free Radical Biol. Med.* **16**:845–850.

Harborne, J.B. 1988. *The Flavonoids: Advances in Research since 1980*. Chapman & Hall, New York.

Harman, D. 1982. The free-radical theory of aging. In *Free Radicals in Biology*, vol. 5. (Pryor, W.A., ed.) pp. 255–275, Academic Press, New York.

Hatate, H. and Kochi, M. 1992. Effect of antioxidants on lipid oxidation in homogenate of Japanese oyster during storage at 4 and −25°C. *Nippon Suisan Gakkaishi* **58**:2397.

Hirayama, O., Nakamura, K., Hamada, S., and Kobayasi, Y. 1994. Singlet oxygen quenching ability of naturally occurring carotenoids. *Lipids* **29**:149–150.

Ho, C.P., Huffman, D.L., Bradford, D.D., Egbert, W.R., Mikel, W.B., and Jones, W.R. 1995. Storage stability of vacuum packaged frozen pork sausage containing soy protein concentrate, carrageenan, or antioxidants. *J. Food Sci.* **60**:257–261.

Hu, J.P., Calomme, M., Lasure, A., De Bruyne, T., Pieters, L., Vlietinck, A., and Vanden Berghe, D.A. 1995. Structure-activity relationship of flavonoids with superoxide scavenging activity. *Biol. Trac. Elem. Res.* **47**:327–331.

Hwang, K.T. and Regenstein, J.M. 1988. Protection of membrane mince lipids from rancidity during frozen storage. *J. Food Sci.* **54**:1120–1124.

Jeremiah, L.E. 1985. Effects of antioxidants on rancidity development and palatability of frozen bacon. *J. Food Protect.* **48**:653–658.

Jitoe, A., Masuda, T., Tengah, I.G.P., Suprapta, D.N., Gara, I.W., and Nakatani, N. 1992. Antioxidant activity of tropical ginger extracts and analysis of the contained curcuminoids. *J. Agric. Food Chem.* **40**:1337–1340.

Jørgensen, K. and Skibsted, L.H. 1993. Carotenoid scavenging of radicals: effect of carotenoid structure and oxygen partial pressure on antioxidative activity. *Z. Lebensm. Unters. Forsch.* **196**:423–429.

Jovanovic, S.V., Steenken, S., Tosic, M., Marjanovic, B., and Simic, M.G. 1994. Flavonoids as antioxidants. *J. Am. Chem. Soc.* **116**:4846–4851.

Jurdi-Haldeman, D., Macneil, J.H., and Yared, D.M. 1987. Antioxidant activity of onion and garlic juices in stored cooked ground lamb. *J. Food Protect.* **50**:411–413, 417.

Kagan, V.E., Serbinova, E.A., and Packer, L. 1990. Generation and recycling of radicals from phenolic antioxidants. *Arch. Biochem. Biophys.* **280**:33–39.

Kaiser, S., Di Mascio, P., Murphy, M.E., and Sies, H. 1990. Physical and chemical scavenging of singlet molecular oxygen by tocopherols. *Arch. Biochem. Biophys.* **277**:101–108.

Kelleher, S.D., Silva, L.A., Hultin, H.O., and Wilhelm, K.A. 1992. Inhibition of lipid oxidation during processing of washed, minced Atlantic mackerel. *J. Food Sci.* **57**:1103–1108, 1119.

Kikuzaki, H. and Nakatani, N. 1993. Antioxidant effects of some ginger constituents. *J. Food Sci.* **58**:1407–1410.

Kim, S.Y., Kim, J.H., Kim, S.K., Oh, M.J., and Jung, M.Y. 1994. Antioxidant activities of selected oriental herb extracts. *J. Am. Oil Chem. Soc.* **71**:633–640.

King, D.L. and Klein, B.P. 1987. Effect of flavonoids and related compounds on soybean lipoxygenase-1 activity. *J. Food Sci.* **52**:220–221.

Kohen, R., Yamamoto, Y., Cundy, K.C., and Ames, B.N. 1988. Antioxidant activity of carnosine, homocarnosine, and anserine present in muscle and brain. *Proc. Natl. Acad. Sci. USA* **85**:3175–3179.

Krahl, L.M., Rhee, K.S., Lin, K.W., Keeton, J.T., and Ziprin, Y.A. 1995. Sodium lactate and sodium tripolyphosphate effects on oxidative stability and sensory properties of precooked reduced-fat pork sausage with carrageenan. *J. Muscle Foods* **6**:243–256.

Krinsky, N.I. 1989. Antioxidant functions of carotenoids. *Free Radical Biol. Med.* **7**:617–635.

Lai, S.-M., Gray, J.I., Smith, D.M., Booren, A.M., Crackel, R.L., and Buckley, D.J. 1991. Effects of oleoresin rosemary, tertiary butylhydroquinone, and sodium tripolyphosphate on the development of oxidative rancidity in restructured chicken nuggets. *J. Food Sci.* **56**:616–620.

Lanari, M.C, Cassens, R.G., Schaefer, D.M., and Scheller, K.K. 1993. Dietary vitamin E enhances color and display life of frozen beef from Holstein steers. *J. Food Sci.* **58**:701–704.

Lanari, M.C., Cassens, R.G., Schaefer, D.M., and Scheller, K.K. 1994. Effect of dietary vitamin E on pigment and lipid stability of frozen beef: a kinetic analysis. *Meat Sci.* **38**:3–15.

Liu, H.F., Booren, A.M., Gray, J.I., and Crackel, R.L. 1992. Antioxidant efficacy of oleoresin rosemary and sodium tripolyphosphate in restructured pork steaks. *J. Food Sci.* **57**:803–806.

Loliger, J. 1991. Use of antioxidants in foods. In *Free Radicals and Food Additives*. (Aruoma, O.I., ed.) pp. 121–150, Taylor & Francis, New York.

Mahoney, J.R., Jr., and Graf, E. 1986. Role of alpha-tocopherol, ascorbic acid, citric acid, and EDTA as oxidants in model systems. *J. Food Sci.* **51**:1293–1296.

Masuda, T. and Jitoe, A. 1994. Antioxidative and antiinflammatory compounds from tropical gingers: isolation, structure determination, and activities of cassumunins A, B, and C, new complex curcuminoids from *Zingiber cassumunar. J. Agric. Food Chem.* **42**:1850–1856.

Matlock, R.G., Terrell, R.N., Savell, J.W., Rhee, K.S., and Dutson, T.R. 1984. Factors affecting properties of precooked-frozen pork sausage patties made with various NaCl/phosphate combinations. *J. Food Sci.* **49**:1372–1375.

McNeill, J., Kakuda, Y., and Findlay, C. 1988. Influence of carcass parts and food additives on the oxidative stability of frozen mechanically separated and hand-deboned chicken meat. *Poultr. Sci.* **67**:270–274.

Michelson, A.M. and Monod, J. 1975. Superoxide dismutases and their applications as oxidation inhibitors. U.S. patent no. 391,0521.

Mikkelsen, A., Bertelsen, G., and Skibsted, L.H. 1991. Polyphosphates as antioxidants in frozen beef patties: lipid oxidation and color quality during retail display. *Z. Lebensm. Unters. Forsch.* **192**:309–318.

Monahan, F.J., Asghar, A., Gray, J.I., Buckley, D.J., and Morrissey, P.A. 1994. Effect of oxidized dietary lipid and vitamin E on the colour stability of pork chops. *Meat Sci.* **37**:205–215.

Moral, A., Tejada, M., and Borderias, A.J. 1986. Stability of frozen trout. I. Treated and untreated minces stored at −12, −18, and −24°C. *J. Food Biochem.* **10**:37–46.

Nakatani, N. 1992. Natural antioxidants from spices. In *Phenolic Compounds in Food and Their Effects on Health. II. Antioxidants and Cancer Prevention.* (Huang, M.-T., Ho, C.-T., and Lee, C.Y., eds.) pp. 72–86, American Chemical Society, Washington, D.C.

Nanni, E.J., Jr., Stallings, M.D., and Sawyer, D.T. 1980. Does superoxide ion oxidize catechol, α-tocopherol, and ascorbic acid by direct electron transfer? *J. Am. Chem. Soc.* **102**:4481–4485.

Packer, J.E., Slater, T.F., and Willson, R.L. 1979. Direct observation of a free radical interaction between vitamin E and vitamin C. *Nature (London)* **278**:737–738.

Pigeolet, E., Corbisier, P., Houbion, A., Lambert, D., Michiels, C., Raes, M., Zachary, M.-D., and Remacle, J. 1990. Glutathione peroxidase, superoxide dismutase, and catalase inactivation by peroxides and oxygen derived free radicals. *Mech. Age Develop.* **51**:283–297.

Porter, W.L., Black, E.D., and Drolet, A.M. 1989. Use of polyamide oxidative fluorescence test on lipid emulsions: contrast in relative effectiveness of antioxidants in bulk versus dispersed systems. *J. Agric. Food Chem.* **37**:615–624.

Pozo, R., Lavety, J., and Love, R.M. 1988. The role of dietary α-tocopherol (vitamin E) in stabilising the canthaxanthin and lipids of rainbow trout muscle. *Aquacult.* **73**:165–175.

Pratt, D.E. 1992. Natural antioxidants from plant material. In *Phenolic Compounds in Food and Their Effects on Health. II. Antioxidants and Cancer Prevention.* (Huang, M.-T., Ho, C.-T., and Lee, C.Y., eds.) pp. 54–71, American Chemical Society, Washington, D.C.

Ramanathan, L. and Das, N.P. 1992. Studies on the control of lipid oxidation in ground fish by some polyphenolic natural products. *J. Agric. Food Chem.* **40**:17–21.

Roginsky, V.A., Barsukova, T.K., Remorova, A.A., and Bors, W. 1996. Moderate antioxidative efficiencies of flavonoids during peroxidation of methyl linoleate in homogeneous and micellar solutions. *J. Am. Oil Chem. Soc.* **73**:777–786.

Salih, A.M., Price, J.F., Smith, D.M., and Dawson, L.E. 1989. Lipid oxidation in turkey meat as influenced by salt, metal cations, and antioxidants. *J. Food Qual.* **12**:71–83.

Santos, E.E.M. and Regenstein, J.M. 1990. Effects of vacuum packaging, glazing, and erythorbic acid on the shelf-life of frozen white hake and mackerel. *J. Food Sci.* **55**:64–70.

Schuler, P. 1990. Natural antioxidants exploited commercially. In *Food Antioxidants.* (Hudson, B.J.F., ed.) pp. 99–170, Elsevier Applied Science, New York.

Schwarz, K., Ernst, H., and Ternes, W. 1996. Evaluation of antioxidative constituents from thyme. *J. Sci. Food Agric.* **70**:217–223.

Selvaraj, P., Jasmine, G., and Jeyachandran, P. 1991. Effect of ascorbic acid dip treatment on frozen storage of squid (*Loligo duvaucelii*, Orbigny). *Fish. Technol.* **28**:117–121.

Serbinova, E., Kagan, V., Han, D., and Packer, L. 1991. Free radical recycling and intramembrane mobility in the antioxidant properties of alpha-tocopherol and alpha-tocotrienol. *Free Radical Biol. Med.* **10**:263–275.

Shahidi, F., Wanasundara, U.N., and Amarowicz, R. 1994. Natural antioxidants from low-pungency mustard flour. *Food Res. Int.* **27**:489–493.

Stodolnik, L. 1994. Effect of xanthan on changes in the lipids of comminuted cod meat during frozen storage. *Chlodnictwo* **29**(4):25–28.

Stoick, S.M., Gray, J.I., Booren, A.M., and Buckley, D.J. 1991. Oxidative stability of restructured beef steaks processed with oleoresin rosemary, tertiary butylhydroquinone, and sodium tripolyphosphate. *J. Food Sci.* **56**:597–600.

Stryer, L. 1975. *Biochemistry*. p. 72, Freeman, San Francisco.

Tabatabaie, T. and Floyd, R.A. 1994. Susceptibility of glutathione peroxidase and glutathione reductase to oxidative damage and the protective effect of spin trapping agents. *Arch. Biochem. Biophys.* **314**:112–119.

Tanaka, M., Kuei, C.W., Nagashima, Y., and Taguchi, T. 1988. Application of antioxidative Maillard reaction products from histidine and glucose to sardine products. *Nippon Suisan Gakkaishi* **54**:1409–1414.

Thed, S.T. and Erickson, M.C. 1994. Alteration in lipid oxidation of channel catfish during frozen storage in response to ascorbate applications. *J. Muscle Foods* **5**:37–48.

Therond, P., Couturier, M., Demelier, J.F., and Lemonnier, F. 1996. Hydroperoxides of erythrocyte phospholipid molecular species formed by lipoxygenase correlate with α-tocopherol levels. *Lipids* **31**:703–708.

Ueda, J.-I., Saito, N., Shimazu, Y., and Ozawa, T. 1996. A comparison of scavenging abilities of antioxidants against hydroxyl radicals. *Arch. Biochem. Biophys.* **333**:377–384.

van Acker, S.A.B.E., van den Berg, D.-J., Tromp, M.N.J.L., Griffioen, D.H., van Bennekom, W.P., van der Vugh, W.J.F., and Bast, A. 1996. Structural aspects of antioxidant activity of flavonoids. *Free Radical Biol. Med.* **20**:331–342.

Vekiari, S.A., Oreopoulou, V., Tzia, C., and Thomopoulos, C.D. 1993. Oregano flavonoids as lipid antioxidants. *J. Am. Oil Chem. Soc.* **70**:483–487.

Waagbø, R., Sandnes, K., Torrissen, O.J., Sandvin, A., and Lie, Ø. 1993. Chemical and sensory evaluation of fillets from Atlantic salmon (*Salmo salar*) fed three levels of N-3 polyunsaturated fatty acids at two levels of vitamin E. *Food Chem.* **46**:361–366.

Wada, S. and Fang, X. 1992. The synergistic antioxidant effect of rosemary extract and α-tocopherol in sardine oil model system and frozen-crushed fish meat. *J. Food Proc. Preserv.* **16**:263–274.

Whang, K., Aberle, E.D., Judge, M.D., and Peng, I.C. 1986. Antioxidative activity of α-tocopherol in cooked and uncooked ground pork. *Meat Sci.* **17**:235–249.

Yen, G.-C. and Duh, P.-D. 1993. Antioxidative properties of methanolic extracts from peanut hulls. *J. Am. Oil Chem. Soc.* **70**:383–386.

Yen, G.-C. and Duh, P.-D. 1994. Scavenging effect of methanolic extracts of peanut hulls on free-radical and active-oxygen species. *J. Agric. Food Chem.* **42**:629–632.

Yen, G.-C. and Duh, P.-D. 1995. Antioxidant activity of methanolic extracts of peanut hulls from various cultivars. *J. Am. Oil Chem. Soc.* **72**:1065–1067.

Yoshimoto, T., Furukawa, M., Yamamoto, S., Horie, T., and Watanabe-Kohno, S. 1983. Flavonoids: potent inhibitors of arachidonate 5-lipoxygenase. *Biochem. Biophys. Res. Commun.* **116**:612–618.

13

Edible Coatings and Films

Yvonne M. Stuchell and John M. Krochta

13.1 Introduction

Frozen storage of foods is an effective means of preservation that prevents or decreases undesirable chemical changes while maintaining many of the characteristics of fresh foods. However, during long-term storage, deterioration may still occur at a much reduced rate. Although much research on edible coatings and films to prevent food quality loss has been accomplished in the past 35 years, only a small portion of the work has been applicable to frozen-food systems.

Edible coatings (formed directly on foods) and films (preformed, then placed on foods or between food components) can function to prevent quality losses in frozen-food systems in numerous ways. Migration of moisture, lipids, and oxygen can occur in frozen foods, resulting in reduced quality of the finished product. Moisture migration is a major problem in both homogeneous and heterogeneous frozen-food systems. An edible coating can reduce the rate of moisture transfer between the food and the surrounding atmosphere, thus slowing the formation of package ice and the appearance of freezer burn. Moisture migration can be reduced between components in heterogeneous food systems, reducing the changes in appearance, texture, and taste that may result. Certain edible films created from hydrophilic materials are highly impermeable to fats and oils and have the potential to reduce migration of oils between components of the food product. Restriction of the movement of oxygen within the food or the availability of oxygen at the food surface can slow oxidative changes in the frozen product. Enrobing a food in an edible coating that is a good oxygen barrier could reduce or eliminate quality changes associated with the oxidation of lipids, vitamins, pigments, and flavor compounds. Edible coatings can also be used to incorporate various food additives into foods at specific locations, for example, they may maintain an antioxidant on the surface of a susceptible food or keep coloring from bleeding from one food component to another.

Additional potential roles of edible coatings and films in frozen foods relate to other structural and sensory aspects. Frozen foods are often susceptible to shattering or fragmentation, and an edible coating could be used to improve the structural integrity and mechanical handling characteristics of such a product. For example, pizza toppings could be held on the surface of a frozen pizza by an appropriate edible coating, which would then become part of the pizza during cooking. Edible coatings could also be used to preserve intact batter and bread layers on frozen products. With regard to sensory attributes, edible coatings can slow the loss of colors, volatile flavors, and aromas from a food product, retaining the freshness of a product for a longer period of time.

There are many challenges to using edible coatings and films on frozen foods. Decisions such as when to coat the food (prior to or after freezing), the method of applying the coating or film (spray, immersion, falling film, wrapping, etc.), and whether to dry or freeze a coating on the food will affect the performance of the coating. The appearance and texture of a coating may be different under freezing conditions or after being thawed or cooked but at all times must be acceptable. In coating formulations with a significant amount of water, moisture sublimation from the coating can be significant (moisture in the coating acts in a sacrificial way for moisture in the product), leaving a porous matrix of material on the food that may not function effectively or look appetizing. The coating or film should be able to withstand the significant temperature fluctuations and handling stresses that processed foods often undergo during manufacturing, storage, transportation from producer to market to home, and storage in home freezers prior to preparation for consumption.

13.2 Materials and Their Properties

Edible films and coatings can be created from components of milk, grains, legumes, and other plant and animal materials. Polysaccharides and their derivatives (alginates, pectins, agars, carrageenans, cellulose derivatives, amylose, starches, chitin, etc.) have been used to develop edible coatings that have good oxygen and carbon dioxide barrier properties at low to intermediate relative humidities (RH) but make poor moisture barriers because of the hydrophilic nature of these polymers. Proteins (casein, whey proteins, wheat gluten, soy proteins, corn zein, gelatin, collagen derivatives, etc.) have not been studied as extensively as polysaccharides for film formation but have similar barrier properties. Lipids (waxes, acetylated monoglycerides, surfactants, oils, etc.) form edible coatings that are good moisture barriers but often have poor structural handling characteristics.

Multicomponent composite films and coatings (which combine polysaccharides and/or proteins with lipids, either as emulsions or laminate systems) attempt to take advantage of the desirable qualities of each component to produce superior barriers. Generally, the polysaccharides and proteins provide structural integrity and (at low RH) oxygen, aroma, and lipid barrier properties, while the lipid component provides moisture barrier attributes.

Most studies on the barrier and mechanical properties of edible coating and film materials have been conducted at 25 ± 5°C and a variety of RH ranges. Because the properties of edible film and coating materials generally have a strong dependence on RH, comparison among studies and materials is difficult. Little data exists on the effect of temperature, especially freezing temperatures.

A bilayer film formed out of an emulsion of stearic and palmitic acid (SA-PA) in an aqueous-ethanolic solution of hydroxypropyl methylcellulose (HPMC) performed well over the temperature range of 40 to −19°C (Kamper and Fennema, 1984). The water vapor permeability (WVP) of this film increased as temperature was either increased or decreased relative to 25°C (table 13.1). Increase of WVP as temperature decreased was attributed to greater film hydration at the lower temperatures. Nonetheless, the WVP of the film was quite low at −19°C, which is comparable to polyethylene. This indicated that the HPMC upon which the otherwise brittle SA-PA layer rested provided good structural integrity to the composite film. Such low WVP at low temperature combined with the low water vapor partial pressure driving force for moisture transport at low temperatures provides an excellent moisture barrier. A related study examined the WVP of a composite film similar to the bilayer film above but with the addition of a thin layer of beeswax (Kester and Fennema, 1989a). This film was studied over the temperature range of 4–40°C. In this case, the WVP increased as temperature was increased or decreased relative to 13°C (table 13.1). Nonetheless, the WVP at 4°C was quite low. The film displayed good resistance to fracture at low temperature. WVP values measured at 25°C showed only a slight increase after 3 and 9 weeks of storage at −40°C, compared to no low-temperature storage. A study of the temperature dependence of wheat gluten and soy protein isolate film WVPs over the range of 5–35°C determined that WVPs increased as temperature decreased (Gennadios et al., 1994) (table 13.1). Again, this was attributed to increased film hydration at lower temperatures. However, the WVPs measured were one to two orders of magnitude greater than the composite films discussed above, reflecting the need for a lipid component to provide moisture barrier properties to an edible film or coating.

A study of the temperature dependence of corn zein, wheat gluten, and wheat gluten–soy protein isolate film oxygen permeability (OP) over the range of 7–35°C at 0% RH determined that the very low OPs became even lower as temperature decreased (Gennadios, Weller and Testin, 1993). However, studies at higher RH values are necessary to assess the impact of greater film hydration on OP values, especially effects that occur at lower temperatures.

The tensile strengths of films based on either corn zein, wheat gluten, methylcellulose, or hydroxypropyl cellulose increased with temperature over the range

Table 13.1. Effect of temperature on water vapor permeabilities of edible films

	Film[a]	Test conditions[b]	Permeability (g mm m^{-2} d^{-1} kPa^{-1})
Kamper and Fennema, 1984	SA:PA:HPMC:PEG	40°C, 83/0% RH	0.095
		25°C, 83/0% RH	0.057
		5°C, 83/0% RH	0.323
		−19°C, 83/0% RH	1.140
Kester and Fennema, 1989a	BW/SA:PA:MC: HPMC:PEG	40°C, 97/0% RH	0.209
		32°C, 97/0% RH	0.086
		25°C, 97/0% RH	0.063
		13°C, 97/0% RH	0.019
		4°C, 97/0% RH	0.044
Gennadios et al., 1994	WG:Gly	35°C, 50/100% RH	1.29
		25°C, 50/100% RH	1.51
		15°C, 50/100% RH	1.72
		5°C, 50/100% RH	2.41
Gennadios et al., 1994	SPI:Gly	35°C, 50/100% RH	1.51
		25°C, 50/100% RH	1.72
		15°C, 50/100% RH	2.11
		5°C, 50/100% RH	3.10

[a]HPMC = hydroxypropylmethyl cellulose, MC = methyl cellulose, SA = stearic acid, PA = palmitic acid, BW = beeswax, PEG = polyethylene glycol, WG = wheat gluten, SPI = soy protein isolate, Gly = glycerol

[b]Relative humidity on top and bottom sides of film (top/bottom)

of 5–45°C and decreased with relative humidity over the range of 23–75% RH (Gennadios and Park, 1993a). These results were attributed to the influence of temperature and RH on film moisture content. Generally, the cellulosic films had greater strength than the protein films, within the film formulations studied.

13.3 Applications

13.3.1 Meats, Poultry, and Seafood

Frozen meats, poultry, and seafood are susceptible to degradation under frozen conditions through moisture loss, lipid oxidation, protein alteration leading to changes in water binding capacity and texture, and flavor, color, and aroma changes. Ice glazes have often been used as a coating, but brittleness of the ice and loss of the glazing due to sublimation require periodic reglazing during storage to maintain the coating. The addition of corn syrup solids, cellulose gums, and pectinates to the glazing solution in an effort to reduce brittleness and sublimation has had limited success and acceptance (Ijichi, 1978).

Polysaccharide Coatings

Alginate and carrageenan coatings have been studied for use in improvement of frozen meat and fish quality. These coatings generally provide only a small barrier to moisture loss. However, the moisture in the coating acts as a sacrificial agent, delaying losses in the product until the coating moisture has been lost.

Earle (1968) patented an edible alginate coating that was applied in a two-step process. First, a solution of soluble sodium alginate was applied by dipping or spraying, followed by a similar application of a second solution containing calcium to set the gel. This was marketed under the trade name Flavor-Tex, and it was described as an edible, clear, odorless, and tasteless film that adapts to any surface and significantly extends shelf life to treated foods (McCormick, 1975). Flavor-Tex was used to coat beef portions and compared with ice glazing and a carrageenan coating (Shaw and Secrist, 1977). Samples were stored at −3°C and −12°C for 8 weeks and 5 months, respectively. When the beef portions were thawed and tested for moisture content, lipid oxidation (using thiobarbituric acid [TBA] value), appearance, and odor, none of the coatings provided significant improvement compared to a simple ice glaze. Ijichi (1978) compared the effectiveness of Flavor-Tex to an ice glaze containing ascorbate and an unglazed control in inhibiting chemical and microbiological deterioration of frozen salmon and rockfish. No significant differences between treatments was noted for total viable microorganisms and trimethylamine or hypoxanthine formation. Moisture analysis showed no significant differences between the ice glaze and Flavor-Tex, but both were superior to the uncoated controls. It was noted that Flavor-Tex may have a slight advantage over glazing because it remains intact upon thawing, thus controlling drip loss. The order of effectiveness (best to worst) with respect to lipid oxidation (TBA value), with each coating significantly different from the others, was ascorbate-ice glaze, Flavor-Tex, and uncoated controls.

Protein Coatings and Films

Klose et al. (1952) dipped frozen turkey steaks in 3% aqueous gelatin solutions containing fine suspensions of various combinations of antioxidants and then stored them at −12°C for 6 months. Rancid off-flavor was significantly reduced in both the meat and skin portions of the coated samples. Development of peroxides was markedly reduced by the gelatin-antioxidant coatings in all cases, with antioxidants at 0.15% total concentration being the most effective. The gelatin coating alone provided some protection against rancidification compared to the uncoated controls. In the same study, turkey halves were coated with a gelatin or gelatin-antioxidant mixture, wrapped in butcher paper, and stored under conditions similar to those used for turkey steaks. Although the level of fat deterioration was greater for the halves than for the steaks, similar results with respect to effectiveness of the coatings was achieved. Again, the gelatin coating provided limited protection against oxidation compared to controls. But, the addition of antioxidant to the gelatin coating increased protection significantly.

Farouk et al. (1990) determined the effect on fluid exudate of beef round steaks wrapped in an edible collagen film (Coffi), followed by either vacuum packaging with high-barrier bags or tray packaging with polyvinyl chloride (PVC) film wrap and storage at −25°C for 1 week. Steaks wrapped in Coffi film and packaged with PVC film had significantly lower amount of exudate than those not wrapped with Coffi film. Steaks wrapped with Coffi film before vacuum packaging also had significantly less exudate than corresponding vacuum-packaged steaks without Coffi film. They suggested that the presence of a film in close contact with the surface of the steaks could have reduced the amount of exudate, and that much of the exudate could have been absorbed by the Coffi film. No significant differences were noted for color or TBA number.

Beef cubes were wrapped and heat sealed in an edible collagen film, wrapped in freezer paper, and stored at −10°C for 5 months (Conca, 1994). When compared to uncoated controls that had been stored under the same conditions in high-barrier plastic bags, the coated samples were given higher ratings after 1.5, 3, and 5 months of storage for odor, flavor, texture, and overall quality. Little difference was found in color or lipid oxidation between the samples. The appearance of the cooked, coated meat samples was judged slightly less acceptable than the uncoated samples in sensory tests after 1.5 months of storage. However, the coated samples gained acceptability in sensory tests after 3 and 5 months of storage, as individuals became used to the appearance.

Stuchell and Krochta (1995) coated King salmon pieces with either an edible whey protein isolate (WPI) solution or a WPI solution followed by 5% antioxidant overspray and then stored the coated salmon pieces for 11 weeks at −23°C. No significant differences were noted with respect to moisture loss for WPI and WPI–antioxidant overspray treatments, and uncoated controls. For samples coated with WPI–antioxidant overspray, the onset of lipid oxidation was delayed for 2 weeks, compared to uncoated samples. In addition, the peak peroxide value of uncoated samples was reduced 55% at 35 days of storage with the WPI–antioxidant overspray coating.

Lipid Coatings

Acetylated monoglycerides were used to protect Thuringer sausages during frozen storage against moisture loss and oxidation of spices (Edible, 1967). Tests showed that after only 28 days of storage, the ordinary sausage in its modified natural casing had lost most of its starting moisture, while the sausages coated with acetylated monoglycerides had lost only 8% of starting moisture. Taste comparisons after a few weeks revealed a definite difference between coated and uncoated sausages.

In another study, chicken breasts were coated with acetylated monoglyceride (Myvacet 7-15) and stored at −18°C for 1- and 8-month periods (Zabik and Dawson, 1963). After cooking, coated samples were comparable in flavor, juiciness, percent moisture, and percent fat when compared to controls wrapped in polyvinylidene chloride (PVDC) wrap. In addition, the coated poultry had significantly less total cooking losses and higher percent press fluid than the control poultry.

Hirasa (1991) coated Silver salmon pieces with Myvacet 5-07 and stored them at −10°C for 10 weeks. Although he noted the formation of cracks and fractures in the coating during frozen storage, the coating was still highly effective at reducing moisture loss compared to uncoated samples. The Myvacet 5-07 coating also retarded the formation of peroxides and TBA in the samples, and the small increase noted when compared to vacuum-packaged controls was thought to be due to oxygen migration through the cracks and fractures of the coating. In order to reduce the development of cracks noted with solid acetylated monoglycerides, Stuchell and Krochta (1995) tested salmon pieces that had been coated with acetylated monoglycerides (Myvacet 9-08 and 9-45) that were liquid at ambient temperature. When stored at −23°C, these coatings were effective at reducing the rate of moisture loss by 42% during the first 3 weeks of storage, compared to uncoated controls. The onset of lipid oxidation was delayed for 2 weeks, and the peak peroxide value was reduced 54% with acetylated monoglyceride coatings when compared to uncoated controls.

Multicomponent Coatings

Multicomponent coatings may be composed of a blend of polysaccharides, proteins, and/or lipids formed from emulsions in a one-step process or formed as laminates in a multistep process. Edible coatings formed from emulsions are desirable in an industrial setting because they are convenient to use and often more resilient than single component (e.g., lipid) coatings. Laminate coatings are more complicated to form (requiring several steps), and as such may not be as desirable.

Two patents were granted to Bauer et al. (1968, 1969) for the following coatings for meat and poultry products: (1) a water-in-oil emulsion coating and (2) a water-in-oil emulsion containing a small amount of a soluble edible gum. The coating consists of fat, water or water-containing soluble gum, and emulsifier, plus other optional ingredients (seasonings, preservatives, flavoring and coloring materials, and viscosity controlling agents) and is applied to the meat at a thickness of 0.01–0.1 inch. One example given was frozen chicken pieces that were dipped in an emulsion consisting of vegetable oil, beeswax, monoglycerides, water, seasonings, and other ingredients and stored at −20°C for several weeks.

Another example involved coating frozen pork chops with an emulsion of vegetable oil, beeswax, monoglycerides, apple juice, seasonings, and other ingredients and storing under conditions similar to those of the chicken study. The coatings were completely absorbed by the meat product during cooking, and the cooked product was found to possess superior eating qualities.

Hirasa (1991) studied the effects of a casein–acetylated monoglyceride (Myvacet 5-07) emulsion coating on lipid oxidation and moisture loss from salmon pieces stored at −10°C for 10 weeks and −18°C for 6 months. Results showed that the emulsion coating had no effect on lipid oxidation. However, the emulsion coating retarded moisture loss from the fish samples, especially when the total solids in the emulsion was high (20%). These coatings were very thick and opaque and were removed prior to analysis of the fish samples. Stuchell and Krochta (1995) tested WPI–acetylated monoglyceride (Myvacet 9-08 and 9-45) emulsion coatings and compared them to a WPI–acetylated monoglyceride bilayer coating applied in two steps on frozen King salmon. The emulsion coatings provided no improvement over WPI-coated samples and were not statistically different from uncoated controls with respect to moisture loss. However, samples with the bilayer coatings had significantly less moisture loss than uncoated samples throughout the 11-week testing period. For samples coated with either emulsion or bilayer coatings, the onset of lipid oxidation was delayed for 2 weeks, compared to uncoated samples. In addition, the peak peroxide value of uncoated samples was reduced 41–62% at 35 days of storage with these coatings.

13.3.2 Fruits and Vegetables

Freezing of fruits and vegetables preserves these foods at their prime in a convenient form that is ready to cook. However, even in the frozen state, these products can lose moisture and acquire a dried-out taste and appearance. Frozen fruits and vegetables no longer respire but are susceptible to oxidative degradation of flavor, aroma, nutrients, and color components during storage. Although literature concerning coatings on fresh or lightly processed fruits and vegetables has been published, there has been no published work to date with respect to edible coatings on frozen fruits and vegetables.

13.3.3 Heterogeneous Foods

Heterogeneous foods have limited shelf life, even under frozen storage, due to moisture transfer between the components or between the product and the surrounding package environment. In foods such as breaded shrimp which are covered by standard of identity requirements, moisture transfer can result in the food product being out of regulation, requiring its removal from the market. In prefried frozen foods, lipid migration can occur between the batter and the food product. Edible coatings have the potential to eliminate or reduce these problems.

An edible coating for frozen seafood that minimizes deterioration during storage was developed by Earle and Snyder (1966). The process generally consists of inactivating the enzymes and bacteria in the seafood, then forming a coating of edible starch-alginate solution that is gelled using a calcium- or aluminum-containing gelling agent. For better protection of seafood during long periods of storage, an edible oil can also be included in the coating. The seafood can then be frozen or breaded and frozen. Shrimp coated with this treatment retained their original flavor, texture, and color after 3 months of frozen storage, while the control shrimp were found to be dry, tough, and generally of poor texture. The breading on shrimp first coated with the starch-alginate solution and then stored under freezing conditions for 6 months before deep-frying was found to have adhered well to the shrimp.

A patent was given to Earle and McKee (1985) for an edible alginate coating that reduced lipid oxidation and lipid and moisture transfer between components in food products having a batter, breading, sauce, or filling. The coatings also protected the food and held the crumbs to the food product. Food products suitable for coating included breaded meat and vegetables, filled dough products such as burritos, and tempura coatings on frozen food products.

Commercially produced sundae ice cream cones have a storage life of about 3 months, after which the cone becomes soggy due to moisture transfer from the ice cream. Rico-Peña and Torres (1990) tested a preformed bilayer film of methylcellulose and palmitic acid as a moisture barrier in simulated sundae ice-cream cones. Samples of sugar cones with these films showed no detectable moisture increase during 10 weeks of storage at −23°C. Those stored at −12°C showed no increase for 4 weeks, and only negligible increases appeared for the duration of the study.

Frozen pizzas and filled pies have a relatively dry crust in contact with a moist sauce or filling. Moisture movement from the sauce or filling to the crust results in a crust with a soggy texture. Kamper and Fennema (1985) tested a preformed, edible bilayer film of stearic-palmitic acid and hydroxypropyl methylcellulose between two food components (saltine crackers and tomato paste) in a model system to determine the film's ability to retard moisture migration. Samples stored at −20°C showed essentially no moisture transfer between components during the 70-day storage period. Kester and Fennema (1989b) tested a preformed stearic-palmitic fatty acid blend and cellulose ether bilayer film as a barrier to moisture migration between bread and a tomato-based sauce stored under abusive conditions of frozen storage (−6.7°C for 9 weeks). The film effectively retarded moisture movement from the sauce to the bread during storage, compared to the samples without film. Desirable sensory properties, measured after cooking the model product, were significantly better in the presence of the film compared to the nonfilm control. Fennema et al. (1990) patented an edible polysaccharide-lipid film for retarding water transfer among components of a multicomponent food product. The preformed film was positioned between the surface of French bread and a commercial pizza sauce, with the hydrophobic side of the film facing the sauce, and stored at −6.7°C for 9 weeks. The edible film provided an effective impediment to movement of water into the bread; and after baking, sensory evaluation showed that desirable sensory attributes of freshly prepared product were maintained.

13.4 Future Work

The limited information available indicates that edible coatings and films can play a significant role in the preservation of frozen foods, extending shelf life through the reduction of moisture, lipid, and aroma transfer; minimization of oxidation of lipids, pigments, flavors, and nutrients; and reduction of mechanical damage. The greatest potential for these films is in heterogeneous food products where the potential for deterioration of the food product is greatest. Additional research is needed on (1) the optimization of coating and film formulations for frozen-food applications; (2) development of systems for the application of coatings and films to foods, both before and after freezing; and (3) assessment of the cost-benefit of edible coatings with respect to frozen-food systems.

References

Bauer, C.D., Neuser, G.L., and Pinkalla, H.A. 1968. Method for preparing a coated meat product. U.S. patent 3,406,081.

Bauer, C.D., Neuser, G.L., and Pinkalla, H.A. 1969. U.S. patent 3,483,004. (As cited by Daniels, R. 1973. *Edible Coatings and Soluble Packaging.* Noyes Data Corporation, Park Ridge, N.J.)

Conca, K.R. 1994. Evaluation of collagen based film as an edible packaging for frozen meat. Presented at the First Multi-Society International Conference: Biodegradable Polymers, June 13–14, 1994, Ann Arbor, Mich. paper 13. (Lecture notes and visual aids published by Executive Conference Management, Plymouth, Mich.)

Earle, R.D. 1968. Method of preserving food by coating same. U.S. patent 3,395,024.

Earle, R.D. and McKee, D.H. 1985. Coated food product and method of making same. U.S. patent 4,504,502.

Earle, R.D. and Snyder, C.E. 1966. U.S. Patent 3,255,021. (As cited by Daniels, R. 1973. *Edible Coatings and Soluble Packaging.* Noyes Data Corporation, Park Ridge, N.J.)

Edible packaging offers pluses for frozen meat, poultry. 1967. *Quick Froz. Foods* **April**:165–67, 213–14.

Farouk, M.M., Price, J.F., and Salih, A.M. 1990. Effect of an edible collagen film overwrap on exudation and lipid oxidation in beef round steak. *J. Food Sci.* **55**:1510–1512, 1563.

Fennema, O.R., Kamper, S.L., and Kester J.J. 1990. Method for making an edible film and for retarding water transfer among multi-component food products. U.S. Patent 4,915,971.

Gennadios, A., Brandenburg, A.H., Park, J.W., Weller, C.L., and Testin, R.F. 1994. Water vapor permeability of wheat gluten and soy protein isolate films. *Ind. Crops Prod.* **2**:189–195.

Gennadios, A., Park, H.J., and Weller, C.L., 1993. Relative humidity and temperature effects on tensile strength of edible protein and cellulose ether films. *Trans. ASAE* **36**:1867–1872.

Gennadios, A., Weller, C.L., and Testin, R.F. 1993. Temperature effect on oxygen permeability of edible protein-based films. *J. Food Sci.* **58**:212–214, 219.

Hirasa, K. 1991. Moisture loss and lipid oxidation in frozen fish-effect of a casein-acetylated monoglyceride edible coating. M.S. Thesis, University of California, Davis.

Ijichi, K.E. 1978. Evaluation of an alginate coating during frozen storage of red snapper and silver salmon. M.S. Thesis, University of California, Davis.

Kamper, S.L. and Fennema, O. 1984. Water vapor permeability of an edible, fatty acid, bilayer film. *J. Food Sci.* **49**:1482–1484.

Kamper, S.L. and Fennema, O. 1985. Use of an edible film to maintain water vapor gradients in foods. *J. Food Sci.* **50**:382–84.

Kester, J.J. and Fennema, O. 1989a. An edible film of lipids and cellulose ethers: barrier properties to moisture vapor transmission and structural evaluation. *J. Food Sci.* **54**:1383–1389.

Kester, J.J. and Fennema, O. 1989b. An edible film of lipids and cellulose ethers: performance in a model frozen-food system. *J. Food Sci.* **54**:1390–1392, 1406.

Klose, A.A., Mecchi, E.P., and Hanson, H.L. 1952. Use of antioxidants in the frozen storage of turkeys. *Food Technol.* **6**(7):308–311.

McCormick, R.D. 1975. Edible coating isolates oxygen and moisture, controls structure—seals in flavor. *Food Prod. Dev.* **9**(4):14–16.

Rico-Peña, D.C. and Torres, J.A. 1990. Edible methylcellulose-based films as moisture-impermeable barriers in sundae ice cream cones. *J. Food Sci.* **55**:1468–1469.

Shaw, C.P. and Secrist, J.L. 1977. Edible coatings for individual frozen meat portions. U.S. Army Natick Research and Development Command, Food Engineering Laboratory, Natick, Mass.

Stuchell, Y.M. and Krochta, J.M. 1995. Edible coatings on frozen king salmon: effect of whey protein isolate and acetylated monoglycerides on moisture loss and lipid oxidation. *J. Food Sci.* **60**:28–31.

Zabik, M.E. and Dawson, L.E. 1963. The acceptability of cooked poultry protected by an edible acetylated monoglyceride coating during fresh and frozen storage. *Food Technol.* **17**(1):87–91.

14

Product Composition and the Quality of Frozen Foods

Norman F. Haard

14.1 Introduction

For the most part, freezing is a very effective way to preserve food-quality indices, namely, color, flavor, nutrient value, and texture. However, freezing and frozen storage of foodstuffs almost always causes physical and chemical changes that lead to some form of quality loss. The extent to which damage occurs depends on several factors including the freezing and storage technology employed, the application of prefreezing treatments, and the nature of the food product. Application of appropriate prefreezing treatments, freezing methods, and storage conditions, as discussed elsewhere in this book, are often effective in minimizing quality deterioration. For example, specific problems that occur in frozen foods, such as gelation of egg yolk, loss of surimi functionality, enzymatic browning of fruit tissue and crustacean meats, and rancidity of plant and animal tissues can be minimized with judicious use of additives or other prefreezing treatments. Freezing damage may be such that the effective frozen storage life of the product is limited, as with the rapid onset of oxidative rancidity in salmonid fishes or the loss of succulent texture during frozen storage of gadoid fishes. In some cases, freezing damage is such that the product is unacceptable. For example, the loss of turgor pressure in raw vegetables like tomato and lettuce yields a flaccid product that would not meet a consumer's expectation of salad components.

Prior to about 1929, foods were frozen with little or no consideration given to the variety or the starting quality of the product. The fish dealer froze any unsold seafood prior to the weekend for sale the following week. Modern food packers select products for freezing with consideration of variety of plant or breed of animal, and optimum structural and chemical makeup. In addition to considering optimal harvest time and freshness of the raw material, the biochemical makeup and stability of the product can be influenced by other factors. Conventional horticultural breeding programs have long been used to select for varietal characteristics that are better suited for specific processing operations such as freezing.

Moreover, with the advent of recombinant DNA techniques it is now possible to engineer transgenic animals or plants to have highly selective traits. In 1994, the first transgenic crop (Flavr Savr tomato) was approved for food use in the United States.

Several factors contribute to the biochemical makeup of the raw material. In addition to the contribution of heritable characteristics, the age of the animal or maturity of the plant, the sex of the animal, and the location of the tissue in the organism may be important. Extrinsic factors related to growing or rearing conditions may also influence biochemical makeup and postharvest changes that can determine suitability of the product for freezing preservation. Two examples, one for a plant tissue and one for an animal tissue, are given here to illustrate this point. "Water core" is a physiological disorder that develops in cold-stored apples. This problem can be minimized by increasing the calcium content of the pulp. Manipulation of calcium content can be achieved by proper nitrogen nutrition of the tree or by treatment of postharvest fruit with Ca^{2+} dips (Saks et al., 1990). Animal muscle cells are equipped with antioxidant enzyme systems that serve to minimize injury caused by free radicals and active oxygen species such as superoxide anions and singlet oxygen. Dietary selenium as well as antemortem exercise of the animal can increase the concentration of antioxidative enzymes in the muscle (Ji et al., 1988). The presence of these protective enzymes in the raw material is expected to minimize the rate of oxidative rancidity during frozen storage of the product.

This chapter explores how (1) the biochemical composition and organization of the sample determines the kind and extent of freezing damage, and (2) the biochemical makeup of the raw material can be engineered by environmental and genetic manipulations. The approach requires a basic understanding of mechanisms responsible for quality change in frozen foods.

14.2 Comparative Aspects of Food Tissues

14.2.1 Nature of Food Tissues

We do not usually think of harvested plant and animal tissues as living organisms, since they do not appear to require external nourishment, exhibit growth, or reproduce. However, we can consider, as did Thomas Aquinas, that "a thing is said to live in so far as it operates of itself and not as moved by another." Indeed, harvested asparagus continues to grow in the refrigerator, a side of beef continues to utilize glycogen and synthesize adenosine triphosphate for cell maintenance, a harvested potato tuber may sprout and reproduce itself after many months of storage, and a grain of wheat or tomato fruit may undergo death if deprived of oxygen. Foods can be classified as (a) intact tissue systems (e.g., fruits, vegetables, cereals, pulses, eggs, and myosystems), (b) disrupted tissue systems (e.g., ground meat, purees, flour, and tomato catsup), (c) noncellular or disrupted cellular systems (e.g., milk, fruit juices, honey, and oils), or (d) combinations thereof (Fennema et al., 1973).

14.2.2 Freezing Damage

With respect to resistance to freezing damage, some basic differences between plant and animal cells have been emphasized (Fennema, 1975). The parenchyma tissue of plants is made up of semirigid, polyhedral cells containing cellulosic cell walls bounded by a pectinaceous middle lamella and often containing an extensive network of air spaces. Mature plant cells contain a number of organelles not found in animal cells, notably chloroplasts, chromoplasts, large vacuoles, protein bodies, amyloplasts, and mature starch granules. The vacuole, which may comprise most of the mature plant cell, contains organic acids and phenols that are released when the fragile membrane is disrupted by freezing damage, which happens readily. In contrast, muscle cells are flexible, elongated fibers that are neatly aligned with minimal air spaces and separated by a glycoprotein-rich extracellular matrix. The texture damage often observed in frozen-thawed plant tissues is attributed to the semirigid nature of the cells and the less orderly packing of the cells in tissue.

Muscle cells are less likely to break as a consequence of freezing and thawing, and structural change is evidenced by cell-cell separation. The nature of the cell contents can influence the propensity for ice nucleation, and the permeability of the cell membrane influences the rate of water movement out of the cell (Reid, 1993). A cell wall or membrane that is not a good barrier to water transfer will favor intracellular dehydration and growth of extracellular ice. Although the muscle sarcolemma tends to have a greater hydraulic permeability than the plant cell wall and membrane, internal cell freezing is more common in animal cells. Considering the generally accepted mechanisms of freezing damage, it is clear that composition and structure of biological cells are key factors (table 14.1).

14.3 Fruits and Vegetables

14.3.1 Optimal Quality for Freezing

Normally, optimum quality for freezing occurs when maturity is near that which is customary for fresh consumption (Fennema, 1975). Other factors (size, color, flavor, etc.) depend on the specific characteristics of the crop and market needs. The basis for cultivar differences in freezing suitability varies with the crop. For example, pea cultivars bred for fresh consumption and canning are not suitable

Table 14.1. Relationship between freezing damage and nature of the biological cell

Mechanism of damage	Nature of damage	Biological variability
Solute concentration: Solute concentration occurs in the unfrozen matrix. Ionic strength increases.	Biopolymer structure and function is influenced by ionic strength; biopolymer aggregation is strongly influenced by concentration.	The kind and amount of solute in the cell is influenced by inter- and intraspecific biological factors.
Solute crystallization: pH can change when buffer components have different eutectic points.	pH influences functional properties of biopolymers, e.g., rates of chemical reactions, water holding capacity, etc.; pH influences color of anthocyanin pigments.	The kind and amount of buffer salts is influenced by inter- and intraspecific biological factors; the initial pH of muscle cells depends on postmortem time prior to freezing.
Dehydration: Osmotic transfer of water from within cell to outside matrix.	Decrease in cell volume may cause rupture or folding of cell membranes; decompartmentation of cell organelles causes mixing of cell components affecting reaction rates, pH, etc.	Inter- and intraspecific biological factors influence composition, plasticity, and permeation properties of biological membranes.
Mechanical Damage: Ice crystal growth can cause membrane distortions by a "ratchet mechanism."	Stress on flexible and rigid structures can cause formation of mechanical damage to structures.	Inter- and intraspecific biological factors influence composition and plasticity of biological membranes.
Biochemical Reactions: Enzyme and nonenzymatic reactions.	Chemical reactions that do not normally occur in the intact cell may occur as a consequence of the above changes.	Inter- and intraspecific biological factors influence the amount and kind of enzymes and chemical reactants present in the cell.

for freezing; the freezing industry employs small, dark-green cultivars that give an attractive appearance in the frozen package (Arthey, 1993). Peas destined for freezing are harvested at a less mature stage (more tender and with higher sugar content) than those used by the canner because freezing does not use the sugar that is included in the canning liqueur, and because of the impact of thermoprocessing on texture. Storage life of frozen peas is determined by stage of ripening and the starch, sugar, and ascorbic acid concentrations (Nacea et al., 1972).

14.3.2 Chemical Composition and Optimal Quality

Factors that influence the chemical composition and freezing characteristics of fruits and vegetables include cultivar, maturity, soil type, sowing date, plant population, pests, disease and weed control, harvesting method and conditions, and handling and transportation (Arthey, 1993). Ideally, the grower of crops destined for the freezer can select for characteristics important for one reason or another. A few examples relating cultivar and intraspecific factors to chemical composition are provided for illustration. The best harvest time for soybeans processed by freezing (86–90 days) appears to be related to low amounts of the amino acids methionine, lysine, and arginine which increase 90–105 days after planting (Islam and Lea, 1981). In frozen kiwi fruit, total acidity and pH are related to diminution of green color (chlorophyll) during frozen storage due to pheophytinization (Cano et al., 1993). The tendency of 40 blackberry cultivars to redden during frozen storage is associated with lower soluble solids, total anthocyanins, and higher titratable acidity; in avoiding this problem, sample maturity is more important than cultivar selection (Sapers et al., 1985). Elevating the amino acid and acid amide contents of sour cherries by N and K fertilization improves taste and keeping quality of the frozen product (Choureitah and Lenz, 1974). Decolorization of frozen strawberries varies with cultivar and appears to depend on the amount of total anthocyanins as well as the type of anthocyanin (Polesello and Rampilli, 1972); the amounts of pelargonidin 3-monoside, pelargonidin, and cyanidin chlorides in different cultivars exhibit good correlation with resistance to bleaching during frozen storage. Differences in the vitamin C content of frozen broad beans depend on cultivar, maturity, and cultivar-dependent losses during blanching and freezing (Kmiecik et al., 1990).

14.3.3 Problems

The two most common problems in frozen fruits and vegetables are (1) chemical reactions that cause changes in flavor, color, texture, and nutrition; and (2) physical damage that causes loss of turgor and texture changes.

Chemical Reactions

Enzymatic and nonenzymatic reactions occur at the temperature and water activity of conventional frozen foods (Karel, 1985). Enzymes known to contribute to deterioration during frozen storage of vegetables include several oxidoreductases such as lipoxygenase, catalase, peroxidase, and polyphenol oxidase as well as lipase and cystine lyase (Williams et al., 1986). The oxidoreductases are of primary importance because their action often leads to off-flavor and pigment bleaching in vegetables and to brown discoloration in some fruits. Lipoxygenase appears to be the main enzyme responsible for pigment bleaching and off-odor in frozen vegetables (Williams et al., 1986; Velasco et al., 1989).

Texture Damage

Susceptibility of fruits and vegetables to texture softening involves several factors (Reeve, 1970). In evaluating strawberry cultivars best suited for freezing, the texture score of the thawed product is a key quality attribute (Garrote and Bertone, 1975). The relationship between turgor and texture is less important in vegetables that are usually eaten after cooking (Grout et al., 1991). Loss of membrane semipermeability and disruption of cellular compartments can be minimized by rapid freezing rate, low storage temperature, and slow thawing. Other strategies that can be used by processors to maintain texture of soft fruits include chemical treatments, blanching, and control of ice crystal nucleation and growth.

14.3.4 Pretreatments to Minimize Chemical Reactions

Blanching

Most vegetables require blanching to destroy problematic enzymes. Exceptions include bell peppers, cucumbers, leeks, herbs, Jerusalem artichokes, onions, swedes, and turnips (Arthey, 1993). Most fruits do not require blanching, although those susceptible to enzymatic browning benefit from inactivation of polyphenoloxidase. The physical appearance of certain fruit juices, notably those of citrus fruits, can also be adversely influenced by the action of pectin methyl esterase and the interaction of divalent ions with demethylated pectin to cause loss of "cloud."

Indicator enzymes to establish blanching efficacy include peroxidase, catalase, and lipoxygenase (Velasco et al., 1989). Moreover, the maturity stage of the commodity can influence the efficacy of blanching, as in the prevention of darkening of frozen banana slices (Cano et al., 1990). Although blanching may serve other functions in addition to enzyme inactivation, the process is energy intensive and may cause undesirable changes such as loss of protein in the green pea (Markh and Zagibalov, 1970), and change the delicate aroma of fruit. Accordingly, other means to control problematic enzymes in frozen fruits and vegetables are also important (Pala, 1982). A general discussion of blanching is provided by Lund (1975).

Additives

Chemicals such as sulfur dioxide, sodium chloride, and malic, citric, and ascorbic acids are commonly used to retard enzymatic browning. For example, quality retention of frozen avocado purees can be achieved by pretreatment with sodium bisulfite and ascorbate to prevent enzymatic browning, and antioxidants to prevent oxidative rancidity.

Transgenic Crops

Another approach to the control of enzyme activity is to produce transgenic crops using antisense technology. For example, enzymatic browning in potato tubers can be prevented by antisense expression of polyphenol oxidase genes (Bachem et al., 1994).

14.3.5 Pretreatments to Prevent Texture Damage

Additives

Softening caused by freezing-thawing can sometimes be minimized by pretreatment of the tissue with $CaCl_2$ and/or sucrose (Polsello and Crivelli, 1971). In peaches, calcium interacts through ionic bonding with the carboxyl groups of pectin, while sucrose forms hydrogen bonds with pectin (Polesello and Maltini, 1970). Sucrose solutions (30–50%) also serve to preserve fruit aroma and prevent enzymatic browning by excluding oxygen. $CaCl_2$ and microwave pretreatment help maintain texture as well as ascorbic acid retention in frozen plantains (Giami, 1991). On the other hand, Ca^{2+} pretreatment is not effective in minimizing softening of frozen tomatoes (Hoeft et al., 1973). Modified gums have also been used as cryoprotectants and are effective in minimizing softening of frozen melon balls (Rio and Miller, 1979).

Low-Temperature Blanching

Heat treatment to activate pectin methyl esterase is sometimes combined with Ca^{2+} to minimize texture softening (Dewdney, 1978; Kaack, 1994; Fuchigami et al., 1995). Stepwise blanching has also been used to preserve the texture of frozen potatoes (Canet et al., 1982).

Antifreeze Proteins and Ice^{+} Genes

With the recent application of rDNA techniques to develop transgenic plants (Knight, 1989), a variety of new approaches to minimization of texture change during freezing can now be envisioned (Lee et al., 1992). At relatively high subzero temperatures, conditions are conducive for the generation of damaging, large ice crystals. Factors that favor supercooling and facilitate efficient ice nucleation can minimize the physical disruption caused by large ice crystals. Hence, the incorporation of antifreeze protein genes from Arctic fish (Johnston, 1990; Hightower et al., 1991), and the bacterial ice nucleating gene (Gurian-Sherman and Lindow, 1993) into the genome of plants may provide products that are more resistant to membrane damage and loss of turgor.

14.4 Eggs

14.4.1 Optimal Quality for Freezing

Frozen eggs and egg products are used in the food industry and in quantity food preparation (Savi, 1973). Whites can be frozen without any additives, but when yolks are frozen and stored below −6°C, an irreversible change in viscosity called *gelation* is observed (Palmer et al., 1970).

14.4.2 Chemical Composition and Optimal Quality

Egg yolk is a dispersion of particles in a clear plasma. Yolk contains 15.7–16.6% protein, 31.8–35.5% lipid, 0.2–1.0% carbohydrate, and 1.1% ash (Powrie and Nakai, 1990). Variability in the lipid content is attributed to strain rather than diet (Marion et al., 1964). The fatty acid composition of yolk lipid is influenced by the type of fat in diets of hens, and cholesterol content of the yolk is influenced by genotype of the bird (Powrie and Nakai, 1990). The low-density lipoproteins (LDLs) of egg yolk contain 84% lipid (3.7% cholesterol, 31% phospholipid, 65% triglyceride). Accordingly, it might be expected that factors such as breed, diet, and age of the hen may influence the composition of LDL particles in such a way as to influence propensity for gelation during freezing.

14.4.3 Problems

Constituents in both plasma and in granules contribute to the gelation problem (Chang et al., 1977). It is not clear how genetic and intraspecific factors may influence the composition of yolk components involved with gelation. An explanation for yolk gelation is that very low density lipoprotein (VLDL) particles become disrupted and form a mesh-type structure (Kurisaki et al., 1980; Nonami et al., 1992). Low-density lipoproteins in granules also form aggregates during freezing (Wakamatu et al., 1982).

14.4.4 Pretreatments to Minimize Gelation

A variety of approaches have been used to minimize the gelation of frozen egg yolk. Sugars (e.g., sucrose, galactose, glucose, and fructose) or NaCl are used as antigelling agents. Because of the high concentration of additive required to remedy the problem, other yolk properties are affected by the treatment. NaCl increases both emulsifying capacity and viscosity, whereas fructose reduces emulsifying capacity of the thawed yolk (Jaax and Travnicek, 1968). Yolk treated with salt (2–10%) and sugar (8–10%) has limited use because of the influence of the treatment on product taste. Treatment of egg yolk with proteolytic enzymes such as papain and trypsin can also prevent gel formation, but the product has reduced emulsifying capacity. Mechanical treatments have also been employed to reduce yolk viscosity. Low levels of salt, sugar, and/or skim milk combined with homogenization stabilizes frozen eggs used for scrambling (Penfield and Campbell, 1990). The reader is referred to Cotterill (1986) for a detailed discussion of this topic.

14.5 Milk

14.5.1 Optimal Quality for Freezing

Although milk is not normally preserved by freezing, a number of dairy products are frozen. Freezing point analysis is done to detect adulteration of milk with water. Lactation stage and time of milking have a significant influence on the freezing point of milk (Buchberger et al., 1986). Freezing point is influenced by the concentration of solutes such as Na, Ca, K, lactose, and so forth. For example, significant lowering of freezing point and increase in Cl^- is observed in milk from cows reared in warm weather (Demott, 1971).

14.5.2 Chemical Composition and Optimal Quality

The composition of dairy products such as cheese, yogurt, and ice cream influences the extent of freezing damage. Detailed discussion of these products is beyond the scope of this chapter. The composition and properties of dairy products is described by Swaisgood (1985). An example relating chemical composition of cheese with freezing damage is provided. Freezing and thawing of Cheddar cheese causes increased fracturability, mealiness, and lower meltability, which is more apparent in low-fat (34%) than normal (38%) and high-fat (44%) cheese and in low-salt cheese (Kasprzak et al., 1994). Aging the cheese after thawing can reverse some of these defects.

14.5.3 Problems

The emulsion stability of milk is decreased due to fat crystallization during freezing and thawing (Oortwijn and Walstra, 1982). This may explain why frozen-thawed milk appears to have a lower fat content than fresh milk (Weese et al., 1969). Frozen storage of UHT milk for 4–8 months produced significant losses in α-tocopherol (Vidal-Valverde et al., 1993) and retinol (Vidal-Valverde et al., 1992). Freezing and thawing of a semihard ewes' milk cheese does not influence sensory characteristics or chemical composition but facilitates proteolysis during subsequent aging (Fontecha et al., 1994).

14.5.4 Pretreatments

Additives of various types are added to stabilize frozen dairy products, but this subject matter is beyond the scope of this chapter. For a general discussion, the reader is referred to the review by Haylock et al. (1995). Salting Coulommier cheese prior to freezing and thawing minimizes texture defects caused by freezing (Richards, 1968). In ice cream, larger ice crystals give a coarse texture, and formation of small ice crystals is facilitated by addition of stabilizing agents such as alginates, vegetable gums, and gelatin (Everington, 1989).

14.6 Food Myosystems

14.6.1 Optimal Quality for Freezing

In general, meat has an excellent frozen storage life. The factors that influence the quality of frozen meat have recently been reviewed (Marsdon and Hendrickson, 1993) and include the intrinsic quality of the raw material, freezing conditions employed, and the protection and storage of the product throughout the cold chain. Some muscles are particularly susceptible to cold shortening and thaw rigor, and as a general practice it is adviseable to allow the muscle to go through rigor mortis prior to freezing.

Although slaughtered meat is normally chilled and held for an appropriate time prior to freezing, there are exceptions to this practice. Muscle to be used in sausage products can be advantageously frozen in the prerigor state, since mixing the frozen meat with NaCl prior to thawing and comminution will prevent thaw rigor. The higher pH of prerigor meat gives the proteins a higher capacity to bind water in sausage products. Muscle pH varies with species and postmortem holding time prior to freezing.

14.6.2 Chemical Composition and Optimal Quality

The freezing point of red meats, as for other tissues, is a function of composition. Factors known to influence freezing point of red meats include species, age, sex, and location of cut (Shafiur-Rahman, 1994). High-quality frozen poultry depends on several factors including maturity of the bird, diet prior to slaughter, and postmortem handling (Fennema, 1975). The breed of bird also influences meat quality (Belitz and Grosch, 1987). The specific heat and latent temperature of freezing for poultry meat depends on the type and chemical composition of the meat, especially water and fat contents (Gromov, 1985). Leg meat contains a higher ratio of unsaturated to saturated fatty acids than chicken breast meat (Miteva and Bakalivanova, 1987). During frozen storage, linoleic, linolenic, and palmitoleic acids decrease more in leg meat than in white meat.

Given the broad taxonomic diversity of fish and shellfish species, it is not surprising that freezing damage varies considerably in fishery products (Haard, 1990). However, freezing is an important method to preserve fish since it is difficult to extend market life at above-freezing temperatures because of the inability of current methods to control autolytic (biochemical) reactions at above-freezing temperatures (Haard, 1992a). The relationship between chemical composition of fish and factors such as biological age, sexual maturation, diet, feeding patterns, and environmental factors has been reviewed (Haard, 1992b). Knowledge of seasonal variation of chemical composition in products from the fishery can affect the choice of raw materials for freezing (Krynowek and Wiggin, 1979). The fish farmer has additional ways to influence fish product quality by manipulation of rearing conditions (Mohr, 1986; Haard, 1992b). For example, off-flavor development in frozen fish can be minimized by keeping the live fish off of feed prior to slaughter (Ross and Love, 1979), or by including antioxidants such as α-tocopherol in their diet (Erickson, 1991; Sekiya et al., 1991). As with land animals, the fatty acid composition of fish lipids can be influenced by diet (Waagbo et al., 1993). On the other hand, freshness of the raw material is more important than other biological factors in obtaining high-quality frozen skipjack (Tanaka et al., 1977). The tolerance limit of time before freezing is 1 h postmortem at 25–36°C.

One of the most important factors determining drip loss in thawed meat is the ultimate pH of the muscle (Lawrie, 1974). As ultimate pH of the beef decreases from 7.3 to 5.5, drip increases from about 1 to 8–15% depending on the cut of meat. *Longissimus dorsi* muscle produces less thaw drip than *psoas* muscle at all pH values. Moreover, a relatively low ultimate pH causing excessive drip in fish meat is often above pH 6. The reason for intrinsic differences in fluid loss in different muscles is not clear.

14.6.3 Problems

The two main causes of quality loss in frozen meat are lipid oxidation and protein denaturation. As mentioned already, freezing and thawing of myosystems often causes disruption of water-holding capacity and this results in drip loss and lowered meat quality. As mentioned, freezing meat after slaughter while the carcass is still warm can result in excessive formation of drip. This is attributed to thaw rigor that results from the sudden high rate of ATP breakdown caused by release of Ca^{2+} from the sarcoplasmic reticulum and mitochondria. Thaw rigor produces excessive shortening or contraction of myofibrils, which can toughen the meat or, when especially severe, causes "sloppy" texture. Extreme sarcomere shortening causes disruption of the normal alignment of the myofibrils and exudation of capillary fluids. In most cases, proper freezing of aged meat results in negligible fluid loss. The structural basis of water-holding capacity in meat is discussed elsewhere (Offer and Knight, 1988). Some fish species exhibit other problems, notably the disruption of connective tissue integrity called *gaping*, and toughening of texture due to denaturation and crosslinking of myofibril proteins. The general aspects of freezing fish (Lavety, 1991; Morrison, 1993) and the importance of biochemical reactions (Haard, 1994) have been reviewed.

Texture

Softening or toughening of meat during freeze-thawing or frozen storage can be caused by several factors including thaw rigor, protein denaturation and cross-linking, and loss in water-holding capacity. Texture toughening during frozen storage of some species of fish is a particularly important problem.

The texture of properly conditioned beef and other red meats is not normally affected by frozen storage. Minor changes in proteins may occur during the slow freezing of beef (Gunar et al., 1973) and pork (Rahelic et al., 1974). There is also evidence that freezing-thawing promotes the natural aging process in beef. Freezing beef inactivates calpastatin activity (Whipple and Koohmaraie, 1992). The decrease in calpastatin after thawing facilitates tenderization of beef by allowing greater proteolysis of key myofibril proteins by calpain. Likewise, prior freezing of meat from the Yorkshire breed of hogs (boars, not gilts or castrates) improves the development of tenderness and juiciness (Anastijevic et al., 1981).

Mechanical disruption of poultry meat prior to freezing influences water-holding capacity, as well as lipid oxidation. Deboned chicken neck and back have lower emulsifying capacity, water-holding capacity, and emulsion stability after frozen storage (Orr and Wogar, 1979). However, the source of necks has a greater influence on the composition and quality characteristics than the deboning machine used.

The loss in water-holding capacity in frozen fish is often attended by development of a dry, fibrous texture. This problem is more severe in gadoid species in which it is associated with denaturation and cross-linking of myosin (Sotelo et al., 1995). Theories to explain this problem include the greater sensitivity of fish myosin to denaturation, and the interaction of myosin with products of lipid hydrolysis, lipid oxidation, and trimethylamine oxide demethylation (Haard, 1994). In this regard, it is interesting that the stability of fish myosin is influenced by rearing temperature (Tsuchimoto et al., 1988).

Oxidative rancidity

Oxidation of unsaturated fatty acids is an important problem in some frozen myosystems, notably pork and fish as well as comminuted poultry, lamb, and beef. The frozen shelf life of ducks, geese, and chickens is determined largely by lipid oxidation (Gutschmidt, 1974). Brown discoloration due to the formation of metmyoglobin in red meats and carotenoid bleaching in fish and poultry tend to parallel fat oxidation. The fat content and composition of red meats vary with species. Beef and lamb have a lower content of polyunsaturated fatty acids than pork. Fatty acid composition of land animals is also influenced by the type of fat in the animals' diet. For example, increasing the amount of fish meal in the diet of hogs results in an increase in long-chain polyunsaturated fatty acids in the meat, and a decrease in frozen storage life (Valaja et al., 1992). Normally, pork has a much shorter frozen storage life than other mammalian meats because of rancidity.

Although beef and lamb are less susceptible to rancidity than pork, the sensory characteristics of frozen beef and lamb may also be influenced by fat oxidation. Muscle disintegration results in a low but significant release of membrane phospholipids and Fe^{2+} ions from myoglobin (Belitz and Grosch, 1987). For lipid oxidation in precooked, frozen ground beef, storage time appears to be a more important factor than storage temperature (Ramamurti, 1986).

Rancidity can develop rapidly during frozen storage of fish (Ortiz and Bello, 1992). Species as well as intraspecific factors influence the propensity for lipid oxidation in frozen fish. Lean fish, like Atlantic cod, may develop specific cold storage off-flavors (e.g., cardboardlike aroma) due to hydrolysis and oxidation of specific membrane phospholipids. In species like salmon, lipid oxidation may also be coupled to bleaching of carotenoid pigments (Dyer et al., 1956; Andersen et al., 1990). Various enzymatic and nonenzymatic factors contribute to lipid oxidation in frozen fish (Haard, 1994).

14.6.4 Pretreatments

Texture Changes

The addition of additives, such as soy protein to beef patties (Berry, 1991) or NaCl and tetrasodium pyrophosphate to buffalo and goat meats (Kondaiah et al., 1985), helps moisture retention during cooking from the frozen state. In pork, cooking loss is greater when NaCl (5%) is added prior to freezing than when the same amount of salt is added after freezing just prior to cooking (Vada et al., 1979). Ground turkey dark meat patties have been formulated with egg white, NaCl, and food grade polyphosphate (King et al., 1986). Cooking drip loss was lowest for samples made with polyphosphate and NaCl. Proteolytic enzymes are sometimes added to tenderize tougher cuts of meat prior to freezing. Some chemical treatments may serve to activate endogenous proteinases. For example, the application of Ca^{2+} after thawing facilitates tenderization of beef by allowing greater proteolysis of key myofibril proteins by calpain. Red meats are often butchered into small portions prior to freezing, thus increasing the tendency for fluid loss during thawing and cooking (Khan and Lenz, 1977). Pretreatment of poultry meat with low doses of electrons (0.1–4 kGy) lowers its superchilling temperature (Dubini et al., 1993). The extent to which the water-ice transition temperature is lowered by irradiation depends on the source of poultry meat.

Poultry is frequently packaged in a skintight film prior to freezing to avoid dehydration (freezer burn) and oxidation. Unfortunately, the exclusion of oxygen from packaged fish appears to promote texture deterioration in gadoids (Lundstrom et al., 1981). Although they do not appear to prevent toughening, polyphosphate dips have long been used to minimize drip loss of fish fillets (Schober and Duerr, 1989). Polyphosphate is more effective than BHT in preventing oxidation and carotenoid bleaching in frozen yellow sea bream (Song et al., 1977). Vacuum pretreatment has been used to minimize texture deterioration of frozen sea cucumbers (Tanaka and Liu, 1975). Although toughening is not normally a serious problem during frozen storage, a patented process to tenderize poultry and other myosystems has been reported (Narushima and Sekino, 1988). Freezing has been reported to alter the recovery of myosin from duck muscle (Smoczkiewiczowa et al., 1978).

Lipid Oxidation

Precooking pork prior to freezing does not appear to influence the increase in peroxide number during frozen storage (Cantoni et al., 1972). Freezing rate also influences the oxidation of lipids and myoglobin in pork sausage products (Wanous et al., 1989). Since denatured heme proteins can promote lipid oxidation, precooking sometimes increases the propensity of meat to undergo oxidation during frozen storage. Comparing changes during frozen storage of lamb muscle frozen directly after slaughter and that cooled prior to freezing, the latter exhibited less increase in free fatty acids and TBA value (Al-Dulaimy and Al-Aswad, 1987). This may be due to the lower pH of meat held prior to freezing. Precooked chicken meat develops rancid flavors and increased TBA values in 2–4 months. Rapid freezing provides a product with less oxidation during storage than does slow freezing (Berry and Cunningham, 1970).

Supplementing the diet of farm animals (hogs, beef, and lamb) with vitamin E (usually as α-tocopherol) to achieve muscle concentrations of 5–7 μg/g delays lipid oxidation and associated discoloration of red meats caused by oxidation of myoglobin (Sherbeck et al., 1995; Wulf et al., 1995). Lipid oxidation in frozen pork is minimized by airtight packaging and the use of antioxidants such as citric acid, butylated hydroxyanisole, and erythorbic acid (Perryman et al., 1989; Licciardello et al., 1982). Cures containing citric acid are not very effective in preventing lipid oxidation in poultry. However, exclusion of oxygen from packaged, deboned poultry meat improves oxidative stability of the frozen product (Barbut et al., 1990). Vacuum packaging and N_2 packaging are better than CO_2 packaging in preventing lipid oxidation in mechanically deboned chicken or turkey meat held for 4 months at −18°C (Jantawat, 1979). Stability of abdominal fat and meat of broilers is influenced by dietary fat and vitamin E supplements (Bartov and Bornstein, 1977).

14.7 Conclusions

Although the kinds of damage associated with freezing of different products are well documented, the molecular mechanisms involved are not always completely understood. Accordingly, the strategy to remedy a given problem based on manipulation of the raw material's biochemical makeup or by pretreatments depends on our understanding of the underlying mechanisms of cellular disruption and chemical reactions that contribute to quality loss. Genetic engineering experiments designed to manipulate the biochemical makeup of plant and animal tissues used for food will prove helpful in both gaining basic understanding of underlying mechanisms and in designing foods more suitable for preservation by freezing.

References

Al-Dulaimy, H.H.A. and Al-Aswad, M.B. 1987. Study of some chemical changes in lamb stored by freezing. *Iraqi J. Agric. Sci.* **5**(1):153–166.

Anastijevic, V., Josipovic, S., Dordevic, M., Tadic, I., Pavlov, I., Bulatovic, Z., and Talijan, R. 1981. Quantitative and certain qualitative properties and sensoric evaluations of meat from young boars, gilts, and castrates of Yorkshire pigs. *32nd Annual Meeting European Association for Animal Production.* p. 7.

Andersen, H.J., Bertelsen, G., Ohlen, G.C.A.A., and Skibsted, L.H. 1990. Development of rancidity in salmonoid steaks during retail display. A comparison of practical storage life of wild salmon and farmed rainbow trout. *Z. Lebensm. Unters. Forsch.* **191**(2):119–122.

Arthey, D. 1993. Freezing vegetables and fruits. In *Frozen Food Technology*. (Mallett, C.P., ed.). pp. 237–269, Chapman and Hall, London.

Bachem, C.W.B., Speckmann, G.-J., Linde, P.C.G., Verheggen, F.T.M., and Hunt, M.D. 1994. Antisense expression of polyphenoloxidase genes inhibits enzymatic browning in potato tubers. *Biotechnology* **12**(11):1101–1105.

Barbut, S., Kakuda, Y., and Chan, D. 1990. Effects of carbon dioxide, freezing, and vacuum packaging on the oxidative stability of mechanically deboned poultry meat. *Poult. Sci.* **69**(10):1813–1815.

Bartov, I. and Bornstein, S. 1977. Stability of abdominal fat and meat of broilers: the interrelationships between the effects of dietary fat and vitamin E supplements. *Br. Poult. Sci.* **18**:47–57.

Belitz, H.-D. and Grosch, W. 1987. *Food Chemistry*. Springer-Verlag, New York.

Berry, B.W. 1991. Effects of soy protein and freezing treatments on cooking loss and composition of beef patties. *J. Muscle Foods* **2**(2):105–118.

Berry, J.B. and Cunningham, F.E. 1970. Factors affecting the flavour of frozen fried chicken. *Poult. Sci.* **49**(5):1236–1242.

Buchberger, J., Graml, R., and Wolfschoon-Pombo, A.F. 1986. Studies of freezing point and composition of milk. Effect of time of milking, lactation stage, lactation number, and herd. *Dtsh Molk. Ztg.* **107**(32):1066–1068.

Canet, W., Espinosa, J., and Ruiz-Altisent, M. 1982. Effects of the stepwise blanching on the texture of frozen potatoes measured by mechanical tests. *Refrig. Sci. Technol.* **4**:284–289.

Cano, M.P., Fuster, C., and Marin, M.A. 1993. Freezing preservation of four Spanish kiwi fruit cultivars (*Actinidia chinensis*, Planch): chemical aspects. *Z. Lebensm. Unters. Forsch.* **196**(2):142–146.

Cano, P., Marin, M.A., and Fuster, C. 1990. Freezing of banana slices. Influence of maturity level and thermal treatment prior to blanching. *J. Food Sci.* **55**(4):1070–1072.

Cantoni, C., Minoccheri, C.B.F., and Merlino, M. 1972. Behavior of peroxides, free fatty acids, and vitamins B1, B2, and B12 during preparation, preservation, and reheating of precooked frozen foods. *Ind. Aliment.* **11**(11):80–82.

Chang, C.M., Powrie, W.D., and Fennema, O. 1977. Studies on the gelation of egg yolk and plasma upon freezing and thawing. *J. Food Sci.* **42**:1658–1665.

Choureitah, A. and Lenz, F. 1974. Fruit quality characteristics of sour cherries in relation to different applications of N and K. *Erwerbsobstbrau* **16**(9):136–138.

Cotterill, O.J. 1986. Freezing egg products. In *Egg Science and Technology*, 3rd ed., (Stadelman, W.J. and Cotterill, O.J., eds.) AVI Publishing, Westport, Conn.

Demott, B.J. 1971. Relationship between weather factors, source of sample, composition of milk, and freezing point depression. *J. Milk Food Technol.* **34**(3):140–141.

Dewdney, P.A. 1978. Plant cells and their role in fruit and vegetable processing. In *Scientific and Technical Surveys*, no. 107. Leatherhead Food RA, Leatherhead, Surrey, U.K.

Dubini, B., Ferrante, L., Leonardi, M., Messina, M., Montecchia, F., and Ponzi-Bossi, M.G. 1993. The freezing and thawing of water in poultry meat and vegetables irradiated by electrons at doses of 0.1–4 kGy. *Int. J. Rad. Biol.* **63**(6):687–692.

Dyer, W.J., Morton, M.L., Fraser, D.J., and Bligh, E.G. 1956. Storage of fish fillets. *J. Fish. Res. Bd. Can.* **13**:569–579.

Erickson, M.C. 1991. Extraction and quantitation of tocopherol in raw and cooked channel catfish. *J. Food Sci.* **56**:1113–1114.

Everington, D.W. 1989. The special problems of freezing ice cream. In *Food Freezing: Today and Tomorrow.* (Bald, W.B., ed.) pp. 133–142, Springer-Verlag, New York.

Fennema, O.R. 1975. Freezing Preservation. In *Physical Principles of Food Preservation (Principles of Food Science. Part III).* (Karel, M., Fennema, O.R., and Lund, D.B., eds) pp. 173–215, Marcel Dekker, New York.

Fennema, O.R., Powrie, W.D., and Marth, E.H. 1973. *Low Temperature Preservation of Foods and Living Matter*. Marcel Dekker, New York.

Fontecha, J., Pelaez, C., Juarez, M., and Martin-Hernandez, M.C. 1994. Effect of freezing and frozen storage on the physicochemical, organoleptic, and microbiological characteristics of a semi-hard ewes' milk cheese. *J. Dairy Sci.* **61**(1):133–142.

Fuchigami, M., Miyazaki, K., and Hyakumoto, N. 1995. Frozen carrots texture and pectic compounds as affected by low temperature blanching and quick freezing. *J. Food Sci.* **60**(1):132–136.

Garrote, R.L. and Bertone, R. 1975. Chemical evaluation and the suitability for freezing of strawberry cultivars. *Rev. ITA* **1**:81–99.

Giami, S.Y. 1991. Effects of pretreatments on the texture and ascorbic acid content of frozen plantain pulp (*Musa paradisiaca*). *J. Sci. Food Agric.* **55**(4):661–666.

Gromov, M.A. 1985. Specific gravity and specific heat of poultry meat. *Myasnaya Ind. SSSR* **10**:42–43.

Grout, W.W., Morris, G.J., and McLellan, M.R. 1991. Freezing of fruits and vegetables. In *Food Freezing: Today and Tomorrow*. (Bald, W.B., ed.) pp. 113–122, Springer-Verlag, New York.

Gunar, E.V., Yakubov, G.Z., Kaminarskaya, A.K., and Derbedeneva, Z.A. 1973. The effect of freezing techniques on the sarcoplasmic and myofibrillar proteins of meat. *Kholodil'naya Tekhnika* **11**:28–32.

Gurian-Sherman, D. and Lindow, S.E. 1993. Bacterial ice nucleation: significance and molecular basis. *FASO J.* **7**(14):1338–1343.

Gutschmidt, J. 1974. The storage life of frozen chicken with regard to the temperature in the cold chain. *Lebensm. Wiss. Technol.* **7**:139–145.

Haard, N.F. 1990. Enzymes from food myosystems. *J. Muscle Foods* **1**:293–338.

Haard, N.F. 1992a. Technological aspects of extending prime quality of seafood: a review. *J. Aquat. Food Prod. Technol.* **1**(3/4):9–27.

Haard, N.F. 1992b. Control of chemical composition and food quality attributes of cultured fish. *Food Res. Int.* **25**:289–307.

Haard, N.F. 1994. Biochemical reactions in fish muscle during frozen storage. In *Seafood Science and Technology*. (Bligh, E.G., ed.) pp.176–209, Blackwell Scientific Publications, Cambridge, Mass.

Haylock, S.J., Towler, C., and Hewitt, S.A. 1995. Dairy component interactions in food products. In *Ingredient Interactions*. (Gaonkar, A.G., ed.) chapter 10, Marcel Dekker, New York.

Hightower, R., Baden, C., Penzes, E. Lund, D.P., and Dunsmuir, P. 1991. Expression of antifreeze proteins in transgenic plants. *Plant Mol. Biol.* **17**(5):1013–1021.

Hoeft, R., Bates, R.P., and Ahmed, E.M. 1973. Cryogenic freezing of tomato slices. *J. Food Sci.* **38**(2):362.

Islam, M.N. and Lea, R.A. 1981. Effect of maturity on the amino acid composition and freezing quality of "Emerald" soybean. *J. Food Sci.* **46**(2):658–659, 663.

Jaax, S. and Travnicek, D. 1968. The effects of pasteurization, selected additives and freezing rate on the gelation of frozen-defrosted egg yolk. *Poult. Sci.* **47**:1013–1022.

Jantawat, P.P. 1979. Composition and stability of lipids from mechanically processed poultry meats. *Diss. Abstr. Int.* **40**(4):1614.

Ji, L.L., Stratman, F.W., and Lardy, H.A. 1988. Antioxidative enzyme systems on rat liver and skeletal systems. Influences of selenium deficiency, chronic training, and acute exercise. *Arch. Biochem. Biophys.* **263**:150–160.

Johnston, I.A. 1990. Cold adaptation in marine organisms. *Phil. Trans. Royal Soc. Lond. Ser. B: Biol. Sci.* **326**(1237):655–666.

Kaack, K. 1994. Blanching of green bean (*Phaseolus vulgaris*). *Plant Foods Hum. Nutr.* **46**(4):353–360.

Karel, M. 1985. Environmental Effects on Chemical Changes in Foods. In *Chemical Changes in Food during Processing*. (Richardson, T. and Finley, J.W., eds.) pp. 483–501, AVI Publishing, Westport, Conn.

Kasprzak, K., Wendorff, W.L., and Chen, C.M. 1994. Freezing qualities of Cheddar-type cheeses containing varied percentages of fat, moisture, and salt. *J. Dairy Sci.* **77**(7):1771–1782.

Khan, A.W. and Lenz, C.P. 1977. Effects of freezing, thawing and storage on some quality factors for portion-size beef cuts. *Meat Sci.* **1**(4):263–270.

King, A.J., Patel, S.B., and Earl, L.A. 1986. Effect of water, egg white, sodium chloride, and polyphosphate on several quality attributes of unfrozen and frozen, cooked dark turkey muscle. *Poult. Sci.* **65**(6):1103–1111.

Kmiecik, W., Lisiewska, Z., and Jaworska, G. 1990. Vitamin C levels in fresh, frozen, and canned broad beans in relation to their varieties and degree of maturity. *Roczniki Panstwowego Zakladu Higieny* **41**(1–2):17–24.

Knight, P. 1989. Engineering fruits and vegetables. *Bio/Technology* **7**(12):1233–1237.

Kondaiah, N., Anjaneyulu, A.S.R., Rao, V.K., Sharma, N., and Joshi, H.B. 1985. Effect of salt and phosphate on the quality of buffalo and goat meat. *Meat Sci.* **15**(3):183–192.

Krynowek, J. and Wiggin, K. 1979. Seasonal variation and frozen storage stability of blue mussels (*Mytilus edulis*). *J. Food Sci.* **44**:1644–1645.

Kurisaki, J.-I., Kaminogawa, S., and Yamauchi, K. 1980. Studies on freeze-thaw gelation of very low density lipoprotein from hen's egg yolk. *J. Food Sci.* **45**:463–465.

Lavety, J. 1991. Physico-chemical problems associated with fish freezing. In *Food Freezing: Today and Tomorrow*. (Bald, W.B., ed.) pp. 123–131, Springer-Verlag, New York.

Lawrie, R.A. 1974. *Meat Science*. Pergamon Press, New York.

Lee, S.P., Zhu, B.L., Chen, T.H.H., and Li, P.H. 1992. Induction of freezing tolerance in potato (*Solanum commersonii*) suspension cultured cells. *Physiol. Plant.* **84**(1): 41–48.

Licciardello, J.J., Ravesi, E.M., and Allsup, M.G. 1982. Stabilization of the flavor of frozen minced whiting. I. Effect of various antioxidants. *Mar. Fish. Rev.* **44**(8):15–21.

Lund, D. 1975. Heat Processing. In *Physical Principles of Food Preservation (Principles of Food Science. Part II)* (Fennema, O.R., ed.) pp. 32–33, Marcel Dekker, New York.

Lundstrom, R.C., Correia, F.F., and Wilhelm, K.A. 1981. Dimethylamine and formaldehyde production in fresh red hake (*Urophysis chuss*): The effect of packaging material oxygen permeability and cellular damage. *Refrig. Sci. Technol.* **4**:457.

Marion, W.W., Nordskog, A.W., Tolman, H.S., and Forsythe, R.H. 1964. Egg composition as influenced by breeding, egg size, age, and season. *Poult. Sci.* **43**:255–264.

Markh, A.T. and Zagibalov, A.F. 1970. Protein changes in green peas during freezing and cold storage. *Kholodil'naya Tekhnika* **47**(2):47–50.

Marsdon, J.L. and Hendrickson, R.L. 1993. Meat and meat products. In *Frozen Food Technology*. (Mallett, C.P., ed.) pp. 168–195, Chapman and Hall, New York.

Miteva, E. and Bakalivanova, T. 1987. Changes in the fatty acid composition of dietetic ready-to-cook chicken meat products during frozen storage. *Nahrung* **31**(10):981–986.

Mohr, V. 1986. Control of nutritional and sensory quality of cultured fish. In *Seafood Quality Determination*. (Kramer, D.E. and Liston, J., eds.) pp. 487–496, Elsevier Science Publishers, Amsterdam.

Morrison, C.R. 1993. Fish and shellfish. In *Frozen Food Technology*. (Mallett, C.P., ed.) pp. 196–236, Chapman and Hall, New York.

Nacea, M., Nicolescu, L., Cristea, S., Vasilescu, S., and Suciu, S. 1972. Keeping quality of frozen stored green peas. *Lucrari de Cercetare* **9**:75–82.

Narushima, N. and Sekino, Y. 1988. Meat tenderization with proteolytic enzyme–containing oil. U.S. patent 4,725,443.

Nonami, Y., Akasawa, M., and Saito, M. 1992. Effect of frozen storage on functional properties of commercial frozen whole egg. *J. Jpn. Soc. Food Sci. Technol.* **39**(1):49–54.

Offer, G. and Knight, P. 1988. The structural basis of water holding capacity in meats. In *Developments in Meat Science*. (Lawrie, R., ed.) pp. 173–244, Elsevier, Amsterdam.

Oortwijn, H. and Walstra, P. 1982. The membranes of recombined fat globules. IV. Effects on properties of the recombined milk. *Nether. Milk Dairy J.* **36**(4):279–290.

Orr, H.L. and Wogar, W.G. 1979. Emulsifying characteristics and composition of mechanically deboned chicken necks and backs from different sources. *Poult. Sci.* **58**(3):577–579.

Ortiz, H. and Bello, R. 1992. Composition and stability of fatty acids from deboned cachama and sardine meat during freezer storage. *Arch. Latinoam. Nutr.* **42**(4):460–466.

Pala, M. 1982. Effect of different pretreatments on the quality of deep frozen green beans and carrots. *Refrig. Sci. Technol.* **4**:224–231.

Palmer, H.H., Ijichi, K., and Roff, H. 1970. Partial thermal reversal of gelation in thawed egg yolk products. *J. Food Sci.* **35**:403.

Penfield, M.P. and Campbell, A.M. 1990. *Experimental Food Science*, 3d ed. pp. 153–155, Academic Press, New York.

Perryman, D.L., Nassif, W.S., and Cairncross, K.D. 1989. The loss of erythorbic acid in minced meat on storage at refrigerated and freezing temperatures. *J. Assoc. Public Anal.* **27**:127–129.

Polsello, A. and Crivelli, G. 1971. Tests on the addition of calcium salts to some fruits as pretreatment for quick freezing. *Ind. Agrar.* **9**(12):413–422.

Polesello, A. and Maltini, E. 1970. Studies on uptake of calcium salts during pretreatment of peaches for freezing. *Ind. Agrar.* **8**(6):199–205.

Polesello, A. and Rampilli, M. 1972. Studies on the behavior and stability of pigments in strawberries. II. Preliminary data on anthocyanin composition in fresh and frozen fruits. *Annali dell'Instituto Sperimentale per la Valorizzazione Tecnologica dei Prodotti Agricola* **3**:41–51.

Powrie, W.D. and Nakai, S. 1990. The chemistry of eggs and egg products. In *Egg Science and Technology*. (Stadelman, W.J. and Cotterill, O.J., eds.) pp. 97–140, Hawthorne Press, Binghamton.

Rahelic, S., Pribis, V., and Skenderovic, B. 1974. The influence of the time of freezing and period of frozen storage on the quality characteristics of beef and pork. II. Influence on changes in nitrogenous substances. *Fleischwirtschaft* **54**(3):497–499, 502.

Ramamurti, K. 1986. Effects of cooking, freezing and frozen storage on the fatty acid pattern of ground beef. *Lebensm. Wiss. Technol.* **19**(6):477–481.

Reeve, R.M. 1970. Relationships of histological structure to texture of fresh and processed fruits and vegetables. *J. Texture Stud.* **1**(3):247–284.

Reid, D.S. 1993. Basic physical phenomena in the freezing and thawing of plant and animal tissues. In *Frozen Food Technology*. (Mallett, C.P., ed.) pp. 1–19, Chapman and Hall, London.

Richards, E. 1968. Storage of Coulommier cheese in a deep freeze cabinet. *Dairy Ind.* **33**(11):755–757.

Rio, M.A.D. and Miller, M.W. 1979. Effect of pretreatment on the quality of frozen melon balls. *Bull. Inst. Int. Froid* **59**(4):1172–1175.

Ross, D.A. and Love, R.M. 1979. Decrease in the cold store flavour developed by frozen fillets of starved cod (*Gadus morhua* L.). *J. Food Technol.* **14**:115–122.

Saks, Y., Sonego, L., and Benarie, R. 1990. Senescent breakdown of Jonathan apples in relation to the water-soluble calcium content of the fruit pulp before and after storage. *J. Am. Soc. Hort. Sci.* **115**(4):615–618.

Sapers, G.M., Burgher, A.M., Phillips, J.G., and Galletta, G.J. 1985. Composition and color of fruit and juice of thornless blackberry cultivars. *J. Am. Soc. Hort. Sci.* **110**(2):243–248.

Savi, P. 1973. Frozen egg products. *Sci. Aliment.* **19**(4):87–93.

Schober, B. and Duerr, F. 1989. Effects of polyphosphates on quality and yield of fish products. *Fisch. Forsch.* **27**(3):71–76.

Sekiya, T., Murata, H., Sakai, T., Yamaguchi, K., Yamashita, K., Ugawa, M., Kanai, M., and Shimada, M. 1991. An attempt to control lipid peroxidation in the tissues of yellowtails and to enhance their biological protective ability by feeding high α-tocopherol brown meals. *Nippon Suisan Gakkaishi* **57**:287–292.

Shafiur-Rahman, M. 1994. The accuracy of prediction of the freezing point of meat from general models. *J. Food Eng.* **21**(1):127–136.

Sherbeck, J.A., Wulf, D.M., Morgan, J.B., Tatum, J.D., Smith, G.C., and Williams, S.N. 1995. Dietary supplementation of vitamin E to feedlot cattle affects beef retail display properties. *J. Food Sci.* **60**(2):250–252.

Smoczkiewiczowa, A., Nowotarska, G., and Wieladek, H. 1978. Determination of myosin (purified using Sephadex G-100) in fresh and frozen duck meat. *Zeszyty-Naukowe Akademia Ekonomiczna w Poznaniu* **73**:28–34.

Song, D.J., Hur, J.W., and Kang, Y.J. 1977. Studies on freezing of yellow sea bream. I. Effects of freezing and storage temperature and chemicals on the quality of yellow sea bream. *Bull. Korean Fish. Soc.* **10**(4):221–226.

Sotelo, C.G., Pineiro, C., and Perez-Martin, R.I. 1995. Denaturation of fish proteins during frozen storage: role of formaldehyde. *Z. Lebensm. Unters. Forsch.* **200**(1): 14–23.

Swaisgood, H.E. 1985. Characteristics of edible fluids of animal origin: milk. In *Food Chemistry*, 2d ed., (Fennema, O.R., ed.) chapter 13, Marcel Dekker, New York.

Tanaka, K. and Liu, J.B. 1975. Freezing preservation of fresh sea cucumber (*Stichopus japonicus* Selenka). *Refrigeration* **50**(567):3–10.

Tanaka, T., Kakuda, K., Nishiwa, K., Nakayama, T., and Otsuka, Y. 1977. Influence of biological conditions of skipjack aboard ship and pretreatment with chilled seawater on the final quality of frozen fish. *Refrigeration* **52**(591):3–24.

Tsuchimoto, M., Tanaka, N., Misa, T., Yada, S., Senta, T., and Yasuda, M. 1988. The influence of rearing temperature on the relative thermostability of myofibrillar Ca^{2+}-ATPases and on the lowering speed of freshness in carp. *Nippon Suisan Gakkaishi* **54**:117–122.

Vada, M., Pusztai, K., Redey, Z., and Lengyel, A. 1979. Variation of quality traits during storage at low temperatures (freezing and cooling). *Proc. Eur. Meet. Meat Res. Workers* **25**:605–610.

Valaja, J., Suomi, K., Alaviuhkola, T., and Immonen, I. 1992. Effect of dietary fish meal on the palatability and fatty acid composition of pork. *Agric. Sci. Finl.* **1**(1):21–26.

Velasco, P.J., Lim, M.H., Pangborn, R.M., and Whitaker, J.R. 1989. Enzymes responsible for off-flavor and off-aroma in blanched and frozen-stored vegetables. *Biotech. Appl. Biochem.* **11**:118–127.

Vidal-Valverde, C., Ruiz, R., and Medrano, A. 1992. Stability of retinol in milk during frozen and other storage conditions. *Z. Lebensm. Unters. Forsch.* **195**(6):562–565.

Vidal-Valverde, C., Ruiz, R., and Medrano, A. 1993. Effects of frozen and other storage conditions on alpha-tocopherol content of cow milk. *J. Dairy Sci.* **76**(6):1520–1525.

Waagbo, R., Sandnes, K., Torrissen, O.J., Sandvin, A., and Lie, O. 1993. Chemical and sensory evaluation of fillets from Atlantic salmon (*Salmo salar*) fed three levels of *n*-3 polyunsaturated fatty acids at two levels of vitamin E. *Food Chem.* **46**(4):361–366.

Wakamatu, T., Sato, Y., and Saito, Y. 1982. Identification of the components responsible for the gelation of egg yolk during freezing. *Agric. Biol. Chem.* **42**(6):1495–1503.

Wanous, M.P., Olson, D.G., and Kraft, A.A. 1989. Pallet location and freezing rate effects the oxidation of lipids and myoglobin in commercial fresh pork sausage. *J. Food Sci.* **54**(3):549–552.

Weese, S.J., Butcher, D.F., and Thomas, R.O. 1969. Effect of freezing and length of storage on milk properties. *J. Dairy Sci.* **52**(11):1724–1726.

Whipple, G. and Koohmaraie, M. 1992. Freezing and calcium chloride marination effects on beef tenderness and calpastatin activity. *J. Anim. Sci.* **70**(10):3081–3085.

Williams, D.C., Lim, M.H., Chen, A.O., Pangborn, R.M., and Whitaker, J.R. 1986. Blanching of vegetables for freezing—which indicator enzyme to choose. *Food Technol.* **40**:130–140.

Wulf, D. M., Morgan, J.B., Sanders, S.K., Tatum, J.D., Smith, G.C., and Williams, S. 1995. Effects of dietary supplementation of vitamin E on storage and caselife properties of lamb retail cuts. *J. Anim. Sci.* **73**:399–405.

15

Role of Packaging in Quality Preservation of Frozen Foods

V. M. Balasubramaniam and M. S. Chinnan

15.1 Introduction

Over the last two decades the American family has changed such that women represent an increasing part of the work force. As a result, in about 62% of families, both husband and wife go to work away from home. Due to changes in life styles, consumers are demanding convenience, variety, quality, healthfulness, and value from the food they consume. According to a survey by Good Housekeeping Institute, nearly 70% of the female respondents expressed interest (ranging from "somewhat" to "high") in frozen foods packaged in microwave-safe materials. The frozen food industry is responding to this challenge by introducing novel food products using improved processing and packaging technologies (Ott, 1988). As a consequence, according to a U.S. Department of Commerce projection, the frozen entrée market has been growing at an annual rate of 15% or greater, and the trend is expected to continue (Rice, 1986a).

The type of package selected by any frozen-food packager is dependent on the particular product, its niche in the marketplace, the company's packaging philosophy, and existing equipment. Therefore, choosing a packaging medium is more complex than merely determining the lowest priced or most consumer-appealing package (Alexander, 1993). In this chapter, the recent advances in frozen packaging technologies and package-product interaction in preserving quality have been reviewed. Emphasis is placed on summarizing packages available to the consumers at the retail level.

15.2 Packaging Material

One of the important catalysts for packaging advancements has been the development of multilayer plastics. These advancements are partially due to the developments in high-barrier plastics, such as the ability to coextrude various polymers. These advances coupled with new food preservation techniques (such as modified atmosphere and dual-ovenable packaging technology) have revolutionized frozen-food packaging sectors. Some important plastics and films used include polyamide (nylon), polyester, polyethylene (PE), low-density polyethylene (LDPE), polypropylene (PP), polystyrene (PS), polyvinylidine chloride (PVDC), polyvinyl chloride (PVC), polyethylene terephthalate (PET), crystallized polyester (CPET), and ethylene vinyl alcohol (EVOH) (Ott, 1988; Rice, 1990). The properties of selected film materials adapted from Parry (1993) are given in table 15.1.

15.3 Frozen-Food Packages

15.3.1 Modified Atmosphere Packaging

Modified atmosphere packaging (MAP) is defined as the enclosure of food products in gas-barrier materials while providing a gaseous environment that has been changed or modified (Young et al., 1988) to inhibit the action of spoilage agents. This results in either maintenance of quality of a perishable food during its natural life, or the actual extension of the shelf life. Vacuum packaging and gas packaging are two major types of packaging available in this category.

Vacuum Packaging

Vacuum packaging involves placing a product in a film of low oxygen permeability, removing air from the package, and applying a hermetic seal (Smith et al., 1990). Vacuum packages are used primarily for shipping frozen food from packers to retail grocery, hotel, restaurant, and institutional outlets.

Under good vacuum conditions, the O_2 level is reduced to less than 1% by vacuum packaging. The method has been shown to be effective in inhibiting bacterial growth at low temperatures, resulting in reduced spoilage and extended product shelf life (Ogrydziak and Brown, 1982; Brewer and Wu, 1993). In addition, it has been shown to be effective in maintaining the quality of a variety of food products including frozen pizza (Piazza, 1988), shrimp (Bramsnaes, 1969), mushrooms (Steinbuch, 1979), ground beef (Lynch et al., 1986; Brewer and Wu, 1993), and pork (Brewer and Harbers, 1991a,b). Vacuum packaging, however, was reported to be less effective than blanching for long-term quality preservation of unblanched frozen spinach (Birnbaum et al., 1979).

Gas Packaging

In this packaging technique, the product is generally stored in an atmosphere containing an appropriate composition of N_2, CO_2, and O_2. N_2 is used primarily

Table 15.1. Properties of selected barrier films

Film[a]	H_2O vapor transmission ($g/m^{-2}\ day^{-1}$ at 38°C and 90% RH)	O_2 permeability ($cm^3 m^{-2}\ day^{-1}\ atm^{-1}$ for 25 μm film at 25°C)
Polyester (oriented)	25–30	50–130
CPET	Variable[b]	Variable[b]
PVDC-PVC copolymer (Saran)	1.5–5.0	8–25
LDPE	18	7,800
HDPE	7–10	2,600
Nylon-6	84–3100	40
Polypropylene (oriented)	6–7	2,000
Plasticized PVC	15–40	500–3,000
Polystyrene (oriented)	100–125	5,000
EVOH	16–18	3–5

[a]Film abbreviations:

PET	Crystallized polyethylene terephthalate
PVC	Polyvinyl chloride
PVDC	Polyvinylidene chloride
LDPE	Low-density polyethylene
HDPE	High-density polyethylene
EVOH	Ethylene vinyl alcohol

[b]Permeabilities change according to level of crystallinity. For each 1% change in crystallinity, there is a 1.5% increase in transmission rate

as a filler to prevent package collapse with foods that absorb the CO_2, or simply to displace O_2 to retard oxidation. The packaging materials used for this type of package should be suited for low-temperature storage as well as have desired gas permeability values.

Gas packaging is often used for frozen storage of precooked, ready-to-eat products (Hwang et al., 1990; Ray et al., 1994). The package containing the product would be distributed in the frozen state. Then the product is usually thawed and sold as chilled product but with extended shelf life (Morris, 1989).

Skin Packaging

Skin packaging is essentially a modified form of vacuum packaging. A preheated film is dropped onto the product which is supported on a lower web of the same film. The air between the two films is withdrawn and an upper web forms around the food to produce a skintight package that is then heat sealed in a vacuum chamber. It is used primarily for packaging frozen meat products. A skin-packaged, frozen red-meat product should be regarded as one that can be stored in the dark until required and then displayed attractively for a few days for retail sale. In a supermarket, skin-packed meat can be sold from frozen display cases just as chilled meat is sold from chilled cases. The advantages would be that the frozen case will require only regular replenishment from frozen dark storage instead of the costly in-store butchery necessary for retailing chilled meat (Taylor, 1979).

15.3.2 Dual-Ovenable Packaging

Due to changing life styles and proliferation of microwave ovens, demand for dual-ovenable packaging has increased in the frozen-food industry. These new packaging materials are made from microwave-penetrable materials and have sufficient heat resistance so that the food may be heated in conventional (gas, electric, and convection) ovens as well. The requirements of a dual-oven tray as a frozen-food package include (1) the ability to protect the food so contained; (2) the ease of heating or cooking the food; and (3) the ability to withstand temperature extremes (ranging from deep-freezing at −40°C to reheating at 200–220°C). Three dual-ovenable tray types are commonly available: paperboard/fiberboard trays, aluminum foil trays, and heat-tolerant plastic trays (Ono, 1990).

Paperboard, molded pulp, and PET-coated paperboard containers form a large and growing segment of the container market. These trays have sufficient heat resistance to resist several minutes of oven temperature in excess of 260°C, which enables dual-oven use for paperboard trays. In addition such packaging lends itself to desirable microwavability, printability, and heat sealability characteristics. Consumer advantages for this type of package include serving-size flexibility (e.g., two-pack and four-pack tear-apart containers) and cost savings due to lower packaging and shipping costs. Dual-ovenable, foil-based packaging has also evolved so that the packages can be used in microwave ovens. The coated film foil is more rigid than a typical paperboard tray, and packaging costs are nearly the same.

Numerous frozen entrées are packaged in crystallized polyester (crystallized polyethylene terephthalate [CPET]) trays. These trays have a variety of uses including packaging, cooking, and serving of food (Ott, 1988). Consumers like the reusability of the CPET tray and tend to perceive foods that are packaged in CPET as being higher in quality. Another type of heat-tolerant plastic packaging is constructed of thermo-set, fiber-reinforced, mineral-filled polyester. These high density trays are sturdy and very durable. Also, packages constructed using polycarbonate and polyethermide coextruded in a five-layer arrangement are used for frozen entrées and side dishes. The tray is dual-ovenable and has consumer-attractive features such as an antistick surface that resists the adhesion of food to the tray interior, and a temperature control that enables it to be removed from ovens without the use of potholders (Rice, 1987).

15.4 Influence of Package Parameters on Quality

The critical package material parameters that influence the quality of foods include gas permeability, water vapor permeability, mechanical strength, and transparency. The functional attributes of the package material such as its microwaveability and ability to protect the contents against temperature abuse should also be considered. A successful package must be tough enough to withstand damage at freezing temperatures from the sharp, unyielding corners of frozen food and should have an appropriate barrier to reduce weight loss from moisture evaporation. Moisture migration may still take place in the frozen food from the interior to the surface, resulting in frost formation between the product and the package. This may lead to desiccation of the surface and development of off-flavors. The internal frosting can be avoided by suitable package design such that the package fits closely around the food. Improperly designed packages may also allow oxygen and/or light to enter, resulting in deterioration of pigments, lipids, proteins, and overall eating quality, especially as frozen storage time increases (Jeremiah, 1980).

15.4.1 Gas Permeability

The exact equilibrium of atmosphere attained within a sealed package depends on many factors of which most importantly on the gas permeability values of the package material. Packaging materials that excluded oxygen forestalled development of oxidative reactions responsible for increases in rancid and acid flavors, and as a result increased the product shelf-life (Brewer and Harbers, 1991a; Bjerkeng and Johnsen, 1995). The formation of thiobarbituric acid–reactive substances (TBARS), a measure of lipid oxidation, was reduced by packaging in materials with a low oxygen transmission rate (Christophersen et al., 1992; Bjerkeng and Johnsen, 1995). Retention of carotenoid and ascorbic acids in oxygen-sensitive frozen foods was reported with oxygen-impermeable pouches (Thakur and Arya, 1988). A package material with very low oxygen permeability, however, may lead to anaerobic conditions followed by tissue breakdown and development of off-flavors and other undesirable metabolic reactions. The permeability of a film is also a function of temperature. Therefore, a film that creates a near-optimum atmosphere at ambient conditions may not necessarily provide optimal atmospheric conditions under the changing storage or handling temperatures encountered during freezing.

15.4.2 Water Vapor Permeability

Water vapor permeability of a package depends on the permeability and surface area of the packaging material. As with gas permeability, water vapor permeability is affected by temperature. Films of various permeabilities are readily available (table 15.1). With vacuum-packaged pork, the moisture was retained, however, PVC-packaged pork had a lower moisture content (Brewer and Harbers, 1991a,b). A choice of film with higher water vapor permeability will result in dehydration of the product and concomitant acceleration of protein denaturation and oxidation of lipid fractions. The water vapor permeability of packaging for vegetables should be low enough to provide protection against dehydration and the quality changes it causes (Ahvenainen and Mälkki, 1985a). On the other hand, a package with a low water permeability value would trap an excessive amount of moisture, which may lead to undesirable microbial growth.

15.4.3 Transparency

Package transparency has the greatest effect on the shelf life of frozen foods when the foods are subjected to light, infrared radiation, and fluctuating temperature conditions. Hawkins et al. (1975) reported that 95% of the energy to which the frozen foods in an open display cabinet are exposed is caused by infrared radiation. Photodegradation of carotenoids depends on light wavelength. Shorter wavelengths result in greater degradation. Therefore, packaging materials that incorporate UV absorbers or light-impermeable materials would likely improve surface color retention of illuminated foods containing carotenoids (Ahvenainen and Mälkki, 1985a,b). Light in conjunction with oxygen availability has been found to accelerate both lipid and pigment oxidation during short-term display (Brewer and Wu, 1993). However, blocking light in the presence of even small amounts of oxygen was insufficient to prevent detrimental effects on color and lipid stability.

15.4.4 Package Environment

In addition to packaging, the quality of the frozen food is significantly affected by the temperature fluctuations during transport, storage, retail cabinet display, and in-home freezer storage. Although stagnant air coupled with packaging material provide an effective buffer to protect against fluctuating storage temperature conditions (Zuritz and Sastry, 1986), the quality of food is affected significantly when the package is stored in the top layer of an open display cabinet (Ahvenainen and Mälkki, 1985a,b). Furthermore, in open retail cabinets, vacuum packaging is not sufficient for photosensitive foods if the package opacity is not high. Therefore, under these conditions, the food should be packaged by vacuum packaging in materials with low oxygen and water vapor permeability and low transparency, and absorptivity at long wavelengths. Temperature fluctuations in home freezers may also be higher than those in commercial storage due to lack of precise controls. Very little work has been done on the quality degradation of frozen foods under these conditions. Nutritional losses and associated food-packaging interactions also merit further investigation.

15.4.5 Special Design Considerations for Microwave Packages

For safe and palatable food, it is often recommended that frozen precooked food should be heated to 70°C or above. However, heating frozen packaged foods according to manufacturer's instructions may not necessarily achieve this minimum temperature range (Bows, 1989). O'Meara and Reilly (1986) and Bows and Richardson (1990) studied the effect of component configuration and packaging materials on microwave reheating of multiportion frozen meals. The reheating rate of multiportion frozen meals is influenced by type of packaging material, type of food, and the orientation of different foods within the package. Improper design may result in extreme temperature variations within a dinner component as well as between components (O'Meara and Reilly, 1986). To ensure proper heating, manufacturers have introduced a number of advanced features in package design.

Susceptors provide specialized layering based on flexible Mylar polyester films and can absorb microwave energy to produce heat in a closely controlled manner (Rice, 1986b). These susceptors can be tailored to the needs of particular food products and particular packages, for example, for paper board trays, plastic bowls, and flexible film bags. Susceptors are applied either in heavy or light layers depending on whether the taste and texture of the food inside the package is enhanced by fast or slow heating. Thicker layers provide faster microwave energy absorption. Films with different heating capabilities can be heat sealed to each compartment according to how much energy various items in the tray require for heating in order to produce a high quality meal. For example, the meat and potato compartments might require more energy input to produce a heated meal, while the dessert compartment might require very small amount of slow heat-through to keep the dessert cool. While all the dinner components are heated in the same oven for the same amount of time at the same heat setting, there need not be any instance of ice in the middle of mashed potatoes or over-heated dessert at the end of microwave cooking (Rice, 1986a). Furthermore, as more and more packaging is sought for crisping and browning the product in addition to heating or reheating the product, there is a need to develop adhesives that withstand the temperature extremes (ranging from deep-freezing at −40°C to reheating at 200–220°C) associated with microwave packaging for frozen foods (Design, 1988).

15.5 Quality Changes in Frozen Foods Associated with Packaging

15.5.1 Physical changes

Color

Consumers associate color with good quality, and it is the most important influence in consumer decision-making regarding acceptability of food products for consumption (Smith, 1981; Brewer and Harbers, 1991a). Freezing or frozen storage generally does not affect the color of cooked meat (MacDougall, 1974). In contrast, in raw meat, Lynch et al. (1986) observed that PVC-packaged beef patties were preferred to vacuum-packaged products in preference testing. The difference in brown discoloration in packages of high and low oxygen permeability is that in the former, metmyoglobin is formed several millimeters below the surface, while in the latter it appears on the surface.

As oxygen permeability of packaging material increased, metmyoglobin increased, and a* (a Hunter color parameter) value and deoxymyoglobin decreased in ground beef studies (Brewer and Wu, 1993). Similar results have been reported by Bhattacharya et al. (1988a). The higher the a* value, the stronger the red color. The lowest a* value usually occurred in pork packaged with PVC, which permitted maximum oxygen penetration and light transmission. Vacuum packaging afforded some protection to a* value but did not always result in the same degree of preservation provided by aluminum foil. Loss of redness and oxymyoglobin in vacuum-packaged ground pork may be attributed to loss of metmyoglobin-reducing ability over time (Bhattacharya et al., 1988a; MacDougall, 1982) or to the reduced temperature encountered in frozen storage (O'Keefe and Hood, 1982). As for the b* values (a second Hunter color parameter), no consistent trends have been found in relation to the surface and interior locations in the meat block, storage time, packaging treatment, or interaction between storage time and packaging treatment (Brewer and Harbers, 1991a). In vegetables, the packaging and freezing does not appear to affect the color of the product during storage (Birnbaum et al., 1979; Forni et al., 1991).

Hwang et al. (1990) reported that vacuum-packaged beef samples were redder than those packaged in N_2/CO_2 and air. Similarly, vacuum packaging treatment led to increased redness compared to meat packaged with 100% N_2, 100% CO_2 or air (Nolan et al., 1989). The authors speculated that the formation of ice crystals on the surface of precooked meat slices packaged in modified atmosphere and ambient air conditions as reasons for difference in color attributes. The major effects of frozen storage time and packaging treatment on instrumental color measurements appeared to be predominantly a surface phenomena (Brewer and Harbers, 1991a; Brewer and Wu, 1993).

Drip Loss

Bhattacharya et al. (1988a) studied the influence of package type (polyethylene bag versus vacuum packaging) and frozen storage temperature on drip and cooking losses and shear strength of frozen beef patties. It was observed that the temperature affected drip loss and cooking loss but not shear strength, whereas the package type affected all these factors. Samples packaged in polyethylene packages had higher drip losses than those that were vacuum packaged. Drip losses for polyethylene packages increased to 14% for high-fat samples and 8.4% for low-fat samples, whereas the corresponding values for vacuum-packaged samples were 10.7 and 6.3% for high- and low-fat samples, respectively.

15.5.2 Nutritional Loss

Frozen foods lose nutrients primarily due to oxidative reactions which tend to accelerate at low moisture levels, that is, when surface dehydration or freezer burn occurs. Therefore, a packaging material with poor water- and oxygen-barrier properties allows moisture to escape and oxygen to enter, resulting in excessive losses of nutrients such as ascorbic acid, provitamin A, thiamin, and folate during frozen storage (Fennema, 1977; Breene, 1994). Use of oxygen-impermeable pouches has been reported to retain carotenoid and ascorbic acids in oxygen-sensitive frozen foods (Thakur and Arya, 1988).

15.5.3 Biochemical and Sensory Changes

During frozen storage, chemical- and enzyme-mediated oxidative degradation and associated off-flavors are the major causes of spoilage. A limiting factor in determining the rate of these oxidative reactions in packaged foods is the availability of oxygen. Packaging materials can thus serve to block out light and oxygen. After long-term frozen storage, oxygen-impermeable packaging maintained color and minimized surface discoloration, off-flavor, and TBA values (Bhattacharya et al., 1988b; Brewer and Harbers, 1991b; Christophersen et al., 1992; Bjerkeng and Johnsen, 1995). Polyethylene-packaged samples had higher TBA values than did vacuum-packaged samples stored at the same temperatures. Samples stored at higher temperature (−12.2°C) had higher TBA values than did samples stored at lower (−23.3 and −34.4°C) temperatures. Packaging treatments that excluded oxygen (e.g., vacuum packaging) were most effective in delaying lipid oxidation, whereas PVC-packaged meat had the highest level of lipid oxidation. Vacuum-packaged samples were also found to have a more constant flavor profile than PVC-packaged samples (Lynch et al., 1986).

Since oxygen exclusion appears critical to maintaining ground pork in a "like fresh" condition, vacuum-packaged pork had lowest scores for rancid, acid, bitter, and sour odors and flavors at all times. Use of aluminum foil, on the other hand, had little impact on preventing the development of rancid and acid flavors and odors unless the foil was used in combination with an oxygen-impermeable packaging material (such as Saran wrap). Since vacuum packaging is not a very viable option for consumers, the ground pork purchased at retail in commercial PVC packaging should be repackaged using Saran wrap or Saran with aluminum foil. This would probably delay changes in oxidative rancidity, preventing waste from consumer rejection of meat that has unacceptable odor and or flavor. Oxidation of lipids may be a surface phenomenon. Removal of discolored surface meat might reveal usable, unoxidized meat. Packaging materials that protect against changes in water-holding capacity and possible evaporative losses also retain juiciness and prevent increase in rubberiness.

15.6 Mathematical Models

15.6.1 Temperature Distribution

The rate of quality loss during the postfreezing phase is primarily temperature dependent. Changes in temperature, such as fluctuations during distribution, storage, retail cabinet display, or in-home freezer storage often result in a reduction in food quality. An understanding of temperature distribution histories in the product as a result of fluctuating storage temperature is therefore important in predicting quality loss. As the available dollars for research shrink, it is necessary to identify critical factors that will have the most impact on product quality without conducting elaborate experimental trials. Under these conditions, mathematical models are useful to describe the influence of packaging on temperature fluctuations in frozen foods. The following researchers have developed such models: Sastry and Kilara (1983) described the thermal response of frozen foods exposed to dithermal storage regimes, Zuritz and Singh (1985) simulated the effect of storage temperature changes through the frozen-food chain, and Mittal and Parkin (1985) studied the response of frozen foods in insulated containers (assuming no loss of heat through the container walls) to a step change in ambient temperatures. These models (analytical or numerical) rely on vigorous mathematical principles and provide solutions to the transient heat conduction problem. Once a suitable model is developed and experimentally verified, the model is useful for predicting the temperature fluctuation for a number of different scenarios.

Many of the earlier simulations (Scott and Heldman, 1984; Mittal and Parkin, 1985) did not consider the influence of packaging materials and particularly their thermal influence. However, the later models of Zuritz and Sastry (1986) and Scott and Heldman (1990) demonstrated that the contribution of packaging materials to the total thermal resistance is very important when the convective resistance of the surrounding air is negligible (high convective heat transfer coefficient value). The packaging boundaries also have significant influence on the temperature differential within the food product. The resistance due to the packaging interface was greater than zero but less than infinity. Thus, the earlier assumption of no resistance to heat flow by the packaging material was invalid, and the resistance to heat transfer due to packaging interface should be taken into account when designing new packages.

15.6.2 Kinetics of Quality Changes

Semiempirical kinetic models are often used to describe the change in quality of frozen foods, and an Arrhenius equation may be used to describe the effect of temperature on the rate constant. These models would facilitate the design of packages that provide products with desired quality attributes.

The changes in a quality attribute P can be modeled as:

$$\frac{dP}{dt} = kP^n$$

where P is the food quality attribute, k is the rate constant, and n is the order of reaction. The quality degradation during frozen storage of foods is often modeled as a first-order reaction. The food-package interactions have been studied using different quality attributes such as change in color (Bhattacharya and Hanna, 1989; Lanari and Zaritzky, 1991; Lanari et al., 1993) and drip and cooking losses (Bhattacharya and Hanna, 1989). Polyethylene-packaged samples had greater rate constants for drip and cooking loss than did vacuum-packaged samples, while vacuum-packaged samples had greater color change than polyethylene-packaged samples (Bhattacharya and Hanna, 1989). Similarly, Lanari and Zaritzky (1991) proposed a first-order kinetic myoglobin interconversion model to explain the effect of vacuum packaging on oxygenation capacity during frozen and partially frozen storage of beef.

Most of the models discussed above were developed either to describe temperature history or study quality degradation during frozen-food storage. Only a limited number of models realistically described the influence of both temperature history and quality degradation (Scott and Heldman, 1990). Future model development should examine the extent of interaction of individual components of a package (as in the case of microwave packaging) and the interaction between the package and the food during frozen storage and subsequent in-container heating. These models would help to predict the likely effect of various packaging material–food system combinations.

15.7 Conclusions

The quality of frozen food products could be preserved (if not improved, in comparison to the original product going through a similar storage period at a temperature above freezing) through careful selection of a product-specific packaging system. In addition, packaging also determines consumers' acceptance of the product. Further developments in plastic lamination, composite packaging, and processing technologies, along with changing family life styles will continue to shape the future improvements in frozen-food packages that meet the consumer needs and requirements.

References

Ahvenainen, R. and Mälkki, Y. 1985a. Influence of packaging on the shelf life of frozen foods. I. Carrot cubes. *J. Food Technol.* **20**:183–192.

Ahvenainen, R. and Mälkki, Y. 1985b. Influence of packaging on the shelf life of frozen foods. II. Baltic herring fillets. *J. Food Technol.* **20**:193–201.

Alexander, G.W. 1993. Preformed containers in a carton and paperboard setup containers for microwaveable food packaging. *Tappi J.* **75**(1):164–168.

Bhattacharya, M. and Hanna, M.A. 1989. Kinetics of drip loss, cooking loss, and color degradation in frozen ground beef during storage. *J. Food Eng.* **9**:83–96.

Bhattacharya, M., Hanna, M.A., and Mandigo, R.W. 1988a. Effect of frozen storage conditions on yields, shear strength, and color of ground beef patties. *J. Food Sci.* **53**:696–700.

Bhattacharya, M., Hanna, M.A., and Mandigo, R.W. 1988b. Lipid oxidation in ground beef patties as affected by time-temperature and product packaging parameters. *J. Food Sci.* **53**:714–717.

Birnbaum, N.R., Hicks, J.R., Tabacchi, M.H., and Brecht. P.E. 1979. Evaluation of evacuated packages as an alternative to blanching for frozen spinach. *J. Food Sci.* **45**:104–106.

Bjerkeng, B. and Johnsen, G. 1995. Frozen storage quality of rainbow trout (*Oncorhynchus mykiss*) as affected by oxygen, illumination, and fillet pigment, *J. Food Sci.* **60**:284–288.

Bows, J.R. 1989. The influence of the physical, thermal, and electrical properties on the quality of food heated by microwaves. (*Technical Memorandum 527*) Campden Food and Drink Research Association, Chipping Campden, UK.

Bows, J.R. and Richardson, P.S. 1990. Effective of component configuration and packaging materials on microwave reheating a frozen three-compartment meat. *Int. J. Food Sci. Technol.* **25**:538–550.

Bramsnaes, F. 1969. Quality and stability of frozen seafood. In *Quality and Stability of Frozen Foods. Time-Temperature Tolerance and its Significance.* (Arsdel, W.B.V., Copley, M. J., and Olson, R.L., eds.) pp 217–236, Wiley-Interscience, New York.

Breene, W.M. 1994. Healthfulness and nutritional quality of fresh versus processed fruits and vegetables: a review. *J. Foodserv. Syst.* **8**:1–45.

Brewer, M.S. and Harbers, C.A.Z. 1991a. Effect of packaging on color and physical characteristics of ground pork in long-term frozen storage. *J. Food Sci.* **56**:363–366.

Brewer, M.S. and Harbers, C.A.Z. 1991b. Effect of packaging on physical and sensory characteristics of ground pork in long-term frozen storage. *J. Food Sci.* **56**:627–631.

Brewer, M.S. and Wu, S.Y. 1993. Display, packaging, and meat block location effects on color and lipid oxidation of frozen lean ground beef. *J. Food Sci.* **58**:1219–1223.

Christophersen, A.G., Bertelsen, G., Anderson, H.J., Knuthesen, P., and Skibsted, L.H. 1992. Storage life of frozen salmonoids. Effect of light and packaging conditions on carotenoid oxidation and lipid oxidation. *Z. Lebensm. Unters. Forsch.* **194**:115–119.

Design element for microwave packaging: adhesives, *A*. 1988. *Food Eng.* **60**(2):48.

Fennema, O. 1977. Loss of vitamins in fresh and frozen foods. *Food Technol.* **31**:32–36.

Forni, E., Crivelli, G., Polesello, A., and Ghezzi, M. 1991. Changes in peas due to freezing and storage. *J. Food Process. Preserv.* **15**:379–389.

Hawkins, A.E., Pearson, C.A., and Raynor, D. 1975. Advantages of low emissivity materials to products in commercial refrigerated open display counters. *Austr. Refrig. Air Cond. Heat.* **29**:15–25.

Hwang, S.Y., Bowers, J.A., and Kropf, D.H. 1990. Flavor, texture, color, and hexanal and TBA values of frozen cooked beef packaged in modified atmosphere. *J. Food Sci.* **55**:26–29.

Jeremiah, L.E. 1980. Effect of frozen storage and protective wrap upon the cooking losses, palatability, and rancidity of fresh and cured pork cuts. *J. Food Sci.* **46**:187–190.

Lanari, M.C. and Zaritzky, N.E. 1991. Effect of packaging and frozen storage temperature on beef pigments. *Int. J. Food Sci. Technol.* **26**:629–640.

Lanari, M.C., Cassens, R.G., Schaefer, D.M., and Scheller, K.K. 1993. Dietary vitamin E enhances color and display life of frozen beef from Holstein steers. *J. Food Sci.* **58**:701–704.

Lynch, N.M., Kastner, C.L., Caul, J.F., and Kropf, D.H. 1986. Flavor profiles of vacuum packaged and polyvinyl chloride packaged ground beef: a comparison of cooked flavor changes occurring during product display. *J. Food Sci.* **51**:258–262, 267.

MacDougall, D.B. 1974. The appearance of frozen meat and its color stability during storage. In *Meat Freezing—Why and How?* (MRI Symposium 3) (Cutting, C.L., ed.) pp. 10.1–10.6, Langford, Bristol, U.K.

MacDougall, D.B. 1982. Changes in color and opacity of meat. *Food Chem.* **9**:75–88.

Mittal, G.S. and Parkin, K.L. 1985. Transportation of frozen foods in insulated containers—Theoretical and experimental results. Presented at the 1985 North Atlantic Region Meeting of the American Society of Agricultural Engineers, Baltimore, M.D. August 11–14. paper NAR85-503.

Morris, C.E. 1989. Convenience for supermarket delis with CAP pizza. *Food Eng.* **61**(6):53–4.

Nolan, N.L., Bowers, J.A., and Kropf, D.H. 1989. Lipid oxidation and sensory analysis of cooked pork and turkey stored under modified atmospheres. *J. Food Sci.* **54**:846–849.

Ogrydziak, D.M. and Brown, W.D. 1982. Temperature effects in modified-atmosphere storage of seafood. *Food Technol.* **36**(5):86, 88, 90–91, 94–96.

O'Keefe, M. and Hood, D.E. 1982. Biochemical factors influencing meat myoglobin formation on beef from muscles of differing stability. *Meat Sci.* **7**:209–227.

O'Meara, J.P. and Reilly, D.B. 1986. The effects of oven parameters on microwave reheating of frozen dinners. *Microwave World* **7**:9–11.

Ono, K. 1990. Frozen food and oven-proof trays. In *Food Packaging*. (Kadoya. T., ed.) pp. 253–268, Academic Press, London.

Ott, D.B. 1988. Trends in food packaging. *J. Home Econ.* **80**(3):36–43.

Parry, R.T. 1993. *Principles and Applications of Modified Atmosphere Packaging of Food*. p. 65, Blackie Academic Professional, London, UK.

Piazza, R. 1988. Crust freezing and vacuum packaging improve quality. *Food Process.* **49**:48–50.

Ray, E.E., Olivas, Y., Vincent, M., and Smith, D. 1994. Effects of modified atmosphere packaging and film on sensory and storage properties of frozen ham strips. *J. Food Qual.* **17**:41–55.

Rice, J. 1986a. Frozen dual-ovenable dinner entrees. *Food Process.* **47**:49–52.

Rice, J. 1986b. Microwave susceptors control energy absorption rates. *Food Process.* **47**:74.

Rice, J. 1987. Stouffer switching from foil to dual-ovenable plastic trays. *Food Process.* **48**:160–161.

Rice, J. 1990. Packaging: past fifty, present, and future. *Food Process.* **51**:110–118.

Sastry, S.K. and Kilara, A. 1983. Temperature response of frozen peas in di-thermal storage regimes. *J. Food Sci.* **48**:77–83.

Scott, E.P. and Heldman, D.R. 1984. Prediction of frozen food quality during fluctuating temperature storage. Presented at 1984 Winter meeting of the American Society of Agricultural Engineers, New Orleans, La., Dec. 11–14. ASAE paper 84-6570.

Scott, E.P. and Heldman, D.R. 1990. Simulation of temperature dependent quality deterioration in frozen foods. *J. Food Eng.* **11**:43–65.

Smith, G.C. 1981. Considerations in meat packaging. *Natl. Provision.* **31**(1):27.

Smith, J., Ramaswamy, H., and Simpson, B. 1990. Developments in food packaging technology. Part 2. Storage aspects. *Trends Food Sci. Technol.* **1**:111–118.

Steinbuch, E. 1979. Quality retention of unblanched frozen vegetables by vacuum packing. 1. Mushrooms. *J. Food Technol.* **14**:321–323.

Taylor, S. 1979. Skin packaging of frozen meat. *Food Manuf.* **(May):**61–63.

Thakur, B.R. and Arya, S.S. 1988. Relative suitability of plastic films for the frozen storage of mango pulp. *J. Food Process. Preserv.* **12**:171–178.

Young, L.L., Reviere, R.D., and Cole, A.B. 1988. Fresh red meats: a place to apply modified atmospheres. *Food Technol.* **42**(9):65–66, 68–69.

Zuritz, C.A. and Sastry, S.K. 1986. Effect of packaging materials on temperature fluctuations in frozen foods: mathematical model and experimental studies. *J. Food Sci.* **51**:1050–1056.

Zuritz, C.A. and Singh, R.P. 1985. Modeling temperature fluctuations in stored frozen foods. *Int. J. Refrig.* **8**:289.

SECTION IV

Monitoring of Quality in Frozen Foods

16

Physical and Ultrastructural Measurements

P. Mallikarjunan and Y.-C. Hung

16.1 Introduction

The frozen-food industry is one of the largest segments in the U.S. and world food markets. According to Mallet (1993), sales of frozen foods grew to \$50 billion in the United States and an additional \$25 billion in Europe in 1990. Frozen foods fill important niches in the overall market because they provide convenient prepared or semiprepared products, can be available all year at consistent cost in markets where fresh or refrigerated products are not, and can be transported greater distances with minimal damage and degradation (Persson and Löndahl, 1993). Frozen foods generally have been perceived by consumers as having better quality than foods processed and preserved by other methods. However, growth of sales of frozen foods have been slowed by the deficiencies in frozen foods as compared to fresh, shelf-stable, and refrigerated alternatives (Dagnoli and Erikson, 1989). The consumers expect the quality of the frozen food to be close to that of a fresh product. Thus, the quality of the frozen foods is most important in determining growth and success.

16.2 Quality Attributes Important to Frozen Foods

Changes in food quality during its frozen and thawed stages play a major role in the success of products to be marketed as frozen foods. The quality changes in the frozen food are due mainly to the rate of freezing, storage temperature and temperature abuse during storage, and length of storage. Thus, it is imperative to determine the quality of frozen foods in all the stages from the fresh product to the thawed product. In addition, for food products consumed after further processing, such as cooking, the quality of frozen food should also be determined after the postprocessing. The following physical quality parameters are studied extensively and considered critical: color, texture, and juiciness. In addition to these parameters, knowledge of the ultrastructural changes in the frozen food is also needed to explain the quality changes occurring in frozen foods by analyzing the ice-crystal structure and cellular damage.

16.3 Color

The first and foremost quality by which a consumer makes a decision is the visual appearance. In addition to influencing the purchase intent, the color of the food also affects the taste perception of the food. Color can be defined as the energy distribution of the light reflected by or transmitted through a particular food product. Depending on how the light reacts with the food, the food can be classified as opaque, translucent, or transparent. Among these reactions, the reflected light determines the color and appearance of the material. The reflected light is affected by the light source, the angle of viewing, and the characteristics of the background; in turn, all these factors affect the color perception. Often, the color is measured using standard CIE (the Commission International de l'Eclariage) scales. In the food industry, the derived color scales such as Hunter *L, a* and *b* or CIE L^*, a^*, and b^* are commonly used. These parameters represent the lightness (L), the degree of redness or greenness ($\pm a$), and the degree of yellowness or blueness ($\pm b$) of the product measured. A value of 0 or 100 for L represents black or white, respectively. These instrumental color measurements correspond to the visual assessment of food color, and the human perception of the color as hue, saturation, and value can be calculated from L^*, a^*, and b^* values.

The hue describes the visual sensation of the color. It represents the appearance of a given area in comparison to one or portions of two or more of the perceived standard colors red, yellow, green, and blue. The hue is measured in angle (degrees or radians) in the color wheel (fig. 16.1).

Chroma, the purity or saturation, describes the intensity of a fundamental color with respect to the amount of white light mixed with it. In other words, converting the coordinate system of a and b to polar coordinates of r and θ gives chroma and hue, respectively. Thus, the hue angle and chroma can be calculated as

$$\text{Chroma or saturation} = [a^2 + b^2]^{0.5}$$

$$\text{Hue angle} = \tan^{-1}\left(\frac{b}{a}\right)$$

The value or lightness is an indication of overall light reflectance of that color. Thus, the color parameter L^* represents the value or color lightness.

In many instances, the differences in color parameters between two stages provide more meaningful interpretation of the processes than absolute color values.

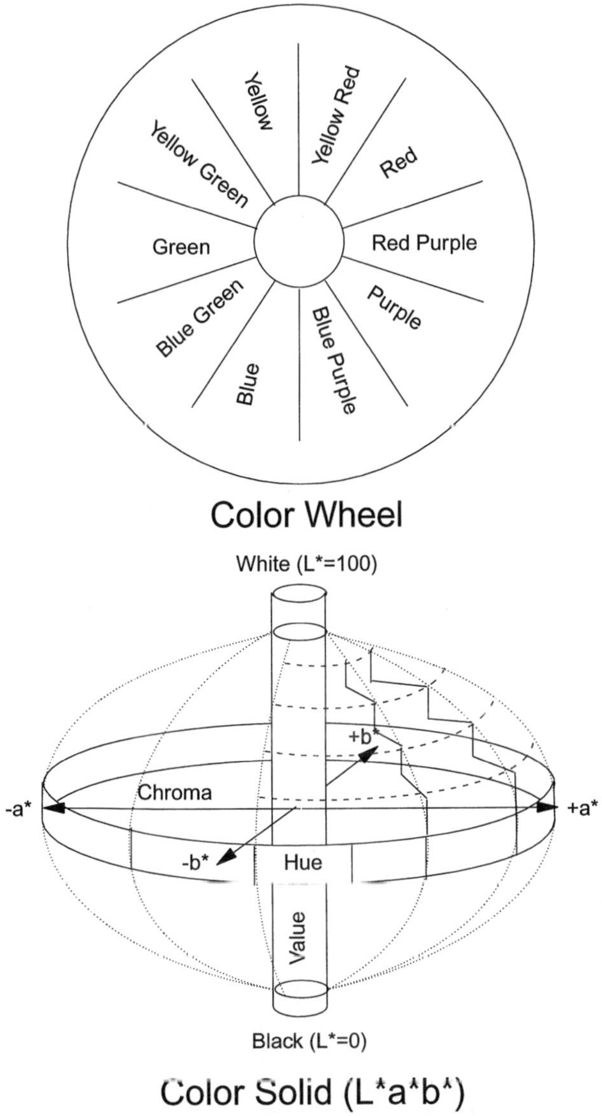

Figure 16.1. Color wheel and solid representing color-measurement systems

The color differences between two stages (e.g., between raw and frozen food following freezing, during storage, or at thawed stages) can be calculated as

$$\text{Hue angle difference} = \tan^{-1}\left(\frac{b}{a}\right) - \tan^{-1}\left(\frac{b_0}{a_0}\right)$$

$$\text{Saturation difference} = [(a - a_0)^2 + (b - b_0)^2]^{0.5}$$

$$\text{Brightness difference} = \text{abs}(L - L_0)$$

$$\text{Total color difference} = [(L - L_0)^2 + (a - a_0)^2 + (b - b_0)^2]^{0.5}$$

where L_0, a_0, and b_0 are the color parameters of the initial stage (or stage 1) and L, a, and b are the corresponding color parameters at other processing stages.

Color can be measured using either a colorimeter or spectrophotometer. To duplicate the responses of the human eye, the colorimeter uses a set of three filters (red, green, and blue) with transmission curves similar to the standard X, Y, and Z curves. The light reflected from the object through each filter is recorded and the tristimulus values are obtained. All tristimulus colorimeters available depend on this principle with individual refinements in photocell response, sensitivity, stability, and reproducibility. Today, the meters also provide the color values in other color spaces. Tristimulus colorimeters usually are small and portable. They come with special attachments and a wide range of aperture openings and are customized for specific applications. The colorimeters are available from Agtron Inc., Sparks, Ne.; BYK-Gardner, Silver Spring, Md.; HF Scientific Inc., Ft. Myers, Fla.; Hunter Lab, Reston, Va.; Minolta Corp., Ramsey, N.J.; and X-Rite Inc., Grandville, Mich.

In recent years, the use of tristimulus colorimeters has increased to include measuring the color of frozen foods, mostly, at the surface. Lanari et al. (1993) measured the surface color of thawed frozen ground beef samples using a colorimeter (Minolta chromameter). They found repeated freezing and thawing decreased the redness (a^*) and color saturation of semimembranosus muscle samples. Andersen and Skibsted (1991) and Akamittath et al. (1990) measured the surface color of frozen meat to study the effect of salt on lipid oxidation. Bhattacharya et al. (1988) also used a colorimeter (Hunter Lab color-difference meter) to measure the surface color of frozen ground beef patties and found an increase in frozen storage time resulted in a loss of color, with products tending to gray. Forni et al. (1991) measured the color changes in peas caused by freezing and frozen storage. The color was measured on 100 g of the pea samples using a Hunter color-difference meter. Along with texture and flavor, the color of the peas was found acceptable ($p < 0.05$) after both freezing and cold storage. Cano and Martin (1992) measured the color of frozen kiwi slices using a Hunter Lab model D25-9 colorimeter and 50 g of pureed sample. Based on the color parameters *L*, *b*, and chroma, the freezing preservation of kiwi fruit slices produced a kiwi fruit product with similar color characteristics and appearance of the fresh fruit ($p < 0.05$).

Spectrophotometers measure a ratio of the light reflected or transmitted from a food product to that from a known reference standard. Using multiple sensors sensitive to a particular wavelength, the tristimulus values are calculated mathematically by applying the energy curves of the light source and that of the standard observer. Hwang et al. (1990) used a spectrophotometer (Hunter Lab) for measuring the color of frozen, precooked beef loin slices. Before the measurements, samples were thawed for 6 min and finish-cooked for 3 min using a microwave oven. Based on redness (a^*), the exclusion of oxygen in the packages by vacuum and N_2/CO_2 produced little adverse effect on product color. Brewer and Harbers (1991) measured the color of frozen ground pork to study the effects of frozen storage on lipid oxidation. Before color measurement using a spectrophotometer (Hunter Lab), all the samples were thawed at 4°C for 48 h. They used a^* for redness and difference in reflectance between 630 and 580 nm for oxymyoglobin content. Ground pork lost its redness and oxymyoglobin during frozen storage, and they concluded that packaging that excluded both oxygen and light was most effective at preventing the loss of red color. Similar results were obtained for frozen ground beef patties over a 12-wk storage at −20°C using the same methodology (Brewer and Wu, 1993). Chen and Trout (1991) measured color (using a Perkin-Elmer spectrophotometer) of restructured beef steaks to study the effect of various binders on color stability during frozen storage. Cano et al. (1990) measured the color of frozen banana slices by determining the reflectance at 485 nm using a spectrophotometer (Perkin-Elmer). The color measurement at 485 nm was correlated with oxidation of p-phenylenediamine. They concluded that the effect of freezing on enzymatic activities of banana slices depended on banana maturity level at the time of processing. A summary of measurement conditions, color measuring devices, and measured color parameters for different frozen foods is listed in table 16.1.

Both tristimulus colorimeters and spectrophotometers provide product color with reasonable accuracy and reproducibility. The choice of which meter to use will depend on the food material and type of application. Tristimulus colorimeters are used primarily in production settings to monitor product quality during each processing step. Spectrophotometers can provide high accuracy and are most suited for laboratory applications. The recent development of portable spectrophotometers enables the use of them in production settings as well. Regardless of the choice of meter, it is important that sample preparation and presentation to the instrument be followed to obtain reproducible measurements. Light scattering and sample uniformity can be altered by grinding, mixing, and blending procedures. It is advisable to turn the sample 90° from the first reading and to use an average of the two readings.

Table 16.1. Color measurement of frozen foods

Product	Device	Freezing condition	Measurement condition	Parameters	Source
Ground beef	Colorimeter	To −20°C in a freezer at −70°C	Thawed at 22°C for 30 min	Saturation index	Lanari et al. (1993)
Ground beef patties	Spectrophotometer	Frozen at −20°C and stored for 12 wk	Thawed	a^*, refl. at 630 and 580 nm	Brewer and Wu (1993)
Ground beef patties	Colorimeter	To −35°C for 4 h and stored at −12, −23, and −34°C for a 20-wk period	Thawed at 6°C for 24 h	Total color difference (ΔE)	Bhattacharya et al. (1988)
Cooked beef loin slices	Spectrophotometer	Frozen at −20°C and stored for 11 wk	Thawed, finish-cooked using a microwave oven	a^*	Hwang et al. (1990)
Ground pork patties	Colorimeter	Frozen at −35°C and stored at −18°C	Not reported	a^*	Andersen and Skibsted (1991)
Ground pork	Spectrophotometer	Frozen at −17°C within 24 h of grinding	Thawed at 4°C for 48 h	a^*, refl. at 630 and 580 nm	Brewer and Harbers (1991)
Restructured meat systems	Colorimeter	Frozen at −30°C for 15–16 h and stored at −10°C	Not reported (or frozen)	a^*	Akamittath et al., (1990)
Restructured steak	Spectrophotometer	Frozen at −34°C for 48 h and stored at −23°C	Scanned from 380 to 760 nm	L^*, a^*, and b^*	Chen and Trout (1991)
Kiwi slices	Colorimeter	An air-blast freezer at −40 to −20°C	Pureed	Hue, chroma, ΔE	Cano and Martin (1992)
Banana slices	Spectrophotometer	Frozen at −24°C	Homogenized and centrifuged; the aliquot of extract was used	Refl. at 485 nm	Cano et al. (1990)
Peas	Colorimeter	Frozen at −35°C for 25 min and stored at −20°C for 6 mo	Not reported	ΔE	Forni et al. (1991)

16.4 Texture

One of the most important quality attributes of frozen food that influences sales is its texture. Food texture depends mainly on the structural constituents. Thus the amount and size of ice crystals in the frozen food influences its texture during frozen storage and especially after thawing. Textural properties are related to the deformation, disintegration, and flow of food under force. The properties can be measured objectively and correlated to sensory textural attributes.

Szczesniak (1983) gave a system of classification of textural characteristics based on fundamental rheological principles. Textural characteristics can be classified into five primary parameters—hardness, cohesiveness, viscosity, elasticity, and adhesiveness—and three secondary parameters—brittleness, chewiness, and gumminess. These properties refer to the behavior of the food in the mouth.

Food materials are viscoelastic in nature, that is, exhibiting properties of both the solid and liquid. Since the majority of frozen foods exhibits more solid characteristics, the texture is measured using applied compression, shear, and/or tensile forces (e.g., Warner-Bratzler [WB] shear test, puncture test, and textural profile analysis). It is best to measure the texture of the food material as close as possible to the stage it is consumed, so that it can be compared with the sensory measurements or consumer acceptance of the product. Thus, in foods such as frozen yogurt and ice creams (Goff et al., 1995), the texture can be measured in its frozen stage. However, the texture of many other frozen foods is determined after thawing and, in some cases, such as meat, after postprocessing (e.g., cooking). The evaluation methods depend on the nature of the product and the sensory property that needs to be compared such as mouth feel, chewiness, or tenderness.

Tenderness is a measure of texture for foods such as meat, fruits, and vegetables. It has been measured as shear, bite, penetration, tensile, and compressive forces. Some objective methods of evaluating tenderness include shear methods using the Warner-Bratzler shear device or Kramer shear press; compression methods (puncture test, texture profile analysis, and stress-relaxation tests); and tensile methods.

In measuring texture, force must be recorded accurately; since rapid force fluctuation and slope changes occur, a high-frequency response capability for the data acquisition is required. In addition, accurate records of the probe position are needed to interpret the data properly. A reproducible deformation rate is critical. To achieve faster data processing and reproducible deformation, use of a universal testing machine for conducting the experiments is necessary. The universal testing machines are available in the market from Instron Corp., Canton, Mass.; TA-Instruments Inc., New Castle, Del.; Testing Machines Inc., Amityville, N.Y.; and Texture Technologies Corp., Scarsdale, N.Y. A summary of measurement conditions, texture-measuring methods, and parameters for different frozen foods is listed in table 16.2.

Table 16.2. Texture measurement of frozen foods

Product	Method	Freezing condition	Measurement condition	Parameters	Source
Green beans	WB shear	Frozen at −18°C and stored for 4 wk	Cooked samples	Shear force	Brewer et al. (1994)
Peas	Kramer shear	IQF at −35°C for 25 min	200 mm/min crosshead speed	Firmness	Forni et al. (1991)
Ground beef patties	Kramer shear	Frozen at −35°C for 4–5 h and stored at −12, −23, or −34°C for 20 wk	Thawed at 6°C for 24 h and cooked in an oven kept at 170°C	Shear strength	Bhattacharya et al. (1988)
Restructured beef steaks	Kramer shear	Frozen at −30°C	Cooked	Tenderness	Lamkey et al. (1986)
Cooked beef loin slices	Puncture (simulated tooth attachment)	Frozen at −20°C and stored for 11 wk	Thawed for 6 min and finish-cooked for 3 min using a microwave oven	Firmness	Hwang et al. (1990)
Cut pears	Puncture (8-mm Magness-Taylor probe)	Frozen at −15°C	Thawed at 21°C	Firmness	Bolin and Huxsoll (1993)
Minced cod fillet	Texture-profile analysis	Frozen at −20°C for 24 h and stored at −14°C for 24 wk	Thawed overnight at 2°C	Hardness	Jahncke et al. (1992)
Black and silver scabbard-fish	Texture-profile analysis	Frozen at −75°C and stored at −18°C for 8 wk after vacuum packaging	Thawed and cooked at 100°C for 2 h	Hardness	Dias et al. (1994)

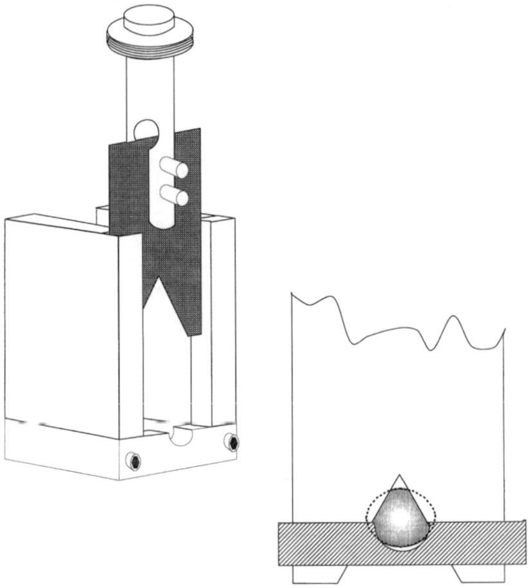

Figure 16.2. Warner-Bratzler shearing device

16.4.1 Shear Test

One of the widely used texture measurements is the Warner-Bratzler shear test. In this test, the sample is sheared using a Warner-Bratzler shear blade. The shear blade has an opening in the shape of an inverted triangle. The sample is placed inside the hole, and the shear blade is moved in an upward direction during testing (fig. 16.2). Detailed discussion on the deformation pattern during testing and interpretation of the testing is widely available in the literature (Voicey, 1976). The following parameters are obtained during a WB shear test:

1. Maximum or initial yield force representing a tensile rupture signifying cohesiveness.
2. Cross-sectional area of the sample at rupture, which indicates compression required to initiate rupture and can be correlated with firmness.
3. Slope of the force-time curve, which provides an index of firmness.
4. Force per unit length of the blade edge cutting the sample at the peak force.

Another shear test used widely is the Kramer shear test. In the Kramer shear test, the sample is placed in a Kramer shear cell (fig. 16.3), which contains a

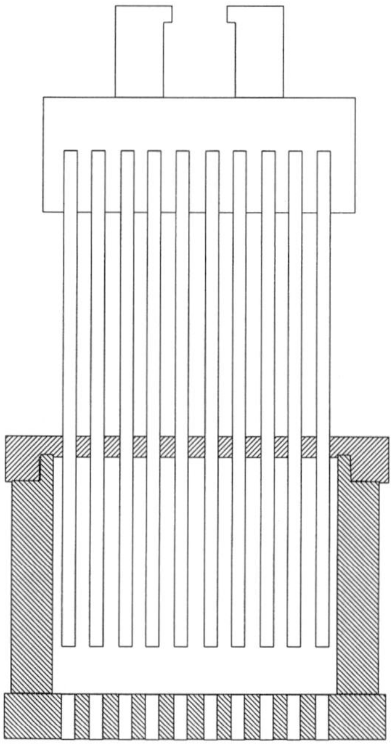

Figure 16.3. Cross-sectional view of Kramer shear compression cell

sample holder and a blade assembly. The Kramer shear blade assembly contains 10 blades, each of which is 0.3175 cm thick. The sample holder is a metal box (6.7 × 6.7 × 6.3 cm) with 10 slots in its bottom. The sample is placed in the sample holder to the desired level (normally to 30% by volume). The sample holder has a matching slotted cover with 10 slots, which is put on the top of the sample holder. The whole assembly is placed in the base of the universal testing machine. The blade assembly is lowered so that the bottom of the set of blades is just touching the slotted cover. After setting the universal testing machine to the desired test speed, the blades are lowered until the product is extruded through the bottom slots. A typical test speed is around 200 mm/min (20 cm/min).

Similarly, the following parameters are obtained during a Kramer shear test:

1. The initial yield force and yield distance.
2. Peak force and peak force distance.
3. Slope and work (area under the force-time curve) required to shear the sample.

Brewer et al. (1994) used a WB shear apparatus to determine the effects of different types of blanching on frozen green beans. The texture measurements were obtained on cooked samples. Texture was softest in beans blanched using boiling water and steam-blanched beans. Microwave blanching for 3 min (at full power) resulted in a similar texture to steam-blanched beans. Forni et al. (1991) used the Kramer shear test to evaluate the firmness of frozen peas. An increase in firmness of peas was observed after frozen storage at −20°C. Bhattacharya et al. (1988) used the Kramer shear test on cooked beef patties. They found an increase in shear strength in ground beef patties stored at −12 to −35°C. Lamkey et al. (1986) used Kramer shear to measure the texture of restructured beef steaks; adding phosphates along with salt increased the tenderness of beef steak frozen at −30°C.

16.4.2 Puncture Test

The puncture test, a compression method, is widely used in texture measurement of fruits and vegetables. The puncture test is popular because of its simplicity since only the force required to push a puncture probe into or through the sample must be determined. Standard penetration techniques with various shapes of probes have been used. Depending on the size and shape of the probe, the puncture test may only give data related to the sample compressibility and/or the force necessary to break down the sample. In addition to using a puncture probe, the sample can be placed on a die, so the probe can penetrate through the sample completely. The following three parameters are obtained during a puncture test:

1. Maximum or initial yield force that can be correlated with firmness.
2. Sample area at rupture, which indicates energy required to initiate rupture.
3. Slope of the force-deformation curve that provides modulus of elasticity.

Cano et al. (1990) studied the effects of thermal treatment before freezing on firmness of banana slices. To evaluate firmness, samples were tested using a Bellevue pressure tester with 5/6-inch plunger. Freezing and thawing without thermal treatment produced an excessively soft product, and a prethermal treatment (blanching) resulted in a firm, acceptable texture. They further concluded that, in addition to thermal treatment, selection of banana maturity level for freezing must compromise between the sensorial quality and processing parameters. Hwang et al. (1990) used the measurements of puncture force to penetrate a meat piece using a simulated tooth attachment. The puncture measurements were obtained on microwave-thawed samples. The frozen, precooked beef samples stored at different modified atmosphere conditions did not show any significant differences in textural parameters. Bolin and Huxsoll (1993) studied the effects of partial drying before freezing on the texture of cut pears. The peak force required to

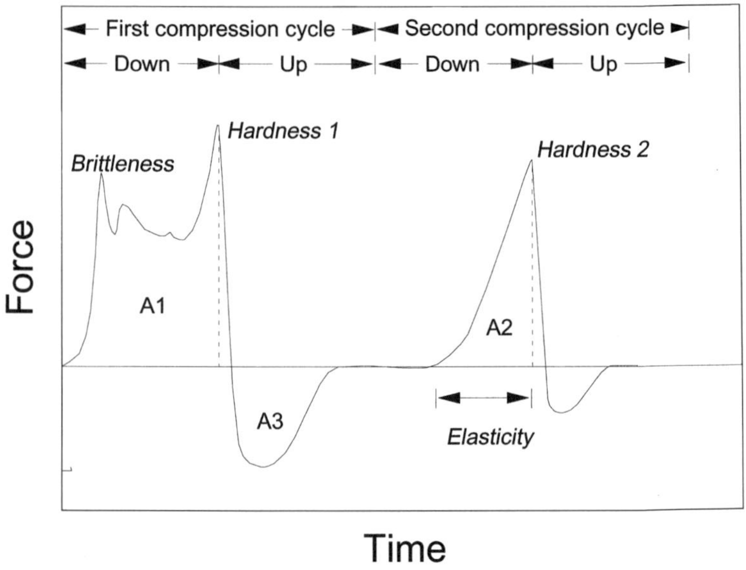

Figure 16.4. Typical force-time curve during a texture-profile analysis

penetrate a side of thawed pear sample using an 8-mm Magness-Taylor probe was correlated with the texture. Concentrating the pears to 50% weight reduction using osmotic drying improved the texture of frozen and thawed pears 1.5 times better than that obtained using hot-air drying.

16.4.3 Textural Profile Analysis

Cyclic loading and unloading to a set deformation during uniaxial compression have been used widely by many investigators to determine several textural parameters. Using texture-profile analysis, six different texture properties can be estimated: hardness, brittleness, cohesiveness, elasticity, adhesiveness, chewiness, and gumminess. Figure 16.4 illustrates a typical texture profile curve obtained using the Instron universal testing machine.

Hardness is measured from the profile as the height of the first peak. Cohesiveness is measured as the ratio of the area under the second peak to the area under the first peak. Springiness or elasticity is defined as the height food recovers during the time that elapses between the end of the first bite and the start of the second bite. Elasticity can be calculated from the width of the second compression curve (*A2*) up to the maximum force. Adhesiveness is measured as the area, *A3*, of the negative peak beneath the baseline of the profile and represents work necessary to pull the probe from the sample. Brittleness is characterized by the multipeak shape of the first bite and measured as the height of the first significant break in the peak. Gumminess is expressed as the product of hardness and cohesiveness. Chewiness is expressed as the product of hardness, cohesiveness, and elasticity.

Jahncke et al. (1992) determined the effects of frozen storage on the texture of minced cod fillet using texture-profile analysis. Before analysis, the samples were thawed overnight at 2°C. Using a crosshead speed of 12.7 cm/min, the samples were compressed to 65% of their original height. In general, with an increase in frozen storage time, hardness values decreased. Jahncke et al. attributed the reduction in hardness to the action of proteases during frozen storage. Dias et al. (1994) used texture-profile analysis to study the effect of frozen storage on black and silver scabbardfish quality. After thawing and cooking the minced samples (25 mm diameter and 40 mm height) at 100°C for 2 h, texture measurements were carried out using an Instron equipped with 10 kg load cell at 10 mm/min crosshead speed. They found a significant increase in load at maximum displacement (hardness) with frozen storage and an inverse correlation with the decrease in salt-soluble proteins.

16.5 Juiciness

Juiciness is another important quality attribute for food products, especially for frozen-food products. In frozen foods, the state of water in the product and the size of ice crystals affect the juiciness of the product. Juiciness can be defined as the amount of juice released during consumption. This affects the texture, flavor, and overall acceptance of the product; the product should not be dry or exudative. The frozen products with larger ice crystals, upon thawing, will experience a higher amount of drip loss (i.e., thawing loss) and will cause an excessive drip that results in a less-appealing product. The other quality loss that affects product juiciness is cooking loss, the amount of juice released during cooking.

In addition to thawing and cooking losses, the food product's ability to hold water plays a crucial role in determining the product juiciness. This capacity of the food product is often called water-holding capacity (WHC). In definition, WHC is the ability of the food product to hold all or part of its own and/or added water (Honikel, 1987). A wide variety of methods for measurement of WHC is reported in the literature. The methods used for measuring WHC of meat systems have been given by Honikel (1987) and Trout (1988). These procedures can be adapted easily to other food systems as well.

There are three basic methods for measuring WHC: methods applying no external force, methods applying external mechanical force, and methods applying thermal force. The most commonly used methods for measuring WHC involve applying external forces such as the filter-paper press method and the centrifugation method. The press method, which is rapid and inexpensive, involves pressing the sample between filter papers and measuring the water released (Grau and Hamm, 1957). The sample can be pressed using a hand-operated screw press, a high-pressure laboratory press (e.g., Carver press), or an Instron universal testing machine. The latter two pieces of equipment offer better control of the applied pressure and, consequently, give more reproducible results. The amount of water released from the sample can be measured either directly by weighing the filter papers or indirectly by measuring the area of filter papers wetter relative to the areas of the pressed sample. Jahncke et al. (1992) used Instron for obtaining expressible moisture from frozen cod. The information on amount of load and time was not reported. Freezing cod at −20°C for 24 h and storing it at −14°C increased the press juice from it. After 3 wk of frozen storage the changes in press juice values were not significantly different ($p > 0.05$). Kenawi (1994) measured the press juice of beef during frozen storage. He also found an increase in press juice with storage time in frozen beef. The conditions used to freeze the samples and the temperature during storage were not reported.

Precise and reproducible results can be obtained with the centrifugal method. The samples, varying in size from 1 to 20 g, are centrifuged at forces between 5,000 and 40,000 × gravity. The amount of water released is determined either directly by weighing the amount of water released or indirectly by weighing the sample after centrifugation. Greer and Murray (1991) used the centrifugal method to evaluate the water-holding capacity of pork during frozen storage. The samples were thawed at 1.7°C for 2 d before Greer and Murray estimated water-holding capacity. They found a decrease in the expressible juice after thawing. Even though—after freezing and thawing—an increase in drip loss was observed, the released juices were reabsorbed by the meat and resulted in a less expressible juice than that in fresh pork.

The methods that use no external forces include measurement of drip losses and thawing losses (Greer and Murray, 1991) from the product. Thermal forces allow measurment of the water-holding capacity in terms of cooking loss (Lamkey et al., 1986; Bhattacharya et al., 1988; Berry, 1992; Brewer et al., 1992). Most of the frozen foods consumed after thawing will be cooked before consumption. Thus the water-holding capacity after heating is of interest. Depending on the nature of the ice crystals and denaturation of proteins and cellular structure in the product, the cooking loss of the product will vary. A summary of measurement conditions, juiciness-measuring methods, and parameters for different frozen foods is listed in table 16.3.

16.6 Ultrastructural Evaluation

Understanding the structure-function relationship is critical for controlling and improving the quality of frozen foods. Cellular structures such as membranes and

Table 16.3. Juiciness measurement of frozen foods

Product	Method	Freezing conditions	Parameters	Source
Frozen cod	Press method	Frozen at −20°C for 24 h and stored at −14°C	Expressible juice	Jahncke et al. (1992)
Beef	Press method	Not reported	Press juice	Kenawi (1994)
Pork	Centrifugal method	Stored at −30°C for 90 d	Drip loss	Greer and Murray (1991)
Kiwi fruit slices	Drip method	Frozen at −40°C and stored at −18°C	Drip loss	Fúster et al. (1994)
Pork	Thawing loss	Stored at −30°C for 90 d	Thaw loss	Greer and Murray (1991)
Restructured beef steaks	Cooking loss	Frozen at −30°C	Cooking loss	Lamkey et al. (1986)
Ground beef patties	Cooking loss	Frozen at −35°C for 4–5 h and stored at −12, −23, or −34°C for 20 wk	Cooking loss	Bhattacharya et al. (1988)
Ground beef patties	Cooking loss	Frozen at −43°C and stored at −30°C	Cooking loss	Berry (1992)
Ground beef patties	Cooking loss	Blast frozen at −20°C	Cooking loss	Brewer et al. (1992)

cell walls are affected by food processing; these changes ultimately lead to changes in quality attributes of the end product.

The primary tool used to study ultrastructure is the microscope. Many light and electron microscope techniques are available for observing the cells. In these techniques, cells must be manipulated to make the cell structure more visible. When cells are stained or fixed, the pretreatment can lead to the creation of artifacts. Artifacts are effects characteristic of a pretreatment and are not normally present in the cell structure. If the structure appears to be the same for each process combination, it can be assumed to be real and without artifacts. Table 16.4 shows a comparison of various types of microscopic techniques.

Conventional light microscopy can be used to study fixed or stained cells, with the magnification of the specimen being limited by the wavelength of the visible light. Light microscopes can outline the shape and size of the cells. Fuchigami et al. (1994) used light microscopy to study the coarse parenchyma of raw, blanched, and frozen-thawed carrot tissues, with very clear results. However, for more structural details, electron microscopy techniques should be employed. In this chapter, emphasis will be given to electron microscopic techniques for ultrastructural evaluation of frozen foods.

16.6.1 Principle of Electron Microscopy

A focused beam of electrons is accelerated by high voltage and interacts with the specimen. All foods must be dehydrated or frozen to a temperature that prevents the generation of measurable amounts of vapor. Two types of electron microscopy are distinguished: scanning electron microscopes (SEM) and transmission electron microscopes (TEM). In SEM, the sample is scanned by an electron beam focused onto a point. On impact, the primary electron beam produces secondary and back-scattered electrons from the specimen surface. These secondary or reflected electrons are processed for display and photography. Scanning electron microscopes operate typically at resolutions on the order of 10 nm. Because accelerating voltage determines the kinetic energy of the electrons, the electrons penetrate more deeply into the specimen at high accelerating voltages. In TEM, the electron beam, accelerated by a voltage of 20–100 kV, penetrates a thin specimen. The enlarged shadow of the specimen is viewed on a fluorescent screen for focusing and selecting the area of interest and then photographed.

This chapter discusses only common procedures related to sample preparation for electron microscopy. With numerous variations in product-specific procedures, more-detailed information on electron microscopy can be obtained from Kalab (1983) or Aguilera and Stanley (1990).

16.6.2 Sample Preparation for Electron Microscopy

With most foods, SEM has been used to study the internal microstructure. Most foods contain water as the major component that will produce volatile vapors in

Table 16.4. Comparison of microscopes

Criterion	LM	TEM	SEM	CLSM
Resolution	200–500	0.2 –1	3–4	Comparable to light microscopy
Magnification (×)	10–1,500	200–300,000	20–100,000	10 –2,000
Lens	Glass/quartz	Electromagnet	Electromagnet	Water immersion/ oil immersion/ Glass/filters
Specimen preparation	Easy	Difficult	Easy	Very easy
General sample-preparation procedure	Fixing, staining, viewing	Fixing, drying, staining, thin sectioning, fracturing, replication, and viewing	Fixing, drying, fracturing, mounting, coating/etching, and viewing	Staining (if needed), sectioning
Specimen thickness	Thin (reflected) Very thin (transmitted)	Very thin	Thin	Thin
Measuring environment	Small, versatile	Small, vacuum	Large, vacuum	Small
Illumination source	White light	Electron beam	Electron beam	Laser
Visualization of samples in frozen stage	Possible if used in reflection mode	No	Yes	Yes
Real-time visualization of a process	Possible	No	No	Yes
Image display	Eye, viewing screen, film	Fluorescent screen, film	CRT screen, computer image processing, film	Computer image processing

vacuum inside the electron microscope. Thus, all the water from the sample should be removed, or the sample should be frozen below −100°C. Conventional drying may severely alter the microstructure of the sample because of the dissolved components in the aqueous phase such as salts, carbohydrates, peptides, and soluble proteins. To minimize unwanted structural changes in the sample, special procedures such as fixation, dehydration, and special drying techniques have been developed. Then, the sample is fractured and the fragments are mounted on metal stubs, rendered conductive by coating with carbon and gold, before examination under the microscope.

Fixation

Fixation renders the structure rigid and resistant to changes during subsequent procedures. Most fixation techniques stabilize proteins in the sample by cross-linking protein molecules using low-molecular aldehydes. Formaldehyde, glutaraldehyde, and acrolein are commonly used for this purpose. In addition to aldehydes, oxidating fixatives such as osmium tetroxide or potassium permanganate are also used as fixing agents. The oxidating fixatives react with the unsaturated lipids. By obscuring the underlying protein matrix or by melting under the electron beam, high concentrations of lipids will interfere during SEM. Thus, if the lipids should not be retained in the sample for SEM, then these oxidative fixatives should be used only on defatted samples. Temperature of fixation may range from 0 to 37°C, and duration may vary from 10 min to several hours with denser samples requiring longer duration.

Drying

Drying is most frequently accomplished by freeze-drying or critical-point drying. In freeze-drying, the sample is frozen rapidly and the ice is subsequently sublimated in vacuum. The samples should be rapidly frozen and maintained at a temperature below −75°C during freeze-drying to prevent recrystallization. In critical-point drying, substitution of liquid CO_2 is used as one of several suitable transitional fluids for the organic solvent with which the dehydrated sample has been impregnated. Subsequently the liquid CO_2 is converted into gaseous CO_2. Critical-point drying is suitable for products based on protein or milk products or for plant tissues, but it produces artifacts in products containing gelatinized starch. Freeze-drying is a better alternative for these products.

Fracturing

The internal microstructure of the sample under study can be revealed by fracturing, which is carried out either with a dried sample (dry-fracturing) or a frozen sample. Dry-fracturing is done under a low-magnification stereoscopic microscope and is suitable for brittle samples such as muscle or cheese. Freeze-fracturing is more suitable for samples that do not require fixation or samples that are frozen.

Mounting

The fractured fragments are mounted on metal SEM stubs. Care should be taken to avoid touching the fracture planes while mounting. Initially, the stubs are sanded and then painted with a conductive silver cement. Then each fragment is positioned on a droplet of freshly applied silver cement. After hardening of the cement, the vertical sides are carefully painted, leaving only the fracture side exposed.

Coating

Stubs with sample fragments are coated with a conducting layer using vacuum evaporation or sputter coating techniques. In vacuum evaporation, the coating material is evaporated consecutively from two to three different angles in an evacuated bell jar. In sputter coating, coating is carried out in an argon atmosphere using the coating material as cathode and the specimen stub as anode.

Viewing

For SEM, the accelerating voltage depends on the nature of the sample and the magnification. Most foods are examined at magnifications between 100× and 20,000×, and most work with food is carried out at 5–20 kV.

16.6.3 Sample Preparation for TEM

Negative Staining

Negative staining is a technique suitable for suspensions of submicroscopic particles translucent to the electron beam. An electron-dense salt solution such as ammonium molybdate is added to the suspension, and a thin layer of the mixture is dried on an electron-transparent film. As the beam passes through the particles under study, it is absorbed by the electron-dense stain. This results in a micrograph with the particles appearing light against a dark background. If protein is present in the suspension, it should be fixed in glutaraldehyde solution.

Metal Shadowing

In this technique, the specimen surface is deposited with metal vapors at a specific angle, resulting in a three-dimensional appearance of micrographs. For deposition, low granularity and high-melting-point metals such as chromium, platinum, palladium, iridium, or tungsten are used. Metal evaporation can be obtained by resistance heating, electron beam heating, or ion beam sputtering in vacuum.

Thin Sectioning

Thin sectioning is a standard TEM technique and is suitable for solid samples. The samples are usually fixed, dehydrated, embedded in an electron-transparent resin, and sectioned using a glass or diamond knife. The image in the electron microscope is formed by the passage of the beam through areas containing varying amounts of stains. A variety of techniques suitable for preparation are available for thin sectioning and can be found in detail elsewhere (Kalab, 1983).

Freeze-Fracturing and Replication

Changes to initial microstructure occur significantly during dehydration (in the previous methods described); to best preserve the original microstructure, the samples should be fixed by rapid freezing (cryofixation). Because of the higher resolutions used in TEM than in SEM, the rate of rapid freezing is critical to prevent ice-crystal formation. All foods can be examined using freeze-fracturing and replication. For suspensions, emulsions, and foams, freeze-fracturing may be the only reliable technique. Procedures related to freeze-fracturing have been discussed in detail by Kalab (1983). The freeze-fractured samples are shadowed at 45° with platinum (or high-melting-point metal), and the metal layer is reinforced with carbon evaporated at an angle of 90°. The replica is separated from the fractured sample by flotation on water. Then the replica is cleaned and lifted on a grid for TEM.

16.6.4 Examples of Ultrastructural Evaluations

Ice Cream

Caldwell et al. (1992) described a low-temperature electron microscopy technique to study ice-cream structure in the frozen and fully hydrated state. The samples for electron microscopy were removed using a pair of stainless-steel rivets to push through the sample. The rivets were then plunged immediately into liquid nitrogen and loaded into a chilled copper stub (−196°C). The stub was attached quickly to the transfer device, and the assembly was immersed in nitrogen slush (−210°C) and delivered to an evacuated preparation chamber. The upper rivet was fractured to reveal a fresh surface. The fractured samples were then transferred to the cold stage of the electron microscope and etched for 10 min at −80°C after being warmed from −160°C. Subsequently the etched sample was transferred to the preparation chamber under vacuum, sputter coated with gold at −150°C, and viewed at 10-kV accelerating voltage in the electron microscope.

Goff et al. (1995) studied the influence of temperature, polysaccharide-stabilizing agents, and overrun on the changes in the rheological properties of ice cream. They used low-temperature electron microscopy as described by Caldwell et al. (1992) to confirm the results they obtained using dynamic rheological measurement techniques. After mixing and aging overnight at 5°C, the ice-cream mixes were frozen to −5°C and 20–100% overruns. In addition, the ice-cream samples were hardened at −35°C for 24 h. For low-temperature electron microscopy, ice-cream samples were taken both at the time of draw from the barrel freezer and after hardening. The structural details provided needed explanation for the changes in the rheological properties of the ice cream. The authors concluded that an increase in storage modulus or higher elasticity may result from a structure increasingly dominated by the air phase.

Other methods for examining ice cream involve TEM (Buchheim, 1974; Berger and White, 1971). For using TEM, the samples were frozen, fractured, and etched, and a replica was prepared with platinum and carbon. The replica was removed and cleaned for examination. Berger and White (1971) studied the fat-destabilization process during ice-cream manufacture as it was influenced by fat composition and processing conditions. TEM provides the high resolution needed for studying the fat-serum interface. Buchheim (1974) was able to observe fat globules, ice crystals, air bubbles, and phase interfaces. However, TEM is not suited for viewing frozen and fully hydrated samples.

Frozen Bread Dough

Frozen bread dough is used by more than 50% of in-store supermarket bakeries, as well as by retail consumers. A number of researchers (Junge et al., 1981; Parades-Lopez and Bushuk, 1982; Berglund et al., 1991) have used electron microscopy to study the ultrastructure of dough and breads. Berglund et al. (1991) used low-temperature SEM to examine the ultrastructure of frozen dough in a fully hydrated state. The prepared dough samples were frozen at −23°C, and samples were examined by LT-SEM shortly after mixing, after frozen storage of 24 h and 24 wk, and after one and three freeze-thaw cycles. Samples were cut from the interior of the frozen loaf with a precooled cutting tool, kept frozen on dry ice, and placed on a cooled specimen holder. For SEM, the sample was rapidly brought to a lower temperature by plunging the specimen holder into nitrogen slush at −196°C. Then, the sample was fractured with a cooled macroknife assembly in a preparation chamber maintained at −160°C and transferred under vacuum onto the cold stage of the cryounit. After being sublimed at −65°C for 6 min, samples were sputter coated with gold at 0.1 Torr and at 30 mA for 3 min. Finally, the coated samples were examined and photographed. Berglund et al. (1991) concluded that the observed changes in the ultrastructure of the starch granules and gluten contributed to the extended proof times and reduced loaf volumes of frozen bread dough.

Frozen Vegetables

The effects of freezing rates on textural damage were studied on frozen blanched and unblanched summer squash by Crivelli and Buonocore (1971). The microscopic measurement details were not provided by these authors. They concluded that blanching made the tissue softer and more sensitive to freezing damage. Mohr (1974) studied the effects of freezing and freezing-thawing on blanched and unblanched vegetables (green peas, spinach, and green beans) and the damage to ultrastructure using SEM. The samples were prepared by fixing them using a 2% glutaraldehyde buffered with 0.02-M sodium phosphate. The method included glutaraldehyde-osmium fixation, staining of the tissue with uranyl acetate during alcohol dehydration, embedding in styrene-metha-crylate, and poststaining with lead citrate. There was less structural disruption in green peas (high starch content) than in tissues of spinach and green bean pods (low starch content). Mohr (1974) concluded that the starch granule component is the major contributor to texture and confirmed these results from consecutive sensory evaluation. Recently, Fuchigami et al. (1995) analyzed cellular structure changes in frozen carrot tissue with SEM and reported that the micrographs of carrot tissue clearly showed the cell damage and plasma membrane separation. They attributed the excessive tissue softening to the severe damage to the membranes by freezing.

Meat

Ultrastructural evaluations of meat tissues have been conducted using SEM by several researchers to investigate the tissue damage resulting from freezing and freeze-thaw cycling. The meat products included pork liver (Jones et al., 1986), beef (Bevilacqua et al., 1979; Bevilacqua and Zaritzky, 1980; Rahelić et al., 1985), and fish (Bello et al., 1982; Kim et al., 1987). In general, for all these studies, the tissue samples were fixed, postfixed, dehydrated, embedded or cryofractured, sputter coated, and finally analyzed using electron microscope. Jones et al. (1986) and Rahelić et al. (1985) used 2% glutaraldehyde, osmium tetroxide, ethanol for fixing, postfixing, and dehydrating, respectively. In addition, Rahelic et al. (1985) stained their tissue samples using uranyl acetate and lead nitrate. For fish samples, glutaraldehyde and potassium ferricyanide were used for fixing by Bello et al. (1982) and Kim et al. (1987), respectively. Because of the wide variations in the selection of fixing agent, postfixing agent, and sample preparation, readers are advised to select a suitable procedure corresponding to the product. In general, the results from these studies on meat tissues concluded that fast freez-

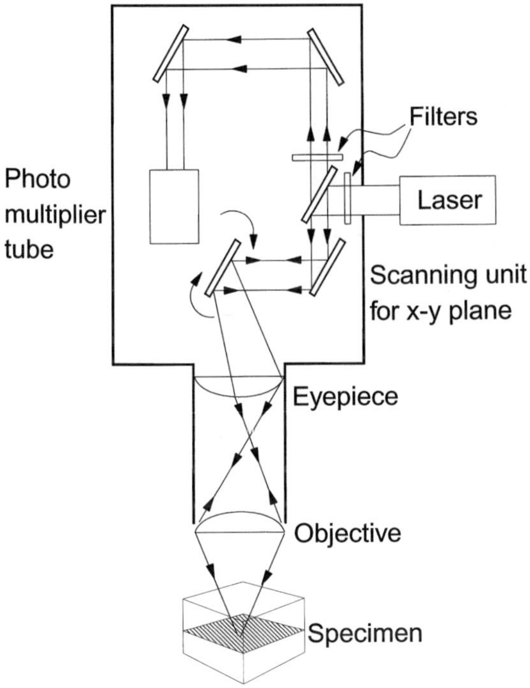

Figure 16.5. Schematic representation of light path in a confocal laser scanning microscope

ing produces smaller ice crystals (Bello et al., 1982), the ultrastructure of the muscle successively changed with decreasing freezing temperature from −10 to −115°C (Rahelić et al., 1985), and severe tissue damage resulted from freeze-thaw cycling compared to refrigerated storage.

16.6.5 Confocal Laser Microscopy

In addition to SEM and TEM techniques to study the microscopic information, another new technique available is laser scanning confocal microscopy (Vodovotz et al., 1996). Confocal laser microscopy is a combination of light microscopy, confocal imaging, and scanning microscopy. In this modified light microscopy, light or a laser light with a narrow range of wavelengths excites a specific fluorescence material. The technique is called confocal because both the illumination and detection (imaging) systems are focused on the same single-volume element of the specimen (fig. 16.5). Thus, confocal microscopy detects only the in-focus region, with out-of-focus parts appearing as black background.

This method provides an image with better resolution than that obtained with conventional light microscopy. In addition, it provides an opportunity to observe a three-dimensional image without the need to physically section the sample. It can also distinguish the spatial location of different components by detecting fluorescence from dyes specific to different chemical species. Confocal microscopy provides many advantages over the other microscopic techniques. One of the main advantages is the minimal sample preparation required. The specimen does not require embedding, sectioning, or fixing because of the optical sectioning capabilities of confocal microscopy. Another advantage is its ability to scan samples at different depths to obtain a *z* series. A stack of *z* series can then be reconstituted using image-rendering software to generate a true three-dimensional image of the object. Confocal microscopy can be applied for real-time studies and enabling the visualization of dynamic processes.

Confocal laser microscopy has not been used widely for studying the ultrastructural information in frozen foods. Recently, attempts were made by Frank (1996) to study the structure of ground beef patties using confocal laser microscopy in frozen, thawed, and cooked stages. They used Nile red and fluorescein isothiocyanate for staining the patties to visualize fat and protein, respectively, and a low magnification (2.5× objective) to provide a high resolution up to 2 mm deep from the surface. Data from confocal microscopy showed fractured fat particles adhering to muscle tissue in thawed raw patties. These results confirm that confocal laser microscopy may be a useful technique for studying the quality of other frozen foods.

16.7 Conclusions

Quality is an important factor in determining the success of frozen foods in the market. Thus, methods to assess these quality attributes should lead the food industry to identify process conditions that will affect acceptance of the end product. With recent improvement in technologies, handheld color meters have become affordable for the food industry. Different types of universal testing machines under $15,000 are also available; they can do specific tests on food materials, making them a tool used by research and development and quality-control departments. The methods described in this chapter have provided the basic information for developing methods specific to a food product.

The developments in ultrastructural evaluations involve digitally processing the images instead of evaluating them by qualitative (visual) comparisons. Also, the use of other imaging techniques such as atomic force microscopy, magnetic resonance imaging, and confocal laser microscopy are evolving to include frozen foods. Developments in measurement techniques will continue to shape future improvements in frozen foods to meet consumers' needs.

References

Aguilera, J.M. and Stanley, D.W. 1990. *Microstructural Principles of Food Processing and Engineering.* Elsevier Science, New York.

Akamittath, J.G., Brekke, C.J., and Schanus, E.G. 1990. Lipid oxidation and color stability in restructured meat systems during frozen storage. *J. Food Sci.* **55**:1513–1517.

Andersen, H.J. and Skibsted, L.H. 1991. Oxidative stability of frozen pork patties: effect of light and added salt. *J. Food Sci.* **56**:1182–1184.

Bello, R.A., Luft, J.H., and Pigott, G.M. 1982. Ultrastructural study of skeletal fish muscle after freezing at different rates. *J. Food Sci.* **47**:1389–1394.

Berger, K.G. and White, G.W. 1971. An electron microscopical investigation of fat destabilization in ice cream. *J. Food Technol.* **6**:285–294.

Berglund, P.T., Shelton, D.R., and Freeman, T.P. 1991. Frozen bread dough ultrastructure as affected by duration of frozen storage and freeze-thaw cycles. *Cer. Chem.* **68**:105–107.

Berry, B.W. 1992. Low fat level effects on sensory, shear, cooking, and chemical properties of ground beef patties. *J. Food Sci.* **57**:537–540.

Bevilacqua, A.E. and Zaritzky, N.E. 1980. Ice morphology in frozen beef. *J. Food Technol.* **15**: 589–597.

Bevilacqua, A.E., Zaritzky, N.E., and Calvelo, A. 1979. Histological measurements of ice in frozen beef. *J. Food Technol.* **14**:237–251.

Bhattacharya, M., Hanna, M.A., and Mandigo, R.W. 1988. Effect of frozen storage conditions on yields, shear strength, and color of ground beef patties. *J. Food Sci.* **53**:696–700.

Bolin, H.R. and Huxsoll, C.C. 1993. Partial drying of cut pears to improve freeze/thaw texture. *J. Food Sci.* **58**:357–360.

Brewer, M.S. and Harbers, C.A.Z. 1991. Effect of packaging on color and physical characteristics of ground pork in long-term frozen storage. *J. Food Sci.* **56**:363–366.

Brewer, M.S., Klein, B.P., Rastogi, B.K., and Perry, A.K. 1994. Microwave blanching effects on chemical, sensory, and color characteristics of frozen green beans. *J. Food Qual.* **17**:245–259.

Brewer, M. S., McKeith, F. K., and Britt, K. 1992. Fat, soy, and carrageenan effects of sensory and physical characteristics of ground beef patties. *J. Food Sci.* **57**:1051–1052, 1055.

Brewer, M.S. and Wu, S.Y. 1993. Display, packaging, and meat block location effects on color and lipid oxidation of frozen lean ground beef. *J. Food Sci.* **58**:1219–1223.

Buchheim, W. 1974. Electron microscopical studies on the structure of whipped cream. *Proceedings of the 19th International Dairy Congress,* 1E, pp. 245–246, Bundesanstalt für Milchforschung, Kiel, Federal Republic of Germany.

Caldwell, K.B., Goff, H.D., and Stanley, D.W. 1992. A low temperature SEM study of ice cream. I. Techniques and general microstructure. *Food Struct.* **11**:1–9.

Cano, M.P. and Martin, M.A. 1992. Pigment composition and color of frozen and canned kiwi fruit slices. *J. Agric. Food Chem.* **40**:2141–2146.

Cano, P., Martin, M.A., and Fuster, C. 1990. Freezing of banana slices. Influence of maturity level and thermal treatment prior to freezing. *J. Food Sci.* **55**:1070–1072.

Chen, C.M. and Trout, G.R. 1991. Color and its stability in restructured beef steaks during frozen storage: effects of various binders. *J. Food Sci.* **56**:1461–1464, 1475.

Crivelli, G. and Buonocore, C. 1971. Rate of freezing and integrity of vegetable tissues. IV. Observations on summer squash (*Cucurbita pepo* L.). *J. Texture Stud.* **2**:89–95.

Dagnoli, J. and Erikson, J.L. 1989. The looming battle for the center of the plate. *Advertising Age* **60**(49):810.

Dias, J., Nunes, M.L., and Mendes, R. 1994. Effect of frozen storage on the chemical and physical properties of black and silver scabbard fish. *J. Sci. Food Agric.* **66**:327–335.

Forni, E., Crivelli, G., Polesello, A., and Ghezzi, M. 1991. Changes in peas due to freezing and storage. *J. Food Proc. Preserv.* **15**:379–389.

Frank, J.F. 1996. Structure of ground beef patties. Presented at the 1996 Annual Meeting of Institute of Food Technologists, New Orleans, La.; abstract no. 70-1.

Fuchigami, M., Hyakumoto, N., and Miyazaki, K. 1995. Programmed freezing affects texture, pectic composition, and electron microscopic structure of carrots. *J. Food Sci.* **60**:137–141.

Fuchigami M., Hyakumoto, N., Miyazaki, K., Nomura, T., and Sasaki, J. 1994. Texture and histological structure of carrots frozen at a programmed rate and thawed in an electrostatic field. *J. Food Sci.* **59**:1162–1167.

Fúster, C., Préastamo, G., and Cano, M.P. 1994. Drip loss, peroxidase, and sensory changes in kiwi fruit slices during frozen storage. *J. Sci. Food. Agric.* **64**: 23–29.

Goff, H.D., Freslon, B., Sahagian, M.E., Hauber, T.D., Stone, A.P., and Stanley, D.W. 1995. Structural development in ice cream: dynamic rheological measurements. *J. Texture Stud.* **26**: 517–536.

Greer, G.G. and Murray, A.C. 1991. Freezing effects on quality, bacteriology, and retail-case life of pork. *J. Food Sci.* **56**:891–894, 912.

Grau, R. and Hamm, R. 1957. Eine einfache Methode zur Bestimmung der Wasserbindung im Muskel. *Naturwissenschaften* **40**:29–30.

Honikel, K.O. 1987. Critical evaluation of methods detecting water holding capacity in meat. In *Accelerated Processing of Meat.* (Romita, A., Valin, C., and Taylor, A.A., eds.) pp. 225–240, Elsevier Applied Science Publishers, London, UK.

Hwang, S.Y., Bowers, J.A., and Kropf, D.H. 1990. Flavor, texture, color, and hexanal and TBA values of frozen cooked beef packaged in modified atmosphere. *J. Food Sci.* **55**:26–29.

Jahncke, M., Baker, R.C., and Regenstein, J.M. 1992. Frozen storage of unwashed cod (*Gadus morhua*) frame mince with and without kidney tissue. *J. Food Sci.* **57**:575–580.

Jones, S.B., Strange, E.D., and Maleeff, B.E. 1986. Ultrastructure of pork liver after freeze-thaw cycling and refrigerated storage. *J. Food Sci.* **51:**761–765.

Junge, R.C., Hoseney, R.C., and Varriano-Martson, E. 1981. Effect of surfactants on air incorporation in dough and the crumb grain in bread. *Cer. Chem.* **58**:338.

Kalab, M. 1983. Electron microscopy of foods. In *Physical Properties of Foods.* (Peleg, M. and Bagley, E.B., eds.) pp. 43–103, AVI Publishing, Westport, Conn.

Kenawi, M.A. 1994. Evaluation of some packaging materials and treatments on some properties of beef during frozen storage. *Food Chem.* **51**:69–74.

Kim, J.M., Lee, C.M., and Hufnagel, L.A. 1987. Textural properties and structure of starch-reinforced surimi gels as affected by heat-setting. *Food Microstructure,* **6**:81–89.

Lamkey, J.W., Mandigo, R.W., and Calkins, C.R. 1986. Effect of salt and phosphate on the texture and color stability of restructured beef steaks. *J. Food Sci.* **51**:873–875, 911.

Lanari, M.C., Cassens, R.G., Schaefer, D.M., and Scheller, K.K. 1993. Dietary vitamin E enhances color and display life of frozen beef from Holstein steers. *J. Food Sci.* **58**:701–704.

Mallet, C.P. (ed.) 1993. *Frozen Food Technology.* Blackie Academic & Professional, London.

Mohr, W.P. 1974. Freeze-thaw (and blanch) damage to vegetable ultrastructure. *J. Texture Stud.* **5**:13–27.

Parades-Lopez, D. and Bushuk, W. 1982. Development and "underdevelopment" of wheat dough by mixing: microscopic structure and its relation to bread-making quality. *Cer. Chem.* **60**:24.

Persson, P.O. and Löndahl, G. 1993. Freezing technology. In *Frozen Food Technology.* (Mallet, C.P., ed.) pp. 20–58, Blackie Academic & Professional, London.

Rahelić, S., Gawwad, A.H., and Puač, S. 1985. Structure of beef *Longissimus dorsi* muscle frozen at various temperatures. Part 2: Ultrastructure of muscles frozen at −10, −22, −33, −78, and −115°C. *Meat Sci.* **14**:73–81.

Szczesniak, A.S. 1983. Physical properties of foods: what they are and their relation to other food properties. In *Physical Properties of Foods.* (Peleg, M. and Bagley, E.B., eds.) pp. 1–37, AVI Publishing, Westport, Conn.

Trout, G.R. 1988. Techniques for measuring water-binding capacity in muscle foods: a review of methodology. *Meat Sci.* **23**:235–252.

Vodovotz, Y., Vittadini, E., Coupland, J., McClements, D.J., and Chinachoti, P. 1996. Bridging the gap: use of confocal microscopy in food research. *Food Technol.* **6**:74–82.

Voicey, P.W. 1976. Engineering assessment and critique of instruments used for meat tenderness evaluation. *J. Texture Stud.* **7**:11–48.

17

Chemical Measurements of Frozen Foods

M. C. Erickson

17.1 Introduction

Chemical reactions occur during freezing and frozen storage of foods. Whereas some of these reactions may be desirable, more often than not they have undesirable consequences. Understanding these reactions is therefore critical in controlling and minimizing the detrimental changes that occur during freezing and subsequent storage. To monitor these reactions, measurement of chemical markers in the food product may be conducted. Chemical markers are compounds whose concentrations often reflect changes in quality. The usefulness of chemical markers is illustrated by their ability to provide quantitative information, to be used in predictive modeling, and to mimic other processes that may take place in the vicinity of the site where the chemical measurements are made. The objective of this chapter is to discuss both the preparation for analysis and the analytical methods used to quantify chemical markers and reactions. Proceeding from discussion of the more traditional techniques, the chapter will also touch on the development and application of new sensor technologies that should have a major impact on quality evaluation of frozen foods.

17.2 Sample Preparation

An important aspect of quality assessment is sampling and preparation of the sample for analysis. The analyst must be aware of the requirements and in many cases the confidence limits of both sampling and analysis. All too often the analyst is prepared to work to very fine limits negated by the sampling procedure used for his or her starting material. As technologies have improved to the point that the use of smaller sample volumes and improved sensitivities have become routine, it has become more critical to protect sample preparation steps against factors that could impinge on the ultimate accuracy and efficient use of these technologies. These factors include sample contamination, recovery losses, and perturbations to the intrinsic chemical behavior of the molecule(s) of interest.

17.2.1 Sampling

In statistical terms, the group of items or units selected from the population of interest and used to represent that population is called the sample (Cochran, 1977). Generally the population of interest is large, while the sample represents a smaller subset of all items or units.

Both manual and continuous sampling procedures exist in the food industry. In continuous sampling procedures, a fraction of the material being analyzed is diverted. The size and number of samples will be dependent on the food item to be analyzed, as well as the statistical accuracy desired. Influencing the size and number of samples drawn is the degree of sample heterogeneity. To minimize heterogeneity, various techniques (e.g., grinding and mixing) are used to ensure that the sample drawn is representative of the entire material and provides a true measure of its overall content. Toward that end, Pettinati et al. (1983) compared the food chopper to the bowl cutter for preparation of meat samples that had been frozen. Based on the precision and accuracy of the results and the simplicity and convenience of the procedures, the bowl-cutter procedure was favored over the food chopper for preparation of meat samples for analysis. Lichon and James (1990) expanded on this study by comparing 12 methods of homogenization on a range of food materials. They found that cryogenic homogenization methods (cryogenic milling; top-drive macerator/dry-ice grind) were effective on the majority of foods tested, allowing milligram subsample masses to be used for analysis. Homogenization in the frozen state is advantageous because food that is normally tough and elastic at room temperature becomes fragile and crystalline, with the lowering of temperature facilitating pulverization of the food to a smaller particle size. Such techniques are especially suited to foods already in the frozen state, but there are limitations in the mass of product that may be inserted into these apparatuses.

17.2.2 Dilution/Filtration

Two of the simplest sample preparation methods used are dilution and filtration. Dilution usually is used to bring the concentration of the analytes into the linear range of the analytical method. Filtration is used when the analyte is in the liquid phase and the interference is in the solid phase. If the analytes are present at trace levels or differ greatly in concentration or if the interference and the analytes are both soluble, neither method is satisfactory.

17.2.3 Extraction

To facilitate the measurement of chemical markers, isolation of the compound from other interfering compounds is often necessary. One of the most common isolation steps is extraction, in which the food material is dispersed in an aqueous or organic solvent, followed by separation of the extract from the food matrix. In the isolation of tocopherols, two basic approaches exist to extract the compounds from food tissues: extraction of tissue with solvents without prior saponification or direct alkaline digestion of tissue to which solvent extraction is applied. The two methods of recovery of tissue tocopherol were compared with five different samples of frozen minced fish muscle. The saponification procedure employed consisted of exposing 1 g muscle to 18% KOH, 25% ethanol, and 3% ascorbic acid during heating at 70°C for 45 min. Following cooling of the saponified mixture, it was extracted three times with 5 ml each of 10% ethyl acetate in hexane. This extract was evaporated to dryness, reconstituted in 1 ml of 98:2 methanol/water, and filtered before quantification by HPLC. An important aspect worth noting in this procedure is the necessity for including ethyl acetate in the extracting solution. In the absence of ethyl acetate, the potassium salts of the fatty acids generated during saponification interfere in the extraction of the tocopherols and lower the recovery. The procedure used for extraction of nonsaponified muscle tissue followed that used by Lakritz and Thayer (1992) with the exception that 1 g rather than 20 g was analyzed and the volume of the extracting solution (9% ethanol in cyclohexane) used per gram tissue was 2.8 times larger. The extract was evaporated to dryness, reconstituted in 98:2 methanol/water, and filtered for quantification by HPLC. Following adjustments for recovery of a spiked standard, tissue samples that had been saponified were calculated to have had 1.4–2.4 times higher levels of α-tocopherol than samples that had not been saponified (table 17.1). In contrast, γ-tocopherol levels in saponified and nonsaponified samples were similar. It is postulated that modification of cellular membranes during freezing restricted access of the solvent to membrane α-tocopherol, whereas exposure of the solvent to γ-tocopherol, which resides in the depot lipids, was not restricted. Extraction efficiency of α-tocopherol, in fact, was dependent on the level of α-tocopherol present in the samples, with a correlation coefficient of -0.98 occurring between the extraction-efficiency ratio and the α-tocopherol levels obtained with saponification. Therefore, since solvent extraction overestimated tocopherol losses in frozen tissues, it is recommended that saponification be employed routinely during tocopherol assays of frozen-food tissue samples.

Procedures used in the isolation of volatiles from foods before gas chromatographic separation and detection include purge and trap or direct sampling of headspace. These procedures, however, exert quantitative and qualitative differences in composition of the isolated sample (Jennings and Filsoof, 1977). Solid-phase microextraction (SPME) of volatiles is a recently developed adsorption/

Table 17.1. Effect of saponification on α- and γ-tocopherol measurements in frozen fish muscle tissue

	α-Tocopherol ($\mu g/g$)[a]			γ-Tocopherol ($\mu g/g$)[a]		
Sample	With saponification	Without saponification	Extraction-efficiency ratio[b]	With saponification	Without saponification	Extraction-efficiency ratio[b]
1	7.15	3.89	1.84	1.68	1.34	1.25
2	0.64	0.27	2.37	0.42	0.29	1.45
3	9.85	6.93	1.42	1.48	1.27	1.16
4	9.46	6.23	1.52	1.42	1.51	0.95
5	8.32	4.74	1.76	1.29	1.14	1.13

[a]Adjusted for recovery of spiked standard

[b]Level obtained with saponification divided by level obtained without saponification

desorption technique that eliminates the need for solvents or complicated approaches in concentrating volatile or nonvolatile compounds in liquid samples of headspace (Zhang and Pawliszyn, 1993; Shirey et al., 1995). By introducing a fused silica fiber coated with a viscous polymer to the headspace above a sample, the volatiles may be adsorbed onto the coating and then transferred to the analytical instrument for desorption and analysis. Because SPME is a new technique, quantitative data specific for food samples are limited. Preliminary assessment of SPME is currently ongoing in this laboratory, but findings to date indicate that it would be an excellent procedure for monitoring the generation of oxidative volatiles in foods that have been stored under frozen conditions.

17.3 Types of Analysis

A variety of analytical tools are used to assess quality in frozen foods including titration, spectrophotometric, spectrofluorometric, chromatographic, and electrophoretic analyses. In the following sections, these analytical tools will be described in greater detail and studies cited where they have been applied to frozen-food systems.

17.3.1 Titration

Measurement of both free fatty acids (FFA) and extractable protein nitrogen has been accomplished by titration methods. In the FFA test, the carboxylic groups are neutralized with the addition of sodium hydroxide and a change in the color of the metacresol indicator signals the end point (Ke and Woyewoda, 1978). Several examples where this method has been applied to frozen foods include the studies of Jonas (1969), Mazeaud and Bilinski (1976), Ke et al. (1978), and de Koning et al. (1987). Correction, however, must be made for the contribution of the acidic phospholipids to the percent FFA value obtained by alkalimetric titration (de Koning et al., 1987; Zhou and Ackman, 1996).

To measure extractable protein nitrogen, the sample is extracted with an aqueous solution containing sodium chloride and the organic nitrogen present in the extract reduced and transformed to ammonium sulfate in the presence of sulfuric acid, catalysts, and hydrogen peroxide. Organic carbon and hydrogen are then oxidized and the boiling solution rendered colorless. After a brief cooling period, the digest is quantitatively transferred to a steam distillation apparatus and strong base added to liberate the ammonia. Application of steam results in complete distillation of the ammonia into boric acid. The ammonia is in turn titrated from the boric acid with a strong acid (at low concentration) and the nitrogen quantified. Since protein is assumed to be 16% nitrogen, the factor 6.25 is used to convert total nitrogen to total protein. With this assay, Love (1956, 1958) showed that the rate of freezing affected protein extractability, and Dias et al. (1994) derived a statistical expression based on the soluble protein content in scabbardfish, permitting the prediction of textural quality during frozen storage.

17.3.2 Spectrophotometry and Spectrofluorimetry

All biochemicals absorb energy from at least one region of the spectrum of electromagnetic radiation. The energies at which this absorption occurs depend on the available electronic, vibrational, and rotational energy levels of the molecule. When absorption is from the ultraviolet/visible region of the spectrum (200–700 nm), transitions occur between electronic energy levels. Spectrophotometry and spectrofluorimetry involve the measurement of these transitions, which permit the characterization and quantification of biological molecules.

Spectrophotometry

Markers for Lipid Oxidation. In shifting the double bond during lipid oxidation, conjugated dienes develop an ability to absorb light in the ultraviolet region with a maximum at 232 nm. Using this absorption property, the level of conjugated dienes was determined in frozen bass, tilapia, and channel catfish by applying the extinction coefficient of conjugated dienes to absorbance values of extracted lipids (Erickson, 1993a,b, 1994). Sensitivity is limited, however, because the diene absorption appears as a rather imprecise shoulder on the strong absorption peak of the nonperoxidized fatty acid itself. By using second-derivative spectroscopy (Corongiu and Milia, 1983; Corongiu et al., 1989) or ultraviolet difference spectroscopy with tandem cuvettes (Vossen et al., 1993), sensitivity can be increased.

Erickson (1993a,b, 1994) also used another spectrophotometric analysis to monitor the extent of oxidation during frozen storage, but in this case a chemical reaction between the compound of interest (malondialdehyde) and thiobarbituric acid preceded the spectrophotometric measurement at 532 nm. Such chemical modifications are typical of many spectrophotometric assays because they improve the sensitivity of the assay and eliminate interference from other compounds that may have been absorbed at the wavelengths of the unmodified compound of interest. Both distillation and aqueous acid extraction were used as preparatory tools for quantification of thiobarbituric acid reactive substances but gave similar responses (Salih et al., 1987; Spanier and Traylor, 1991). In either case, addition of chelator and an antioxidant, such as propyl gallate, to the samples before homogenization is recommended to retard further lipid oxidation.

Markers for Protein Denaturation. Studies of dimethylamine (DMA) and formaldehyde contents in frozen gadoid fish have also been conducted using spectrophotometric measurements (Landolt and Hultin, 1982; Ciarlo et al., 1985; Samson et al., 1985). In the case of DMA, a trichloroacetic acid extract of fish muscle is reacted with a copper ammonium reagent and carbon disulfide at 40–50°C to produce a yellow-colored copper dimethyl-dithiocarbamate (Castell et al., 1974). The solution is acidified with acetic acid and shaken vigorously to facilitate extraction of the colored complex into a benzene layer. After drying with sodium sulfate, absorbance of the benzene layer is measured at 440 nm. The spectrophotometric analysis most commonly employed for formaldehyde, on the other hand, is based on that of Nash (1953), in which free or loosely bound formaldehyde in a TCA or perchloric acid extract is reacted at 60°C with Nash color reagent containing acetylacetone with an excess of ammonium acetate to produce diacetyldihydrolutidine, measurable spectrophotometrically at 415 nm.

Enzyme Activity Assays. As a prefreezing operation, blanching is the primary means of inactivating enzymes present in vegetables. Residual activity of enzymes can therefore serve as an indicator of the adequacy of blanching. In this regard, the most frequently conducted assay is for peroxidase activity and involves monitoring spectrophotometrically the increase in absorbance at 470 nm when an extract from a vegetable product is incubated with guiaicol and hydrogen peroxide (Whitaker, 1994). Another enzyme that has been suggested as a blanching indicator is lipoxygenase. Its activity may be followed spectrophotometrically by observing the increase in conjugated dienes or hydroperoxides when an extract of the product is incubated with linoleic acid. A coupled reaction involving the bleaching of the yellow color of carotene by lipoxygenase may also be used (Whitaker, 1994).

On the basis that freezing and thawing leads to cellular disruption and release of bound enzymes into the cellular fluid, various enzymes have been proposed for distinguishing fresh from frozen-thawed muscle. Several of these enyzmes may be assayed spectrophotometrically and include ATPase, lactate dehydrogenase, cytochrome oxidase, α-glucosidase, and cathepsin D (Barbagli and Crescenzi, 1981; Shimomura et al., 1987; Nambudiri and Gopakumar, 1992; Nilsson and Ekstrand, 1993).

Spectrofluorimetry

Fluorescence is the emission of light by a sample that absorbs light (of higher energy) at a lower wavelength. While all molecules absorb photons, relatively few fluoresce significantly at room temperature, so these latter properties are especially useful for resolving minor components in complex mixtures. Furthermore, the high sensitivity of photodetectors and the ability of monochromators or filters to resolve incident from emitted light makes fluorescence a sensitive analytical procedure with nanogram amounts capable of being detected.

Spectrofluorimetry is a useful tool in the analysis of fluorescent pigments, a chemical marker for later stages of lipid oxidation. These pigments arise when lipid oxidation products react with proteins, amino acids, and nitrogen-containing phospholipids (Gardner, 1979; Gutteridge et al., 1982; Leake and Karel, 1985).

These fluorescent products form two distinct groups based on their fluorescent wavelength maxima. The one group formed between aldehydes (other than malondialdehyde) and a variety of amino compounds exhibits an excitation maximum between 320 and 360 nm and an emission maximum between 410 and 460 nm. The second group, with an excitation maximum between 370 and 405 nm and an emission maximum between 450 and 470 nm, is due to the reaction of malondialdehyde with amino compounds (Leake and Karel, 1985). Conjugated Schiff bases are proposed to be responsible for the second group's fluorescence while the mechanism of the first group's fluorescence is speculated to involve aldol condensation of alkanals to alk-2-enals before cross-linking with the amine group (Kikugawa and Beppu, 1987; Iio and Yoden, 1988).

Fluorescent pigment measurements have been employed on both organic and aqueous extracts from frozen chicken and fish (Pikul et al., 1984; Erickson, 1993a,b, 1994). Pikul et al. (1984) found that frozen storage generated a relatively greater increase of fluorescence in the organic compared to the aqueous extract. Erickson (1994) noted, however, that compounds other than fluorescent pigments were present in fish extracts and limited the sensitivity of the fluorescent pigment measurement for monitoring oxidative changes. More recently, solid-sample fluorescence spectrophotometry has been applied to dried foods to enable measurement of protein-bound fluorescence (Fujimoto et al., 1990; Endo and Fujimoto, 1993). While fluorescence intensity varied with the food type, it was positively correlated with the carbonyl value of extracted lipids, demonstrating its value as a simple method for determination of oxidative deterioration. Its utility in measurement of oxidative deterioration of frozen foods, however, has yet to be demonstrated and reported on in the scientific literature.

As briefly discussed in the spectrophotometer section, enzyme activity assays have been used to distinguish fresh from frozen-thawed products. One of these enzyme assays, the measurement of β-N-acetylglucosaminidase, utilizes a fluorometric measurement of free methylumbellierone (excitation 360 nm, emission 448 nm), which is released upon incubation of an aqueous extract with the substrate 4-methylumbelliferyl glycoside (Barrett, 1972). Kitamikado et al. (1990) found this fluorometric assay to be a reliable and convenient method for testing the freezing history of edible fish. One drawback is that it is not applicable to mammalian meats such as beef, pork, and mutton because the blood of these animals contains no neutral β-N-acetylglucosaminidase.

Nuclear Magnetic Resonance Spectroscopy

Nuclear magnetic resonance (NMR) spectroscopy is based on the interaction of a nucleus with an applied external magnetic field and provides information on the macroscopic spatial distribution of nuclei within a sample by recording the influence of linear magnetic field gradients on the NMR signal (Morris, 1986). The noninvasive quality of this imaging technique allows the visualization of body features in many food items without compromising the structural integrity of the food system. For example, this form of spectroscopy has been used to visualize the freezing of water in a salmon steak (McCarthy and Kauten, 1990), the freezing interface of peaches (McCarthy et al., 1989), and the presence of frozen and unfrozen areas in fish (Howell et al., 1996). Nuclear magnetic resonance spectroscopy has also been advocated as a useful tool for determining quality deterioration during frozen storage of fish muscle (Lambelet et al., 1995). In frozen storage of cod, a supplemental component appeared in the water proton relaxation, which was presumably associated with juice exudation. The relaxation time associated with this additional component was dependent on the temperature and length of frozen storage.

Near-Infrared Spectroscopy

Near-infrared (NIR) spectroscopy reflectance instruments scan in the spectral region of 700–2,500 nm and are based on the principle that fundamental molecular vibrations excited in the midinfrared region give information on molecular structure. Hong et al. (1994) showed that NIR spectroscopy could be used to quantify oxidative deterioration of oils. The absorbance at 2,076 nm was attributed to the hydroxy group of hydroperoxides and the absorbance at 2,052 nm to the ester bond of triacylglycerols. Correlation between the NIR method and the traditional iodometric method for peroxide quantification was 0.98. NIR spectroscopy could therefore serve as a simple and rapid method for measurement of peroxides in lipids extracted from frozen foods.

On-line NIR instruments have been designed to scan a wide range of food types, from liquids and emulsions to powdered, crushed, and extruded solids (Giese, 1993). Directing a filtered beam of infrared light onto the surface of the food, the amount of energy absorbed at several frequencies is measured and transformed to concentration data using a calibration stored in the instrument's central processor. Applied to foods before freezing, NIR instruments may serve to monitor variations in chemical parameters, such as moisture, that ultimately affect the quality of the frozen product. Since the measurement is taken from the surface of the material, misleading data may be obtained if the surface is not representative of the entire product.

17.3.3 Electrophoresis

Electrophoresis is commonly used to separate proteins by exposing the molecules to an electric field. Several studies have employed electrophoresis to determine the proteins involved in denaturation (Owusu-Ansah and Hultin, 1986; Leblanc et al., 1987; Ragnarsson and Regenstein, 1989; Dias et al., 1994). Routine application of this procedure is not warranted but may provide valuable information when interpretation of textural deterioration is desired. Electrophoresis has proven more valuable than enzyme activity measurements in differentiating frozen from fresh poultry meat (Niemann et al., 1995). Using polyacrylamide gel isoelectric focusing, a band was found at pI 9.45 on the electropherogram of frozen/thawed products but not for fresh poultry breast and thigh muscle.

17.3.4 Chromatography

Thin-Layer Chromatography

In thin-layer chromatography (TLC), substances are separated by partitioning between a solid support (silica gel, paper, or alumina) and a moving liquid phase. Exchange of molecules between the solid and liquid phases occurs by adsorption-desorption as the solvent front moves over the surface. To identify thermal inactivation of lipid-acyl-hydrolase and adequacy of blanching in spinach destined to be frozen, decomposition of lecithin and mono- and digalactosyl diglyceride has been monitored by TLC (Park et al., 1996). In this study, lipid extracts were applied to activated Kieselgel-G plates and developed in a chromatography chamber with chloroform/methanol/water (75:25:4).

Gas-Liquid Chromatography

Gas-liquid chromatography (GLC) is a technique for separating volatile substances. In GLC the components to be separated are carried through the column by an inert gas. The sample mixture is partitioned between the carrier gas and a nonvolatile solvent supported on an inert size-graded solid. The solvent selectively retards the sample components, according to their distribution coefficient, until they form separate bands in the carrier gas. These component bands leave the column in the gas stream and are recorded as a function of time by a detector.

Deterioration in flavor and aroma that occurs during frozen storage of food products is commonly evaluated by GLC. Using this technique, changes in volatile constituents have been noted in frozen fruits (Etievant et al., 1986; Douillard and Guichard, 1990; Ueda and Iwata, 1992), frozen vegetables (Bengtsson et al., 1967; Murray et al., 1976), frozen pork (Brewer et al., 1992), and frozen fish (Erickson, 1993a,b; Freeman and Hearnsberger, 1993; Erickson, 1994; Brannan and Erickson, 1996). When only a few volatiles are of concern, packed columns are sufficient for separation and quantification. The advantages to separation on these columns are the reduced run times required ($<$15 min). Hence, more samples may be analyzed within one day. Using a packed column of Porapak P (80–100 mesh), generation of the oxidative volatiles propanal, pentanal, and hexanal were monitored in frozen bass by Erickson (1993a). To generate more-complex profiles of volatiles, capillary columns should be used. Using a cross-linked 5% phenylmethylsilicone capillary column, volatile profiles of frozen cooked chicken stored for 0, 2, and 6 mo at $-20°C$ were generated (fig. 17.1). The major differences between these profiles are in the larger size of many of the

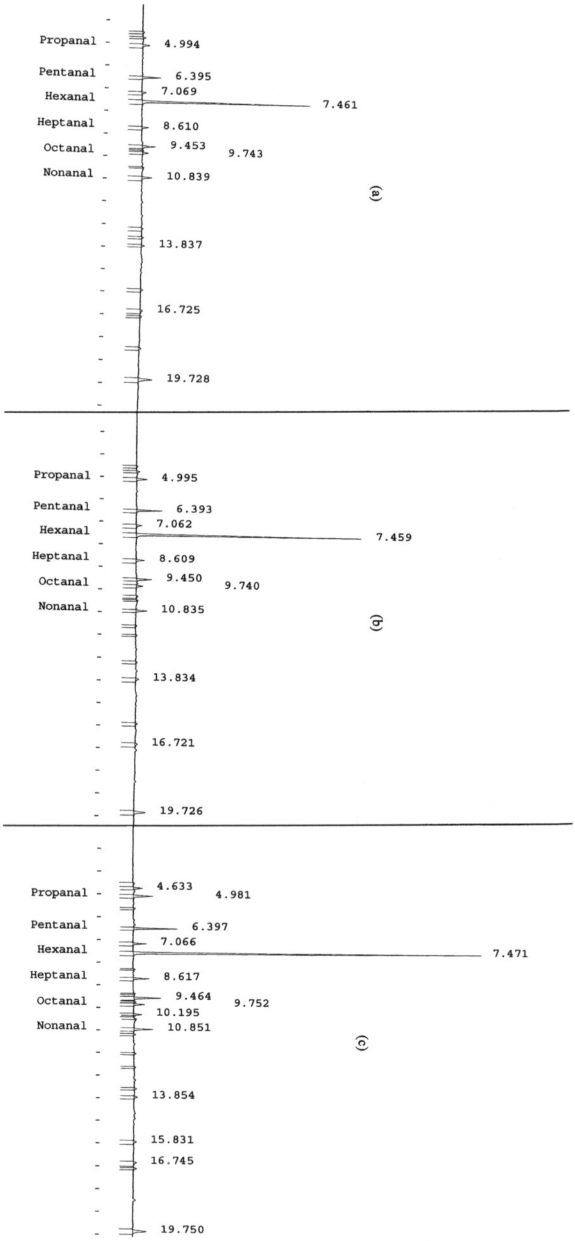

Figure 17.1. Utilization of solid-phase microextraction (SPME) and capillary gas chromatography for comparison of volatiles in cooked chicken stored at $-20°C$ for (a) 0 mo, (b) 2 mo, and (c) 6 mo. SPME fibers with 100-μM polydimethylsiloxane coating were exposed to headspace of cooked chicken (2 g) for 30 min at 70°C. Volatiles were desorbed (225°C) onto the column (40°C) and the oven heated at 20°C/min for 10.5 min, then held at 250°C for 15 min. Eluted volatiles were detected using a flame ionization detector.

peaks as storage progresses from 0 to 6 mo. Decreases in volatiles contributing to fresh flavors, however, have also been detected in frozen samples using capillary columns (Douillard and Guichard, 1990; Ueda and Iwata, 1992).

High-Performance Liquid Chromatography

High-performance liquid chromatography (HPLC) is an instrumental method that easily can separate and detect a multitude of chemical substances. It is used primarily for the separation of nonvolatile compounds, whereas GLC is used to separate volatile substances. An HPLC system consists of a stationary phase, a mobile phase (liquid), a pumping system, and a detector. The liquid mobile phase, carrying the sample, is pumped through the stationary phase contained in a steel column. Interaction in the sample molecules between the mobile and stationary phases affects sample preparation. Detectors used in an HPLC system may include ultraviolet spectrometers, fluorometers, light scattering, refractive index, and electrochemical detectors.

HPLC has served as a useful tool in the detection of degradation of the antioxidant vitamins, tocopherol, and ascorbic acid in frozen fish (Erickson, 1993a,b, 1994; Brannan and Erickson, 1996). Separation and quantification of tocopherols were accomplished using a Zorbax ODS column (4.6 mm $\times$ 2.5 cm, 5-μm particle size) and a mobile phase of 98:2 (methanol/water, v/v). Separation and quantification of ascorbic acid were achieved using a Brownlee Spheri-5-ODS analytical column (4.6 $\times$ 220 mm, 5-μm packing) and a mobile phase of 40-mM sodium acetate and 1-mM decylamine in methanol/water (15:85, v/v) at pH 4.0. In both the ascorbic acid and tocopherol analyses, the compounds were detected using an electrochemical detector.

17.3.5 Sensors

A sensor is a device that converts, through a transducer, a chemical parameter to a detectable signal (Nylander, 1985). A recent addition to this technological field is electronic nose systems that are capable of predicting sensory-like properties of complex flavors from their chemical constituents. The principal component of these systems is an array of sensors (metal oxide, conducting polymer, surface acoustic wave, or hybrid systems). The sensors respond to a wide variety of chemical classes but vary in magnitude of response to any single component, giving them the broad resolution and discriminatory ability necessary to perform aroma analysis on a wide variety of food products. In analysis, the sample vapor is exposed to the sensors. The output signal is produced from the change in resistance as the organic vapor reacts on the surface of the sensor. Adding another dimension of information is the rate of response and return to initial conditions of the sensor, which will vary with sample type, volatility, and complexity. The signal produced by the sensor is then analyzed by computer using sophisticated chemometric and artificial neural network software. Using such a system, qualitative and quantitative attributes may be assigned to those aroma patterns so that the system learns to recognize the desired characteristics of the sample. Once trained, the system can identify products that have changed significantly in aroma and flavor. Because this technology is applicable to a wide range of food products, it will no doubt be a major tool in the future for evaluation of quality of frozen foods.

17.4 Conclusion

Chemical measurements are an important tool for evaluation of quality in frozen foods. Spectrophotometric, titrimetric, fluorometric, chromatographic, and electrophoretic analyses have all been used to assess degradative reactions that occur during freezing and frozen storage. While varying in sophistication and apparatus required, the usefulness of an analysis is dependent on the specific component that needs to be measured and in the proper preparation of a sample for analysis. Often, more than one type of chemical measurement is required to adequately assess the magnitude of change that has occurred during the freezing and storage processes. Complementary sensory and physical measurements will also aid in the evaluation. The recent development of sensor systems to measure aromatic and flavor profiles of food products will help to link the chemical and sensory assessment of frozen-food quality.

References

Barbagli, C. and Crescenzi, G.S. 1981. Influence of freezing and thawing on the release of cytochrome oxidase from chicken's liver and from beef and trout muscle. *J. Food Sci.* **46**:491–493.

Barrett, A.J. 1972. Lysosomal enzymes. In *Lysosomes: A Laboratory Handbook.* (Dingle, J.T., ed.) pp. 46–135, American Elsevier Publishing, New York.

Bengtsson, B.L., Bosund, I., and Rasmussen, I. 1967. Hexanal and ethanol formation in peas in relation to off-flavor development. *Food Technol.* **21**:478.

Brannan, R.G. and Erickson, M.C. 1996. Sensory assessment of frozen stored channel catfish in relation to lipid oxidation. *J. Aquat. Food Prod. Technol.* **5**:67–80.

Brewer, M.S., Ikins, W.G., and Harbers, C.A.Z. 1992. TBA values, sensory characteristics, and volatiles in ground pork during long-term frozen storage: effects of packaging. *J. Food Sci.* **57**:558–563, 580.

Castell, C.H., Smith, B., and Dyer, W.J. 1974. Simultaneous measurements of trimethylamine and dimethylamine in fish, and their use for estimating quality of frozen stored gadoid fillets. *J. Fish. Res. Board Can.* **31**:383–389.

Ciarlo, A.S., Boeri, R.L., and Giannini, D.H. 1985. Storage life of frozen blocks of Patagonian hake (*Merluccius hubbsi*) filleted and minced. *J. Food Sci.* **50**:723–726, 738.

Cochran, W.G. 1977. *Sampling Techniques,* 3d ed. John Wiley & Sons, New York.

Corongiu, F.P., Banni, S., and Dessi, M.A. 1989. Conjugated dienes detected in tissue lipid extracts by second derivative spectrophotometry. *Free Radical Biol. Med.* **7**:183–186.

Corongiu, F.P. and Milia, A. 1983. An improved and simple method for determining diene conjugation in polyunsaturated fatty acids. *Chem.-Biol. Interact.* **44**:289–297.

Dias, J., Nunes, M.L., and Mendes, R. 1994. Effect of frozen storage on the chemical and physical properties of black and silver scabbardfish. *J. Sci. Food Agric.* **66**:327–335.

Douillard, C. and Guichard, E. 1990. The aroma of strawberry (*Fragaria ananassa*): characterization of some cultivars and influence of freezing. *J. Sci. Food Agric.* **50**:517–531.

Endo, Y. and Fujimoto, K. 1993. Evaluation of oxidative deterioration of lipid in dried foods using solid sample fluorescence spectrophotometry. *Nippon Shokuhin Kogyo Gakkaishi* **40**:176–180.

Erickson, M.C. 1993a. Ability of chemical measurements to differentiate oxidative stabilities of frozen minced muscle tissue from farm-raised striped bass and hybrid striped bass. *Food Chem.* **48**:381–385.

Erickson, M.C. 1993b. Compositional parameters and their relationship to oxidative stability of channel catfish. *J. Agric. Food Chem.* **41**:1213–1218.

Erickson, M.C. 1994. Comparison of chemical measurements to differentiate oxidative stability of frozen minced tilapia fish muscle. *Int. J. Food Sci. Technol.* **29**:585–591.

Etievant, P.X., Guichard, E.A., and Issanchou, S.N. 1986. The flavour components of Mirabelle plums. Examination of the aroma constituents of fresh fruits: variation of head-space composition induced by deep-freezing and thawing. *Sci. Aliment.* **6**:417–432.

Freeman, D.W. and Hearnsberger, J.O. 1993. An instrumental method for determining rancidity in frozen catfish fillets. *J. Aquat. Food Prod. Technol.* **2**:35–50.

Fujimoto, K., Mohri, S., Hasegawa, K., and Endo, Y. 1990. Oxidative deterioration of fish meat. *Food Rev. Int.* **6**:603–616.

Gardner, H.W. 1979. Lipid hydroperoxide reactivity with proteins and amino acids: a review. *J. Agric. Food Chem.* **27**:220–229.

Giese, J. 1993. On-line sensors for food processing. *Food Technol.* **47**(5):88–95.

Gutteridge, J.M.C., Kerry, P.J., and Armstrong, D. 1982. Autoxidized and lipoxidase-treated polyunsaturated fatty acids autofluorescence associated with the decomposition of lipid peroxides. *Biochim. Biophys. Acta* **711**:460–465.

Hong, J.-H., Yamaoka-Koseki, S., and Yasumoto, K. 1994. Analysis of peroxide values in edible oils by near-infrared spectroscopy. *Nippon Shokuhin Kogyo Gakkaishi* **41**:277–280.

Howell, N., Shavila, Y., Grootveld, M., and Williams, S. 1996. High-resolution NMR and magnetic resonance imaging (MRI) studies on fresh and frozen cod (*Gadus morhua*) and haddock (*Melanogrammus aeglefinus*). *J. Sci. Food Agric.* **72**:49–56.

Iio, T. and Yoden, K. 1988. Formation of fluorescent substances from degradation products of methyl linoleate hydroperoxides with amino compounds. *Lipids* **23**:1069–1072.

Jennings, W.G. and Filsoof, M. 1977. Comparison of sample preparation techniques for gas chromatographic analysis. *J. Agric. Food Chem.* **25**:440–445.

Jonas, R.E.E. 1969. Effect of magnesium sulphate and cryoprotective agents on free fatty acid formation in cold-stored fish muscle. *J. Fish. Res. Board Can.* **26**:2237–2240.

Ke, P.J., Linke, B.A., and Ackman, R.G. 1978. Acceleration of lipid oxidation in frozen mackerel fillet by pretreatment with microwave heating. *J. Food Sci.* **43**:38–40.

Ke, P.J. and Woyewoda, A.D. 1978. A titrimetric method for determination of free fatty acids in tissues and lipids with ternary solvents and m-cresol purple indicator. *Anal. Chim. Acta* **99**:347.

Kikugawa, K. and Beppu, M. 1987. Involvement of lipid oxidation products in the formation of fluorescent and cross-linked proteins. *Chem. Phys. Lipids* **44**:277–296.

Kitamikado, M., Yuan, C.-S., and Ueno, R. 1990. An enzymatic method designed to differentiate between fresh and frozen-thawed fish. *J. Food Sci.* **55**:74–76.

de Koning, A.J., Milkovitch, S., and Mol, T.H. 1987. The origin of free fatty acids formed in frozen cape hake mince (*Merluccius capensis*, Castelnau) during cold storage at −18°C. *J. Sci. Food Agric.* **39**:79–84.

Lakritz, L. and Thayer, D.W. 1992. Effect of ionizing radiation on unesterified tocopherols in fresh chicken breast muscle. *Meat Sci.* **32**:257–265.

Lambelet, P., Renevey, F., Kaabi, C., and Raemy, A. 1995. Low-field nuclear magnetic resonance relaxation study of stored or processed cod. *J. Agric. Food Chem.* **43**:1462–1466.

Landolt, L.A. and Hultin, H.O. 1982. Inhibition of dimethylamine formation in frozen red hake muscle after removal of trimethylamine oxide and soluble proteins. *J. Food Biochem.* **6**:111–125.

Leake, L. and Karel, M. 1985. Nature of fluorescent compounds generated by exposure of protein to oxidizing lipids. *J. Food Biochem.* **9**:117–136.

Leblanc, E.L, Leblanc, R.J., and Gill, T.A. 1987. Effects of pressure processing on frozen stored muscle proteins of Atlantic cod (*Gadus morhua*) fillets. *J. Food Proc. Preserv.* **11**:209–235.

Lichon, M.J. and James, K.W. 1990. Homogenization methods for analysis of foodstuffs. *J. Assoc. Off. Anal. Chem.* **73**:820–825.

Love, R.M. 1956. Influence of freezing-rate on denaturation of cold-stored fish. *Nature (London)* **178**:988–989.

Love, R.M. 1958. Studies on protein denaturation in frozen fish. III. The mechanism and site of denaturation at low temperatures. *J. Sci. Food Agric.* **9**:609–617.

Mazeaud, F. and Bilinski, E. 1976. Free fatty acids and the onset of rancidity in rainbow trout (*Salmo gairdneri*) flesh: effect of phospholipase A. *J. Fish. Res. Board Can.* **33**:1297–1302.

McCarthy, M.J., Charoenrein, S., Heil, J.R., and Reid, D.S. 1989. MRI measurements in frozen systems. In *Proceedings of the International Conference on Technical Innovations in Freezing and Refrigeration of Fruits and Vegetables.* (Reid, D.S., ed.) pp. 136–140, University of California–Davis.

McCarthy, M.J. and Kauten, R.J. 1990. Magnetic resonance imaging applications in food research. *Trends Food Sci. Technol.* **12**:134–139.

Morris, P.G. 1986. *Nuclear Magnetic Resonance Imaging in Medicine and Biology*. Claredon Press, Oxford.

Murray, K.E., Shipton, J., Whitfield, F.B., and Last, J.H. 1976. The volatiles of off-flavored unblanched green peas. *J. Sci. Food Agric.* **27**:1093–1107.

Nambudiri, D.D. and Gopakumar, K. 1992. ATPase and lactate dehydrogenase activities in frozen stored fish muscle as indices of cold storage deterioration. *J. Food Sci.* **57**:72–76.

Nash, T. 1953. The colorimetric estimation of formaldehyde by means of the antzsch reaction. *Biochem. J.* **55**:416.

Niemann, C., Beneke, B., and Bentler, W. 1995. Poultry meat: determining freezing methods. *Fleischwirtschaft* **75**:323–326.

Nilsson, K. and Ekstrand, B. 1993. The effect of storage on ice and various freezing treatments on enzyme leakage in muscle tissue of rainbow trout (*Oncorhynchus mykiss*). *Z. Lebensm. Unters. Forsch.* **197**:3–7.

Nylander, C. 1985. Chemical and biological sensors. *J. Phys. E: Sci. Instrum.* **18**:736–749.

Owusu-Ansah, Y. and Hultin, H.O. 1986. Chemical and physical changes in red hake fillets during frozen storage. *J. Food Sci.* **51**:1402–1406.

Park, K.H., Kim, J.W., and Kim, M.J. 1996. Application of enzymatic reactions for evaluation of quality changes in frozen vegetables. In *Chemical Markers for Processed and Stored Foods*. (Lee, T.-C. and Ho, H.-J., eds.) pp. 159–178, American Chemical Society, Washington, D.C.

Pettinati, J.D., Ackerman, S.A., Jenkins, R.K., Happich, M.L., and Phillips, J.G. 1983. Comparative analysis of meat samples prepared with food chopper and bowl cutter. *J. Assoc. Off. Anal. Chem.* **66**:759–765.

Pikul, J., Leszczynski, D.E., Niewiarowicz, A., and Kummerow, F.A. 1984. Lipid oxidation in chicken breast and leg meat after sequential treatments of frozen storage, cooking, refrigerated storage, and reheating. *J. Food Technol.* **19**:575–584.

Ragnarsson, K. and Regenstein, J.M. 1989. Changes in electrophoretic patterns of gadoid and non-gadoid fish muscle during frozen storage. *J. Food Sci.* **54**:819–823.

Salih, A.M., Smith, D.M., Price, J.F., and Dawson, L.E. 1987. Modified extraction 2-thiobarbituric acid method for measuring lipid oxidation in poultry. *Poultr. Sci.* **66**:1483–1488.

Samson, A., Regenstein, J.M., and Laird, W.M. 1985. Measuring textural changes in frozen minced cod flesh. *J. Food Biochem.* **9**:147–159.

Shimomura, H., Takahashi, T., Morishita, T., and Ueno, R. 1987. Investigation of the differentiation of frozen-thawed fish from unfrozen fish by comparison of lysosomal enzyme activity. *Nippon Suisan Gakkaishi* **53**:1841–1845.

Shirey, R.E., Woolley, C.L., and Mindrup, R.F. 1995. Solventless extraction of flavors and other food components. *Food Test. Anal.* **1**(1):39–42.

Spanier, A.M. and Traylor, R.D. 1991. A rapid, direct chemical assay for the quantitative determination of thiobarbituric acid reactive substances in raw, cooked, and cooked/stored muscle foods. *J. Muscle Foods* **2**:165–176.

Ueda, Y. and Iwata, T. 1992. Undesirable odour of frozen strawberries. *J. Jpn. Soc. Hort. Sci.* **51**:219.

Vossen, R.C.R.M., van Dam-Mieras, M.C.E., Hornstra, G., and Zwaal, R.F.A. 1993. Continuous monitoring of lipid peroxidation by measuring conjugated diene formation in an aqueous liposome suspension. *Lipids* **28**:857–861.

Whitaker, J.R. 1994. *Principles of Enzymology for the Food Sciences*, 2d ed. Marcel Dekker, New York.

Zhang, Z. and Pawliszyn, J. 1993. Headspace solid-phase microextraction. *Anal. Chem.* **65**:1843–1852.

Zhou, S. and Ackman, R.G. 1996. Interference of polar lipids with the alkalimetric determination of free fatty acid in fish lipids. *J. Am. Oil Chem. Soc.* **73**:1019–1023.

18

Sensory Evaluation Methods to Measure Quality of Frozen Food

A. V. A. Resurreccion

18.1 Introduction

Before the quality of frozen food can be evaluated, it must be defined. The definition of the sensory quality of a food or food product is "the acceptance of the sensory characteristics of a product by consumers who are regular users of the product category, or who comprise the target market for the product" (Galvez and Resurreccion, 1992). The measurement of sensory quality of frozen food involves consumer affective tests wherein only regular consumers of frozen foods participate as panelists to determine acceptance or preference for the frozen food. By definition, sensory quality cannot be measured by any other means except through consumer affective tests. However, although consumers are able to rate their acceptance of the attributes of a product, they are unable to reliably quantify intensities of specific attributes of the food. Sensory descriptive analysis tests, using highly trained panelists to rate attributes of food much like an instrument would, and instrumental measurements serve to quantify intensities of those attributes. This chapter will discuss consumer affective tests and also mention the sensory descriptive analysis tests.

Consumer tests are necessary in several stages throughout the frozen-food product cycle (ASTM, 1979). Sensory evaluation by consumers is important in the development of the product itself for new-product development guidance, product improvement and optimization, and product maintenance. The testing methods employ untrained individuals and larger numbers of panelists than required in sensory analytical test methods. Consumer affective tests appear to be relatively simple to plan and conduct. Unfortunately, the seemingly easy task is often invalidated by improper data collection methodology, which invariably leads to faulty interpretation of results. Validity and reliability of the consumer-testing methodologies are extremely important.

Affective testing is one of the most important activities in a sensory evaluation program. The primary purpose of affective tests is to assess personal responses by current and regular users of a product or potential customers of a product prototype to the product or specific product characteristics. Consumer evaluation concerns itself with testing certain products using untrained people who are or will become the ultimate users of the product (ASTM, 1979). These products are evaluated on the basis of visual differences, taste, smell, touch, and hearing.

18.2 Strategic Approach to Product Development

Most products created today are designed to satisfy the needs of consumers (ATSM, 1979). A strategic approach to the development of a food product is shown in figure 18.1. Product concepts may arise from a variety of sources including the marketing and research departments, management input, consumer research, and idea generation. Some traditional approaches have their origin in quality grading systems such as those developed for dairy products, meat, fruit, and so forth. Unfortunately, quality grades do not always agree with consumer-acceptance data. Most product research relies on the technical knowledge of the researcher or, in industrial situations, of the product specialist. Often, if consumer data are available these are not used systematically.

The strategic approach to frozen-food development and the development of food products in general involves as the initial step the identification of the product and then the identification of critical attributes of the product. One of the recommended strategies to assess a concept early in the development stage is by using a focus group.

In a focus group, 8 to 12 (usually 10) consumers representing the target market participate in a discussion led by a moderator. The consumers' responses are used to qualitatively assess the concept or to identify critical attributes of a product or concept. Beyond concept development, initial prototype products are developed that possess the critical attributes valued by the consumers and that are devoid of the attributes consumers consider unacceptable. At this stage, consumer sensory tests are a useful tool for product guidance. The objective of consumer sensory tests at this stage is to assess performance potential and to provide product guidance for further development with the goal of maximizing acceptance (ASTM, 1979).

Quantification occurs after the critical quality attributes have been identified. Consumer affective tests are conducted to quantify overall acceptance or consumer responses to critical attributes defined as factors that determine a product's preference or acceptance. Test participants may be composed of 100 consumers for product development guidance or 100 to 500 consumers in three or four cities throughout the country to determine acceptance through central-location or home-use tests. Quantification of acceptability should not involve trained or "experienced" panelists, as were used in studies on frozen pork chops (Miller et al., 1985) or consumer panels with fewer than 24 participants as used in studies on frozen strawberries (Gallander and Hill, 1982).

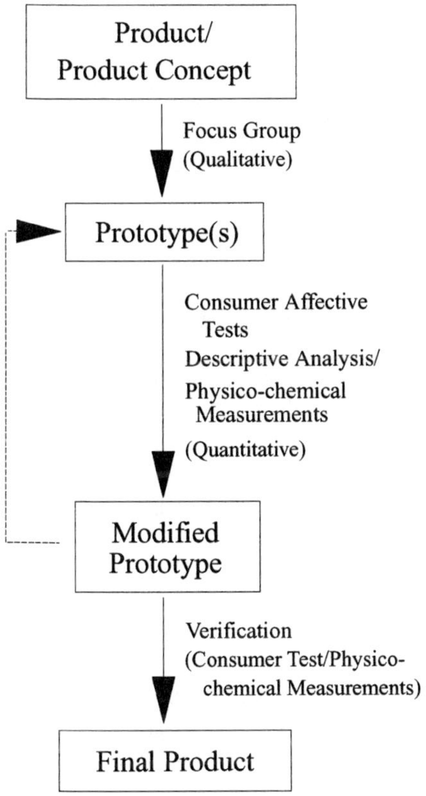

Figure 18.1. A strategic approach to the development of a food product

A successful acceptance test will not guarantee success of a product in the marketplace and is not a substitute for large-scale market tests. If prior testing has shown adequate acceptance, the decision may be made at this point on whether the product will proceed to the test market stage. At this point, execution of test market research is the function of the market research department. Consumer tests are necessary because product developers will need to increase production capability using large-scale manufacturing equipment. Formulation or process changes may require sensory tests to ensure that these changes do not compromise consumer acceptability.

The quantification of quality for assessing or monitoring quality of frozen foods will often involve sensory descriptive analysis and/or physicochemical measurements. Descriptive analysis and physicochemical measurements are used to characterize and establish quantitative limits for the critical attributes that determine consumer acceptance.

18.3 Applications of Affective Tests in the Product Cycle

The implicit goal behind any and all sensory evaluation efforts in the food industry is to improve quality in order to influence food choices (purchases) at the point of sale. It is therefore important to measure appearance, flavor, and texture as perceived by consumers.

18.3.1 Product Maintenance

Consumer sensory evaluation is essential in product maintenance for monitoring product acceptance to establish and maintain position within a market. Furthermore, consumer tests may be important in establishing the length of time that a product will remain stable and to establish open-dating (sell-by or use-by dates) parameters. Home-use tests are often used to monitor relative acceptance or preference. Consumer tests conducted in the sensory laboratory or central location may be used to explain changes in market share caused by competition (ASTM, 1979). If market share is eroding, consumer sensory tests may provide direction for changes that need to be implemented to increase consumer acceptance.

18.3.2 Product Improvement

Consumer affective tests are used in product improvement. After a period of rapidly increasing sales and profits, it is common to observe a leveling of sales increase (ASTM, 1979). Processors may improve the product by adding new flavors or introduce line extensions of the basic product. These should be tested as one would any new product. Laboratory tests, central-location tests, and home-use tests are used to assess acceptability of the new products against the original and competing products.

18.3.3 Product Optimization

Because of profit decline that accompanies the maturity of a product, cost reductions that do not compromise the acceptance of the product are vital. These changes may be arrived at after product optimization studies. Changes in the formulation and processing of raw materials should not decrease acceptance of the product.

18.4 Qualitative and Quantitative Tests for Determining Quality

Consumer sensory research can be classified into two major categories, qualitative and quantitative. Qualitative research is descriptive and does not involve measurements, whereas quantitative research involves measurements.

18.4.1 Qualitative Measurements

Several qualitative methods exist, including one-on-one in-depth interviews, group interviews, and focus groups. The most commonly used qualitative research method is the focus group. The focus group is a method in which small groups of consumers are convened for 90 min to discuss their reactions to products or product concepts. They are useful in gaining insight into consumers' preferences and defining critical attributes of a product (Galvez and Resurreccion, 1992). This method is used to determine product attributes that consumers think are important and should be maximized in the product and characteristics that consumers do not like and should be minimized or eliminated from the product. If this knowledge does not exist, a great deal of time and money can be wasted on research that yields no actionable data. The distinguishing feature of this method is the unstructured approach.

Although quantitative research may provide answers that are projectable to a larger population, it may not be able to provide the answers that focus groups can. Qualitative research usually needs to precede quantitative testing to help establish criteria for data collection and follow quantitative testing to aid the explanation of quantitative data (Chambers and Smith, 1991). Focus groups may also be useful in gathering extensive information about the product category. For example, a producer of frozen baked goods may be interested in extending its product line to include a frozen snack food for children's school lunches but has only sketchy ideas on how this product category is perceived; consequently, it conducts focus groups to help generate information on the language that consumers use to talk about the product category. A focus group may be used to interpret new questions raised by quantitative research. Finally, focus groups can be useful in designing a better questionnaire to help determine the most important questions and appropriate wording. While focus groups may offer, in some instances, faster turnaround time and lower costs, one must remember that although qualitative research results can prove to be useful, they should not be used as a substitute for quantitative research.

Advantages

There are several advantages for using focus groups to obtain consumer information not easily obtained with quantitative research. Focus groups offer flexibility and provide the ability for marketing, research and development, and management personnel to observe real consumers in an interactive setting. In a focus group, observers can easily view interactions between the consumer and the product. Furthermore, observers can see how participants interact with other participants as they build upon each other's observations and input. Focus groups involve fewer participants compared to quantitative methods and can be arranged on short notice and at a lower cost relative to a major field study. Because the method is qualitative, statistical analysis is inappropriate and unnecessary. Because of their flexibility, focus groups are able to meet many diverse marketing and research and development objectives including testing new products and new product concepts and questionnaire development for quantitative research. Furthermore, the flexibility of the method allows for new objectives and redirection of the topics of discussion as it progresses.

Disadvantages

Focus group results are not quantitative. No measurements are made and statistical treatment of qualitative data is inappropriate. The sample size is small and results are not projectable to a population. The participants do not represent the target market. Although a serious attempt should be made to screen for participants who represent the target market, the small sample size does not allow the target market to be fully represented. Results cannot be used independently of quantitative research for strategic decision support. The topics and direction of the discussion are moderator dependent. Careful interpretation of the data is required.

18.4.2 Quantitative Measurements

Acceptance and Preference Tests

Acceptance may be defined as (1) an experience, or feature of experience, characterized by a positive attitude toward the food and/or (2) actual utilization (such as purchase or eating) of the food. Acceptance may be measured by preference or liking of a specific food item (Amerine et al., 1965). In its simplest form, acceptability is inferred from scale ratings. These tests measure acceptability or liking for a food. The question asked is: "How much do you like the product?" (Stone and Sidel, 1993) or "How acceptable is the product?" (Stone and Sidel, 1993; Meilgaard et al., 1991).

The acceptance test provides an estimate of product acceptance based on its sensory properties. This measure of acceptance does not usually consider the influence of packaging, price, or advertising and as such cannot guarantee success in the marketplace.

Preference tests refer to all affective tests based on a measurement of preference or a measurement from which relative preference may be determined (IFT/SED, 1981). Preference may be defined as the measurement of the appeal of one food or food product over another (Stone and Sidel, 1993). The question asked is: "Which sample do you prefer?" or "Which sample do you like better?"

Preference may be defined as (1) an expression of higher degree of liking; (2) choice of one object over others; or (3) psychological continuum of affectivity (pleasantness/unpleasantness) upon which such choices are based (Amerine et al., 1965). It may include the choice of one sample over another, a ranked order of liking, or an expression of opinion on a hedonic (like/dislike) scale. Scaling methods allow us to directly measure degree of liking and to compute preferences from these data.

The measurement of preference from paired comparison or ranking tests is direct, whereas preference from hedonic ratings is implied. Preference is determined by knowing which product is scored significantly higher (more liked) than another product in the sample set or which product is scored higher than another by significantly more people. Stone and Sidel (1993) state that "there is an obvious and direct relationship between measuring product liking/acceptance and preference."

Preference methods can be used to determine differences in preference but not product differences determined using discriminative tests. In some instances, difference tests have been used separately before preference tests to determine whether perceptible difference exists between two samples. If a perceptible difference between two samples does not exist, subsequent preference testing is considered unnecessary (IFT/SED, 1981).

The methods most frequently used to determine preference and quantify acceptance are the paired preference tests and tests employing the nine-point hedonic scale, respectively. Ranking and ratio scaling have likewise been used.

Paired Preference Test. In this test, two samples are usually presented simultaneously, or sequentially if samples have a characteristic that would prevent their being presented at the same time. The panelists are asked to indicate which of the two products is preferred. The instructions may or may not force the panelist to make a decision. If a forced choice is imposed, the panelist may not give a no preference response (ASTM, 1996) and must indicate a preference for one sample over another. The method may also be used for multiple paired preferences within a series of paired sample sets, such as a standard product versus each of several experimental products. One or more sample pairs of products may be tested per sensory evaluation session; the maximum number of samples that can be evaluated is determined by physiological and psychological constraints.

The paired-preference test is simple and relatively easy to organize and conduct. The two coded products are served together to a naive panelist who is asked to indicate which one is preferred after sampling both products (fig. 18.2). A frequently used option is "no preference," and a fourth option may be "dislike both equally."

Advantages of the paired-preference test are its ease in organization and implementation. Only two orders of presentation, A-B and B-A, are possible. However, the order of presentation must be balanced in preference tests, or position bias can easily occur. The panelist usually evaluates only one pair of products in a test without a replication. The no-preference category may be used when 100 or more respondents participate in the test. If the dislike-both-equally category is used, the analysis should be based only on the preference responses. If one obtains >5% in

PAIRED PREFERENCE TEST

Code __________ Date ____________

In front of you are two samples. Beginning with the sample on the left, evaluate each and **circle the sample that you prefer.** You must choose one. You may retaste as often as you need to.

Thank you.

362 **179**

Figure 18.2. An example of a ballot to be used for the paired-preference test

this category the appropriateness of products for the tests or the subjects used for the product should be questioned (Stone and Sidel, 1993).

Disadvantages of the paired-preference test are that it is less informative because there is no direct measure of the magnitude of preference. If the fourth category (no preference) is not used, there will be no way to determine whether both products were disliked. Because preference testing asks the respondent to evaluate the product on a total basis, the testing may be especially susceptible to unintentional bias that can arise from slight differences in placement on the serving tray, serving temperatures, sample volume, and variability of the sample from the manufacturing process, among other factors (ASTM, 1996). The test is less efficient because there is only one response per product pair as opposed to one response per product.

In a multiproduct test, there will be considerable interaction because of flavor carryover from one product to the next. In addition, memory can be a confounding variable if a sequential presentation is selected for the paired test.

Rating Tests. Scale ratings reflect respondents' perceived intensity of a specified attribute under a given set of conditions. Various rating scales, the hedonic scale, and the food action scale have been developed and used in hedonic tests and food action rating.

Hedonic test. This method relies on naive or untrained panelists to indicate their feeling toward a sample using a hedonic scale (fig. 18.3). Samples are presented monadically, sequentially, or in groups, and panelists are asked how much they like or dislike each sample. Penfield et al. (1989) used a seven-point hedonic scale to rate the consumer acceptance of vacuum-packaged frozen restructured

The Nine-point Hedonic Scale

Like extremely	________
Like Very Much	________
Like Moderately	________
Like Slightly	________
Neither Like Nor Dislike	________
Dislike Slightly	________
Dislike Moderately	________
Dislike Very Much	________
Dislike Extremely	________

Figure 18.3. The hedonic scale

beef steaks. Hedonic ratings of the 36-member consumer panel indicated no differences (3.4 and 3.6 or neither like nor dislike) between steaks cooked in a conventional or microwave oven. The same panel rated frozen restructured steaks containing 15, 20, and 25% fat equally. When identical samples were evaluated by a panel that was given orientation sessions until they became familiar with attribute terms such as appearance, softness, chewiness, moistness, greasiness, and off-flavor, they found that steaks cooked in the microwave were more well-done, harder to tooth pressure, required more effort in preparation for swallowing, and were less juicy than those cooked in a conventional oven. These results indicate that although differences in intensity of the different attributes were found by the trained panel, these were not large enough to affect the sensory quality of steaks as determined in the consumer-acceptance tests.

The nine-point hedonic scale has been used for many years and has gained special consideration because of its suitability in the measurement of product acceptance and preference. It was developed by Jones et al. (1955) and Peryam and Pilgrim (1957). Further research resulted in a nine-point bipolar scale with a neutral point in the middle and nine statements that describe each of the points or categories. The scale is easily understood by panelists and simple to use. Although many different forms of hedonic scales such as five- and seven-point scales have been used, the reliability and validity of the nine-point hedonic scale in the assessment of several hundred food items has been confirmed (Stone and Sidel, 1985).

Modifications to the nine-point scale have been suggested, such as eliminating the neutral (neither like nor dislike), like-moderately, and dislike-moderately points, but they were either unsuccessful or had no practical value. Another concern was the bipolar scale and the analysis of the results as unidirectional, but there is no evidence that consumers experience difficulty with the scale and that the statistical analysis presents a problem (Stone and Sidel, 1993).

Stone and Sidel (1993) state that the stability of responses and the extent to which the data can be used as a sensory benchmark for any particular product category are of particular value. Acceptance of a product can be quantified by mean and standard deviation. When tested against a wide array of competitive products, an ordering of means of the different products can be made. It likewise allows the comparison of a product with the highest possible scores that the product category may achieve. Furthermore, parametric statistical analysis, such as analysis of variance (ANOVA) of acceptance data from tests using the nine-point hedonic scale, can provide useful information about product differences (Stone and Sidel, 1993).

Consumer responses from the use of a hedonic scale can likewise be converted to ranks or paired-preference data. To convert to paired-preference data, it is necessary to count the number of subjects who scored one product higher and analyze the result using $p = 1/2$, or binomial distribution. The nine-point hedonic scale has yielded reliable and valid results. Efforts to improve the scale have been unsuccessful, and it should continue to be used with confidence (Stone and Sidel, 1993).

Food action (FACT) rating scale. The FACT scale by Schutz (1965) is used to measure acceptance of a product by a population and the general attitude toward a product. The eating scale includes action and affective statements. Nine categories are represented, and one or more samples may be tested. Samples are presented sequentially in a balanced order, and the panelist is asked to decide which of the statements on the scale best represents his or her attitude. The ratings are converted to numerical scores to facilitate statistical analysis of data. Penfield et al. (1989) and Collins et al. (1995) used a six-point FACT scale for the frozen restructured beef steaks discussed earlier and for frozen, baked sweet potatoes. The six-point FACT scale they used had a rating of 1 corresponding to "I would hardly ever buy this," and a rating of 6 for "I would buy this very often." Results for microwave- and conventional-oven cooked steaks were neutral and did not vary with cooking method.

Ranking. Ranking is an extension of the paired-preference test. One or more coded samples are presented simultaneously and in sufficient amounts to allow the panelist to retaste the sample. The number of samples that can be included in the test largely depends on the panelists' attention spans and memories, as well as physiological considerations. With untrained or naive panelists, no more than four to six samples may be included in a test (ASTM, 1996). The panelist is asked to assign an order to the samples according to his or her preference. Ranking tests evaluate samples only in relation to one another. The amount of liking (or disliking) of individual samples cannot be determined adequately by this method (IFT/SED, 1981).

Ratio Scaling or Magnitude Estimation. In ratio scaling or magnitude estimation (Moskowitz, 1974) panelists employ their own scales based on ratios. A ratio scale implies a constant ratio between fixed points or a zero point. Magnitude estimation is the most common ratio scale used in consumer testing. For example, in estimating acceptance of a sample, if panelists like the second sample twice as much as they did the previous sample, they should give it a value of two times the value given to the first sample.

Magnitude estimation has been applied to a wide range of products. ASTM (1996) indicates its usefulness in evaluation of moderate to large suprathreshold differences, stating that "the measurement of very small differences among similar products or sensations is more efficiently accomplished using other sensory techniques."

Consumer Test Methods by Location

Consumer responses needed for the quantification of acceptance or preference can be conducted in a sensory laboratory setting, in central-location tests (CLT), or in home-use tests (HUT), also known as home-placement tests. A specialized form of CLT that has been used is one conducted in a mobile laboratory.

Sensory Laboratory Tests. Sensory laboratory tests, as the name implies, are conducted in the sensory laboratory of the food company, consulting firm, or research organization. This type of test is the most frequently used method for obtaining consumer responses and was used by Calhoun et al. (1996) in studying the effects of connective tissue/acidic phosphate preblends on reduced-fat frankfurters.

The panel consists of consumers who are recruited from a consumer database consisting of prerecruited consumers or company employees and screened for eligibility to participate in the tests. Usually 25–50 responses are obtained, but this number is small and represents only a rough screening (ASTM, 1996). Using as few as 24 panelists may make it difficult to establish a statistically significant difference in a test. In general, 100 responses are considered adequate (ASTM, 1996) and guidelines (IFT/SED, 1981) indicate 50–100 desirable. The recommended number of products per sitting is two to five.

Advantages. The major advantage of using a sensory laboratory location to conduct a consumer-affective test is the convenience of location. The sensory laboratory is usually located within the building of the food company, consulting firm, or research organization. The sensory laboratory provides a convenient location for the sensory research team but may be less convenient for consumer panelists. Although use of employee panels is generally not recommended in affective tests, in instances when the panelists used are employees of the company, the sensory laboratory would be a convenient location. If the employees have participated previously in consumer tests, they would be familiar with the testing procedures and thus save the researcher time. In the sensory laboratory, the researchers are able to control the conditions of the testing environment such as lighting, noise, and other distractions including odors and to conduct the test in individually partitioned booths where panelists are prevented from distracting each other. Furthermore, because the sensory laboratory is usually adjacent to a fully equipped kitchen, sample preparation steps including cooking, reheating, slicing, and serving can be conducted under highly controlled conditions. One of the advantages of conducting a consumer test in the sensory laboratory is the rapid turnaround time for results to be obtained because of the proximity to the data-processing facilities.

Disadvantages. The disadvantages of conducting consumer tests in a sensory laboratory are the limitations on the consumer sample being used. If consumers screened from a prerecruited consumer database are used, most who would be willing to participate are those whose homes or places of employment are situated close to the sensory laboratory. Another disadvantage is a source of bias when company employees are used; therefore, in such instances, care must be taken to recruit only employees who are not familiar with the production, testing, or marketing of the product and whose attitudes toward a list of commonly consumed foods are similar to those of a consumer population consisting of nonemployees. One other disadvantage is the limited information that may be obtained during a laboratory test compared to a home-use test where samples will be used in normal consumption practices such as drinking from a full glass compared to sipping.

Central-Location Tests. Central-location tests (CLT) are conducted in a location away from the sensory laboratory that is accessible to large numbers of consumers—such as a shopping mall, hotel, or food-service establishment (Miller et al., 1995). The panel may comprise consumers who are recruited from a consumer database consisting of prerecruited consumers or those intercepted at the central location and screened for eligibility to participate in the tests. Usually 100 or more responses are obtained and several central locations may be used in the evaluation of a product. The recommended number of products to be evaluated in a central-location test is four.

Advantages. The major advantage of the central-location test is the ability to recruit and test a large number of participants and obtain a large number of responses when access to a consumer database is not available. In addition, actual consumers participate in the test; no company employees are used. The results have considerable impact and validity because actual consumers are used. Several more products may be tested than would be possible in a home-use test.

Disadvantages. A disadvantage of the central-location test is the distance from the food company, consulting firm, or research organization. Often, the central location has limited facilities, equipment, and resources for food preparation and testing. Facilities that may be lacking are space for a registration area, kitchen, environmentally controlled conditions, and individually partitioned booths for the sensory test. The inadequacy of facilities would limit the type of tasks that can be performed by panelists and may pose a severe limitation on the ability to conduct the test under the experimentally controlled conditions required. The researchers should consider the possibility of walkouts and the high potential for distraction in a central location. The large number of panelists who can be used for this type of test is a disadvantage as it is an advantage because of the time and staff requirements to conduct the test. Finally, as in the sensory laboratory test, the time of exposure is a disadvantage compared to the home-use test where the consumer will have more and longer exposure to a product (Stone and Sidel, 1993).

Mobile Laboratory. A special form of central-location test provides the advantages found in both the sensory laboratory and the central-location tests. A fully equipped mobile laboratory with complete facilities for food preparation and sensory testing is parked at a central location where a large number of consumers can be intercepted; the tests are conducted within the mobile laboratory.

The panel consists of consumers who are intercepted to participate in the tests. There is no prerecruitment or screening except for age. Usually 75 to 100 or more responses can be obtained in one location, and the mobile laboratory test is generally conducted in more than one location. Several central locations may be used in the evaluation of a product. Data are collected by trained interviewers. The recommended number of products to be evaluated is two to four.

Advantages. The major advantage of using a mobile laboratory for a consumer-affective test is the ability to use a large number of real consumers for the test and the availability of facilities to maintain experimentally controlled facilities and environmental conditions for food preparation and testing. There is unlimited access to consumers because a mobile laboratory can be taken almost everywhere a specific target population can be intercepted for the test; no employees are used as participants in the test.

Disadvantages. The disadvantages of the mobile laboratory test are the expense of maintaining the laboratory and logistical arrangements for parking and power supply that have to be made before the test. When conducting a mobile laboratory test, the planning and preparation are more rigorous compared to a laboratory test where the facilities, equipment, and supplies are located within walking distance from the food preparation test site.

Not all foods can be tested in a mobile laboratory because food preparation can be limited and testing conditions likewise limited by space constraints. Usually a mobile laboratory test is conducted by trained interviewers rather than using a self-administered questionnaire, adding to personnel and training costs.

Home-Use Tests. Home-use tests (HUT) are also referred to as home-placement tests. These tests, as the name implies, require that the test be conducted in the participants' own homes. Samples are either mailed to the consumers or consumers are asked to pick them up from the sensory laboratory or a central location.

The panel consists of consumers who are usually prerecruited, part of a consumer database, and screened for their eligibility to participate in the tests. However, in some cases employees of a company who have little or no responsibility for production, testing, or marketing of a product are used. Usually more than 100 responses are obtained per product. The number of products to be tested should be limited to one or two. The home-use test was used in determining the acceptance of frozen and unfrozen beef steaks (Johnson et al., 1990; Miller et al., 1995) and frozen, irradiated poultry (Hashim et al., 1995).

Advantages. The major advantage of the home-use test is that the products are tested under actual home-use conditions. In addition, one can not only test a product for acceptance or preference but also for product performance. Furthermore, the products are tested in the actual home environment and under normal consumption conditions compared with those in the sensory laboratory or central-location tests. Responses may be obtained from the respondent, who is usually the major shopper and purchaser of food, as well as the other members of the household. In many instances, other pieces of information such as the types of competitive products found in the home during the test, usage patterns, and other information that would be useful in marketing the product may be obtained. If the participants of the tests are prerecruited and screened from an existing database, they are aware of their role and the importance of the data collected, which will likely result in a high response rate.

Disadvantages. The main disadvantage of the home-use test is that it provides testing conditions that are not researcher controlled and as such could yield variable results. In addition, it requires a considerable amount of time to implement, distribute samples to participants, and collect participants' responses. Little can be done to exert any control over the testing conditions once the product is in the home of the respondent. The home-use test can be expensive in terms of product cost because sample sizes are much larger than one would serve in a sensory laboratory or central-location test. Mailing or distributing the products to participants will add to the cost of the test. If participants have not been prerecruited and screened from an existing database, they will likely be less aware of the importance of their role and of the data being collected. In such cases, response rates may be lower than expected.

18.5 Product Characterization

Once consumer acceptance for the product has been quantified, the researchers should define the lower boundary for consumer acceptance. This may be done with input from company management. Some may be unwilling to set the limit at "like slightly" (=6) or higher on a nine-point hedonic scale but may opt for a higher limit such as "like moderately" (=7) or higher, for example.

Simultaneously, characterization of the same products used in the consumer-affective tests may be conducted using various established instrumental or physicochemical methods, when these exist, for the critical attributes of the product. Sensory descriptive analysis tests are often used to characterize products when there are no instrumental or physicochemical tests that can be used to validly and reliably characterize product attributes.

18.5.1 Descriptive Analysis

Descriptive analysis is a sensory method by which attributes of a food or food product are identified and quantified using human subjects specifically trained for this sensory method (ASTM, 1993). Several descriptive analysis methods have been described, including the flavor profile, quantitative descriptive analysis (QDA), spectrum method, and texture profile. Variations of these methods are used, and sensory researchers adapt these to fit their testing situations. In studies on frozen pork roasts, Siemens et al. (1990) found that consumers preferred roasts with no fat cover more than roasts with 0.6-cm fat cover. However, in descriptive analysis tests on the same samples a trained panel found that fat cover affected only chewiness among nine attributes (two color, four texture, and three flavor) evaluated.

Flavor Profile

The flavor profile is a method that describes the aroma and flavor of products or ingredients (ASTM, 1996). The method consists of formal procedures for describing and assessing the flavor of a product in a reproducible manner. It involves a minimum of four assessors selected on the basis of screening tests, a personal interview to determine availability and personality traits, and then training over a period of 6 mo. The assessors evaluate the food or food product in a quiet, well-lit, odor-free room. It is suggested that a round table be used to facilitate discussion. Assessments take approximately 15 min per sample. One to three sessions are held. Assessors make independent evaluations and rate character note intensities using a seven-point scale from threshold to strong.

The panel leader compiles the individual results, and discussions are held until a consensus is reached. The panel leader prepares a final profile in tabular and/or graphical format.

Quantitative Descriptive Analysis

Quantitative descriptive analysis (QDA) is an analysis method that describes all sensory properties of a product and quantifies their intensities (ASTM, 1996). It is led by a descriptive-analysis moderator. It requires 10 to 12 qualified panelists, although in some tests 8 to 15 panelists may be involved. Training is conducted in a conference-style room. During training, panelists, under the direction of the panel leader, develop terminology to describe all of the product's sensory properties and definitions and a standardized evaluation procedure. The training requires 2 wk of training or approximately 6–10 h, organized into 90-min sessions. The panel leader organizes the training, provides the products, introduces references when needed, and facilitates activities but does not participate in the evaluation of samples as an assessor. Products are evaluated in partitioned booths. Assessors take 3–20 min per product using a repeated-trials design. For most tests, a four-replicate design is sufficient to provide for analysis (ASTM, 1996) and a minimum of three replications is recommended. References are provided as needed. Fifteen-centimeter line scales are used in rating samples.

Spectrum Descriptive-Analysis Method

The spectrum method utilizes universal scales that are constant for all attributes and products (ASTM, 1996). Perceived intensities are judged according to these universal scales. The training is led by a moderator trained in the spectrum descriptive-analysis method. Twelve to fifteen assessors are trained and ten to twelve are used in a given project.

The assessors are trained in a room with a round table. Training consists of orientation and practice phases. They develop the terminology, definitions, evaluation techniques, and agreed on references to be used. The practice phase involves the application of procedures learned during orientation. A total of 10 to 12 exercises is conducted during the three or more months of panel work. The panel is ready for evaluation after performance has been assessed by the panel leader.

Panelists evaluate food in partitioned booths. Evaluation is approximately 15 min per product. Evaluations are conducted in duplicate or triplicate in separate evaluation sessions.

Texture Profile

Texture-profile analysis is led by a moderator trained in that area. In general, texture profiling follows descriptive procedures for texture with emphasis on using specific references for each specific attribute scale. The texture-profile analysis was used to determine the effect of added collagen fibers and water on hardness, cohesiveness, springiness, chewiness, and juiciness of chicken frankfurters (Meullenet et al., 1994). A round table is used to facilitate discussion and evaluation. Assessors are trained on texture definitions, evaluation procedures, and standard reference scales and develop terminology, definitions, and evaluation procedures. The way the product is manipulated in the mouth during evaluation is standardized. Evaluation takes 5–15 min per product; a minimum of three replications is recommended. References are provided as needed and graphical rating scales are used.

Hybrid Descriptive-Analysis Methods

Descriptive-analysis methods employ a combination of some aspects of both QDA and spectrum analysis methods. These are often referred to as *hybrid* methods and are widely used in sensory descriptive analysis.

18.5.2 Physicochemical Measurements

Physicochemical measurements are likewise used to characterize attributes such as color, flavor, taste, texture, and viscosity. Instrumental color measurements of lightness, chroma, and hue angle can be obtained using a colorimeter. Some flavors such as oxidized flavors have been correlated with peak areas obtained from gas chromatographic analysis of headspace volatile components such as hexanal. The basic tastes can be quantified by analysis of chloride, sucrose, or acid. Texture can be quantified using measurements of shear, cutting force, work to cut, or parameters calculated from instrumental texture-profile analysis. Viscosity can be characterized using a viscometer.

18.6 Modeling of Consumer Acceptance

Overall consumer acceptance, or acceptance of sensory attributes of appearance, flavor, and texture, can be plotted against descriptive-analysis or physicochemical measurements. The relationships found, if any, may be linear or curvilinear (fig. 18.4). For example, Cardello et al. (1982), in studying relationships between judgments of perceived texture of frozen fish by trained and consumer panelists, observed good linear correlations between scalar judgments of texture, although a broader perceptual range was seen in trained panelists' responses. They concluded that regression equations can be developed to relate trained-panel data to consumer data. The regression equation for consumer ratings for juiciness, rated on a nine-point hedonic scale, of a meat product (Munoz and Chambers, 1993) was based on descriptive-analysis scores ranging from 0 to 150 for moisture release where juiciness $= -0.658 + 0.098$ (moisture release). This equation had

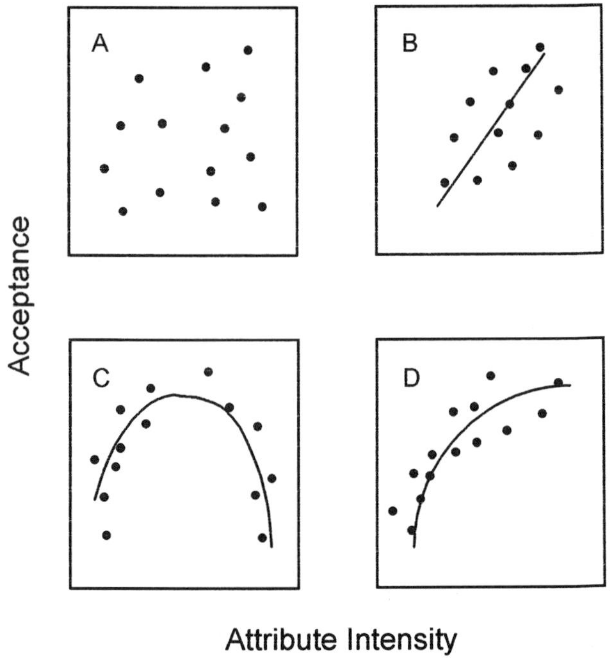

Figure 18.4. Plots of consumer affective scores versus attribute intensity

an R^2 of 0.9. Cardello et al. (1982) developed mathematical regression models with the acceptance scores as the dependent variables and physicochemical measurements as the independent variables. A similar procedure may be used for obtaining models to quantify overall acceptance.

18.7 Summary and Conclusion

In the development of high-quality frozen-food products, a strategic approach that involves the identification of critical attributes is recommended. Quality or consumer acceptance of the food or food product by regular consumers of the product can be quantified by using consumer-affective tests and characterizing sensory properties of the product using sensory descriptive analysis ratings or physicochemical measurements. The relationships can be determined by plotting acceptance ratings against the descriptive analysis ratings or physicochemical measurements. Mathematical models may be developed to predict consumer acceptance scores from descriptive analysis ratings or physicochemical measurements. These equations may be used to establish specifications for frozen food that correspond to a predetermined degree of quality.

References

Amerine, M.A., Pangborn, R.M., and Roesler, E.B. 1965. *Principles of Sensory Evaluation of Food.* Academic Press, New York.

ASTM. 1979. *Manual on Consumer Sensory Evaluation.* (Schaefer, E.E., ed.) American Society for Testing and Materials, Philadelphia.

ASTM. 1993. *Manual on Descriptive Analysis Testing for Sensory Evaluation.* American Society for Testing and Materials, Philadelphia.

ASTM. 1996. *Sensory Testing Methods,* 2nd ed. (Chambers, E. and Wolf, M.B., eds.) American Society for Testing and Materials, Philadelphia.

Calhoun, C.M., Eilert, S.J., and Mandigo, R.W. 1996. Connective tissue/acidic phosphate preblend effects on reduced fat frankfurters. *J. Food Sci.* **61**:459–464.

Cardello, A.V., Maller, O., Kapsalis, J.G., Segars, R.A., Sawyer, F.M., Murphy, C., and Moskowitz, H.R. 1982. Perception of texture by trained and consumer panelists. *J. Food Sci.* **47**:1186–1197.

Chambers, E., IV, and Smith, E.A. 1991. The uses of qualitative research in product research and development. In *Sensory Science Theory and Applications in Foods.* (Lawless, H.T. and Klein, B.P., eds.) Marcel Dekker, New York.

Collins, J.L., Liao, J.-Y., and Penfield, M.P. 1995. Chemical, physical, and sensory attributes of formed and frozen, baked sweetpotato. *J. Food Sci.* **60**:465–467.

Gallander, J.F. and Hill, R.G., Jr. 1982. Effect of variety and harvest date on the quality of frozen strawberries. *Food Process. Technol. Res. Circle* **271**:43–45.

Galvez, F.C.F. and Resurreccion, A.V.A. 1992. Reliability of the focus group technique in determining the quality characteristics of mungbean (*Vigna radiata* (L). *Wilczec*) noodles. *J. Sens. Stud.* **7**:315–326.

Hashim, I.B., Resurreccion, A.V.A., and McWatters, K.H. 1995. Consumer acceptance of irradiated poultry. *Poultr. Sci.* **74**:1287–1294.

IFT/SED. 1981. Guidelines for the preparation and review of papers reporting sensory evaluation data. *Food Technol.* **35**(11):50–59.

Johnson, R.C., Jr., Romans, J.R., Muller, T.S., Costello, W.J., and Jones, K.W. 1990. Physical, chemical, and sensory characteristics of four types of beef steak. *J. Food Sci.* **55**:1264–1267, 1273.

Jones, L.V., Peryam, D.R., and Thurstone, L.L. 1955. Development of a scale for measuring soldiers' food preferences. *Food Res.* **20**:512–520.

Meilgaard, M., Civille, G.V., and Carr, B.T. 1991. *Sensory Evaluation Techniques,* 2d ed. pp. 188–211, CRC Press, Boca Raton, Fla.

Meullenet, J.-F., Chang, H.C., Carpenter, J.A., and Resurreccion, A.V.A. 1994. Textural properties of chicken frankfurters with added collagen fibers. *J. Food Sci.* **59**:729–733.

Miller, L.F., Hedrick, H.B., and Bailey, M.E. 1985. Sensory and chemical characteristics of pork chops as affected by precooking, curing, and frozen storage. *J. Food Sci.* **50**:478–481.

Miller, M.F., Hoover, L.C., Cook, K.D., Guerra, A.L., Huffman, K.L., Tinney, K.S., Ramsey, C.B., Brittin, H.C., and Huffman, L.M. 1995. Consumer acceptability of beef steak tenderness in the home and restaurant. *J. Food Sci.* **60**:963–965.

Moskowitz, H. 1974. Sensory evaluation by magnitude estimation. *Food Technol.* **28**(11):16.

Munoz, A.M. and Chambers, E., IV. 1993. Relating sensory measurements to consumer acceptance of meat products. *Food Technol.* **47**(11):128–131, 134.

Penfield, M.P., Costello, C.A., McNeil, M.A., and Riemann, M.J. 1989. Effects of fat level and cooking methods on physical and sensory characteristics of restructured beef steaks. *J. Food Qual.* **11**:349–356.

Peryam, D.R. and Pilgrim, F.J. 1957. Hedonic scale method of measuring food preferences. *Food Technol.* **11**(9):9–14.

Schutz, H. 1965. A food action scale for measuring food acceptance. *J. Food Sci.* **30**:365–374.

Siemens, A.L., Heymann, H., Hedrick, H.B., Karrasch, M.A., Ellersieck, M., and Eggeman, M.K. 1990. Effects of external fat cover, bone removal, and endpoint cooking temperature on sensory attributes and composition of pork center loin chops and roasts. *J. Sens. Stud.* **4**:179–188.

Stone, H. and Sidel, J.L. 1985. Affective testing. In *Sensory Evaluation Practices*. Academic Press, Orlando, Fla.

Stone, H. and Sidel, J.L. 1993. Affective testing. In *Sensory Evaluation Practices,* 2d ed. Academic Press, New York.

19

Shelf-Life Testing: Procedures and Prediction Methods

Bin Fu and Theodore P. Labuza

19.1 Introduction

The shelf life of a food can be defined as the time period within which the food is safe to consume and/or has an acceptable quality to consumers. Just like any other food, frozen foods deteriorate during storage by different modes or mechanisms, as summarized in table 19.1. Microbes usually are not a problem since they generally cannot grow at freezing temperatures unless subjected to extensive temperature abuse above the freezing point. Enzymes, which can cause flavor change (lipoxygenase) in nonblanched fruits and vegetables and accelerated deterioration reactions in meat and poultry (enzymes released following disruption to organelle membranes during precooking) are a big concern for frozen foods. Cell damage or protein and starch interactions during freezing cause drip and mushiness upon thawing. Discoloration could occur by nonenzymatic browning, bleaching, and freezer burn. Vitamin C loss is often a major concern for frozen vegetables. Physical changes, such as package ice formation, moisture loss, emulsion destabilization, and recrystallization of sugars and ice in frozen desserts are often accelerated by fluctuating temperatures.

For any specific frozen product, the mode determining its shelf life depends on the product characteristics (raw materials, ingredients, and formulation), prefreezing treatment, freezing process, packaging film and processes, and of course storage conditions. All of the quality deterioration and potential hazards are usually exaggerated or complicated by a fluctuating time-temperature environment (e.g., freeze/thaw cycle) during storage. On the other hand, the shelf life of a frozen food can be extended through ingredient selection, process modification, and change of package or storage conditions, as discussed in section 3 of this book.

This chapter will focus on shelf-life testing of frozen foods for product development and market practices. Shelf-life testing consists basically of selecting the quality characteristics that deteriorate most rapidly in time and the mathematical modeling of the change. Table 19.1 can be used as a reference for the selection of

Table 19.1. Deterioration modes of frozen foods

Frozen foods	Deterioration modes
Frozen meats, poultry, and seafood	Rancidity
	Toughening (protein denaturation)
	Discoloration
	Desiccation (freezer burn)
Frozen fruits and vegetables	Loss of nutrients (vitamins)
	Loss of texture (temperature abuse)
	Loss of flavor (lipoxygenase, peroxidase)
	Loss of tissue moisture (forming package ice)
	Discoloration
Frozen concentrated juices	Loss of nutrients (vitamins)
	Loss of flavor
	Loss of cloudiness
	Discoloration
	Yeast growth (upon temperature abuse)
Frozen dairy products (e.g., ice cream and yogurt)	Iciness (recrystallization of ice crystals)
	Sandiness (lactose crystallization)
	Loss of flavor
	Disruption of emulsion system
Frozen convenience foods	Rancidity in meat portions
	Weeping and curdling of sauces
	Loss of flavor
	Discoloration
	Package ice
Frozen bakery products (e.g., raw dough, bread, and croissants)	Burst can (upon temperature abuse) (dough)
	Loss of fermentation capability (dough)
	Staling (becoming leathery)
	Loss of fresh aroma

quality characteristics, which depends on the specific product and usually requires professional judgment. Mathematical modeling of quality deterioration will be discussed next.

19.2 Modeling of Quality Deterioration

19.2.1 Basic Equation

A frozen food starts to degrade once it is produced (fig. 19.1). The rate and degree of degradation depend on both the composition and the environmental conditions during storage and distribution. In general, the loss of food quality or shelf life is evaluated by measuring a characteristic quality index, A. The change of quality index A with time (dA/dt) usually can be represented by the following kinetic equation:

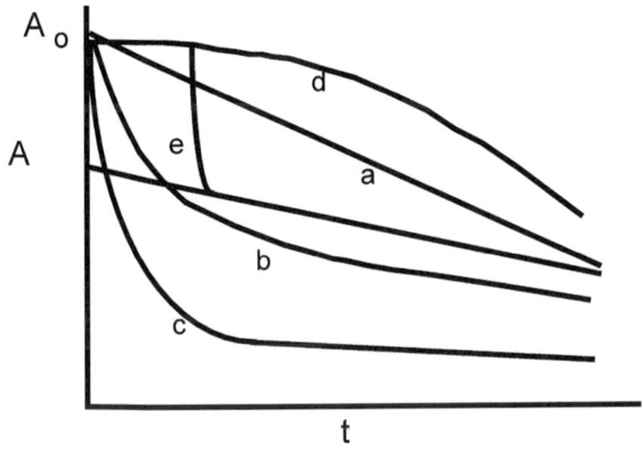

Figure 19.1. Quality deterioration curves: (a) linear; (b) exponential, (c) hyperbolic; (d) quadratic; and (e) complex

$$-dA/dt = k\,A^n \tag{19.1}$$

where k is called a rate constant depending on temperature, product, and packaging characteristics; n is a power factor called reaction order that defines whether the rate of change is dependent on the amount of A present. If environmental factors are held constant, n also determines the shape of deterioration curve.

19.2.2 Zero- and First-Order Kinetics

Equation 19.1 can also be written as

$$f(A) = k\,t \tag{19.2}$$

where f(A) is the quality function, and k and t are the same as in equation 19.1. The form of f(A) depends on the value of n. When n is equal to zero it is called zero-order reaction kinetics, which implies that the rate of quality loss is constant under constant environmental conditions (curve (a) in figure 19.1). If n is equal to 1 it is called first-order reaction kinetics, which results in an exponential decrease in rate of loss as quality decreases (curve b in figure 19.1, which becomes a straight line if plotted on a semilog plot). These quality functions can be expressed as follows:

$$f(A) = A_o - A = k_z t \qquad \text{zero order} \tag{19.3a}$$

$$f(A) = \ln A_o - \ln A = k_f t \qquad \text{first order} \tag{19.3b}$$

where A_o is the initial quality value. If A_e corresponds to the quality value at the end of shelf life, the shelf life (θ) of the food is inversely proportional to the rate constant

$$\theta = (A_o - A_e) / k_z \qquad \text{zero order} \qquad \textbf{(19.4a)}$$

$$\theta = \ln (A_o/A_e) / k_f \qquad \text{first order} \qquad \textbf{(19.4b)}$$

It should be noted that most chemical reactions leading to quality loss in frozen-food systems are much more complex. However, the reaction kinetics can be simplified into either pseudo–zero order or pseudo–first order kinetics. In the case of complex reaction kinetics with respect to reactants, an intermediate or a final product (e.g., peroxides or hexanal in lipid oxidation) could be used as a quality index. There are few cases where neither zero- nor first-order kinetics apply. Curve *c* in figure 19.1 shows the degradation curve for a second-order reaction (with single reactant), which also shows a straight line on a semilog paper. A fractional order should be used to describe curve *d* in figure 19.1.

Sometimes there is an induction period or lag time before the quality deterioration begins (e.g., browning pigment formation in the Maillard reaction or a microbial growth lag phase, as shown in curve *e* in figure 19.1. The length of the lag depends on many issues, but temperature is a predominant factor. Given this, modeling of both the induction or lag period and deterioration phase is necessary for accurate prediction of quality loss or shelf life remaining. An example of such work has been demonstrated by Fu et al. (1991) for the growth of bacteria in milk.

In certain circumstances (e.g., *A* represents a sensory hedonic score), a nonkinetic approach, such as a statistical data fitting technique, can also be used to describe the deterioration curves. Varsanyi and Somogyi (1983) found that the change in quality characteristics as a function of time could be approximately described with linear, quadratic, and hyperbolic functions and that storage temperature and packing conditions affected the shape of the deterioration curves. However, except for the linear curve, the parameters determined by data fitting are difficult to use for prediction under variable storage conditions.

19.2.3 Temperature Dependence of Deterioration Rate

Arrhenius Kinetics

Once a frozen product is made and packaged and starts its journey from the manufacturer's plant to warehouse, distribution center, retail store, and finally consumer's freezer, the rate of quality loss is primarily temperature dependent (Zaritzky, 1982). The Arrhenius relationship is often used to describe the temperature dependence of deterioration rate where for either zero or first order:

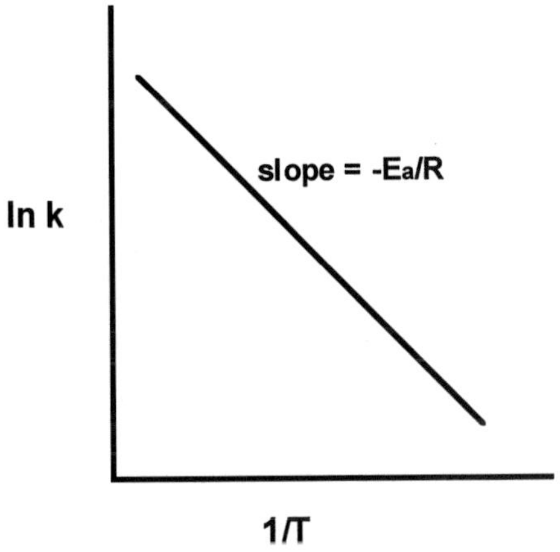

Figure 19.2. Arrhenius plot

$$k = k_o \exp(-E_a/RT) \tag{19.5a}$$

or

$$\ln k = \ln k_o - E_a/(RT) \tag{19.5b}$$

where k_o is a preexponential factor; E_a is an activation energy in cal/mol; R is the gas constant in cal/mol K and equal to 1.986; and T is an absolute temperature in K (273 + °C). Thus, a plot of the rate constant on semilog paper as a function of reciprocal absolute temperature ($1/T$) gives a straight line as shown in figure 19.2. The activation energy is determined from the slope of the line (divided by the gas constant R). A steeper slope means the reaction is more temperature sensitive (i.e., a small change in T produces a large change in rate).

Thus, by studying a deterioration process and measuring the rate of loss at two or three temperatures (higher than storage temperature), one could then extrapolate on an Arrhenius plot with a straight line to predict the deterioration rate at the desired storage temperature. This is the basis for accelerated shelf-life testing (ASLT), which will be discussed later. One should note, however, that in some cases a straight line will not ensue for a variety of reasons, especially if a phase change occurs (Labuza and Riboh, 1982). Thus for frozen foods, extrapolation from temperatures above 0°C are meaningless for shelf-life prediction.

WLF Kinetics

Besides the Arrhenius equation, another popular equation at least in the more recent food literature is the Williams-Landel-Ferry (WLF) model (Williams et al., 1955). Its original form was based on the variation of the viscosity in the temperature range above T_g as addressed in chapter 3. When the rate constant at T_g' is substituted for T_g (T_g' is the T_g of a maximally freeze-concentrated system), the WLF model can be written as follows:

$$\log (k_T/k_g) = C_1(T - T_g')/[(C_2 + (T - T_g')] \qquad \textbf{(19.6a)}$$

or

$$[\log (k_T/k_g)]^{-1} = (C_2/C_1)/(T - T_g') + 1/C_1 \qquad \textbf{(19.6b)}$$

where C_1 and C_2 are constants. Thus a plot of $[\log (k_T/k_g)]^{-1}$ versus $(T - T_g)^{-1}$ will be a straight line with the slope equal to C_2/C_1 and the intercept equal to $1/C_1$. As can be seen, this is a two-parameter temperature-dependent model as is the Arrhenius equation.

Frozen foods stored below T_g' are stable to ice recrystallization and other physical changes. Levine and Slade (1988) postulated that stability is related to the temperature difference between storage temperature and T_g'. This cryostabilization of foods assumes stability below T_g' and rapid decrease of stability above T_g' according to the WLF relationship, exhibiting an increase in reaction rate much higher than expected from the Arrhenius kinetics. However, this may not be true since the rate of chemical reactions can be expected to be influenced by temperature increase in a complex way: (1) an increase of the rate constant, resulting from both the viscosity decrease and the increased molecular mobility (Fennema 1996); and (2) a decrease of the reaction rate as a consequence of the increasing dilution of the reactants (Roos et al., 1996). For these reasons, it seems that the WLF model overpredicts the temperature effect of rate constant (Simatos et al., 1989). As noted by Nelson and Labuza (1994), because of the small temperature range over which foods are stored (e.g., about 30°C for dry foods and about 20°C for frozen foods), both the Arrhenius and the WLF model give good correlations as long as one does not use the universal coefficients suggested by Slade and Levine (1991). In fact, Lim and Reid (1991) assumed WLF applied for their data of enzymic activity at different temperatures in the frozen state as shown in figure 19.3a. They based this on the fact that there was no reaction below −10°C. However, as shown by Nelson and Labuza (1994) and represented in figure 19.3b, the data fit the Arrhenius relationship with an r^2 of 0.999. The challenge in applying the WLF model for stability or shelf-life prediction is that (1) T_g is not known; (2) T_g is difficult to determine; and (3) the universal coefficients of Levine and Slade (1986) are not applicable.

Shelf-Life Model

Most published data related to quality deterioration do not give rates or rate constants but rather are in the form of an overall shelf life (end-point analysis) as a function of storage temperature. Since the temperature range used is usually

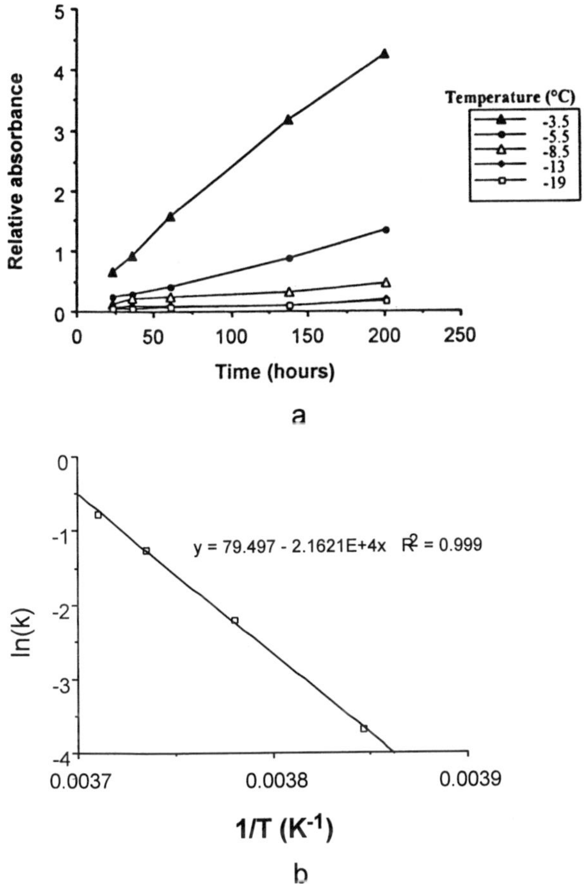

Figure 19.3. Hydrolysis of maltodextrin in the frozen state (Lim and Reid, 1991) (a) rate as a function of temperature (note T_g is -10°C); (b) Arrhenius plot

quite narrow, the following exponential relationship exists between shelf life and storage temperature:

$$\theta = \exp(-bT + c) \qquad \textbf{(19.7a)}$$

$$\text{or} \qquad \ln \theta = -bT + c \qquad \textbf{(19.7b)}$$

where q is shelf life at temperature T in °C, b is the slope of the semilog plot of θ versus T, and c is the intercept or reference temperature as shown in figure 19.4. Practically, this is used frequently for shelf-life determination and prediction because of its simplicity and straightforwardness.

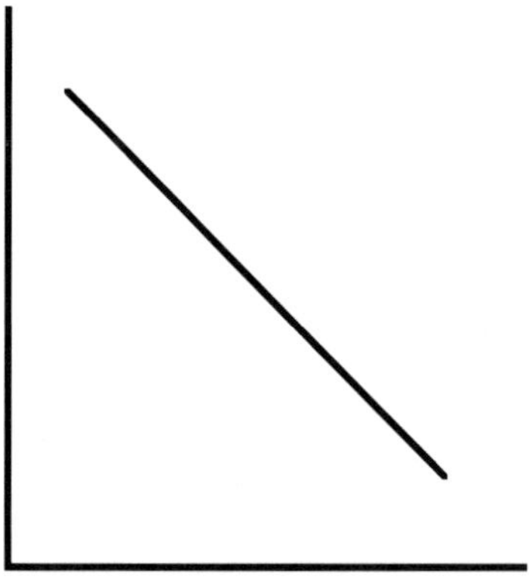

Figure 19.4. Shelf-life plot

Q_{10} or q_{10}

The Q_{10} approach is also often used for estimation of the temperature acceleration of shelf life, which is defined as

$$Q_{10} = \text{rate @ } T_1 + 10\ °\text{C / rate @ } T_1 \quad \textbf{(19.8a)}$$

$$Q_{10} = \text{shelf life @ } T_1 \text{ / shelf life @ } T_1 + 10\ °\text{C} \quad \textbf{(19.8b)}$$

$$Q_{10} = (q_{10})^{1.8} \quad \textbf{(19.8c)}$$

where T_1 is temperature in degrees Celsius. If the temperature unit is in degrees Fahrenheit, then the term q_{10} is used, which in fact is more common than Q_{10} in the frozen-food literature.

The magnitude of Q_{10} depends on the food system, the temperature, and the absolute range. Q_{10} values from 2 to 20 have been found for frozen foods (Labuza, 1982; Labuza and Schmidl, 1985). Q_{10} can be shown to be related to the Arrhenius equation and the shelf-life model through the following expression:

$$Q_{10} = \exp\{10\,E_a/[R\,T\,(T + 10)]\} \quad \textbf{(19.9a)}$$

$$Q_{10} = \exp(10\,b) \quad \textbf{(19.9b)}$$

Thus Q_{10} is not constant but depends on E_a and the absolute temperature T. Some data gleaned from Jul (1984) and Labuza (1982) are shown in table 19.2.

Table 19.2. Estimate of the Q_{10} for shelf life of selected frozen foods

	Days of high-quality life (HQL)		
Item	−10°C	−20°C	Q_{10}
Pork sausage	20	120	6
Pork	50	400	8
Beef	60	200	3.3
Ground hamburger	250	800	3.2
Fried hamburger	35	250	7
Raw poultry	200	700	3.5
Fried poultry	25	700	28
Fatty fish	7	60	9

Other Models

The following models have also been proposed to describe the temperature dependence of the rate constant (Kwolek and Bookwalter, 1971) for frozen systems

$$k_T = a + b\,T \tag{19.10a}$$

$$k_T = a\,T^b \tag{19.10b}$$

$$k_T = a\,/\,(b - T) \tag{19.10c}$$

where a and b are constants. In most cases, equation 19.10c fits data better. However, all of these equations have very limited practical application.

19.2.4 Time-Temperature Tolerance

Frozen foods are often exposed to a variable temperature environment, for example, during distribution or the freezing/defrosting cycle in retail or home freezers. In general, the value of the quality function, f(A), at time t under changing environmental conditions can be estimated from

$$f(A) = \int k_{T(t)}\,dt \tag{19.11}$$

where $T(t)$ is the temperature as a function of time. The form of f(A) depends on the reaction order as discussed previously. If an effective temperature, T_{eff}, is defined as the constant temperature exposure that causes the same quality change as the variable temperature condition, as proposed by Schwimmer et al. (1955), then

$$f(A) = k_{eff}\,t \tag{19.12}$$

The rate constant at that defined temperature is termed the effective rate constant, that is, k_{eff}.

To estimate the quality change under variable temperature conditions, one needs to either solve for f(A) numerically or know the value of T_{eff} or k_{eff} that corresponds to the variable conditions. The numerical approach for a randomly variable temperature history is essentially the same as the time/temperature/tolerance (TTT) approach initiated by Van Arsdel et al. (1969) and derived empirically in the 1960s for the prediction of shelf life of frozen foods (Jul, 1984). It is assumed that the temperature history of the product is known. Thus the fraction of shelf life consumed, f_{con}, was calculated as the sum of the times at each temperature interval, t_i, divided by the shelf life at that temperature, θ_i

$$f_{con} = \Sigma\,(t_i / \theta_i) \qquad \textbf{(19.13)}$$

Thus the remaining shelf life at a reference temperature is equivalent to $(1 - f_{con})^* \theta$.

Equation 19.13 assumes that the rule of additivity is valid for frozen foods (Jul, 1984), which means that the loss of remaining storage life or quality can be calculated from knowledge of the prior time-temperature episodes the product has been exposed to. This also implies that the prior sequence of the time-temperature episodes is of no importance except to calculate the amount of quality remaining up to that time—that is, there is no history effect. If the rule of additivity is valid with reasonable accuracy, the use of time-temperature integrators (TTI) should provide reliable results with respect to prediction of shelf life remaining, which will be discussed later.

However, there are some cases where the total effect of various temperature experiences may not be independent of the order in which they occur or of the nature of temperature history. For example, widely fluctuating temperatures may cause freezer burn or in-package desiccation, which is not additive (Jul, 1984). Where the colloidal nature of a product is affected, the effect of time-temperature history may not be additive either, especially with freeze/thaw cycles. This is also true when growth of microorganisms occurs (Fu et al., 1991). Certain chemical reactions, enzymatic as well as nonenzymatic, could proceed even more rapidly at temperatures below freezing. This is called a negative effect of temperature (Singh and Wang, 1977), and could be caused by one or more of the following factors: (1) a freeze-concentration effect; (2) the catalytic effect of ice crystals; (3) a greater mobility of protons in ice than in water; (4) an increase or decrease in pH with freezing; (5) a favorable orientation of reactants in the partially frozen state; (6) a salting in or out of proteins; (7) a decrease in the dielectric constant; and (8) the development of antioxidants at higher temperatures. As has been shown by Fennema (1975), the freeze-concentration effect can cause rates of chemical reactions to increase dramatically just below the freezing point (fig. 19.5) (e.g., ascorbic acid loss at −3°C can be faster than at higher temperatures), so one should not use data in the −4 to 0°C range or above as part of an accel-

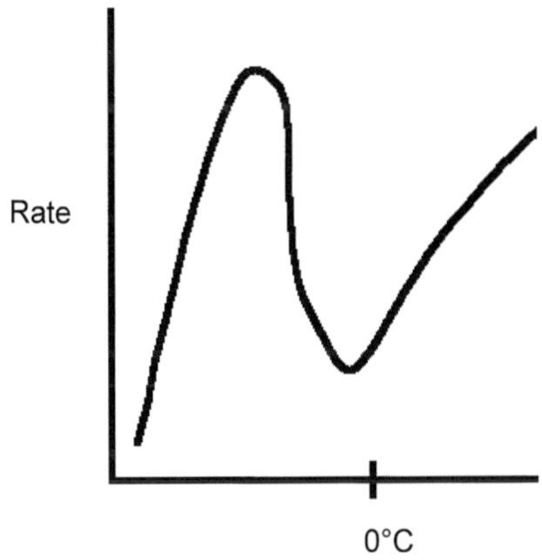

Figure 19.5. Rate of chemical reaction as a function of temperature above and below the freezing point of a food

erated shelf-life test to predict rates at lower temperatures. Fennema (1975) showed that the time to 50% loss of vitamin C in broccoli was 44 d at −5°C, 120 d at −2°C, and 162 d at +2°C. This concentration effect is evident in the shelf-life plot of frozen strawberries, as shown in figure 19.6, using the data of Guadagni (1968). If the data collected only at 25 and 30°F (−3.9 and −1.1°C) are used, the predicted shelf life at 0°F (−17.8°C) is more than 27 years; if data are collected at only 20 and 25°F (−6.7 and −3.9°C), the shelf life predicted at 0°F is 40 d while data below 20°F extrapolated to the true expected shelf life is about 280 d.

The response ratio of the food to changes in environmental temperature (R_T) is dependent on the fluctuating temperature conditions as well as the heat-transfer properties of the food and the package (Cairns et al., 1974; Dagerskog, 1974). In the analysis of food shelf life, an inherent assumption is made that the food responds instantaneously to the environmental temperature changes (i.e., $R_T = 1$). This may be acceptable if a surface deterioration process is the deterministic factor for shelf life, such as mold growth in some foods. Freeze-defrost cycles generally can be considered as sinusoidal oscillations. The amplitude of the effect is reduced inside the package by some factor, thus $R_T < 1$. It can be expected that the shorter the period of the ambient variation the smaller the R_T and hence the smaller the amplitude of the cyclic temperature variation in the package. Zuritz and Sastry (1986) also studied the effect of packaging materials on temperature fluctuations for frozen ice cream and found that packaging materials coupled with a layer of stagnant air were effective barriers against thermal fluctuations.

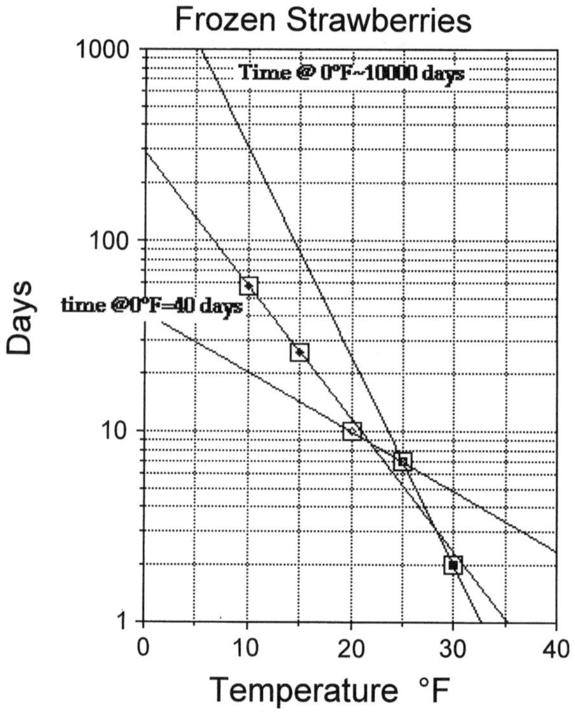

Figure 19.6. Shelf-life plot of frozen strawberries showing the influence of some selected shelf-life data at temperatures just below the freezing point on prediction of shelf life at 0°F. (Data from Guadagni, 1968.)

19.2.5 Hazard Function

After the product is produced, it may fail at any point in time in accordance with its life distribution (Nelson, 1972). The hazard function h(t) of a distribution is defined for $t \geq 0$ by

$$h(t) = f(t)/[1 - F(t)] \quad \textbf{(19.14)}$$

where f(t) is a probability density function and F(t) is a cumulative distribution function. The h(t) is the conditional probability of failure at time t, given that failure has not occurred before.

The behavior of a hazard function for studying the shelf life of food products can be easily understood by examining the bathtub-shaped curve in figure 19.7. Note that at time t_o, a frozen-food product begins its journey to many distribution outlets for consumption. During the time between t_o and t_1, early failures may occur owing to a failure in the process itself, faulty packaging, extreme initial

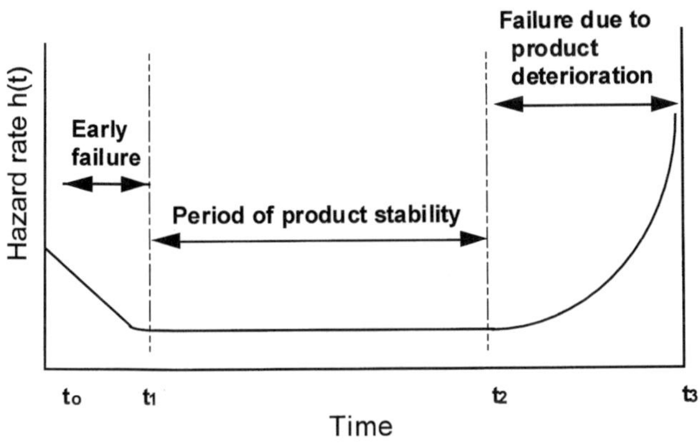

Figure 19.7. Failure rate as a function of time

product abuse, and many other environmental stresses to which the product is subjected. Early failure should not be taken as a true failure relative to the shelf life of the product unless it represents the normal condition. From t_1 to t_2 one can expect, barring chance major temperature fluctuations, no failures. This interval represents the true period of the product's stability. The failure rate is small and almost constant during this time. The hazard or failure rate increases from time t_2 to the termination point t_3 because of the true deteriorative changes occurring within the product. The concept of hazard function is important in the analysis and interpretation of the failure times of a product.

A fundamental assumption underlying statistical analysis of shelf-life testing is that the shelf-life distribution of a food product belongs to a family of probability distributions and that observations are statistically independent. Parameters of a shelf-life distribution are estimated by use of shelf-life testing experimental data. Once the parameters of a shelf-life model have been estimated, they can be used to predict the probabilities of various events, such as future failures (Nelson, 1972). Five statistical models—normal, log normal, exponential, Weibull, and extreme-value distributions—were tested for a few food products (Gacula and Kubala, 1975; Labuza and Schmidl, 1988), with the Weibull distribution (described later) fitting best.

19.3 Shelf-Life Testing: Overall Aspects

19.3.1 Purpose

In the development of any new food product including reformulation, change of packaging, or storage/distribution condition (to penetrate into a new market), one important aspect is the knowledge of shelf life. The shelf life of a food product is

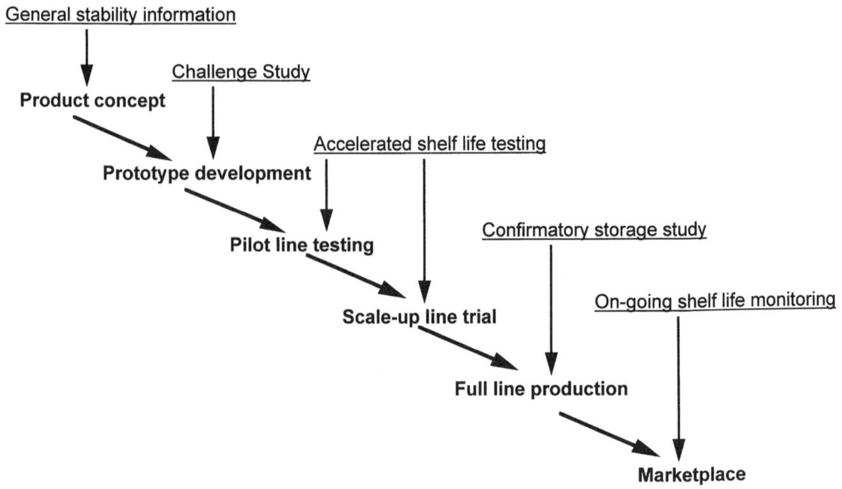

Figure 19.8. Shelf-life testing strategy at different product-development stages

vital to its success in the marketplace. This life must at least exceed the minimum distribution time required from the processor to the consumer. Shelf-life testing can assess problems that the product has in the development stage, following a fail small–fail early philosophy, thereby eliminating large disasters later. Marketing/brand managers also need reliable shelf life data to position the products and establish the brand. Periodic determination of shelf-life helps to provide assurance that the product remains consistent over time with respect to quality.

Different shelf-life testing strategies are necessary at different stages, as illustrated in figure 19.8. If the objective is to identify whether pathogens and spoilage microbes will grow in the case of temperature abuse, then a challenge study is necessary. If the objective is to quickly estimate the approximate shelf life of the product, then an ASLT can be used, as long as the proper temperature range is chosen. A confirmatory shelf-life test may be conducted at the last stage with simulated distribution chain conditions, although in today's research and development environment, this may be skipped.

19.3.2 Shelf-Life Criteria

The criterion for the end of shelf life may be variable depending on the definition of product quality grade, so the shelf life of a product may also be variable. The shelf life of most perishable and semiperishable foods is based almost solely on sensory quality (Fu and Labuza, 1993). For example, fresh meat degrades mainly by bacterial activity and rapid chemical oxidations that cause an off-flavor development and loss of color. This is readily recognizable by consumers. In contrast, many longer-shelf-life foods including most frozen foods degrade mainly by slow chemical reactions such as loss of nutritional value. For example, the vitamin C content of some frozen fruits and vegetables may fall below the required standard as listed on the label before sensory quality becomes inadequate.

The criteria for shelf life may also vary depending on the sensitivity of the consumer. For consumers, taste, odor, and appearance are the most obvious criteria; in academia and in the industry, sensory evaluation correlated with instrumental measurements of a given quality index (e.g., vitamin C level) are usually conducted. In general, the criteria level corresponding to the end of shelf life of a product depends on (1) any legal requirement, such as zero tolerance for botulinum toxin; (2) consumer preferences or marketing requirements; and (3) cost. In essence, the end of shelf life depends on the percentage of consumers a company is willing to displease. If 100% acceptance is required, then high-cost ingredients and absolute control of distribution up to point of consumption are necessary or there will always be some people who get foods beyond their shelf life. The aim is to keep this number as small as possible.

Just-Noticeable Difference (JND)

Sensory (organoleptic) examination of foods was a general procedure used by the human race to evaluate wholesomeness of foods long before the discovery of microorganisms. Sensory evaluation of foods by scientific methods can be used to evaluate such attributes as taste, odor, body, texture, color, and appearance. Changes in these attributes may be brought out by microbial or nonmicrobial actions, usually the latter for frozen foods.

The methods used to evaluate sensory shelf-life data include difference testing and hedonic scoring. Difference testing can involve paired comparisons, duo-trio tests, or triangle tests. The paired-comparison procedure determines the time when a measurable difference in quality occurs between two test samples at a certain level of probability. When applied to frozen foods, this method is often referred to as the *just-noticeable difference* (JND) test or *high-quality life* (HQL) test (Jul, 1984), which is usually based on flavor changes. Duo-trio testing compares two unknowns to an unabused control sample and asks the question of whether the quality of the unknowns is the same as or different from the identified control. Triangle testing determines the one different product among three test samples presented randomly to a set of judges (at least 10). Probability plots are used to predict shelf life at a given probability level. The difference method can result in finding a difference when none really exists (type I error) or not finding one when indeed there is a true difference (type II error). Labuza and Schmidl (1988) have discussed this topic more thoroughly in relationship to shelf-life testing, which is not commonly found in sensory textbooks. Table 19.3 shows some data from Guadagni (1968) for HQL of frozen foods.

Table 19.3. Days of high-quality life for fruits and vegetables

Product	Type	0°F	10°F	20°F
Apples	Pie filling	360	250	60
Blueberries	Pie filling	175	77	18
Cherries	Pie filling	490	260	60
Peaches	Retail syrup	360	45	6
Blackberries	Bulk, no sugar	630	280	50
Raspberries	Bulk, no sugar	720	315	70
	Retail, syrup	720	110	18
Strawberries	Bulk, sugar	630	90	18
	Retail	360	60	10
Green beans	Retail	296	94	30
Cauliflower	Retail	291	61	13
Peas	Retail	305	90	27
Spinach	Retail	187	57	23
Corn	Retail	720	360	
Corn on the cob	Retail	275	150	

Source: Guadagni (1968)

Hedonic Scoring

Hedonic scoring—which indicates acceptance on a numerical scale (e.g., a 1–9 point scale labeled from “dislike extremely” to “like extremely”)—is typically used for shelf-life evaluation. The test can be designed to evaluate not only the overall acceptance of the product but also that of specific characteristics such as flavor, texture, appearance, aftertaste, and so forth. Trained panels can also use this technique on a line scale, which can be converted to numerical equivalents.

If the hedonic method is used to evaluate shelf life, one can simply use the score as quality index A and plot the score versus storage time, run a linear regression, and choose the end of shelf life as the time when the progressed value drops below a preset level (Waltzeko and Labuza, 1976; Gacula, 1975). The shelf life determined in this way is called the practical shelf life (PSL) for frozen foods (Jul, 1984) and is longer than the HQL or JND. The use of hedonic rating scales may be of limited use in shelf-life testing, yet it is probably the most used method. Many food companies use a loss in hedonic score equal to $\Delta = 0.5$ for HQL and $\Delta = 1.5$ for PSL as the end of shelf life (Labuza, 1982). Objective measurements and professional judgment are often required to determine the end point. Data in table 19.4 are from a report published by the former Refrigerated and Frozen Foods Institute (1973). Unfortunately there were no methods given, but the data suggest that the PSL is about two to three times longer than the HQL value. This in itself suggests that the HQL methods can be used to shorten shelf-life testing times.

Table 19.4 Relationship between practical shelf life (PSL) and high-quality life (HQL) for frozen foods

Frozen food	*PSL/HQL ratio*
Lean meat	1.9–2
Fatty meat	2.0–2.4
Lean fish	1.9–2.2
Fatty fish	2.4–2.7
Precooked foods	2.8–3.0
Fruit	2.8–3.1
Vegetables	3.1–3.5

Instrumental Analysis

Chemical or instrumental analyses, such as moisture, nutrient loss, free fatty acids, or color measurement, that closely correlate to sensory attributes can supplement sensory techniques. They are usually less expensive and time-consuming than sensory approaches. A correlation between a physical or chemical test can increase the confidence level of the sensory results. For example, the following constituents or properties can be considered for monitoring chemical changes of pizza quality during frozen storage: total free fatty acids, specific volatile free fatty acids by HPLC, peroxides, oxidative volatiles (e.g., hexanal) by gas chromatography (GC), spice volatiles by GC, lysine, color (decrease in red color or increase in brown), in addition to sensory evaluation of taste and flavor (Labuza, 1986). Most sensory experts agree that analytical methods should complement the sensory tests. Conversely, the end point determined by objective measurements should be confirmed by sensory techniques as well.

Weibull Hazard Analysis

The Weibull hazard procedure requires one to first make an estimation of the time to the end of shelf life. This becomes the initial estimated time limit for the study. The time limit is then divided into several segments where panelists grade the product. Additional panelists are added at a constant number for each subsequent time period to maximize the number of testers near the end of the test. The panelist is asked to grade the food as good (acceptable) or bad (unacceptable), that is, no ranking on a hedonic score. When the product is identified as unacceptable by 50% of the panelists, the number of testers for the next period is increased by the number of failed samples plus the constant number. The interval between sample times is also shortened as the end of shelf life nears. The test ends when no more samples or panelists are available. The scores are ranked and the cumulative hazard calculated. The critical probability of failure, P_c, can then be calculated from the following equation:

$$P_c = 100\,\{1 - \exp[-\Sigma(H/100)]\} \qquad \textbf{(19.15)}$$

where H is the hazard value equal to 100/rank. Choosing $P_c = 50\%$ corresponds to an accumulated hazard value of 69.3%.

The relationship between the logarithm of storage time (log t) and the logarithm of hazard value (log H) is linear

$$\log t = (1/\beta)\log H + \log \alpha \qquad \textbf{(19.16)}$$

where β is the shape parameter and α is the scale parameter. The shelf life can then be determined based on the desired probability level allowed for product failure. The lower this probability, the shorter the shelf life. This plot then allows one to make a management decision with respect to the probability of displeasing a certain fraction of consumers. It is hoped that the distribution time is such that greater than 99% of the product is consumed before the end of shelf life based on displeasing less than X% of consumers where X% is an economic determination (i.e., what you can afford to do). A detailed example was given by Labuza and Schmidl (1988). It should be noted that this process can also be used for simple analytical tests such as plate counts or vitamin C. In these cases the number of panelists are replaced with the number of samples tested. Some criteria, such as 20% vitamin C loss, are used as the negative response. Figure 19.9 shows an example of a Weibull plot for a frozen food based on assumed data. A shelf life of 16 mo is found at $P_c = 50\%$ from the graph. From this graph, if 95% of the food was distributed and consumed in 3 wk, only 1% of the consumers would be displeased (or 0.95% of the product is out of compliance). If the rest were held and consumed at 10.5 wk, 50% of those eating it would have out-of-quality food or another $0.5 \times 5\% = 2.5\%$ of the total number of original product. Thus in this distribution model about 3.5% of the product is unacceptable. To improve on this, the product must either move faster or one must distribute it at a lower temperature. Wittinger and Smith (1986) used this approach to determine sensory shelf life of ice cream based on iciness and found a shelf life of 5 wk at 0°F (−17.8°C), which fits the general data for iciness in ice cream as shown in figure 19.10 (Labuza, 1982). It should be noted that this gives a Q_{10} of about 12.

19.3.3 General Procedures

Shelf-life testing experiments are designed to measure the average shelf life of a product under given conditions. General procedures for shelf-life testing of foods were proposed by Labuza and Schmidl (1985), including

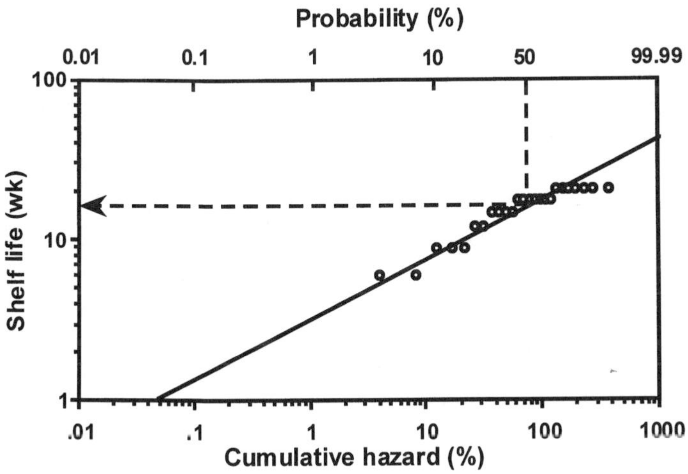

Figure 19.9. An example of a Weibull plot for a frozen food; a shelf life of 16 wk was determined at $P_c = 50\%$.

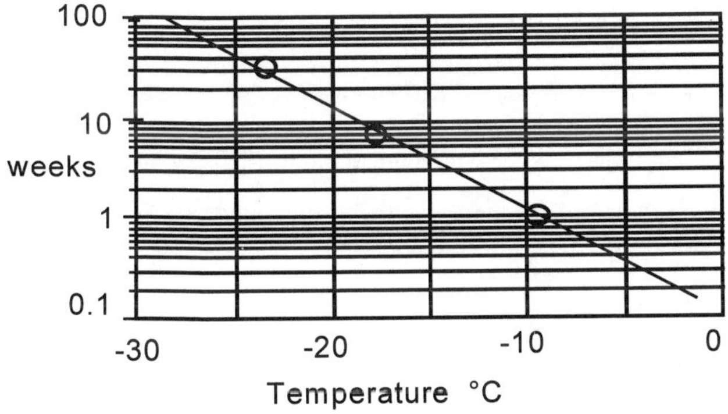

Figure 19.10. Shelf-life plot for ice cream based on iciness perception from data of Labuza (1982)

Step 1. Develop testing protocol. The protocol should consist of (1) specific objective; (2) detailed test design in terms of product, package, and storage condition; (3) execution procedures in terms of time, space, and resource availability; and (4) cost estimation.

Step 2. Identify key quality indicator. Any previous shelf-life data and kinetic parameters of food deterioration available in the literature (Labuza, 1982; Man and Jones, 1994) or the distribution turnover time of a similar or competitive product in the market place, if any, would be very helpful in this preliminary identification and in determining the shelf-life requirement.

Step 3. Estimate product sample and control needs. The number of samples and controls required should be based on the detailed experimental design. If sufficient product is available, extra samples should be placed into each storage condition. Now and then it may be necessary to recheck a sample, especially if a value is not in line with other data. It would be disastrous to be out of sample before failure has occurred or the predetermined termination of the test is reached. Extra controls should also be prepared and stored. When the samples are placed in storage rooms, they should be positioned so that the complete package is exposed to the external atmosphere unless otherwise specified. The specific location of the test sample should be recorded. Temperature controllers should be checked for accuracy periodically. In addition, removal of all unused samples from the storage room to make space for future studies is a must.

There are various thoughts when it comes to using a control product. Some sensory experts prefer an actual physical control; others are satisfied to just use the numbers obtained in the zero-time evaluation. There are three alternatives when using a physical example as a control: (1) making the control from scratch each time using the same ingredients, procedures, and so forth; (2) deep-freezing the control (e.g., pizza held at −70°C) and accepting that it might have changed slightly but minimally compared to the product in shelf life; (3) using a fresh batch of product that may not be identical.

Step 4. Select proper package materials and package size. This is largely dependent on shelf-life requirements, packaging costs and availability, and consumer information. Factors such as vacuum packaging, nitrogen flushing, or use of antioxidants are often considered in combination with packaging materials.

Step 5. Choose storage conditions. Storage conditions are chosen based on the type of shelf-life testing. For example, the intended commercial storage/distribution temperature range should be used in confirmatory shelf-life testing. Elevated temperatures are often used in accelerated shelf-life testing to obtain data for prediction of shelf life at lower temperatures or for prediction of shelf life under variable time-temperature distributions. Humidity control and/or monitoring is less important for frozen foods as compared to other foods (e.g., snacks, cakes, pies, and pastries). Light in the room should be properly controlled depending on the package.

Step 6. Estimate sampling frequency and duration of testing. The sampling frequency is generally an estimation based on experience from prior studies with similar foods. However, once one knows an interval at one temperature, then the intervals at other temperatures can be estimated using a Q_{10} value, that is, if Q_{10} is 3 then for a 10°C lower temperature, the sampling periods can be three times longer. If the interval between sampling is too long, the risk of under- or overestimating shelf life increases. The more analyses completed, the more accurate will be the shelf-life determination.

The question as to when one should end the experiment must be based on some preset criterion for failure. One criterion could be the minimum shelf-life requirement driven by product category, distribution chain, and the benchmark's product stability. If there is an accompanying sensory test, the end time can be based on some organoleptic inferior quality criterion from which one then can get a microbial or chemical index limit. For frozen products, several weeks to months are usually needed. If the shelf life can be estimated with any accuracy, the test intervals can be lengthened and clustered around the expected failure period. Most of the experts only require about six evaluations to provide reliable results.

Step 7. Schedule for execution. Before scheduling the starting date for a shelf-life test, one must check for the availability of ingredients, packaging materials, and storage space, and the time and resource available in the pilot plant or processing plant to prepare the samples. One should also check for the time and resources available in the microbial lab, the analytical lab, and/or the sensory support staff throughout the test period. A copy of the test request and schedule should be sent in advance to those who will be doing the work. The courtesy of providing those involved with this advance information always pays dividends. Holidays should be marked on the scheduling calendar, since scheduling too many evaluations near major holidays or on Friday afternoons is not recommended. However, once scheduled, sample observations on weekends and holidays should not be skipped, since important data points could be missed.

Step 8. Take sample and evaluate quality. Samples should be taken and evaluated following predetermined schedules. Sampling plans should be feasible administratively and economically, taking into account the heterogeneity of the food. Maxcy and Wallen (1983) pointed out the problem of heterogeneity of samples in shelf-life prediction. Multiple subsamples (≥ 3) should be done for nonhomogeneous samples. A single package is usually used as an experimental unit. Replication of three or four units is desired for each measurement. For frozen foods, a thawing process is often involved in the sampling procedure. Proper thawing or microwave heating is critical to the product quality. All samples should be thawed or microwaved in the same way to minimize bias.

The intended analyses should be based on the specific mode of deterioration, which was discussed earlier. Whatever the choice, the tests should be reasonable and logical. The key is to ensure measurement of the right thing. If the wrong quality factor is measured, the test starts out as a failure. Unfortunately, in many cases this cannot be established initially, so sensory evaluation is necessary in almost all shelf-life tests. Key sensory evaluation techniques for frozen foods have been discussed briefly in this chapter and are presented in greater detail in chapter 18.

At the time of each pull, one unit of the sample should be evaluated (informally by a minimum of two to three people) for changes in flavor and texture. This should be done in addition to the final tasting before a consumer sensory test. This is necessary since it helps the developer know approximately how the product is doing during the progress of the shelf life, helping to prevent any surprises in the results. Control samples may need to be prepared fresh.

Step 9. Analyze data. Shelf life is the predicted day at which the stored product (test pull) is X% less than the control at day zero. The data should be plotted and regressed to determine that point using the proper model (zero or first). All too often the data are not analyzed until the experiment is over and then the scientist finds nothing can be concluded because of lack of points, a poor fit, or some surprises. Statistical curve fitting should be consistent with the chosen model based on a theoretical mechanism. The amount of change and number of data points are related to the coefficient of variation (CV) of the test. A weighting factor may be used in estimating the rate constant and its statistical limits. When the data for an attribute do not fit the regression model well (adjusted R square of <0.8), scientific judgment should be used to decide whether the data are applicable.

When in doubt, a rerun on retention samples might help clarify the results. Error analysis could be performed before experiments are run by first finding inherent errors in time, temperature, and quality index measurements, then calculating an expected standard deviation for the plot being used to determine a rate constant. If the experimental data have a standard deviation much higher than the expected value, either the functional form of the rate expression is incorrect or the data contain errors from unanticipated sources.

Step 10. Prepare shelf-life report. Depending on the type of shelf-life determination, the results should either throw light on the technical viability of the product or provide answers to the questions about the maximum safe shelf life and maximum quality shelf life of the product. Before a shelf life is finally set, factors in the scale-up of shelf-life data will need to be taken into consideration. Based on results from ASLT, the provisional shelf life will be set for the product. There is no government regulation that defines the product end point except for that related to nutrient levels (vitamin C and vitamin A) in 21 CFR 101.9(g)(1)(ii), which states that for the vitamins listed, the analysis level cannot be below 80% of the label value if it is a natural food with no added nutrients and cannot be below 100% (21 CFR 101.9[g][1][i]) if the product has any added vitamin or nutrient regardless of whether it is the nutrient under test. Thus one must base the label value on some predicted initial variability and some predicted loss during distribution and storage. The FDA usually takes samples at the supermarket level (where it can purchase them) for compliance testing, not from the end of the process line, so distribution losses must be factored in.

The end point of shelf life is thus dependent on your corporate objectives and how much risk the company is willing to take with the brand. No shelf-life test is completed until a termination summary has been written. All termination summaries should include the objective of the test, product description, package description, conditions and length of storage, methods of evaluation, results (in the form of graphs, shelf-life plots, and Q_{10} values), and conclusions. Termination summaries should become a permanent record in the company library for future reference and preferably indexed well on a computer database for later retrieval. The final shelf life should also be set to give a clear margin of safety. In any case, the shelf life of a new product, particularly of the high-risk category, should be set based on data that relate to the worst-case manufacturing and storage scenario. The shelf life can then be reviewed and if necessary reset in light of further experience in manufacturing and control after the product has been launched.

Step 11. Implementation. One should get top management's approval of the test results so that they can be implemented. Management must believe and support those test results. It is important for production, sales, distribution, purchasing, and quality control to work together to ensure that the production is properly handled from the time of manufacture until the product is consumed.

19.4 Challenge Study

19.4.1 Basis

Freezing reduces the microbial population of foods, but considerable numbers usually survive even prolonged frozen storage. A challenge study is often used in the laboratory to determine the factors and factor interactions as they affect the shelf life of the product. Such simulated experiments enable the researcher to better control the study. A challenge study is necessary for frozen foods for two reasons: (1) to predict microbial growth and potential risk of the product upon temperature abuse in a distribution chain; and (2) to assess the relative stability and the relative risk of different formulas, processes, or packaging materials, which is crucial in new-product development. A challenge study may also be considered as a preliminary shelf-life determination in terms of microbiological safety. It is often used in the early stage of development since if microbial safety is a concern at this stage, reformulating can be done quickly.

19.4.2 Microbial Abuse Procedures

Step 1. Identify barriers. A composition/ingredient analysis should be done to identify any barrier(s) against spoilage microbes and pathogens in case of temperature abuse.

Step 2. Choose types of organisms/strains and inoculation level. One principle is to use an organism or strain previously isolated from the product or similar foods that is responsible for spoilage or risk. The more isolates in the study, the greater the confidence in the accuracy of the shelf-life assessment. An inoculation level, which is generally much higher than the normal contamination level in a product, must also be determined. If the average contamination level for a particular product is known, then the inoculation level should be as close as possible to that level. Sometimes several inoculation levels are used.

Step 3. Determine temperature abuse conditions. After inoculation, products should be packaged using the desired commercial packaging conditions, and subjected to temperature abuse. Factorial design and response surface methodology are often used in designing a challenge study. A typical temperature abuse condition used by some food companies is provided in table 19.5. It starts out with five sets of test packages placed at −18°C to begin the cycle. At the end of the first 24 h, one set of packages is removed and tested for microbiological

Table 19.5. A typical temperature abuse test sequence for microbial challenge studies

Day	Abuse temperature cycle	Number of package sets remaining
1	24 h at −18°C	5
2	20 h at −18°C 4 h at 38°C	4
3	20 h at −18°C 4 h at 38°C	3
4	20 h at −18°C 4 h at 38°C	2
5	20 h at −18°C 4 h at 38 °C	1

Source: Labuza and Schmidl (1985)

indicators to establish a zero-time level. All the other packages are kept at −18°C for the next 20 h, then removed and abused by placing them at 38°C for 4 h. Another set of packages is then removed for microbiological testing, and the cycle is repeated for the remaining packages (i.e., they are all returned to −18°C for at least 20 h, then abused at 38°C for 4 h). This procedure is repeated so that one set goes through at least four freeze-thaw cycles. If there is no significant increase in spoilage organisms or pathogenic organisms after the fourth cycle, the food is deemed safe microbiologically.

Step 4. Do microbial survival analysis. This analysis is to discover whether there is any microbial growth upon temperature abuse or whether the inoculated microbes survived the process. Appropriate detection and enumeration techniques should be used.

19.4.3 Applicability

The use of inoculated pack studies conducted by independent laboratories allows a food processor to assess the relative risks that can occur under conditions of temperature abuse of the food product in question. Taking frozen pizza as an example, both the cheese and sausage, if naturally fermented, will have high total counts of bacteria. Since the product is usually partially prebaked and then frozen, the numbers of vegetative microorganisms will decrease until thawing occurs. Unfortunately, pathogens such as *Staphylococcus aureus* will not be totally inactivated by these treatments. If the product is abused during distribution so severely that the temperature near the surface reaches about 7°C, pathogens may grow. A challenge study with *Staphylococcus aureus* will verify the microbial safety of the product.

It should be noted that inoculated pack studies with pathogens should not be conducted in food industry laboratories that are located close to the food-processing facilities because of the possible transfer of pathogens to food products. No sensory panel can be applied to evaluate the inoculated samples other than visual observation.

19.5 Accelerated Shelf-Life Testing

19.5.1 Basis

During product development, preliminary shelf-life knowledge is often needed in addition to microbiological safety. Shelf-life testing experiments at this stage are often accelerated to evaluate the effects of various formulation and processing parameters on shelf-life stability of the product being developed periodically since one cannot afford the relatively long shelf-life period for a frozen food stored under normal freezing conditions. In addition, temperature fluctuations may occur in distribution and retail holding for frozen storage. Thus kinetic studies at several temperatures within that range are necessary to predict its shelf life. Accelerated shelf-life testing conducted at elevated isothermal temperatures and/or with freeze/thaw cycles for frozen products have been used extensively for several decades by industry and government agencies (Labuza and Schmidl, 1985). The Arrhenius relation and Q_{10} approach are used to extrapolate the results to the expected lower storage temperature. Acceleration factors other than temperature have also been studied for some other deterioration modes such as moisture gain or loss and lipid oxidation (Labuza, 1984) but rarely for frozen foods.

19.5.2 Unique Procedures

Step 1. Clarify test objectives. In general, there are two occasions where ASLT applies: (1) to estimate approximate shelf life quickly during development stage and (2) to collect kinetic parameters for actual shelf-life prediction as in the marketplace, which generally is conducted near the launch phase.

Step 2. Select accelerating temperature conditions. Suggested isothermal accelerating conditions for frozen foods are -15, -10, and -5°C with a control stored at < -40°C (Labuza and Schmidl, 1985). The inherent assumption is that the deterioration mechanism is the same across the temperature range, although as noted earlier there is concern about how close to freezing one can go.

Moisture migration from the food into the surrounding air with resulting desiccation of the food and ice-crystal formation in the package is a major mode of deterioration of frozen foods under fluctuating temperature conditions. Cycling temperature storage is used to test for this (i.e., from 0°F or 10°F up to 20°F, with one day at each temperature and then repeated several times). A freeze-thaw cycling study is also needed to determine its effect on sensory quality. Usually, the high temperature can be much lower than that used in a microbial challenge study unless microbial survival is still a concern. Typically, cycling temperature/time can be three to five 24-h cycles between −18 and −7°C or between −18 and 7°C, depending on the product.

Step 3. Estimate testing time and sampling frequency. Testing times are dependent on a desired shelf life at target storage conditions. For example, given that a shelf life of 12 mo at −18°C is desired, a shelf-life plot can be constructed. Figure 19.11 indicates the test time at −4°C that equates to 12 mo at −18°C for various Q_{10} values. Sampling times at −4°C should thus be 1 wk, 2 wk, 1 mo, 3 mo, and 4.5–5 mo. Most published results suggest that Q_{10} values for vitamin C and quality loss in frozen vegetables range from 2 to 20 and that the shelf life of vegetables is only 6–8 mo at −18°C (Labuza, 1982). Considering these Q_{10} values, a product that does not retain good quality for 4.5 months at −4°C may not retain good quality for 12 mo at −18°C. This also suggests the sampling frequency shown in table 19.6. All simple tests should be conducted at each sampling time, whereas sensory testing should be concentrated mainly toward the end of the test sequence with a few near the beginning.

Step 4. Determine end point. Figure 19.12 shows a comparison of times to various levels for the loss of vitamin C in frozen spinach as a function of temperature (Kramer, 1974). The dotted line represents the 80/80 rule; that is, from a legal standpoint, for natural products, 80% of the tested sample must have no more than a loss of 20% (i.e., 80% of the label value). Consumer sensory testing will not always give such a clear shelf-life result since different shelf-life times can result using different quality attributes. Often professional judgment has to be made to decide which factor to use as the base for the end of the product's shelf life. When shelf life is unacceptably short, adjustments should be made to the food, its environment, packaging, process, and hygienic conditions until a suitable extension of shelf life can be achieved. For some products, the test results may demonstrate that the target shelf life is not attainable. At this point, the question of whether to launch the new product with a shorter shelf life or to abandon the entire project becomes a marketing decision.

Step 5. Estimate kinetic parameters. From each test storage condition, estimation of k or q is needed to make the appropriate shelf-life plot. From this information the potential shelf life and confidence interval for the storage condition can be estimated. Then parameters for the Arrhenius relation and the shelf-life plot, which are used for shelf-life prediction, are determined by linear regression.

Step 6. Extrapolate to normal freezing storage condition. The most useful shelf-life information is obtained for product kept at its intended storage temperature, which is about −18°C for retail frozen products and −23°C for distribution of frozen foods. Figure 19.13 demonstrates how the shelf-life plot is used for extrapolation. It is always a good practice to compare a model's prediction against actual experimental results because of the potential for errors from using the higher-temperature data as noted earlier, along with the other errors suggested by Labuza and Riboh (1982). In addition, the existence of a glass transition at a temperature between the test temperature and the prediction temperature would lead to error as shown by Nelson and Labuza (1994). In the case of frozen foods, most likely the error would be an underprediction of the shelf life.

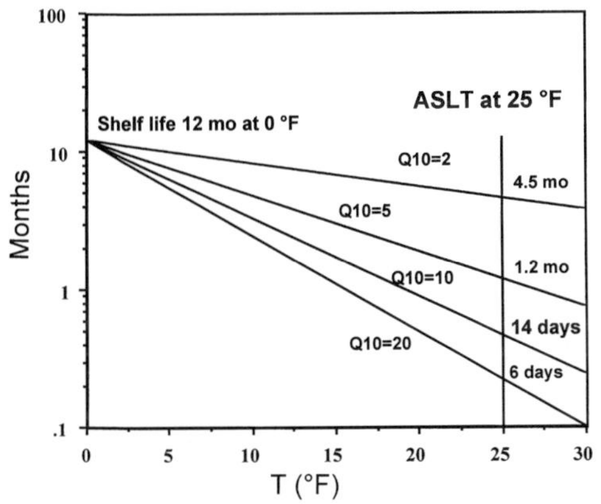

Figure 19.11. Shelf-life testing times at 25°F equivalent to 12 mo at 0°F for various Q_{10} values

Table 19.6. Sampling frequency for frozen pizza ASLT

Temperature (°C)	Sampling times (wk)
−4	1, 2[a], 3, 4, 5, 8, 12, 14, 16[a], 20[a]
−7	2, 4[a], 10, 15[a], 20[a]
−10	4[a], 10, 15[a], 20[a]

[a]Sensory test times

Source: Labuza (1986)

Step 7. Predict quality loss for a fluctuating time-temperature distribution. The prediction is based on two assumptions: (1) that there is no history effect from the time-temperature variation and (2) that the key deterioration mode does not change as a function of temperature. The frozen spinach data shown in figure 19.12 are used in the following example in table 19.7 for a time-temperature distribution. The line equivalent to 20% loss is set as the end of shelf life limit; that is, if A_0 = 36 mg/100 g then A at the end of shelf life is 0.8 × 36 = 28.8 mg, or 7.2 mg vitamin C can be lost. For each temperature of exposure, the time on the 80/80 line is the time for 20% loss; thus, at −10°F, the 20% loss time (equivalent to 100% shelf life) is 16.5 mo. For 6-mo storage at −10°F, there is 6/16.5 or 36.3% of the shelf life lost, and the amount left is 36 − (0.36 × 7.2) = 33.4 mg. Since as noted 80% of A_0 is equal to 28.8 mg/100 g at the end of shelf life, this product is still acceptable at the end of the set of three different time/temperature exposures. In fact, the shelf life left at 5°F = (1 − 0.775) × 3.3 = 0.74 mo = 22 d.

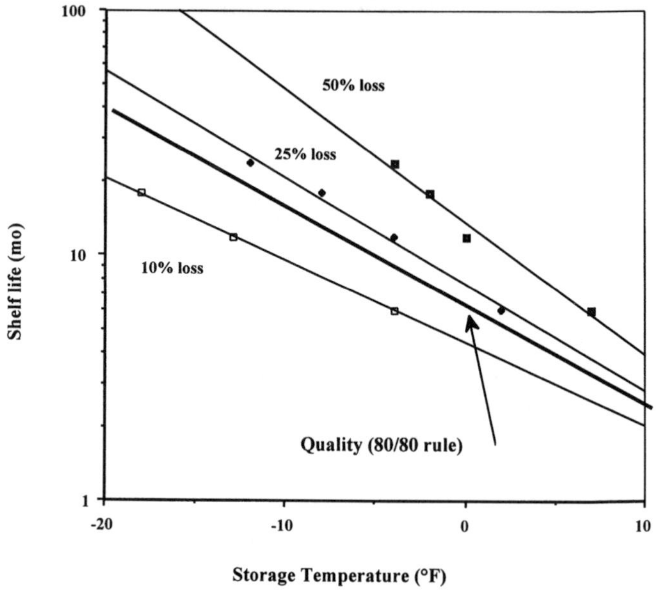

Figure 19.12. Shelf life of frozen spinach as a function of vitamin-loss level

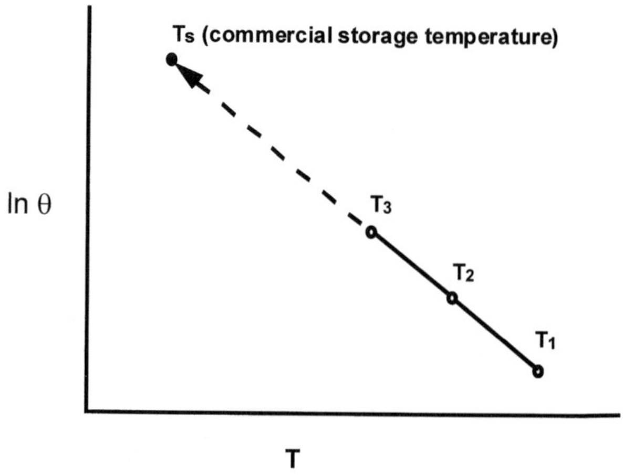

Figure 19.13. Extrapolation from ASLT

Table 19.7. Estimation of quality remaining of frozen spinach after exposure to a variable time-temperature history with A_o = 36 mg/100 g spinach

Temperature (°F)	Time t (mo)	θ shelf life (mo)	f_{con} (t/θ)	Σf_{con}	$A_{remaining}$ (mg/100 g)
−10	6	16.5	0.363	0.363	33.4
+3	1	4.5	0.256	0.619	31.5
+12	0.25	1.6	0.156	0.775	30.4

19.5.3 Applicability

Because of the relatively long shelf life for frozen foods and the unique feature of freezing, the degree of temperature elevation is largely limited. Prediction of actual shelf life from ASLT may be limited severely except in very simple food systems. Frozen foods such as pizzas may present problems with moisture migration: The moisture may diffuse from the pizza sauce, which has a higher a_w, into the crust containing a lower a_w, creating a pizza crust that is limp and soggy. Product-development scientists should use the results only as a guideline and must include as many storage conditions as possible to minimize prediction errors.

ASLT is a quick method that cannot replace the normal storage tests discussed next. Once it is verified that the extrapolation may be wrong (i.e., too large an error), then a careful look should be taken at the deterioration mode, the experiment design and procedure, the data collected, and the model developed. If the extrapolation underpredicts the true shelf life, then it becomes an economic concern; if it overpredicts, then reformulating may be necessary. If the shelf-life prediction indicates that the product meets the stability expectation, then the product has a chance of performing satisfactorily in the marketplace.

19.6 Confirmatory Storage Study

19.6.1 Basis

The difference in potential shelf life should be considered when scaling up from experimental test batches to pilot plant and then to full-scale production. Experience has shown that results of small-scale experiments in the laboratory may not be of much use for large-scale production (Graf and Saguy, 1991). Scale-up not only affects the processibility and quality of a food product, but it often alters its shelf life. Depending on the mode of failure and the food scientist's approach to inhibiting microbial growth and chemical reactions leading to deterioration, scale-up may increase or decrease shelf life. Because of the difficulty in predicting the precise effect of scale-up on food stability, a confirmatory shelf-life study

(Graf and Saguy, 1991) simulating real distribution conditions (if available) and using fluctuating temperatures or freeze/thaw cycles is often needed. Ideally the products to be tested are produced by the plant under typical steady-state manufacturing conditions. In this way the test can confirm the ability to achieve an acceptable product quality and shelf life of a realistic food (not hand-picked showcase samples) at its intended commercial storage temperature conditions.

19.6.2 Unique Procedures

Step 1. Identify real distribution chain conditions. The final decision as to the likely shelf life of a given frozen product must take into account that the product is not going to be stored at a steady temperature of −18°C (0°F) from the end of freezing until thawing for consumption or during end cooking. A typical frozen product will spend part of its shelf life in a bulk cold store, a refrigerated vehicle or container, a distribution store, a retail display cabinet or institutional frozen-food storage cabinet, out of refrigeration during the journey from the retail outlet to home, and in a home freezer or star-marked frozen-food storage compartment before being consumed in the frozen state, thawed, or end cooked. An essential part of any shelf-life determination in markets where marked (European Star) systems are in use is to assess the shelf life at −6°C (21°F) for Star-1 and −12°C (10°F) for Star-2 (Symons, 1994).

Step 2. Determine shelf life. Shelf-life determination can be done by using either a kinetic modeling study or a sensory evaluation. Figure 19.14 is the same shelf-life plot for frozen strawberries using the same data as presented by Guadagni (1968). Based on those data, a demonstrative calculation for a known time-temperature distribution can be conducted and is summarized in table 19.8. The calculations are based on the following: It is assumed that the order is zero and that at time zero the quality is 100%. At the end of shelf life the quality level is set to 0%. Thus the rate of loss is 100% divided by the shelf-life time as found in figure 19.14. For the first period, which is holding in the producer's locker for 250 d at −22°C, the rate of loss is 100/660 or 0.15% per day based on a shelf life of 660 d at the temperature. Given that the holding time was 250 d, the accumulated loss is 250 × 0.15 or 38% loss. As seen, there is a total of 66% loss up to the point of storage in the consumer's freezer. If consumed within 3 wk there is still 22% of the shelf life remaining. Another way of looking at it is with a loss of 66% of the total shelf life, there is 0.34 × 180 or about 2 mo of shelf life left at −13°C. The question would be whether that is enough.

Step 3. Correlate sensory with instrumental analysis. Since at this stage effective methods intended for routine quality control must be established, a correlation to some simply measured objective index must be made.

19.6.3 Applicability

This type of test is usually carried out toward the end of the product-development process, most certainly using product samples obtained during prelaunch runs.

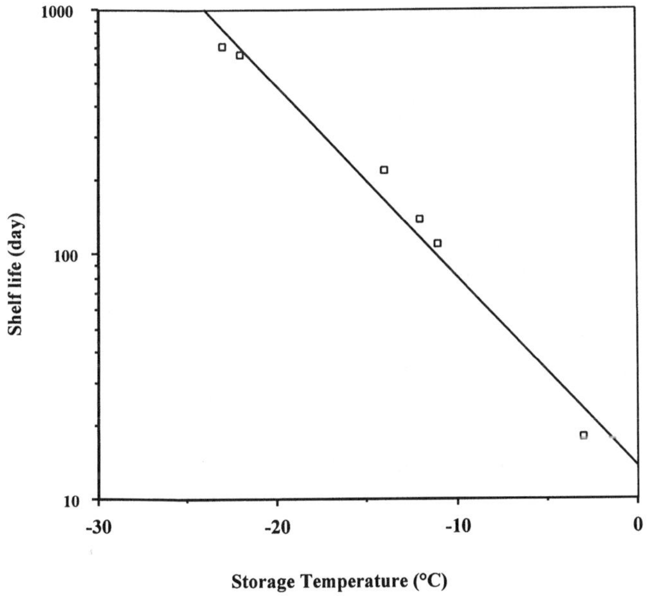

Figure 19.14. Shelf-life plot for frozen strawberries from Guadagni (1968)

Table 19.8. Loss of shelf life of frozen strawberries

Stage	Time (d)	Temperature (°C)	Shelf life θs	Loss rate per day = 100/θs	% Loss
Producer	250	−22	660	0.15	38
Transport	2	−14	220	0.45	1
Wholesale	50	−23	710	0.14	7
Transport	1	−12	140	0.71	0.7
Retail	21	−11	110	0.91	19
Transport	0.1	−3	18	5.56	1
Subtotal	324				66
Home freezer	21	−13	180	0.56	12
Total	344				78

Source: Jul (1984)

This will finally enable the shelf life to be set and the material, product, process, and packaging specifications to be finalized before a full-scale product launch. However, for frozen products, it is likely that product launch may occur before any confirmatory shelf-life determination is completed. In this case, an ASLT for products manufactured on the scale-up line can be employed simultaneously. Without ASLT, no prediction can be made for other variable storage conditions (e.g., another distribution chain as the market expands). In many cases sensory techniques must be used simultaneously to determine the practical shelf life. These tests require a large number of samples and panelists and are thus relatively expensive and may be passed over.

As discussed earlier, individual packages will experience a series of different environments corresponding to different locations during distribution and storage. Other packages from the same consignment will meet conditions that differ both in duration of storage and by traversing different parts of the distribution system, different temperatures, and varied humidities. Thus, one cannot assume that all packages will be stored for the same period under identical conditions or will react in the same way, which is why methods or devices that integrate time-temperature exposure are required. In 1996 the USDA and FDA both began to examine the potential for using such devices for monitoring safety during distribution (Transportation, 1996).

19.7 Ongoing Shelf-Life Monitoring

19.7.1 Basis

It is very important that the shelf life of every existing product be monitored on a regular basis and in a way compatible with the shelf life in question, production volume, and environmental conditions to which the product is exposed (or even abused) up to the point of consumption. During storage and transportation distribution phases, temperature conditions are often less than ideal and temperature abuses can occur. Reaction rates that are related to quality-loss characteristics are strongly temperature dependent, even when all the other factors are controlled through effective packaging and maintained at the desirable levels. Ideally, a cost-effective way to either maintain temperature or individually monitor the temperature conditions of food products throughout distribution is necessary in order to indicate their real quality state. If either one is achieved, it could lead to an effective quality control of the distribution, optimized stock rotation, and reduction of waste, as well as some meaningful information on the remaining shelf life of the products. This calls for control over pallets, the cases of product, and, finally, the individual packages at every step in distribution. Control over pallets generally is achieved by the computerized recording system of the producer in concert with the bulk storage facility. Control over cases is achieved in Europe by printing the mandatory open shelf-life date demanded on individual packets on the outer case in standard format. Time-temperature indicators (TTI) were developed to monitor individual packages but could also serve in both other functions (Taoukis et al., 1991).

19.7.2 Time-Temperature Indicators

Principle

Broadly speaking, a time-temperature integrator/indicator is a device or tag that can keep track of an accumulated time-temperature distribution function to which a perishable product is subjected from the point of manufacture to the display shelf of the retail outlet or to the consumer. The operation of a TTI is based on mechanical, chemical, or enzymatic systems that change irreversibly from the time of their activation. The rate increases at higher temperatures in a manner similar to most chemical reactions. The change is usually expressed as a visible response in the form of mechanical deformation, color development, and color movement.

TTIs can be classified into (1) Temperature indicators, which would indicate whether the package is below a predetermined temperature (Billet, 1983). (2) Temperature abuse indicators. For guidance on whether a frozen product has been exposed to temperature abuse, a small ball or cube of ice has often been placed in the frozen food pack; if it has disappeared or lost its shape, one would know that the product has been above 0°C. (3) Temperature abuse integrators, devices that integrate the time and severity of the abuse (i.e., the sum of the multiples of degrees in degrees Fahrenheit) over a critical point and the time of excess in minutes. In the frozen-food range, these devices are available for 0°, 10°, and 20°F but may be supplied with other melting points (Jul, 1984). (4) Electronic or mechanical time-temperature integrators. Much effort has been devoted to developing sophisticated electronic time-temperature integrators from which data could be downloaded and the accumulated effect on shelf life calculated for a certain food after its passage through the freezer chain (Olley, 1976). Many companies supply such instruments (Ryan, Cox, and Sensitech), but no company presently has put together spreadsheets to do the analysis as we have shown previously. In fact some researchers talk about the area under the *T*-*t* curve as being important (i.e., number of °C/d) when in fact it is not. One company (Remosys, UK) has an electronic integrator that displays the days of shelf life left for fresh fish. The square-root temperature dependence model is used. (5) Chemical time-temperature integrators show the physical or color changes as noted earlier and respond with the same temperature sensitivity (Q_{10} or E_a) as the food they are monitoring. Companies that supply such devices are 3M (St. Paul, Minn.), Lifelines (Morristown, N.J.) and VITSAB (Malmo, Sweden). Note that the latter company marketed TTIs formerly as the i-Point Company. A general review has been done by Taoukis et al. (1991).

Correlation of a TTI with a Food

Figure 19.15 shows a schematic of how a TTI can be used to predict the shelf life of a food product based on mathematical modeling. The left side of this scheme indicates that tag kinetics are required. As noted, these kinetics have been

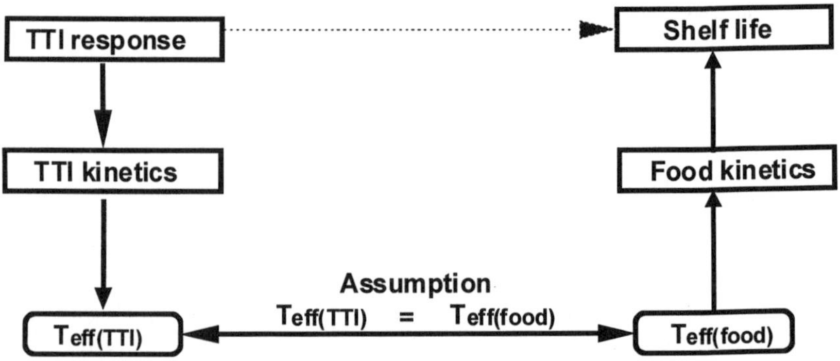

Figure 19.15. Application scheme of time-temperature indicators

determined for several types of TTIs. The tag basically predicts an effective temperature for a variable distribution, which is then used to predict the shelf life of the food using the right side of the scheme. This effective temperature is defined as the constant temperature that—if the tag remained there—would give the same measured response as for the variable temperature exposure. The assumption is made that the effective temperature of the TTI response is equal to that for the food deterioration based on the Arrhenius equation; the TTI response can then be integrated for a variable time-temperature distribution. The rate constant for a TTI response must also follow the Arrhenius theory. The $E_{A(TTI)}$ values of the commercially available indicators cover the range of the most important deteriorative reactions in foods, including those frozen (Fu and Labuza, 1992).

Other approaches such as equivalent time and equivalent point can also be used to correlate TTI response with food reaction based on the Arrhenius relation (Fu and Labuza, 1992). These approaches will be able to provide information on remaining quality or shelf life of the food at a targeted storage temperature after exposure to a variable time-temperature history. They can also be used to compare the effectiveness of different distribution chains.

Applicability

The usefulness of these sensors in monitoring quality depends on how well they can mimic changes in quality. It has been proven mathematically that the effective temperature of the TTI will not be equal to that of a food reaction unless the storage temperature is constant or the activation energy of the TTI is the same as that of the quality-loss reaction. Thus, the device will be able to indicate the shelf life remaining only for reactions whose activation energy is the same as or close ($\Delta E_a < 5$ Kcal/mol) to that of the TTI or some correction factor is needed (Taoukis et al., 1991). The reliability for TTI applications depends on the vari-

Table 19.9. Suggested technical standards and procedures of TTI for frozen foods

Test procedure	Technical standard
Temperature response test −25, −15, −10, −5, +5°C	Reproducible end point Maximum tolerance: ±6 d or 2.5% of the life span of TTI
Evaluation of kinetic parameters of TTI. Make an Arrhenius plot based on the data collected in the above temperature response test	Arrhenius plot and the values of activation energy and pre-exponential factor
Temperature cycling test Cycling range: −25° to −15°C, −15° to −10°C, −10° to −5°C, −25° to +5°C Cycling period: 10 times the interned life span of a TTI	The predicted rate constant at the reference temperature from the Arrhenius equation differs from the actual measured value by <10%.
Simulated field test	Applicable and reliable

Source: George and Shaw (1992)

ability in responses within the temperature range encountered and the confidence on the determined kinetic parameters, as well as the difference between $E_{A(food)}$ and $E_{A(TTI)}$. Table 19.9 lists some testing procedures for TTI to be used for monitoring frozen-food distribution.

In the case of frozen strawberries, a decrease in firmness and the measured amount of ascorbic acid was well correlated with the response of the i-POINT model 2340 (Singh and Wells, 1987; Wells and Singh, 1988). Singh and Wells also studied the potential applications of TTI for monitoring quality changes during frozen storage of hamburger (Singh and Wells, 1985) and other frozen foods (Wells et al., 1987). It should be noted here that they did not use the tags to predict shelf life, but merely a correlation between quality change and tag change. In fact there are no published data giving a comparison of a TTI prediction to the actual quality change of a food. Despite this, some food companies are using the Lifeline-type tags on some frozen foods such as turkey rolls (Taoukis et al., 1991).

The applicability problem involves deviations from the Arrhenius relationship for both the TTI and food, as well as the heat transfer problem, since a TTI tag is usually applied on a package surface. As a result, it does not reflect the temperature response in the center of a pallet load and the chemical and light sensitivity of TTIs. The cost of a TTI also depends on the quantity required. All of these aspects and their potential solutions have been discussed in detail by Taoukis and Labuza (1989a,b).

19.7.3 Other Devices

Alternatives to monitoring temperature during food distribution include the use of flexible, miniature electronic temperature-recording devices (LeBlanc, 1988). They record time-temperature information that can be displayed and processed at the receiving end by interfacing with a microcomputer. In 1989 a satellite tracking system (Geostar Satellite Tracking Service, Geostar Corp., Washington, D.C.) was introduced (Labuza and Taoukis, 1990), but it is only used for drug shipments because of the cost. McMeekin and Olley (1986) gave a list of other time-temperature recording systems. This will provide a valuable record for future use and contribute toward evidence of due diligence.

19.8 Conclusions

The monitoring and prediction of the shelf life of frozen foods is more complex than that of most other foods because of the introduction of a phase change, that is, the freezing out of water, which varies as a function of temperature. Thus application of typical physical chemical principals such as the Arrhenius relationship (log rate versus inverse absolute temperature) is limited. A second limitation is that the experimental temperature range is small, because one cannot use data above the freezing point. In addition the phase change of water to ice induces physical changes in the food, further complicating the situation and for which there are no mathematical models. Despite this, some simple mathematical models can be used to treat the experimental data in order to predict shelf life for a typical frozen-food distribution. This chapter has shown how the simple log shelf life versus temperature plot can be generated by choosing the right storage conditions and the right methods of analysis.

As noted, in order to predict, one needs analytical tools. From the consumer's standpoint, the ultimate tool is sensory acceptability. Because all of the consumers cannot be pleased 100% of the time, the end of shelf life becomes an economic decision-making process. Of all the sensory scaling tests reviewed, it was shown that the use of a Weibull hazard analysis test procedure (prediction of probability of failure) is recommended as the best method for prediction of end of shelf life.

References

Billet, K.G. 1983. The I-point TTM: A versatile biochemical time-temperature integrator. IIR Commission C2 Preprints, Proceedings of the 16th International Congress on Refrigeration. pp. 629–631, International Institute of Refrigeration, Paris.

Cairns, J.A., Oswin, C.R., and Paine, S.A. 1974. *Packaging for Climatic Protection.* Butterworth, London.

Dagerskog, M. 1974. Development of a computer program to simulate temperature distribution and time-temperature exposure of packed frozen food during handling in the distribution chain. Swedish Institute for Food Preservation Research (SIK), report TSBN 91–7200–014–8.

Fennema, O. 1975. Freezing preservation. In *Physical Principles of Food Preservation.* (Fennema, O.R., ed.) pp. 173–215, Marcel Dekker, New York.

Fennema, O. 1996. Water and molecular mobility. In *Food Chemistry*, 3d ed. (Fennema, O.R., ed.) Marcel Dekker, New York.

Fu, B. and Labuza, T.P. 1992. Considerations for the application of time-temperature integrators in food distribution. *J. Food Distri. Res.* **23**(1):9–17.

Fu, B. and Labuza, T.P. 1993. Shelf life prediction: theory and application. *Food Control* **4**(3):125–133.

Fu, B., Taoukis, P.S., and Labuza, T.P. 1991. Predictive microbiology for monitoring spoilage of dairy products with time-temperature integrators. *J. Food Sci.* **56**:1209–1215.

Gacula, M.C., Jr. 1975. The design of equipments for shelf life study. *J. Food Sci.* **40**:399–403.

Gacula, M.C., Jr. and Kubala, J.J. 1975. Statistical models for shelf life failures. *J. Food Sci.* **40**:404–409.

George, R.M. and Shaw, R. 1992. A food industry specification for defining the technical standard and procedures for the evaluation of temperature and time temperature indicators. Technical Manual no. 35. Campden Food & Drink Research Association, London.

Graf, E. and Saguy, I.S. (eds.) 1991. *Food Product Development: From Concept to the Marketplace*. Van Nostrand Reinhold, New York.

Guadagni, D.G. 1968. Cold storage life of frozen foods and vegetables as a function of time and temperature. In *Low Temperature Biology of Food Stuffs*. (Hawthorne, J. and Rolfe, E., eds.) pp. 399–412, Pergamon Press, New York.

Jul, M. 1984. *The Quality of Frozen Foods*. Academic Press, London.

Kramer, A. 1974. Storage retention of nutrients. *Food Technol.* **28**(1):50–60.

Kwolek, W. and Bookwalter, G.N. 1971. Predicting storage stability from time-temperature data. *Food Technol.* **25**(10):51.

Labuza, T.P. 1982. *Open Shelf Life Dating of Foods*. Food and Nutrition Press, Westport, Conn.

Labuza, T.P. 1984. Application of chemical kinetics to deterioration of foods. *J. Chem. Edu.* **61**:348–358.

Labuza, T.P. 1986. An integrated approach to food chemistry: illustrative cases. In *Food Chemistry*, 2d ed. (Fennema, O.R., ed.) pp. 913–938, Marcel Dekker, New York.

Labuza, T.P. and Riboh, D. 1982. Theory and application of Arrhenius kinetics to deterioration of foods. *Food Technol.* **36**(10):66–74.

Labuza, T.P. and Schmidl, M.K. 1985. Accelerated shelf-life testing of foods. *Food Technol.* **39**(9):57–62, 64, 134.

Labuza, T.P. and Schmidl, M.K. 1988. Use of sensory data in the shelf life testing of foods: principles and graphical methods for evaluation. *Cereal Foods World* **33**:193–206.

Labuza, T.P. and Taoukis, P.S. 1990. The relationship between processing and shelf-life. In *Foods for the 90s.* (Birch, G., ed.) p. 73, Elsevier Press, London.

LeBlanc, D.I. 1988. Time-temperature indicating devices for frozen foods. *J. Inst. Can. Sci. Technol. Aliment.* **21**:236.

Levine, H. and Slade, L. 1986. A polymer physico-chemical approach to the study of commercial starch hydrolysis products (SHPs). *Carbohydr. Polymer* **6**:213.

Levine, H. and Slade, L. 1988. Principles of "cryostabilization" technology from structure/property relationships of carbohydrate/water systems: a review. *Cryo-Lett.* **9**:21.

Lim, M.H. and Reid, D.S. 1991. Studies of reaction kinetics in relation to the Tg′ of polymers in frozen model systems. In *Water Relationships in Foods.* (Levine, H. and Slade, L., eds.) pp. 103–122, Plenum Press, New York.

Man, C.M.D. and Jones, A.A. (eds.) 1994. *Shelf Life Evaluation of Foods.* Blackie Academic & Professional, London.

Maxcy, R.B. and Wallen, S.E. 1983. Heterogeneity of samples as a problem in shelf-life prediction. *J. Food Prot.* **46**:542–544.

McMeekin, T.A. and Olley, J. 1986. Predictive microbiology. *Food Technol. Aust.* **38**:331–334.

Nelson, K.A. and Labuza, T.P. 1994. Water activity and food polymer science: implications of state on Arrhenius and WLF models in predicting shelf life. *J. Food Eng.* **22**:271–289.

Nelson, W. 1972. Theory and application of hazard plotting for censored failure data. *Technometrics* **14**:945–966.

Olley, J. 1976. Temperature indicators, temperature integrators, temperature function integrators, and the food spoilage chain. Australia National Committee Joint Meeting of Commissions C2, D1, D2, D3, and E1. International Institute of Refrigeration, Paris.

Refrigerated and Frozen Food Institute. 1973. Information bulletin 73-1.

Roos, Y.H., Karel, M., and Kokini, J. 1996. Glass transition in low moisture and frozen foods. *Food Technol.* **50**:95–107.

Schwimmer, S., Ingraham, L.L., and Hughes, H.M. 1955. Temperature tolerance in frozen food processing: effective temperatures in thermally fluctuating systems. *Ind. Eng. Chem.* **47**(6):1149–1151.

Simatos, D., Blond, G., and Le Meste, M. 1989. Relation between glass transition and stability of a frozen product. *Cryo-Lett.* **10**:77–84.

Singh, R.P. and Wang, C.Y. 1977. Quality of frozen foods: a review. *J. Food Proc. Eng.* **1**:97–127.

Singh, R.P. and Wells, J.H. 1985. Use of time-temperature indicators to monitor quality of frozen hamburger. *Food Technol.* **39**(12):42–50.

Singh, R.P. and Wells, J.H. 1987. Monitoring quality changes in stored frozen strawberries with time-temperature indicators. *Int. J. Refrig.* **10**:296–300.

Slade, L. and Levine, H. 1991. Beyond water activity: recent advances based on an alternative approach to the assessment of food quality and safety. *Crit. Rev. Food Sci. Nutr.* **30**:115–360.

Symons, H. 1994. Frozen foods. In *Shelf Life Evaluation of Foods.* (Man, C.M.D. and Jones, A.A., eds.) pp. 296–316, Blackie Academic & Professional, London.

Taoukis, P. and Labuza, T.P. 1989a. Applicability of time temperature indicators as shelf life monitors of food products. *J. Food Sci.* **54**:783–788.

Taoukis, P. and Labuza, T.P. 1989b. Reliability of time-temperature indicators as food quality monitors under non-isothermal conditions. *J. Food Sci.* **54**:789–792.

Taoukis, P.S., Fu, B. and Labuza, T.P. 1991. Time-temperature indicators. *Food Technol.* **45**(10):70–82.

Transportation and storage requirements for potentially hazardous foods. 1996. 61 Federal Register 59, 372–59, 382.

Van Arsdel, W.B., Coply, M.J., and Olson, R.L. (eds.) 1969. *Quality and Stability of Frozen Foods*. Wiley-Interscience, New York.

Varsanyi, I. and Somogyi, L. 1983. Determination of the shelf life of food products. *Acta Alimentaria* **12**(2):73–100.

Waltzeko, P. and Labuza, T.P. 1976. Accelerated shelf life testing of an intermediate moisture food. *J. Food Sci.* **41**:1338.

Wells, J.H. and Singh, R.P. 1988. Application of time-temperature indicators in monitoring changes in quality attributes of perishable and semiperishable foods. *J. Food Sci.* **53**(1):148–156.

Wells, J.H., Singh, R.P., and Noble, A.C. 1987. A graphical interpretation of time-temperature related quality changes in frozen foods. *J. Food Sci.* **52** (2):435–444.

Williams, M.K., Landel, R.F., and Ferry, J.D. 1955. The temperature dependence of relaxation mechanisms in amorphous polymers and other glass-forming liquids. *J. Chem. Eng.* **77**:3701–3707.

Wittinger, S.A. and Smith, D.E. 1986. Effect of sweeteners and stabilizers on selected sensory attributes and shelf life of ice cream. *J. Food Sci.* **51**(6):1463–1466, 1470.

Zaritzky, N.E. 1982. Mathematical simulation of the thermal behavior of frozen meat during its storage and distribution. *J. Food Proc. Eng.* **6**:15–36.

Zuritz, C.A. and Sastry, S.K. 1986. Effect of packaging materials on temperature fluctuations in frozen foods: mathematical model and experimental studies. *J. Food Sci.* **51**(4):1050–1056.

SECTION V

Strategies to Ensure Frozen Product Quality Today and Tomorrow

20

Consumer Acceptance of Frozen Foods

Denise W. Lam and Fredric Caporaso

20.1 Introduction

Over the last several decades, the American lifestyle has evolved from the traditional nuclear family consisting of one wage-earning adult, a homemaker, and children, to more complex situations involving two working parents, single working parents, and single, career-oriented adults. Americans today appear to be busier than ever, striving to maintain hectic schedules, all the while demanding their lives be made more convenient and simpler with new time-saving techniques. The frozen-foods industry has been quick to recognize this burgeoning and fast-growing niche and has taken advantage of the continuing need for convenience.

Development of frozen foods in the form of meals and entrées, snacks, side dishes, desserts, and so forth flew to new heights beginning in the late 1980s. Food processors scrambled to dominate the market by launching a wide array of similar frozen products which then proceeded to over saturate the market. By 1991, fierce competition between huge frozen-food marketers, such as ConAgra, Tyson Foods, and others, coupled with the recession and rampant promotional strategies, caused sales to fizzle (Tackling, 1992; Dyslin, 1993; Liesse, 1993). However, consumer demand for the convenience of frozen foods has not ceased, and the market is as promising as ever, albeit still very competitive. Processors are changing their tactics and looking at new approaches to satisfying this particular niche by studying trends, becoming more innovative, and listening closely to consumers.

20.2 Market Segmentation

Right from the start, frozen-food manufacturers identify their consumers and place them into target groups. Frozen foods are developed to cater to these target groups that comprise busy working men and women who have little or no time to cook. Yet food companies are aware that frozen-food consumers are no longer only women between the ages of 35 and 50 (Mullich, 1994). In fact, several groups have sprouted from one large target group to create many different niches, each with its own potential to lodge itself securely into a specific category of the frozen-food industry.

Male consumers, for example, actually make up about 30% of the consumers who do the shopping and prepare food (Mullich, 1994). Some companies have focused on this group by increasing the portion sizes of frozen entrées. Other targeted consumers include those who are considered mature (over 50 years of age). They are currently 25% of the American population (Gerson, 1991) and are continuing to grow as the baby boomer generation approaches the mature age. Healthy and more flavorful meals tend to appeal to these consumers who found Healthy Choice frozen entrées launched by ConAgra in 1989 quite acceptable (Gerson, 1991). Although this particular line of frozen food did not initially target the mature age group, it brought attention to that marketing possibility for processors.

Still other areas of focus include children. However, that focus has switched directly to parents who are more concerned with nutrition than fun and variety for their kids (Market, 1995). On the other hand, frozen items such as microwavable pizzas do not necessarily need the extra push toward parents. These will be purchased anyway, primarily for teenagers who consume them as snacks (Despite, 1994). Apparently, 90% of the consumers of frozen pizzas are children, and the only factor that may sway parents into buying alternative pizza products could be the consumer perception of freshness and better quality (Despite, 1994).

Furthermore, target groups should be sorted according to regional food preferences. Such preferences were brought about by an incredible ethnic diversity throughout all areas of the United States, by variety and availability of local produce, and also by economic and political events (Cousminer and Hartman, 1996). In a study done by Peavey et al. (1994), even consumer attitudes and perception of quality toward frozen fish differed between easterners and people living in the Midwest. Generally, those from the East, who had better access to fresh fish than midwesterners, tended to view frozen fish negatively whether or not they had had experience in using it. Some midwesterners, on the other hand, were not certain they could tell the difference between fresh and frozen fish and found them equally acceptable.

20.3 Quality in Frozen Foods from the Consumer's Perspective

The quality of any marketable item, whether it be food or something else, is determined by what consumers perceive as quality. In the case of frozen foods, quality is established by how well these foods meet consumer needs (Kimbrell, 1994a). Besides obvious issues relating the appearance and freshness of the product to quality, other factors exist. The key to defining these quality factors is in pinpointing what consumers want from a frozen-food product.

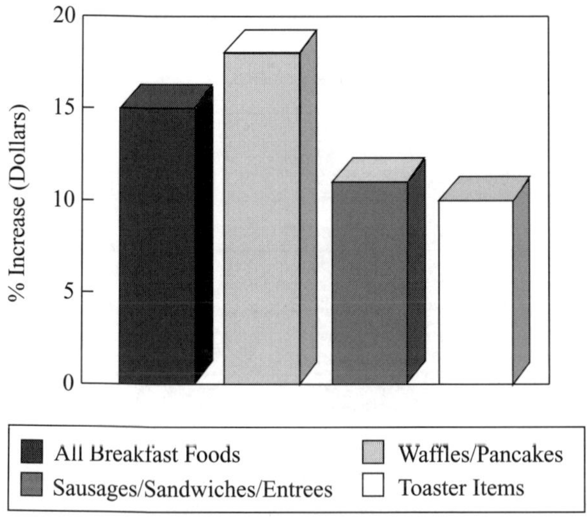

Figure 20.1. Frozen breakfast food sales (1989–1990) in the United States. (Adapted from Hollingsworth, 1991.)

First, consumers want convenience (Kimbrell, 1994a). Americans live life on the fast track today and barely have the time to grab a decent full meal. As shown in figure 20.1, the frozen breakfast food category has been one of the fastest growing in the market since 1980 (Kimbrell, 1991; Davis, 1992).

Because of time constraints, Americans have a greater tendency to eat breakfast at home rather than in restaurants. Also, increased sales in frozen breakfast foods are helped along by consumer recognition that breakfast is an important meal especially for children who go to school. Households with young families are especially keen on nutrition, and parents know their kids ought to eat a good breakfast (Kimbrell, 1991).

Lunchtime has also been made more convenient for consumers. Tight schedules and tight budgets result in more people lunching at their desks in the office (Vrzal, 1994). Microwaves located in the workplace allow consumers to have a hot meal in just minutes whether they tote homemade meals or store-bought frozen ones. Some workplaces even make available frozen products in vending machines at reasonable prices (Sloan, 1993).

Another quality factor is variety. Consumers constantly want to try new products (Kimbrell, 1994a), and food processors continue to search for new food ideals that will tantalize consumers and win their interest. During the frozen-food price wars of the early 1990s, variety was certainly not emphasized; every company in the running tried to promote the same types of product. Now these companies have grown wiser and concentrate on developing innovative products different from those sold by competitors. It is important to study current trends and

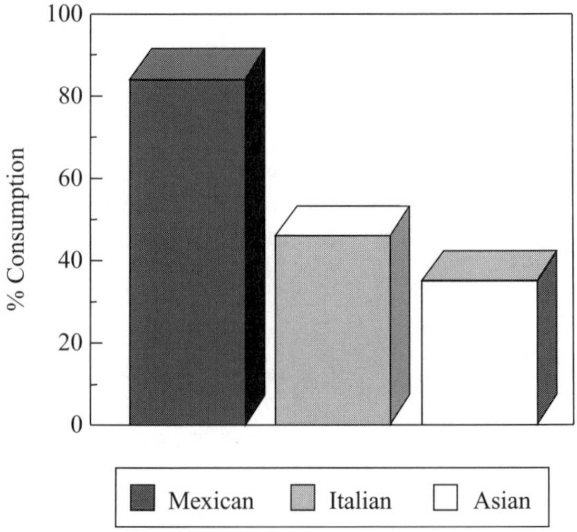

Figure 20.2. Ethnic food trends (1986–1996) in the United States. (Adapted from Postlewaite, 1996.)

ascertain what is popular and what consumer taste preferences are at the moment (Kimbrell, 1994b). Ethnic foods, for example, are an excellent place to begin, as this category has grown increasingly popular. Hence, the key is to be different and to offer the consumer more choices. Figure 20.2 shows the consumption of products in the three largest ethnic categories since 1986 (Postlewaite, 1996). In the Asian food segment, Thai cuisine alone has exhibited an annual growth rate of approximately 25% (Sloan, 1996).

Alongside a great variety of frozen products, companies are also focusing on quality by adding value to their foods. Value can be added to products in a variety of ways including (1) increasing the use of choice ingredients, for example, adding more white meat to frozen chicken pot pies (Kimbrell, 1994a); (2) cross-branding or cobranding, that is, marketing a frozen product made with a well-known and trusted name brand (Stilwell, 1995); (3) modification of the method of preparation, for example, handmade versus mass produced; (4) the use of all-natural flavors and/or ingredients; (5) selection of a product favorably perceived by the consumer, such as wild fish rather than farmed fish (Peavy et al., 1994), or untreated animal products rather than chemically treated products; and (6) improvement of nutrition through reduced fat or reduced calorie options (Kimbrell, 1994a). Due to increasing concerns about weight control, nutrition, and health, food processors are particularly interested in developing good-tasting frozen products with reduced fat, cholesterol, sodium, and calories. At this point, food science and creativity are truly challenged. Fat replacers have found their way into many frozen foods. These include vegetable proteins, meat analogs, and extenders in frozen meat products; bulking agents such as maltodextrins in frozen dairy desserts; and gums and microcrystalline cellulose in frozen sauces, to name a few. Products containing such ingredients must not only hold up against freeze-thaw stability tests, but must also win consumer acceptance and approval as well.

Environmental concerns may or may not be criteria for determining the quality of a frozen product in the eye of the consumer. Much of the concern for the environment has given way to convenience (Blalock, 1995). Now consumers are more occupied with whether they find the packaging to be easy to open and readily disposable after they have heated the food and eaten it directly from the container. Food processors have, on the other hand, discovered that keeping an eye out for the environment is well worth their while. By decreasing the amount of and improving packaging materials, they have been able to cut back on costs. Use of recyclable, lighter materials to manufacture eat-in-box containers eliminates the need to hassle with boil-in-bag packaging, decreases the amount of overall packaging waste, and presents the consumer with a nifty and convenient eat-out-of-and-toss container (Blalock, 1995). Environmental issues ultimately take a back seat to all the cost-cutting and convenience, but in the end, all three parties, consumer, company, and environment, come out ahead.

20.4 Consumer Education

The postprocess handling of frozen foods obviously plays a major role in maintaining the quality and acceptability of such products. Oftentimes, frozen products must withstand freeze-thaw cycles before even reaching the freezer in the consumer's home. According to Bannar (1991), a frozen product may undergo the following series of seven steps before being properly stored in the freezer: (1) the frozen product is packaged and stored in a freezer until ready to be shipped, (2) cases are hauled into trucks and delivered to a distribution warehouse loading dock where they are stored again in another freezer, (3) cases wait at the loading dock before being sent to a retail store, (4) the frozen items are displayed in the freezer cases of the store, (5) the consumer places the frozen item into a shopping cart, and (6) the consumer loads the groceries into the car where they may remain through several errands under heated conditions or hot weather before finally reaching the consumer's freezer. Interspersed between these six steps there may be glitches that prevent or delay proper freezer storage.

If none of the scenarios of improper handling can be effectively prevented before a product reaches a consumer, it is up to the consumer to ensure the remaining quality of the frozen goods. Usually shoppers are aware of the most basic handling practices and generally leave frozen foods as the last items to pick up before going to the cash register (Thayer, 1994). However, changes in American demographics and lifestyles raise concerns about safe food-handling practices. Families have decreased in size while the American population has increased. More families consist of two working parents or single working parents who often leave their children alone to shop and prepare food on their own (Knabel, 1995). Proper handling techniques are more likely to be compromised due to complacency and lack of training in correct procedures. Both adults and children need to be educated in frozen-food storage and handling. Frozen items, particularly meats, should be thawed in the refrigerator or microwave (Knabel, 1995); frozen entrées, snacks, and so forth should be placed immediately into the freezer before they have time to thaw and start to spoil or degrade; frozen desserts should remain outside the freezer for as short a time as possible to minimize crystallization; and items to be stored frozen for long periods of time should be tightly wrapped to prevent loss of moisture and freezer burn. According to Peavy et al. (1994), ice crystals in and around a package denote a product of lower quality.

Consumers need to be educated and must understand that the quality of their frozen goods might be adversely affected in their own kitchens. Currently consumers can refer to a number of resources that offer food safety information and give helpful tips and advice on proper handling techniques. Federal establishments such as the Centers for Disease Control (CDC), FDA, and USDA provide literature and/or audiovisual materials readily available to the public at little or no cost (Knabel, 1995). Furthermore, means of reaching consumers include training educators on food safety and targeting grade-school children. Education at an early age may be most effective because it takes place prior to development of poor food-handling habits. Also, distribution of information on food labels, leaflets, and recipe cards at grocery stores and computer networks, as well as educational programs intended for public television may prove to be worthy strategies of spreading food safety awareness (Knabel, 1995).

20.5 Future Trends

After having suffered through price wars and heavy competition in the frozen-food sector, processors are certainly reluctant to experience anything such as that again. By achieving a balance in their merchandising strategies, growth of individual frozen food categories is attainable (Kimbrell, 1994b). Companies are moving toward discovering new niches in the frozen-food industry in order to avoid oversaturation of the market and bombardment of consumers with a slew of products each of which is indistinguishable from the next. Development of a single category should be focused and managed from the perspective of a product development team. This team should consist of representatives from all working areas of a company—marketing, research and development, operations, purchasing, finance, and so forth—who will push forward to deliver what a consumer wants (Kimbrell, 1994c). Balance is maintained and a profitable product that meets consumer quality standards can be launched.

Food processors have begun to recover from the frozen-food price wars and are striving to find ways of regaining their profit losses. With the wars behind them, certain companies are reinvesting in brand equity. By focusing on the brands and introducing new and unique products, companies hope to be well received by consumers whose tastes are constantly changing. Overall, the future of frozen foods shows promise and growth, particularly since the busy lifestyles of the American population currently show no signs of slowing down. However, even if lifestyles were to adjust to a slower pace, the frozen-food industry would, one hopes, profit from the consumer's trust in quality brand names.

References

Bannar, R.J. 1991. The freeze/thaw rollercoaster. *Dairy Frozen Foods* **2**(4):37–38, 40.

Blalock, C. 1995. Lean and green. *Refrig. Frozen Foods* **6**(4):58, 60.

Cousminer, J. and Hartman, G. 1996. Understanding America's regional taste preferences. *Food Technol.* **50**(7):73–77.

Davis, S. 1992. Supper solutions. *Prepared Foods* **161**(4):89–90.

Despite some pressure, frozen pizza remains a hot category. 1994. *Food R&D* **1**(1):20.

Dyslin, J. 1993. Kids meal meltdown. *Prepared Foods* **162**(11):13.

Gerson, V. 1991. Creating foods for the mature. *Refrig. Frozen Foods* **2**(10):26–28, 30–32.

Hollingsworth, P. 1991. Frozen breakfast sales go sky high. *Dairy Frozen Foods* **2**(4):14–18.

Kimbrell, W. 1991. The frozen factor. *Dairy Frozen Foods* **2**(4):20–22, 24–25.

Kimbrell, W. 1994a. Healthy changes. *Food R&D* **1**(4):32–33.

Kimbrell, W. 1994b. Staying ahead of the game. *Food R&D* **1**(7):20.

Kimbrell, W. 1994c. The power of brand equity. *Food R&D* **1**(8):22, 24.

Knabel, S.J. 1995. Foodborne illness: role of home food handling practices. *Food Technol.* **49**(4):119–131.

Liesse, J. 1993. The rise and fall of frozen kids: how a game of war killed a category. *Advertis. Age* **64**(32):1.

Market watch: kids foods. *Refrig. Frozen Foods* **6**(2):10, 13.

Mullich, J. 1994. Going for the guys. *Food R&D* **1**(5):1, 8.

Peavey, S., Work, T., and Riley, J. 1994. Consumer attitudes toward fresh and frozen fish. *J. Aquat. Food Prod. Technol.* **3**:71–87.

Postlewaite, K. 1996. World of wonders: ethnic food processors offer a new world of regional cuisines, specialty niche products, and healthier foods. *Refrig. Frozen Foods* **7**(9):16, 18, 23.

Sloan, E. 1993. Food with get up and go! *Food Technol.* **47**(10):26.

Sloan, E. 1996. The top 10 trends to watch and work on. *Food Technol.* **50**(7):55–62, 64–68, 70–71.

Stilwell, S.B. 1995. Targeting trends. *Refrig. Frozen Foods* **6**(1): 46, 48, 50.

Tackling tough times. 1992. *Refrig. Frozen Foods* **3**(2):40–42.

Thayer, W. 1994. Frozen foods get mixed reviews. *Frozen Food Age* **42**(8):1, 20.

Vrzal, A. 1994. Savoring the main course. *Prepared Foods* **163**(5):89–90.

21

Marketing Frozen Foods

Eugene Jones

21.1 Introduction

Technological innovations as early as 1869 led to the commercial development and marketing of some frozen foods. Pigeon was one of the first commercially marketed frozen products and was soon followed by fish, poultry, other meats, liquid eggs, and fruits (Montgomery, 1995). These early products saw limited distribution through retail establishments because mechanical refrigeration was in short supply in required distribution channels. As late as 1938, about 60% of frozen foods was distributed in bulk form to ice cream manufacturers, bakeries, and preservers; about 30% went to institutional establishments; and the remaining 10% went to retail stores (Beck, 1993). Much of the bulk production was subsequently processed into other products, and many of these final products were sold as nonfrozen foods. Frozen fruits represented almost half of all frozen foods in 1938, and fruits also constituted the majority of bulk frozen foods.

Retail distribution of frozen foods gained its impetus from the development of commercially frozen vegetables in 1929. Food scientists worked closely with Clarence Birdseye, considered the father of frozen foods, to develop the scientific methods for blanching and processing frozen vegetables. These vegetables had strong retail and institutional appeal and they grew to overtake frozen fruits by 1943 (Frozen, 1985). Of course, the government rationing system for food during World War II aided the growth of frozen foods, especially that of frozen vegetables. Other major segments of the frozen-food industry during this time period consisted of poultry, seafood, and other meats. Prepared foods and concentrated fruit juices also became significant segments of the frozen-food industry by the late 1940s (table 21.1). All segments of the industry have achieved remarkable growth, and today the industry has an economic value of over $55 billion (Compendium, 1995).

While the frozen-food industry today is mature and has tremendous economic value, its growth and development present an interesting study in product market-

Table 21.1. Major frozen food sectors (millions of pounds), 1942–1994

Year	Fruits	Vegetables	Poultry	Meats	Seafood	Prepared foods	Frozen juices	Total
1942	275	220	70	12	41	5	–	623
1946	540	475	150	12	74	40	–	1291
1950	475	590	316	75	133	60	300	1949
1954	523	974	676	200	267	280	783	3703
1958	610	1702	1445	300	300	540	760	5657
1962	668	2261	2224	450	596	1152	1762	9113
1966	756	3415	2528	842	971	2354	1702	12568
1970	743	4294	2645	971	1265	3696	1775	15389
1974	756	5405	2342	1222	1405	3981	2307	17418
1978	631	6278	2428	1370	1740	5387	2699	20533
1982	818	7221	2895	1623	1686	5296	3054	22593
1986	862	7758	3358	1490	2185	5955	3008	24616
1990	1118	9493	4406	1372	2396	6879	2794	28458
1994	1094	12025	4932	1480	2387	6738	3640	32296

Source: *Quick Frozen Foods International,* various issues

ing. Development of frozen foods, unlike that of canned or dehydrated foods, required major capital investments at all levels of the marketing channel. Expensive and functional refrigeration was required in processing plants, warehouse storage facilities, transportation networks, retail stores, and institutional establishments. Moreover, successful marketing of frozen foods required each stage of the marketing channel to be able to maintain product temperatures at or near −18°C as products moved from one stage to the next (Symons, 1994). Adequate refrigeration throughout the marketing system, of course, was just the first obstacle to frozen-food marketing. The industry was further confronted with consumer resistance to frozen foods because of perceptions of poor quality and high prices (Beck, 1993). Further, even in the absence of consumer resistance, high energy costs, especially at retail establishments, placed a realistic constraint on the growth and expansion of frozen foods.

As suggested by the preceding paragraph, marketing frozen foods required industry officials to go beyond the traditional marketing focus of product, place, price, and promotion. These latter factors are critical focus points for firms wishing to carve out a market share from within an existing industry, but they are just a small part of a larger marketing mix for firms wishing to establish an industry. To be sure, the scope of marketing is extremely wide. Among the many factors firms focused on in the development of the frozen-food industry include product planning, warehousing, packaging, branding, selling, transportation, wholesaling, retailing, sales-force management, credit, transportation, advertising, pricing, and consumer analysis. During early phases of development, frozen-food marketers were most concerned about the adequacy of the refrigeration system throughout the marketing channels. Little was done to address consumers' perceptions of poor quality and high prices because producers lacked control over quality assurance measures, and consumers' perceptions of value were directly related to the price/quality factor. In other words, the value of frozen foods could improve only as the quality improved. As requisite investments were made to stabilize quality throughout the distribution system, advertising and product merchandising gained greater prominence among the marketing mix.

The enhanced role of advertising and merchandising in the marketing mix resulted partly from the rapid development and introduction of new products. Many of the aforementioned factors also played significant roles in the marketing process, but advertising and promotion provided the most effective way of communicating product attributes to potential buyers. Specifically, marketers attempted to communicate the ability of their products to provide consumer satisfaction. Such satisfaction is invariably related to a product's actual or perceived quality. For most food products, consumers have no way of assessing quality before purchase and consumption. Thus, to initiate product trial and possibly repeat purchases, marketers attempted to provide consumers some information about product attributes through advertising and promotion. For some products, a consumer's perception of quality and value are derived from advertising messages about product attributes. As an example, marketers of frozen foods attempted to address the high price image of frozen foods by emphasizing their convenience attributes. Indeed, studies show that the convenience attribute of frozen foods has now become the leading factor precipitating consumer purchases (Boehning, 1996). Advertising expenditures through many of the leading media (TV, radio, magazines, etc.) have grown over time, and many of the messages are now focused on the convenience, quality, nutritional, and value attributes of frozen foods (Martin, 1975). Equally important to the frozen-food industry, however, is the extensive merchandising and promotion that are cooperatively sponsored by manufacturers, wholesalers, and retailers (Boehning, 1996). In many respects, this cooperative promotion during a specified month of each year is the most successful effort of the frozen-food industry.

In addition to advertising, merchandising, and cooperative promotion, frozen foods have benefitted from a positive image of the freezing technique of food preserving. For quite some time, consumers have recognized that freezing is an excellent and often superior method of preserving foods. Indeed, frozen foods have enjoyed considerably more growth than other forms of preserved foods—canned and dehydrated. Given the growth and development of the frozen food market, the primary objective of this chapter is to describe the marketing methods that have led to the success of this segment of the food industry.

21.2 Early Marketing and Infrastructure Development (1929–1947)

Classical or early definitions of food marketing placed considerable emphasis on the need for firms to develop and enlarge consumer demand for their product offerings (Evans and Berman, 1992). Such emphasis was most appropriate for the early phases of the frozen-food industry. The simple description of food as frozen was perceived by many consumers as being synonymous with spoiled. Thus, early processors of frozen foods had to be concerned not only with developing the requisite freezing processes to preserve foods, but also with the use of acceptable and descriptive terminology to induce consumer trial. Descriptive terms like "frosted foods" and "quick-frozen foods" gained widespread usage during the early phases of development of frozen foods. Such terminology was necessary to distinguish frozen foods as developed by the scientific process of freezing rather than emanating from the simple process of changing an unfrozen product to its frozen state. For example, many consumers had observed fresh fruits and vegetables undergo weather-related freezing that caused them to become inedible in their thawed or unfrozen state. Terms such as "quick frozen" and "frosted" were intended to convey the message that the overall quality of frozen foods is also a function of the speed at which products are frozen (Symons, 1994).

Frozen-food processors attempted to address the quality perception problem by gaining control over the initial selection of foods for freezing and then controlling each subsequent step of handling, preparation, freezing, storage, and distribution (Montgomery, 1995). Making a quality product available, however, did not solve the consumer resistance problem. This problem was dealt with most directly by enterprising retailers. To induce trial and generate an enhanced prospect of continued purchases, retailers developed the practice of easing a small pack of frozen food into the shopping bags of consumers (Martin, 1974). This practice evolved from the well-known fact that many desirable product attributes are best communicated to consumers through the process of allowing them to experience the goods (Carlton and Perloff, 1990). Processors encouraged this retail practice, as they surmised that products initially demanded at retail would eventually be demanded at and carried by institutional establishments.

Frozen foods gained greater acceptance in the United States as the processing and distribution channels offered higher product quality, greater product selections, and expanded purchasing options for consumers. Product quality was enhanced as frozen-food processors developed and preselected food varieties for freezing, most notably fruits and vegetables. These initial product selections enhanced the quality image of frozen foods throughout the marketing channel. A few processors, particularly Birdseye, gained early and complete control over its product distribution. The most common distribution channels were institutional establishments, frozen-food stores, major department stores, and a few retail food stores (Beck, 1993). As refrigeration technology improved, retail food stores expanded their utilization of freezers and made them more accessible to consumers. Consumer purchases of frozen foods, however, were still constrained by insufficient in-home storage. To ameliorate this problem, independent firms offered specialized locker plants with consumer rental units for storing frozen foods.

Continued technological advancements for mechanical refrigeration in the early 1940s led to lower cost, increased use, and ownership of freezers in retail stores and consumer households. Coupled with these developments was an endorsement of some frozen foods by the American Medical Association as being highly nutritious, and a plea from the U.S. government for a 100% increase in frozen-food production (Beck, 1993). Additionally, food scientists concluded that some frozen foods could actually contain more vitamins than their fresh counterparts. For example, a fresh potato stored for 6 months could contain less vitamin C than an equivalent amount of frozen potatoes. This probability exists because, like all foods, stored potatoes lose nutrients over time. The confluence of these developments served to weaken consumer resistance to frozen foods and renew processors' expectations about prospects for future growth and development.

Developments in frozen-food marketing continued as freezer manufacturers produced self-service storage cabinets for retailers, and frozen-food processors developed expanded food lines and new product offerings. Chief among these new products were frozen concentrated orange juice and frozen potato products in 1945 (Montgomery, 1995). Additionally, prepared foods gained greater acceptance among consumers during the mid 1940s. Indeed many new processors of prepared foods entered this market as the government lifted price controls on most frozen foods in 1946. Unfortunately, these new entrants as well as many existing firms severely underestimated the product characteristics desired by consumers. In short, the prepared food sector placed little emphasis on product quality during this time period. No amount of advertising and promotion can be effective in changing consumers' quality perceptions when these perceptions result from consumption of low-quality products. Indeed, studies suggest that advertisement of branded products in the absence of quality control serves to diminish, not increase sales. A contributing factor to processors' underestimation of consumer product preferences was the strong, but war-related, growth of prepared foods during 1945–1946. This growth, however, was artificially induced by large government purchases and a rationing system that favored frozen foods (Beck, 1993).

Sales of prepared frozen foods thrived during the war period partly because consumers had limited access to other competing forms of foods, namely, fresh and canned. As competing forms of food became more readily available, consumers substituted these supposedly higher-quality foods for the lower quality (either real or perceived) frozen prepared foods. As evidence of this substitution, product sales of prepared foods fell from $40 million in 1946 to $5 million in 1947 (Compendium, 1995). Martin (1968) offers a very succinct explanation for this drastic decline. According to him, products were made available "without any scientific experimentation to ascertain proper methods for freezing, without any regard for quality, without any notion of proper packaging, pricing, product size, or marketing." As previously stated, such disregard for the development of products with attributes desired by consumers led to a near decimation of prepared foods by 1947. Moreover, product sales declined across all segments of the frozen-food industry as many consumers inferred the quality standards of other frozen foods from their experience with prepared frozen foods. Thus, by the end of 1947, the frozen-food industry faced large inventories of unsold foods, drastic price cutting among firms to stimulate sales, unfavorable consumer perceptions of product quality, and an overall uncertain future. Fortunately, some of the very products introduced during the mid 1940s would serve to propel the industry forward during the 1950s and even into the 1990s.

21.3 Market Expansion and Infrastructure Maturity (1948–1970)

Drastic declines in frozen-food sales during the late 1940s prompted the industry to reevaluate its production and marketing strategies. Specifically, the industry gained a greater appreciation for the interrelationship of production and marketing, recognizing that the ultimate objective of both should be consumer satisfaction. To this end, the industry attempted to address the issues of quality, packaging, pricing, product size, and overall marketing as identified by Martin (1968). Stated differently, the industry elevated consumer satisfaction to a much higher position in its marketing equation, recognizing that consumer quality perception problems were real, as opposed to being mere perceptions.

Marketing methods implemented to help reverse the decline in frozen-food sales included (1) improved product quality, (2) better merchandising, (3) more advertising and promotion, (4) better packaging and display, and (5) new product introductions (Beck, 1993). In addition, the industry benefitted from economic, social, and technological changes in the larger U.S. economy. These factors, together with others, served to generate an annual growth rate of over 13% for the 1948–1970 time period (fig. 21.1). A better perspective on this growth rate is gained when it is considered relative to the 3% rate of frozen-food growth since 1970.

Faced with a negative product image, accumulating inventories, and declining prices, many frozen-food packers were forced out of the industry as prices fell below their production costs. Firms surviving the short-run shakeout broadened their marketing approach to achieve long-run growth. Improved product quality resulted from efforts by firms to preselect high-quality products for freezing and then use appropriate freezing technology in the production process. Coupled with the production and selection process, of course, was improved packaging, handling, and merchandising of products. Attractive, lifelike color pictures replaced

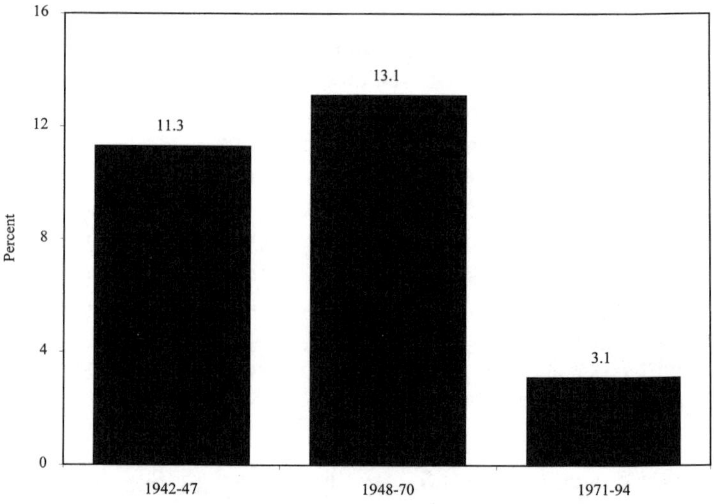

Figure 21.1. Growth (annual percentage) in production of frozen foods. (Source: *Quick Frozen Foods* and *Quick Frozen Foods International,* 1942–1994.)

the ever-present polar bears on many frozen-food products, and instructions for appropriate in-home use were printed on many of these packages. From a marketing standpoint, it may be said that consumers gained information that influenced their overall satisfaction and the total utility of frozen foods. Frozen-food processors and packers increased their utilization of distributors, allowing for wider distribution as well as greater control over the handling of their products. Lowering distribution costs, of course, also was a factor that prompted changes in distribution methods. To its credit, the industry accepted the challenge of weeding out production and marketing inefficiencies that developed and thrived during the war period (Beck, 1993).

Large retailers attempted to gain greater control over their cost by allocating a sizable share of their retail space to their own brands. During the early 1950s, three of the larger chain stores—A & P, First National, and Safeway—stocked their frozen-food departments up to 75% with their own brands (Williams, 1958). Such product displays made it difficult for smaller packers and even some larger ones to secure sufficient shelf space in retail stores. To operate their plants at or near production capacity, several frozen-food packers of national brands were forced to pack private label or chain store brands. This reliance on chain brands was part of an accepted view that the industry could grow and expand only through volume and consistently low prices. An alternative view that encompassed high margins and slow growth was perceived to be more accommodating to a larger number of firms in frozen-food marketing but less promising in enabling the industry to compete effectively with other food sectors, namely, the canned, dehydrated, and fresh markets (Williams, 1958). It should also be recog-

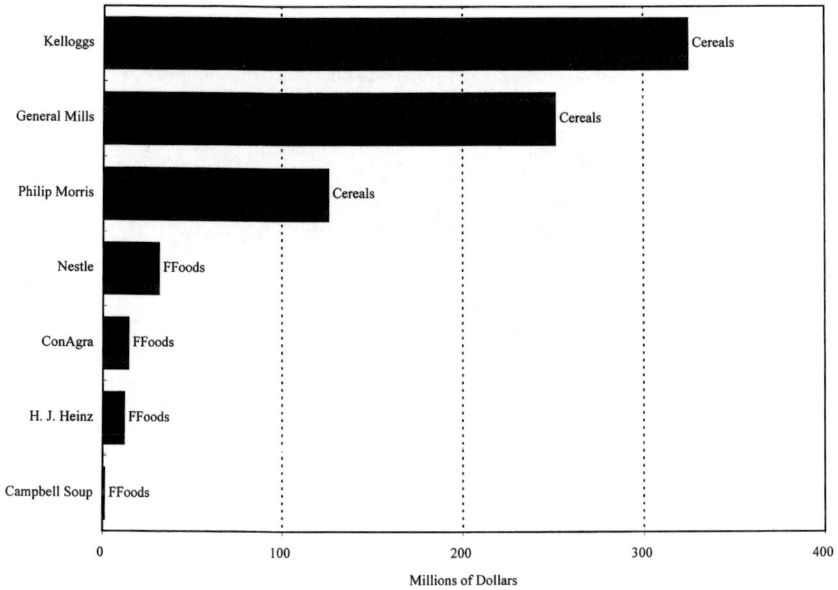

Figure 21.2. Advertising expenditures on frozen foods (*Ffoods*) and cereals by selected firms in 1992. (Source: *Leading National Advertisers Ad Dollar Summary,* 1993.)

nized that high energy costs for the display of frozen foods limit the amount of available space and therefore the extent of product differentiation among frozen foods as compared to other less costly product categories. With a relatively smaller amount of product differentiation, advertising expenditures on frozen foods are also likely to be smaller relative to other product categories. For example, as shown in figures 21.2 and 21.3, the leading breakfast cereal and snack food producers spend more than three times as much on advertising as the leading frozen-food producers.

Improved marketing techniques also included more in-store demonstrations, increased availability of economy packs for larger families, introduction of many new products for retail establishments, and the introduction of portion-control products for institutional establishments. Undergirding all of these marketing changes was an increase in advertising and promotion of frozen foods, especially by larger companies. Indeed, advertising messages, particularly during the 1950s, are credited with being the key marketing element that allowed frozen foods to gain widespread consumer acceptance. As one example, Campbell Soup gained almost instant acceptance of several frozen soups that it introduced and heavily advertised during the mid 1950s (Williams, 1958). These products excelled partly because they did indeed possess the attributes promoted in their advertisements, namely, quality and convenience. Acceptance of frozen foods at institutional establishments, however, was due less to advertising and more to considerations of

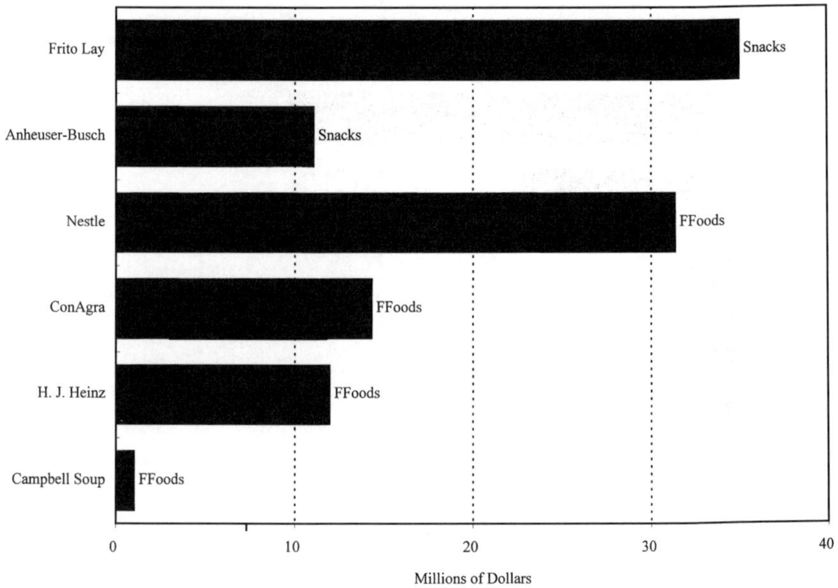

Figure 21.3. Advertising expenditures on frozen and snack foods by selected firms in 1992. (Source: *Leading National Advertisers Ad Dollar Summary,* 1993.)

potential savings on labor costs, and uncertain prices and yields for fresh commodities.

Further aiding the growth of frozen-food marketing during this period was the growing acceptance of frozen concentrated orange juice at retail establishments, and the near instant acceptance of frozen potatoes at institutional establishments (Beck, 1993). Product advertising of frozen concentrated juice, its year-round availability, and increased ownership of in-home freezer space served to stimulate frozen orange juice sales. Further, emphasis on quality and the convenience attributes of frozen concentrated orange juice served to convince consumers that the product was indeed of high quality and good value. Frozen potatoes proved to be a natural product for institutional establishments, yielding satisfying taste and reduced labor costs for food service establishments. The establishment of the first McDonald's in 1955 gave birth to a fast-food industry that would prove instrumental in the success of frozen potatoes, especially frozen french fries. Indeed frozen french fries are given much of the credit for revitalizing the frozen-food industry after the 1947 debacle (Williams, 1958). Of course, two of the driving forces behind the growth of fast foods were the increase of women in the labor force and rising income. Growth in the numbers of women in the labor force increased the demand for convenience foods, and growing income made these foods more affordable.

Technological innovations occurred at all levels of the marketing channel. Processors and frozen-food packers gained access to more efficient and larger freezers; distributors and transporters of frozen foods could select from a range of new shipping containers that greatly reduced packing costs and eliminated many potential product damages; retailers gained access to more efficient and better-designed coolers for displaying their products; institutional establishments and frozen-food packers gained access to freezers that could maintain complete stability in product temperatures; and consumers could select from an expanding range of more efficient and better-designed freezers (Beck, 1993). Further, technological advances made refrigerated trucks more affordable; and this allowed for expanded territory coverage by many distributors. Mishandling of products was greatly reduced or eliminated entirely, which increased the benefits to the industry. In essence, all of these factors served to enhance quality, and this enhanced quality served to substantiate advertisers' claim that frozen foods possess convenience and quality attributes.

21.4 Product Segmentation and Proliferation (1971–Present)

Despite several years of marketing focus on volume and low margins, frozen foods entered the decade of the 1970s with a high-price product image. Market research studies continued to confirm that most consumers considered frozen food to be more expensive, relative to quantity, than canned, dry, or fresh foods (Martin, 1974). To alter or alleviate this perception and thereby expand frozen food use, the industry implemented a broader marketing approach that attempted to educate consumers and food service operators on the benefits of frozen foods. General areas of focus for this marketing education included convenience, nutritional value, freshness and quality, ease of storage and preparation, satisfaction of taste, and an economical price. Indeed, with respect to price, many frozen-food processors introduced product lines with differing prices for various market segments (Scientific, 1987). Ironically, the highest priced and highest quality product segment invariably enjoyed the highest sales, despite the high-price image for frozen foods. This trend suggests that products offering the best value are not necessarily those offering the lowest price. Equally important, it suggests that marketers have a role to play in communicating the value of products to consumers.

The convenience image of frozen foods was enhanced greatly during this period as many prepared foods were introduced in response to technological advances in the production of microwave ovens. New products introduced in response to a 21-fold increase in household ownership of microwave ovens during 1970–1975 included a host of microwaveable vegetables, pizzas, breakfast entrées, popcorn, meat pies, and many low-calorie products (Scientific, 1987). Competing with the marketing image of frozen foods as convenience products, of course, were the growth of fast-food restaurant sales and the growth of prepared foods at retail supermarkets. In response to this competition, frozen-food producers attempted to market their products as convenience foods with good value.

As technological advances for in-home preparation of frozen foods continued to provide opportunities for new product developments, the production of frozen foods became increasingly concentrated in the hands of fewer and fewer firms. While competition among these firms generated new product offerings, it also generated considerable product proliferation as firms competed among themselves for market shares. As firms marketed products with similar attributes, price promotion became a common marketing theme (Liesse, 1992). Although such price promotion often generated desired sales, it sometimes did so with less than satisfactory profits. As an example, ConAgra gained a 40% market share of the healthy prepared foods sector during the 1988–1992 period, but its profits as well as that of its competitors were less than satisfactory (Liesse, 1992). Again, limited display space for frozen foods often forces retailers to replace slower-moving products with faster-moving ones.

Further complicating the marketing of frozen foods during this time period was the growing health consciousness of consumers, and an increasing desire to add more fresh fruits and vegetables to their diets. In response to this changing environment, the frozen-food industry attempted to market frozen foods as being as fresh as, and perhaps even fresher than, fresh fruits and vegetables. Additionally, the frozen-food industry convinced the National Cancer Institute and the Produce for Better Health Foundation that frozen fruits and vegetables should also come under an approved umbrella of nutritional fruits and vegetables. As such, the 5-a-Day for better health program now includes frozen as well as fresh fruits and vegetables.

Although new prepared food products and the addition of frozen fruits and vegetables to the 5-a-Day promotional program have aided the growth of frozen foods, frozen potatoes have been the leading growth product during this period. During the 1970s, frozen potatoes enjoyed a growth rate of over 7% (fig. 21.4). Much of this growth was due to the popularity of frozen french fries at institutional establishments. As early as 1960, frozen potatoes replaced peas as the number one frozen product in the institutional market. By 1977, frozen potatoes had propelled the dollar value of frozen foods at institutional establishments ahead of that at retail establishments (fig. 21.5); by 1979, poundage use of frozen foods at institutional establishments was essentially equal to that at retail establishments (fig. 21.6). Continued growth in the use of frozen potatoes and other frozen foods at institutional establishments has completely reversed the 60:40 retail:institutional distribution of frozen foods in 1970 to a 40:60 retail:institutional distribution in 1994. New marketing emphasis on convenience undoubtedly contributed to this phenomenal growth of frozen foods at institutional establishments. Other contributing factors included growth of fast-food establishments, increased labor force participation of women, and higher income.

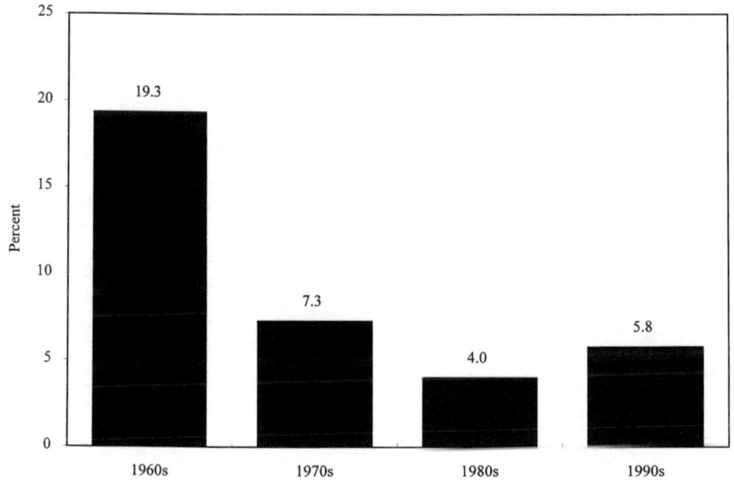

Figure 21.4. Growth (annual percentage) in production of frozen potatoes. (Source: *USDA Potato Facts,* 1996.)

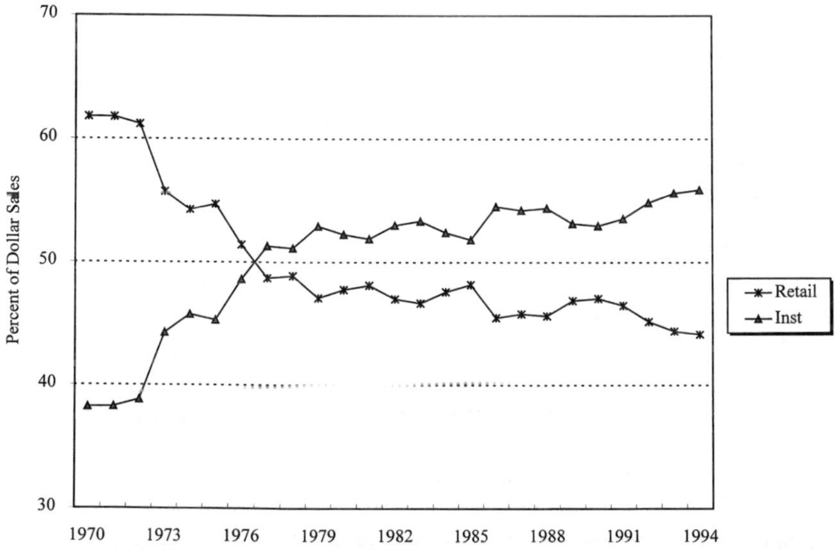

Figure 21.5. Retail and institutional distribution of frozen foods by dollar value. (Source: *Quick Frozen Foods International,* 1970–1994.)

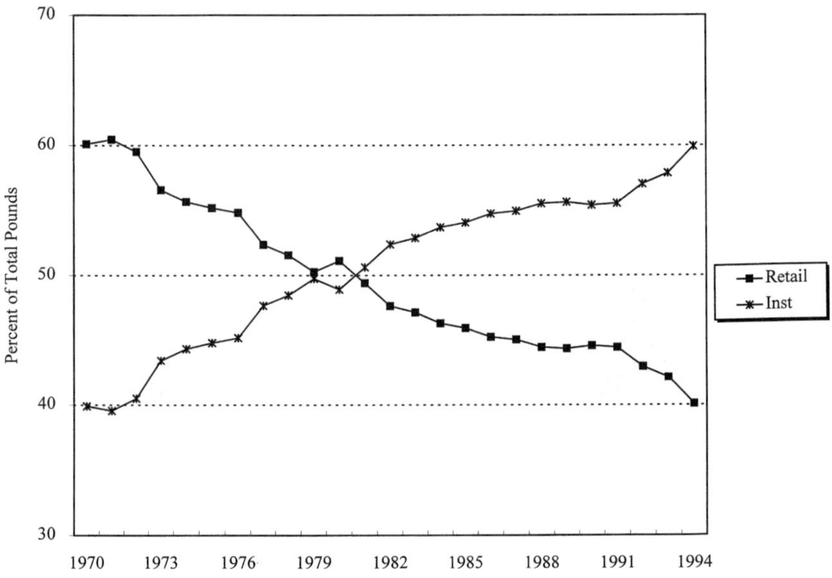

Figure 21.6. Retail and institutional distribution of frozen foods by poundage. (Source: *Quick Frozen Foods International,* 1970–1994.)

While frozen potato products were a near-perfect match for institutional establishments, other frozen products were less well received at both institutional and retail establishments. At the retail level, many frozen foods, especially prepared foods, performed below consumers' expectations. Some of these problems revolved around improper distribution of otherwise quality products, but others related to poor taste, texture, flavor, and overall quality. As an example, frozen entrées and dinners consisting of two or more food products would often heat unevenly. Again, it must be emphasized that advertising and promotion are not effective in addressing real quality problems. Institutional establishments were reluctant to accept some frozen-food products because they were perceived to be nothing more than retail products in larger packages. These establishments demanded higher-quality products, greater variety, portion control, more flexibility in final product preparation, and expanded suggestions for menu planning (Angione, 1973).

To provide better service to the retail and food service markets, the frozen-food industry recognized differences between these two market segments and attempted to meet these specialized needs and preferences. Processors serving both sectors developed product lines and marketing information for the respective sectors. In many instances, a frozen-food processor specialized in a particular sector. For example, many frozen potato processors specialized in servicing the food service sector, with limited distribution of retail potato products. Indeed, one potato processor, Ore-Ida, controls over 54% of the retail potato market (Spethmann, 1993). Such marketing specialization allows firms to more easily provide consumers with quality products possessing the desired product attributes. Likewise, firms specializing in supplying the institutional market can more easily provide food-service operators with consistent quality and other desired product characteristics. Of course, products containing the attributes desired by end users can be marketed easily and effectively. Indeed, the overall success of the frozen-food industry is undoubtedly due to the fact that manufacturers have endowed their products with quality attributes that can be marketed successfully.

21.5 Summary and Conclusions

The frozen-food industry has evolved from one in which frozen fruits represented the dominant sector to one in which fruits is now the smallest of seven frozen-food sectors. Unlike frozen fruits, frozen fruit juices have enjoyed considerable growth, with frozen concentrated orange juice leading this growth. Indeed, frozen concentrated orange juice played a crucial role in the revitalization of the frozen-food industry after a sharp downturn following the war. Growth of frozen fruits was undoubtedly limited by the poor freezing ability of many fruits. Yet, the early experience of frozen-food packers with fruits and vegetables played a significant role in the ultimate implementation of quality control measures for frozen foods. By preselecting fruits and vegetables for freezing, packers and processors discovered a direct relationship between the quality of the initial products and the quality of the final products. To help erase a poor quality image for frozen foods, frozen-food packers developed marketing messages that emphasized the quality control measures utilized by the industry to produce high-quality foods. These messages, however, were not effective until the products actually possessed these quality attributes.

Some of the quality problems for frozen foods resulted from the inability of frozen-food packers to control the distribution or marketing of frozen foods. To preserve quality products, storage facilities are required that can keep products at temperatures of 0°F (−18°C) or below at all levels of the marketing channel. Unfortunately, the magnitude of required capital investments at all levels of the marketing channel sometimes generated some weak links along the channel. In many ways, these significant capital investments serve to distinguish the frozen-food industry from other, less capital-intensive industries.

As products moved through processors' plants, warehouse storage facilities, transportation networks, distribution centers, retail stores, institutional establishments, and even consumers' storage facilities, quality could easily deteriorate at some point along the channel and the packers would have no way of knowing when and where this process occurred. Of course, some of the quality image for frozen foods resulted from factors that packers could control. Some frozen-food packers failed to implement quality control standards for their products during the strong, war-related demand for frozen foods in the mid 1940s, and this production problem served to prolong a quality perception problem for the industry. Drastic declines in sales after the war period served to remind producers of the importance quality plays in the preference function of consumers. This period also highlighted the fact that no amount of advertising can influence consumers' perception of quality when these perceptions emanate from personal experiences with low-quality products.

Several factors played significant roles in helping the industry overcome its quality perception problems and achieve significant growth. Key among these factors were (1) improved quality of frozen foods, (2) a marketing educational program that stressed the convenience and nutritional aspects of frozen foods, (3) increased advertising and promotion of frozen foods, and (4) the introduction of many new frozen-food products. In addition, the growth of fast-food restaurants and an increased recognition of the role that frozen potatoes could play in reducing labor cost served to stimulate demand for a product that was advertised and promoted for its convenience attributes. Further, frozen concentrated orange juice, heavily advertised as a convenience product in retail markets, gained strong consumer acceptance and helped to provide parallel growth in the retail market as frozen potatoes were providing strong growth in the institutional market. Further aiding the growth of frozen foods and diminishing their quality perception problem was the introduction of many high-quality prepared food products in the early 1970s. Perhaps a final plus for the image of frozen foods occurred in 1994 when frozen fruits and vegetables were accepted as part of a national nutritional program. This program encourages all Americans to eat a minimum of 5–9 servings of fruits and vegetables per day.

Continued growth is expected for frozen foods, but high energy costs for display space are likely to keep advertising expenditures for frozen foods below that of many other food categories. Yet, if frozen-food manufacturers can continue to add convenience, quality, nutritional, and value attributes to frozen-food products, an accelerated rate of growth is quite possible.

References

Angione, J. 1973. The serendipity in new institutional frozens. *Quick Frozen Foods* **August**:15.

Beck, S. 1993. Events, happenings, news, and photos of the frozen food industry. *Frozen Food Dig.* **April/May**:73–193.

Boehning, J.C. 1996. Consumers: frozen are on quality par with fresh foods. *Supermarket News* **May 6**, p. 128.

Carlton, D.W. and Perloff, J.M. 1990. *Modern Industrial Organization.* HarperCollins Publishers, New York.

Compendium of leading global frozen food facts and figures, A. 1995. *Quick Frozen Foods Int.* **October**:A1–A20.

Evans, J.R. and Berman, B. 1992. *Marketing,* 5th ed. Macmillan Publishing Company, New York.

Frozen food almanac. 1985. *Quick Frozen Foods* **October**:31–52.

Liesse, J. 1992. 100 Leading national advertiser. *Advertis. Age* **September**:24.

Martin, S. 1968. Scientific ingenuity, marketing skill spur frozen foods. In *The Freezing Preservation of Foods*. (Tressler, D.K., Van Arsdel, W.B., and Copley, M.J., eds.) p. 6, AVI Publishing, Westport, Conn.

Martin, S. 1974. Ending frozen food illiteracy. *Quick Frozen Foods* **March**:16.

Martin, S. 1975. The first 50 years of frozen foods. *Quick Frozen Foods* **July**:12.

Montgomery, N.D. 1995. NFFA, frozen food industry celebrate anniversaries in 1995. *Frozen Food Digest* **February/March**:10–72.

Scientific Ingenuity, Marketing Skill Spur Frozen Food Developments. 1987. *Quick Frozen Foods Int.* **October**:183–189.

Spethmann, B. 1993. Ore-Ida ups spud share bid. *Brandweek* **May 24**:8.

Symons, H. 1994. Frozen foods. In *Shelf Life Evaluation of Foods.* (Man, C.M.D. and Jones, A.A., eds.) Blackie Academic and Professional, Glasgow.

Williams, E.W. 1958. A biography of frozen foods covering 20 years. *Quick Frozen Foods* **August**:135–286.

22

Total Quality Management for the Frozen-Food Industry

John G. Surak and John F. Haury

22.1 Introduction

The primary focus of this book is frozen foods and the inherent product characteristics associated with quality in frozen foods. This chapter focuses on methods businesses may use to achieve product quality. The approach serves as a method to link the customers' perception of quality to the actual quality attributes of a frozen-food product. Many corporations are achieving this objective by implementing the concepts of Total Quality Management (TQM).[1]

22.2 The Total Quality Management Process

TQM describes an overall business management strategy and process that aligns employees with the following objectives:

- Increase value to customers or increase perceived quality to delight customers now and in the future.
- Continuously improve the quality and productivity of *all* business processes and reduce or eliminate waste.

Perceived quality[2] or customer-defined quality is the number-one attribute that drives market share. Perceived quality differs from objective quality[3] or technical

[1]The implementation of TQM has not been completely successful in the United States. The appendix to this chapter (section 22.6) discusses the successes and failures of the TQM process in businesses.

[2]Perceived quality is the total package of product attributes and/or services that influence customers in making their purchasing decisions.

[3]Objective quality is the quality that is typically measured by the company. It is composed of a number of attributes which include conformance, features, performance, reliability, durability, and serviceability (Garvin, 1988).

excellence. The PIMS (Profit Impact of Market Strategy) (Buzzell and Gale, 1987) program identified the following benefits for achieving superior perceived quality:

- Stronger customer loyalty
- Greater product repurchase
- Reduced vulnerability to product price wars
- Ability to command higher relative price
- Reduced marketing costs
- Improved market share

The amount of waste in an organization can be great. H. James Harrington, past president of the American Society for Quality Control (1987) estimates that the typical poor-quality costs in a manufacturing company that does not have a TQM process are 20–30% of sales dollars. In a service company these figures are 30–50% of sales dollars.

The seven categories of the Malcolm Baldrige National Quality Award describe the components of the TQM process (fig. 22.1 and table 22.1) (NIST, 1997). These seven categories interlink to form a dynamic and circular process.

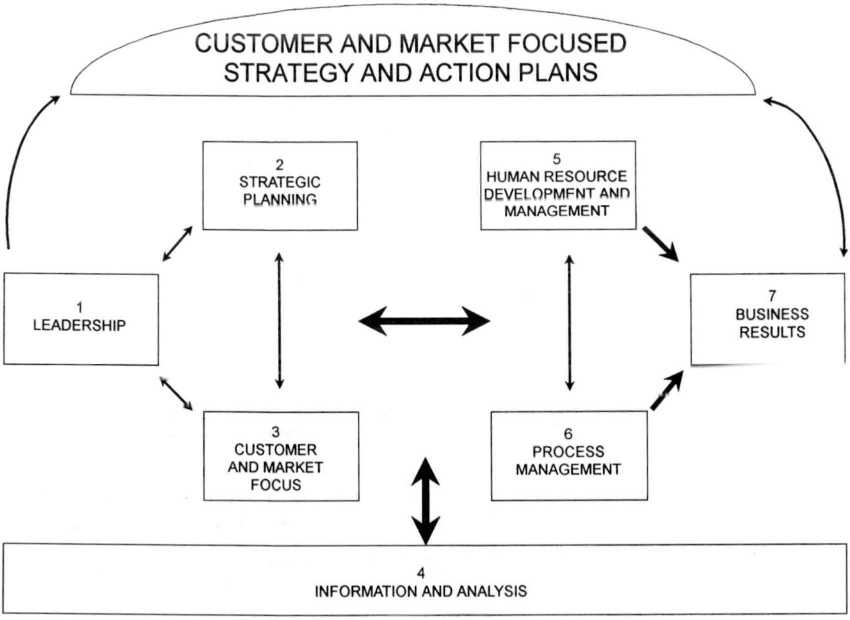

Figure 22.1. The Baldrige framework for TQM. (Source: NIST, 1997.)

Table 22.1. Description of the categories for the Malcolm Baldrige National Quality Award

Number	Category	Description
1.0	Leadership	The Leadership category examines senior leaders' personal leadership in creating and sustaining values, company direction, performance expectations, customer focus, and a leadership system that promotes performance excellence. Also examined is how the values and expectations are integrated into the company's leadership system, including how the company continuously learns and improves, and addresses its societal responsibilities and community involvement.
2.0	Strategic Planning	The Strategic Planning category examines how the company sets strategic directions and how it determines key action plans. Also examined is how the plans are translated into an effective performance-management system.
3.0	Customer and Market Focus	The Customer and Market Focus category examines how the company determines requirements and expectations of customers and markets. Also examined is how the company enhances relationships with customers and determines their satisfaction.
4.0	Information and Analysis	The Information and Analysis category examines the management and effectiveness of the use of data and information to support key company processes and the company's performance management system.
5.0	Human Resource Development and Management	The Human Resource Development and Management category examines how the work force is enabled to develop and utilize its full potential, aligned with the company's performance objectives. Also examined are the company's efforts to build and maintain an environment conducive to performance excellence, full participation, and personal and organizational growth.
6.0	Process Management	The Process Management category examines the key aspects of process management,

		including customer-focused design, product and service delivery processes, support processes, and supply and partnering processes involving all work units. The category examines how key processes are designed, effectively managed, and improved to achieve higher performance.
7.0	Business Results	The Business Results category examines the company's performance and improvement in key business areas: customer satisfaction, financial and marketplace performance, human resources, supplier and partnership performance, and operational performance. Also examined are performance levels relative to competitors.

Source: NIST (1997)

Leadership is the driving force for the implementation of TQM. TQM achieves a mature level only when the executives lead by example or by "walking the talk." A critical part of the leadership process is to make the vision, mission, and guiding principles a day-to-day guide that assists all employees in their decision-making processes. In addition, executives must focus their quality improvement activities on Big Q quality issues rather than Little Q quality issues (table 22.2). The categories of the Baldrige criteria form a dynamic framework for business that is composed of three basic elements: (1) a customer- and market-focused strategy and action plan; (2) a system that comprises leadership, strategic planning, customer and market focus, human resources development and management, process management, and business results; and (3) information and analysis that allow management to use a fact-based decision-making process. As part of the development of the management information system, operational measures and quality improvement measures need to be linked to the customer. These measures must also provide real-time information to determine if the quality improvement process is aligned with business strategies and operational planning to produce the desired results. TQM also focuses on using a systematic approach to increase the individual's potential and creativity and, subsequently, employee empowerment. In addition, TQM allows companies to develop a new relationship with suppliers. Instead of seeing customers and/or suppliers as adversaries, the company develops partnerships with both the customers and suppliers. This is done to effectively and efficiently meet the final customer's needs. Progress is measured by assessing quality and productivity improvement, operational performance, waste reduction, and financial results. Finally, the Baldrige criteria stress that all of the company's actions must point toward business results. Business results are a composite of customer, financial, and nonfinancial performance results.

Table 22.2. Big Q vs. Little Q

Big Q	Little Q
• All products and services produced by the company whether or not they are sold.	• Just the manufactured goods that are sold.
• All processes including business support processes.	• Processes directly related to manufacturing.
• Quality is a business problem.	• Quality is a technical problem.
• All stakeholders in the corporation.	• Customers who buy the product.
• Quality goals are part of the business plan.	• Quality goals are part of the production plan.
• Improvement is directed at company improvement. Evaluation is measured by responsiveness to customers' needs.	• Improvement is directed at departmental improvement.
	• Evaluation is measured as conformance to specifications, procedures, or standards.
• Training for quality management is company wide.	• Training for quality management is concentrated in the Quality department.
• Quality is coordinated by a council of senior executives.	• Quality is coordinated by the quality manager.

Source: Juran (1989)

TQM emphasizes a process-focused management as a means to achieve increased business results, rather than a results-focused management system (Easton, 1993). In the frozen-food industry, TQM emphasizes assessment of the development, production, and delivery of frozen foods as a single process rather than as a series of independent or individual steps. Results-focused management can be defined as setting objectives, obtaining feedback, and creating incentives so that the company can achieve its goals. The results belong to the individual. The individual's role is to determine and execute the best way of achieving the company's goals. Management's purpose is to hire the best people, define the goals, establish incentives, allocate resources, and evaluate results. Process-focused management stresses that results belong to the process and the process belongs to the organization. The individual's role, at all levels in the organization, is to develop, improve, manage, and operate the process. The specific mix of responsibilities is dependent on the individuals' job responsibilities and assignments and level within the organization. Thus, TQM emphasizes the development of an approach that consists of plans,[4] strategies, and methods. For example, if a

[4]William A. Golomski compared and contrasted the definition of plans and strategies in the following manner: Plans are the activities a person undertakes when the individual knows exactly how to

frozen-food company has limited warehousing capacity, the company must understand the actual customers' changing needs and expectations. This knowledge allows the company to communicate this information to employees so the company can respond to the customers' needs by altering production and delivery schedules. When this approach is employed and executed throughout the organization, the organization will achieve the desired objectives of increased customer satisfaction, increased customer satisfaction relative to competitors, increased customer retention, and market share gain. These results will in turn lead to improved financial objectives.

22.3 Continuous Improvement of All Processes

TQM expands on the concept of continuous improvement. Continuous improvement is not confined to just the manufacturing portion of the company. It encompasses the continuous improvement of *all* business processes. The nonmanufacturing processes such as research and development, marketing, and legal services tend to be difficult to improve because the processes are poorly defined.

Companies have five resources (money, physical assets, technology, people, and time) available to improve quality and to meet customer needs. Many companies have effectively used only the first three assets. However, competitive advantages can be gained in the marketplace if all five resources are applied to meeting customer needs. Effective utilization of the company's human resource potential is a difficult task for the traditionally managed company. These companies take a hierarchical approach to management. Workers are hired to supply muscle and are encouraged to "park their brains at the door." Managers are responsible for telling the workers what, when, and how to do the work. This system creates animosity and antagonism between the workers and the managers. The effective implementation of TQM requires the development of a high-performance work force. A high-performance workforce assumes that all employees are creative and desire and are able to do an excellent job. For example, in a plant manufacturing a frozen product, all steps from purchasing of the raw product and ingredients to processing, freezing, packaging, and storage will influence the final quality of the product. Employees on the production line see firsthand where bottlenecks develop and where inferior quality product slips through the inspection steps. For instance, if blemishes are a problem in the final product, the line employees'

---

complete a job. Since the person specifically knows the requirements of future steps, the individual takes into account those future requirements during the process design of the current process step. Plans should be developed with built-in flexibility so that appropriate action can be executed when an unforseen problem occurs. Strategies are those activities a person takes when the individual knows what must be done but does not know exactly how to accomplish the task. The individual cannot identify the process for completion of the next step until the previous step is completed. Therefore, learning must occur as one step is completed and the next step is started (Tribus, 1992).

insights may show where a product should be handled quickly and where it should not be handled to minimize the blemishes. The manager's primary role would then be to remove the roadblocks that have inhibited the line employees from doing their jobs in the most efficient manner. In some food-processing companies, high-performance work teams need only a limited amount of management. In these companies, line employees are responsible for the hiring and disciplining of team members and for developing the production schedules.

The development of high-performance work systems is critical for the success of the TQM process. Management and employees must commit to the development of a process that allows the changing of the corporate culture and individual behavior. It is essential that managers lead by example and provide opportunities that allow for the growth of employees. Former Secretary of Labor Robert Reich (1991) emphasized the importance of a company's human resources talent when he stated that the only resource a company has that differentiates it from its competitors is the employees.

Time is a nonrenewable resource. All organizations are allocated an equal amount of time. Therefore, companies can gain a strategic advantage by compressing cycle time. This allows companies to increase their ability to respond rapidly to the ever-changing customer's needs. Cycle time can be decreased through process improvement, productivity improvement, and process simplification. The effective use of time and human resource potential can give the small- and medium-sized processing companies a competitive advantage and level the playing field for small food processing companies as they compete in the global marketplace.

Greene (1993) lists about 100 different quality improvement tools. It is critical that all employees learn to effectively use the appropriate tools to study a process from both the qualitative and quantitative perspective. The two primary tools of quality improvement are flow diagrams and cause-and-effect charts (or Ishikawa charts, fish-bone diagrams, or C&E diagrams). The flow diagram is considered to be the most important tool of quality improvement. The development of flow diagrams is essential in solving chronic problems since the actual source of the problems is usually further upstream than the suspected step in the process. Flow diagrams aid in the identification of the specific steps of the process (fig. 22.2). Once the basic steps are incorporated into a flow diagram, details can be added to the diagram to allow understanding of specific subprocesses.

The cause-and-effect diagram helps to identify causal factors of a problem by focusing on multiple sources of variation that can affect a system (materials, methods, machines, measurements, personnel, management, and environment) (fig. 22.3). In addition, cause-and-effect diagrams help individuals look beyond symptoms to aid in the identification of the root causes of the problem. For example, such a diagram would be useful in minimizing the inevitable condensation of water that occurs on both food and equipment during freezing and during handling of nonpackaged frozen products.

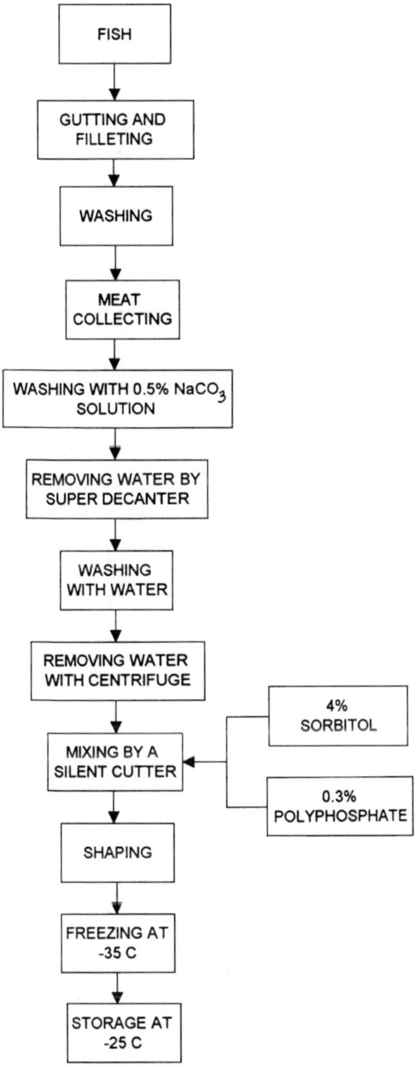

Figure 22.2. Flow diagram for the production of frozen surimi

Once the root cause of a problem is identified, appropriate actions can be taken to eliminate it. World-class companies consider continuous improvement as a never-ending process. These companies utilize an ongoing process to ensure a continual refinement of all controlled processes, and a process to identify and eliminate problems before the problems occur.

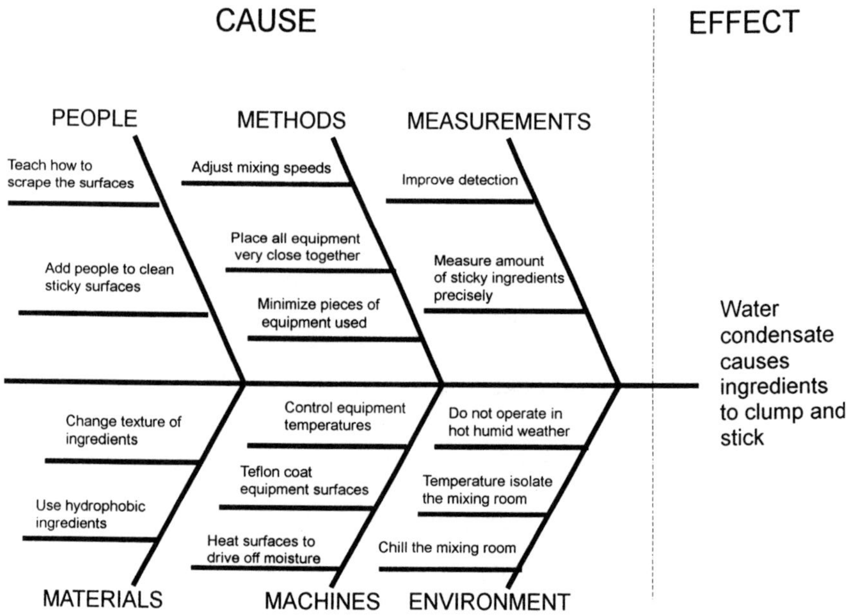

Figure 22.3. The cause-and-effect diagram for water condensate

If continuous improvement is to provide a return on investment, appropriate project selection is critical (Juran, 1989). Initial selection of projects for quality improvement is easy because most companies have chronic problems that cause obvious waste. Some individuals describe this stage as "picking the low-hanging fruit." Once these problems are solved, other criteria need to be used to identify areas for improvement. The criteria used to select projects include the following:

- Increase in value to customers
- Urgency to solve the problem
- Reduction of poor quality costs
- Ease of solving the problem, such as readily available technical solutions or perceived minimal resistance to change
- Health and future of product line

22.4 Increasing Value to Customers

To develop a strategy that focuses on increasing value to the customers, companies need to understand the Voice of the Customer. Drucker (1974) wrote indirectly about the Voice of the Customer in 1974 when he stated that customers

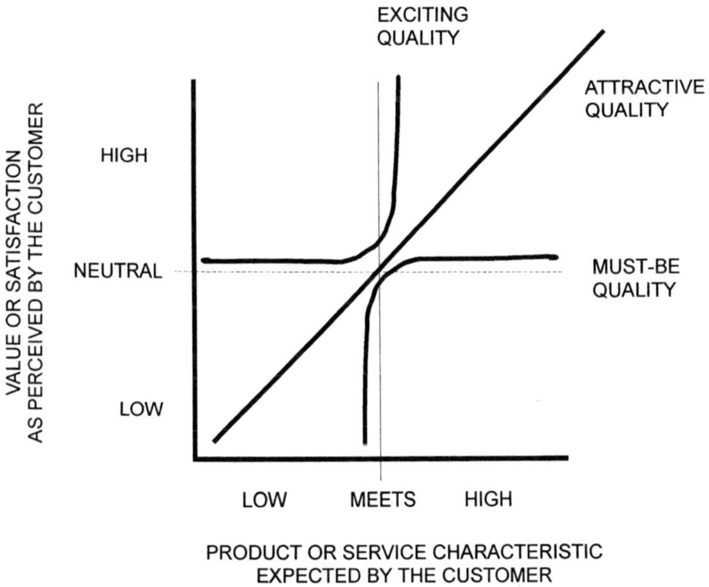

Figure 22.4. The Kano model that correlates perceived quality with expected product or service characteristic. (Adapted from Kano et al., 1984.)

purchase a want or customers buy value. In contrast, the company only manufactures and sells a product. Drucker further states that companies may decide to add product characteristics in order to gain a competitive advantage and develop an "excellent" product from the company's perspective. If the customer considers these characteristics irrelevant, the company is adding waste and expense to the product. Therefore, a company needs to understand the Voice of the Customer and effectively correlate this information into product specifications and the process requirements.

Kano and his colleagues of the Univeristy of Tokyo (1984) further expanded this concept when they correlated the customers' evaluation of the product/service characteristic to the customers' perception of the value that is provided by the product/service characteristic. From this correlation, Kano et al. defined three types of quality (fig. 22.4) as follows:

Must-be quality Must-be quality characteristics or implicit quality are those characteristics that the customer takes for granted. When companies gather consumer information about the product or service, the customer will not identify these quality characteristics. The customer does not notice these characteristics unless the characteristics are either poorly presented or absent. The absence or poor presentation of any of these characteristics causes a significant decline in customer value or perceived quality of the product or service. An example of must-be quality is that customers demand safe food.

Attractive quality These are characteristics that the customer identifies and considers important. As the characteristics are added to the product and service in a consistent manner that meets or exceeds customer expectations, the customer value or perceived quality increases. To delight a customer, nearly all of the characteristics must be presented from the customers perspective in a perfect or near-perfect manner. As a rule of thumb, if customer satisfaction is measured on a five-point scale, these characteristics must be presented at a level of four points or higher to achieve customer delight. For example, customers desire that the actual presentation of a food after its preparation match the graphics that are presented on the package. This presents a technical challenge in the development of multiportion frozen foods since the meat item may have a different heating requirement when compared to the vegetable. By applying different films with different heating capabilities to the food compartments, the industry assures that both the meat and vegetable will have acceptable product quality under conditions in which only one heating time and heat setting is applied to the product.

Exciting quality The customer will not identify these characteristics. These characteristics fill an unstated need or solve a problem. The customer does not know nor expect the company to solve this problem or meet the need. Therefore, there is no negative effect when these characteristics are not present in a product. However, when the characteristics are present in a product or service, these characteristics cause customer delight, and the perceived value skyrockets. For example, a frozen dessert could be developed with the following characteristics: tasty, reduced calories, and added nutrient value. The company can achieve customer delight by focusing its resources to understand and meet unstated customer needs. Information regarding problems customers face should be collected as part of the process used to determine the Voice of the Customer.

Determining if customers are satisfied or delighted, Michael Moon, a management consultant, suggests that a company ask themselves questions such as the following (Whiting, 1992):

- Will the customers repurchase the product?
- If so, will these customers be willing to pay a 35% premium above first-time purchase costs?
- Do the customers recommend the product or service to potential customers who will also purchase the product or service?

Kano et al. (1984) also defined the following types of "quality" that should be avoided in products or services:

Indifferent quality	These characteristics neither add nor subtract from customers' perception of value when the characteristics are present. These characteristics should be avoided because they add needless cost to the product.
Reverse quality	Characteristics that decrease the perceived quality of a product when present in the product/service.

Understanding the customer is critical for a company that wishes to become a customer-driven organization. Companies need to consistently monitor the Voice of the Customer because the customer's needs are changing as a result of the entry of new products, and positive and negative exposure to existing products and advertizing. World-class companies use multiple techniques or listening posts to gather this information. These techniques include the following:

- Focus groups
- Surveys
- Interviews (both one-on-one and group interviews)
- Mail questionnaires
- Product clinics

Questions should be developed in a manner that captures information from the customer's perspective. Unfortunately, this can lead to collection of messy data. Customers tend to mix needs, solutions, and concerns in their responses. Data analysis must be conducted in a manner that allows for the separation of this information and for the identification of the customers' root wants.

Quality Function Deployment (QFD) provides a systematic approach to link the Voice of the Customer to product specifications and process characteristics (Day, 1993). This provides a means to ensure that the primary customer wants are addressed during product development. The first step of the process is to identify technical requirements or product characteristics that will satisfy each of the customer root wants. The analysis is done by using QFD's House of Quality (fig. 22.5). The House of Quality has two parts, the customer part (horizontal) and the

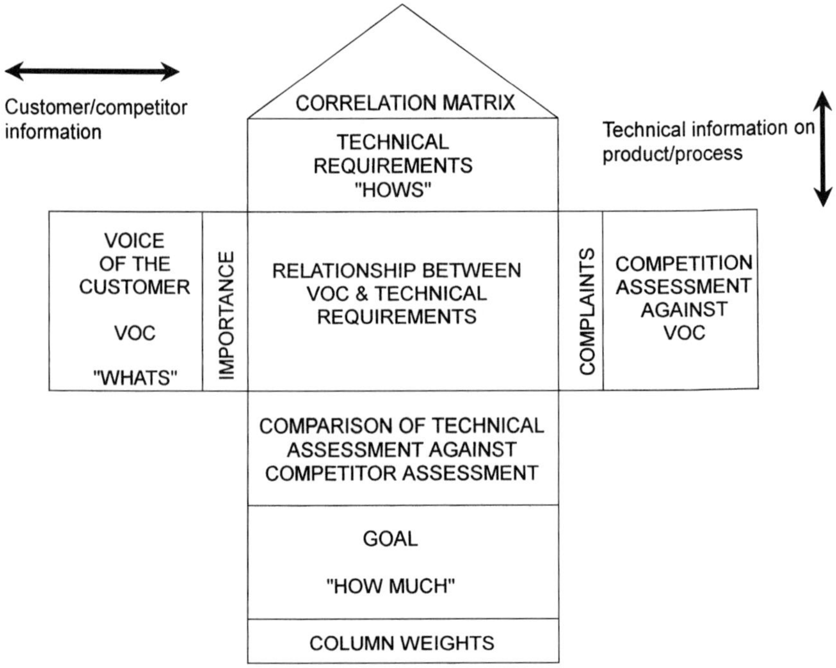

Figure 22.5. The QFD house of quality

technical part (vertical). The customer part expresses customer requirements from the customers' perspective (Voice of the Customer), the importance rating of the need from the customers' perspective, the assessment of customer complaints, and the customer's evaluation of competitive products. The technical part identifies how the company will respond to each of the Voices of the Customer. Each response needs to be measurable and actionable. Also included in this part is a comparison of the company's product or service against the competitors' product or service. In the center of the House of Quality is a matrix that identifies both strengths and negative relations between the Voices of the Customer and technical requirements. The roof of the House of Quality provides an opportunity to assess both positive and negative relations between the various technical requirements. The analysis allows the company to rapidly develop products that both are superior to competitors' products and exceed customers' needs. Figure 22.6 shows a basic House of Quality for the development of mashed potatoes.

A series of QFD matrices can be used as part of the product development process. When this is done, a series of House of Quality matrices are developed in which the input (or the Whats) for the next matrix is the output (or the How and How Much) from the previous matrix (fig. 22.7). At the first matrix level, the

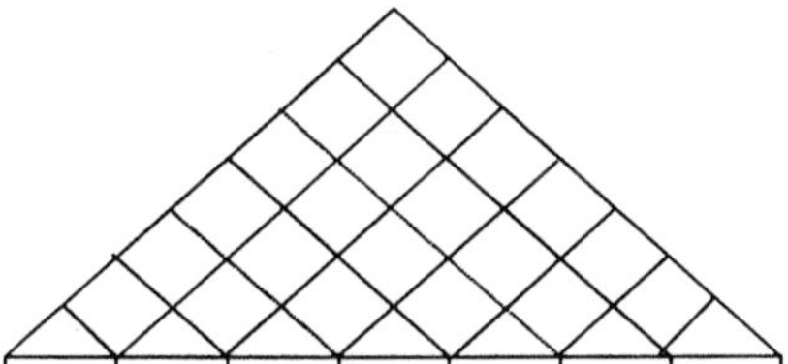

TECHNICAL REQUIREMENTS OR HOWS / VOICE OF THE CUSTOMER OR WHATS	STATISTICAL WEIGHT CONTROL	LOW TEMPERATURE CONTROL	PEEL, SCRUB, TRIM CONTROL	GOOD MANUFACTURING PRACTICES	HACCP	INGREDIENT ADDITIONS	SANITATION AND CLEANUP PROCEDURES
CORRECT FLAVORS	1	1	3	6	6	9	6
POURABLE FROM PACKAGE (NO FROZEN BLOCKS)	1	9	1	1	1	1	1
VISUALLY PURE (NO POTATO SKIN)	1	3	9	6	1	6	6
CORRECT WEIGHT OF PACKAGE	9	1	1	1	1	1	1
SANITARY PRODUCT	1	9	3	9	9	3	9
OTHERS							

CORRELATIONS ABOVE INDICATE THE FOLLOWING:
9 = STRONG, 6 = GOOD, 3 = WEAK, 1 = NONE

Figure 22.6. A basic House of Quality diagram for the production of mashed potatoes

Voice of the Customer is used to generate the technical requirements for the product. Technical requirements address product characteristics rather than product specifications. For example, the customers may want a "healthy" product. This can translate into the technical requirements that identify the nutritional composition of the product. During the development of the second-level matrix, the technical requirements become the input for the development of product requirements and product specifications such as levels of casein, corn oil, minerals, starch, sucrose, and so forth. During the development of the final matrix, the product specifications become the input for the development of process characteristics such as methods that control the addition of corn oil to a batch. When the QFD is properly conducted, it is possible to link process characteristics directly to product requirements that meet or exceed customer root wants.

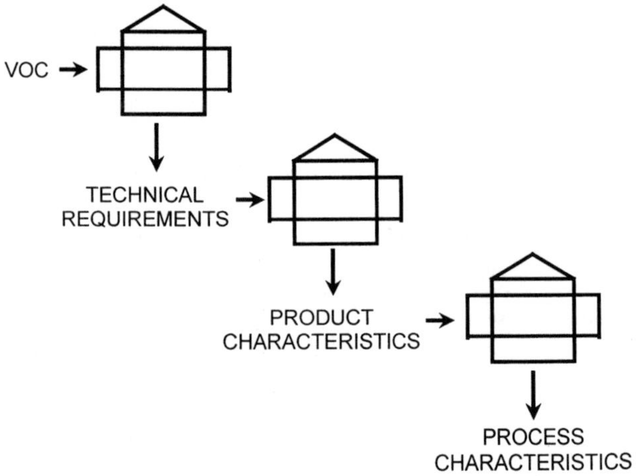

Figure 22.7. Use of a series of QFD matricies as part of the product-development process

As a company's TQM process matures, the company becomes customer driven. The characteristics of the customer-driven company include the following:

- Setting high standards that are based on increasing value to customers
- Knowing the customer[5]
- Creating an image of customer expectations
- Designing products and services that maximize customer satisfaction
- Allocating resources to ensure customer satisfaction
- Ensuring that customer satisfaction is everyone's business

22.5 Summary

TQM provides a paradigm change that allows a company to compete in the global marketplace. TQM is not a "silver bullet" or the ultimate answer to business success and profits. The TQM strategy must be integrated with other strategies such as financial, sales, and marketing strategies.

TQM is a cyclical process rather than a linear process that has a beginning and end. This can pose some problems to food-processing companies that try to im-

[5]These companies are obsessed with knowing the customer to such an extent that the company knows the customer better than the customer knows him or herself.

plement TQM since it is not obvious where to begin this process. Many companies will start on the journey toward world-class quality by implementing quality improvement and statistical process control in the manufacturing section of the business. As experience is generated, the continuous improvement process and customer focus can be expanded to all other business processes. In addition, the leadership/management function also changes to ensure that the TQM process will thrive. These changes include revising the reward and recognition system, removing roadblocks that inhibit workers from doing their jobs, effective training, encouraging cross-functional teams, and developing new relations with customers and suppliers. In addition, senior executives need to ensure that the vision, mission, and guiding principles become a living document that guides the day-to-day decision-making process at all levels in the company. Then issues of quality, safety, and customer value become priority issues for the company. This in turn aligns all employees with the objectives of increasing value to customers and improving the quality of all processes.

22.6 Appendix: TQM—Success or Failure?

Articles have been written for the business and quality literature that describe both the success and failure of the TQM process. The A. D. Little, Ltd. surveyed the top 5,000 manufacturing companies in the United States. TQM had a significant impact in improving the performance of only one-third of these companies. Rath and Strong surveyed 95 senior executives. Thirty-eight percent of these executives gave their quality improvement processes a failing grade (Harrington and Harrington, 1995).

One reason for the lack of success of the TQM effort is that many competitors are using TQM as a process to improve the effectiveness of their companies. Therefore, there is a substantial increase in competition in the industrial sector. Each year, companies may have to achieve substantial gains in quality and productivity just to maintain their current status. However, there are also a number of reasons for the lack of success of the TQM process. Harrington and Harrington (1995) attributed the lack of success to the following factors:

- There was lack of trust between management and employees.
- Upper management did not believe they needed to change.
- The wrong person was selected as the internal champion for the continuous improvement process.
- The wrong consultant was selected or the company terminated the relationship with the proper consultant too soon.
- The organization did not include middle management in the continuous improvement process.

In a second study, Easton (1993) reported that the lack of success of the TQM process could be attributed to one or more of the following reasons:

- Incomplete understanding of TQM by executives which leads to partial implementation of TQM
- Inability to link together the components of TQM into a single process and subsequently link the TQM process to other business processes
- Inability to link the quality improvement efforts to either an increase in the perceived quality attributes or an increase in the return on investment
- Impatience in implementing the TQM process which leads to false expectations that a mature TQM process can be rapidly achieved
- Equating the TQM implementation with the implementation process of a single quality-improvement technique such as Statistical Process Control (SPC)
- Inability to expand the continuous improvement process to include quantitative analysis of process parameters, and subsequent reliance on the use of brainstorming and other qualitative tools to drive the quality-improvement process
- Failure to use appropriate measures to assess the quality-improvement process and link operational performance measures to customers
- Failure to develop all employees to their fullest extent
- Use of a narrow definition of TQM, such as equating TQM with the philosophy of a single quality guru.

Many companies have achieved significant financial improvements as a result of implementing TQM. The U.S. Government Accounting Office (GAO) (1991) conducted and published information on "bottom-line" performance of 20 of the highest-scoring applicants for the Malcolm Baldrige National Quality Award. TQM helped these companies achieve increases in a number of critical performance measures including the following average annual increases:

- 13.7% in market share
- 8.6% in sales per employee
- 1.3% in return on assets
- 0.4% in return on sales

One of the GAO analysts concluded that "the data from the 20 companies studied suggested that TQM strategies, if properly designed and implemented, can significantly improve company performance."

Easton and Jarrell (1994) surveyed 108 firms and correlated the maturity of the TQM process with financial performance. They reported that companies that had a more mature TQM process had a stronger financial performance than companies with a less mature TQM performance. The investigators also reported that downsizing due to corporate restructuring did not significantly improve corporate performance.

References

Buzzell, R.D. and Gale, B.T. 1987. *The PIMS Principles: Linking Strategy to Performances*. Free Press, New York.

Day, R.G. 1993. *Quality Function Deployment: Linking a Company with its Customers*. ASQC Quality Press, Milwaukee, Wis.

Drucker, P.F. 1974. *Management: Tasks, Responsibilities, Practices*. Harper Row Publishers, New York.

Easton G.S. 1993. The 1993 State of U.S. Total Quality Management: A Baldrige Examiner's Perspective. *Calif. Manage. Rev.* **Spring**:32–54.

Easton, G.S. and Jarrell, S.L. 1994. Private communication.

Garvin, D.A. 1988. *Managing Quality*. Free Press, New York.

Greene, R.T. 1993. *Global Quality: A Synthesis of the World's Best Management Methods*. ASQC Quality Press, Milwaukee, Wis.

Harrington, H.J. 1987. *Poor Quality Costs*. ASQC Quality Press, Milwaukee, Wis.

Harrington, H.J. and Harrington, J.S. 1995. *Total Improvement Management: The Next Generation in Performance Improvement.* McGraw Hill, New York.

Juran, J.M. 1989. *Juran on Leadership for Quality*. Free Press, New York.

Kano, N., Serake, N., Takahashi, F., and Tsuji, S. 1984. Attractive quality and must be quality. *Jpn. Soc. Qual. Control* **14**:38–48. (English translation published in *Best in Quality Series*, vol. 7. Hromni, J.D., ed.) ASQC Quality Press, Milwaukee, Wis.

NIST. 1997. *Criteria for Perfomance Excellence*. National Institute of Standards and Technology, Gaitherburg, Md.

Reich, R. 1991. Statement made on *Quality or Else*, a television show produced by the Corporation for Public Broadcasting, October 13, 20, and 27.

Tribus, M. 1992. *Quality First,* 4th ed. (Publication Number 1459) National Institute for Engineering Management and Systems, Alexandria, Va.

U.S. General Accounting Office. 1991. *Management Practices: Companies Improve Performance Through Quality Efforts*. (GAO/NSIAD 91-90) Washington, D.C.

Whiting, R. 1992. Do you really know what your customers want. *Electron. Business* **October**:73.

23

Quality Enhancement

T. M. M. Malundo, R. L. Shewfelt, K. H. McWatters, and Y.-C. Hung

23.1 Quality is Defined by Consumers

The concept of quality is changing (Golomski, 1993; Grijspaardt-Vink, 1993). Three decades ago, the quality of a commodity was based on the technical judgment of experts on specific product properties (Koster, 1990; Sidel and Stone, 1993). Today, industrial approaches define quality more in terms of consumer acceptance than expert opinion (Hildrum et al., 1984).

Quality is defined as the assumed totality of those characteristics that (1) differentiate individual units of a product, (2) are significant in determining a product's degree of acceptability to buyers (Kramer and Twigg, 1970), and (3) can satisfy given consumer needs (von Alvensleben and Meier, 1990). Since quality is best defined by the buyer, criteria for judging quality changes as a product travels through the marketing chain (Shewfelt, 1993). Criteria that consumers ultimately use as bases for purchase and consumption decisions are the most significant (Lyon, 1990). The definition of quality may also differ according to the needs and perceptions of consumers (von Alvensleben and Meier, 1990). Thus, marketing research constantly attempts to understand the needs and preferences of consumers so that quality standards can be set accordingly (Schutz, 1988; Sidel and Stone, 1993; van Trijp and Schifferstein, 1995).

23.2 Quality Research Must be Consumer Driven

Delivering products of consistently superior quality is important in the successful marketing of agricultural products (Lidror and Prussia, 1990; Brumfield et al., 1993). In handling and marketing of food products, management strategies that emphasize consumer needs and preferences when setting quality standards have been recommended (Shewfelt, 1993, 1994; Crothers, 1994; Ingram, 1994; Sansolo, 1994). Quality research compatible with the goals of a consumer-driven management strategy have, therefore, been encouraged (Fishken, 1983; Schutz, 1983; Koster, 1990; Shewfelt and Prussia, 1993; Shewfelt, 1994; Shewfelt et al., 1997).

Fishken (1983) outlined a consumer-oriented approach to identifying optimum product formulations. The approach determines the effects of ingredients on consumer acceptability in four basic steps: ingredient screening, identification of high-impact ingredients, designing a test product, and consumer testing.

The multiple regression approach to optimization described by Schutz (1983) is based on the premise that, for each product, acceptance is determined by a set of product properties. Therefore, this approach focuses mainly on identifying critical product attributes, although the effect of ingredients may indirectly be studied. Briefly, the procedure necessitates the evaluation of 20 or more product samples using a battery of chemical and instrumental methods, descriptive analysis, and large-scale consumer testing (involving at least 100 panelists). Data reduction is conducted using multivariate statistical techniques. A mathematical model that relates consumer response to significant product attributes is generated through regression analysis.

Koster (1990) proposed a strategy similar to the multiple regression approach that translates consumer preferences into product properties. However, only sensory attributes perceived by a trained descriptive panel constitute the critical product properties. Chemical and physical measurements are then related to critical sensory attributes. In order to increase the probability of market success, Koster's method recommends that at least two rounds of consumer and descriptive tests be conducted. An innovative feature of this method is the use of tests that study specific aspects of consumer preference behavior such as product boredom, product irritation, and authentic liking for a product. These tests have been evaluated to varying degrees for sensitivity, reliability, and external validity.

Quality Enhancement (QE) is a new approach to quality research (Shewfelt and Prussia, 1993; Shewfelt, 1994; Shewfelt et al., 1997). It allows for the continuous measurement or improvement of product quality consistent with consumer expectations without repeated consumer tests. Similar to the multiple regression approach to product optimization, it is based on the premise that there is a finite number of product attributes critical to consumer acceptance. However, these attributes are both product- and target-specific, and are determined in a manner more strategic than the procedures described by Schutz (1983) and Koster (1990). Also, like the multiple regression approach, QE allows for the possibility of producing goods of a significantly better quality than those presently available through the aid of a mathematical model that expresses consumer acceptance as a function of critical quality attributes. Like Fishken (1983), Schutz (1983), and Koster (1990), the proponents of QE acknowledge the limitations of their approach in guaranteeing market success. Those recommending QE realize that other factors external to the product affect consumer acceptability, but these factors are either ignored or built into constraints during the evaluation of acceptability.

23.3 QE Has Four Operational Modes

The QE model presented in figure 23.1 has four operational modes—strategic, basic, applied, and developmental, each interlinked by the evaluation of consumer acceptability. In contrast to the linear process of other research strategies (Pangborn, 1979; Koster 1990), QE is interactive with regard to its operational modes. Although the completion of the strategic mode followed by the evaluation phase is required before operation in the other modes is possible, research can be simultaneously performed in the remaining modes once work in the evaluation phase is concluded (Shewfelt, 1994).

Strategic QE is concerned with identifying product attributes important to consumer acceptance and objective methods for quantifying these. Consistent with the dogma of QE, the product and target market are first selected. Critical product attributes are identified through qualitative methods that allow consultation with members of the target population. This approach is more strategic and provides a better insight into the basis for consumer acceptance than those described by Schutz (1983) and Koster (1990). For logistical reasons, QE limits the number of critical quality attributes to a maximum of six for each target population. Once these attributes have been determined, procedures for objectively measuring them are either identified or developed (Shewfelt, 1994; Shewfelt et al., 1997).

The highlight of the evaluation phase is the consumer acceptability function

$$C_t = f(q_1, q_2, \ldots, q_n) \qquad \textbf{(23.1)}$$

where C_t is the percentage of consumers in the target population that finds the product acceptable, q_i is the critical product attribute i ($i = 1, 2, \ldots, n$), and n is the number of critical attributes (≤ 6). The generation of the mathematical model necessitates consumer testing, within the target population, a set of products with attribute levels varying within and outside the expected range of acceptability. The critical product attributes need to be quantified using the objective methods identified during strategic QE. The evaluation of acceptability should be based on an intent-to-consume question. Hedonic scaling or product ranking are not options. The consumer acceptability equation, which must be validated, is generated by regressing acceptability as a function of critical quality measures (Shewfelt et al., 1997).

When basic research such as mechanistic studies is needed to understand the bases for existing technology or formulation, QE may operate in the basic mode. Present process or product may then be modified as a result of a better appreciation of underlying principles. Applied research such as formulation or process optimization studies can be conducted during applied QE. Developmental QE investigates the effects of external product characteristics such as serving size and packaging on acceptability. During the basic, applied, and developmental modes, product quality may be assessed and consumer acceptability predicted without repeated consumer tests, using the consumer acceptability equation (Shewfelt et al., 1997).

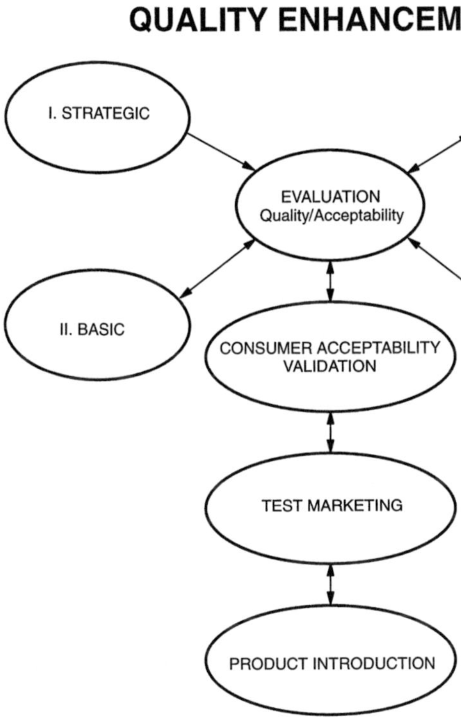

Figure 23.1. Quality enhancement model and linkage to new product introduction

QE is a relatively new strategy to product research and, admittedly, no complete application of this approach has been published. However, specific components have been effectively used in other research (Shewfelt, 1994), and, as with most new concepts, QE embodies elements of research approaches that have been previously employed (Golomski, 1993). The next few sections will review literature pertinent to the implementation of QE, with emphasis on research involving frozen products.

23.4 Qualitative Research Offers Detailed Insights into Consumer Attitudes

Researchers attempting to improve the quality of frozen foods must know the product properties that consumers desire and expect, and translate these into technically viable options to improve the product (Lyon, 1990; Chambers and Smith, 1991; Brumfield et al., 1993). Quality research that fails to identify critical product attributes may prove to be an exercise in futility (Schutz, 1983;

Chambers and Smith, 1991). For instance, it is clear that flavor is sacrificed in the development of low-fat and no-fat frozen dessert products such as ice cream (Hatchwell, 1994). Lipids function in many ways in these products, including direct contribution to flavor, enrobing of other flavor molecules, lubrication, and mouth feel. In addition, lipid substitutes can produce off-flavors. It is not clear which of these functions are critical to acceptability by consumers in the target population, which consists of those wishing to consume frozen desserts while reducing calorie and/or fat consumption.

Qualitative research techniques allow for a more detailed insight into the product perception process experienced by consumers (van Trijp and Schifferstein, 1995). In contrast to quantitative methods, these techniques use words and observations to explain reality and attempt to describe people in natural situations (Krueger, 1994). Used properly, qualitative methods can detail information about consumers' attitudes, opinions, perceptions, behavior, habits, and practices (Chambers and Smith, 1991). Qualitative methods commonly employed in the food industry are focus groups (Kraft, 1981; Galvez and Resurreccion, 1992), in-depth interviews/surveys, or a combination of both. Other less frequently used methods include minigroups, maxigroups, dual-moderated groups, and respondent-plus-client groups (Sokolow, 1988).

23.5 Focus Groups Can Give In-Depth Information on Consumer Attitudes

Surveys and one-on-one interviews have been used to understand consumer attitudes toward food products (Aaker and Keller, 1990; Diaz-Knauf et al., 1993; George, 1993). These qualitative methods have the disadvantage of using closed-ended questions in questionnaires developed by the investigator. Information from respondents is limited to the choices offered, and therefore findings could be unintentionally influenced by the biases of the researcher through oversight or omission (Krueger, 1994).

In contrast, the focus group interview is a nondirective technique that begins with limited assumptions, uses open-ended questions, and allows individuals to respond without providing clues for potential response categories. This approach is in opposition to the structured and directive process of a survey or one-on-one interview which is usually dominated by the investigator or interviewer (Krueger, 1994).

A focus group is a carefully planned discussion designed to obtain perceptions on a defined area of interest. It is made up of 7–12 participants who are similar in at least one specific characteristic. This similarity is determined by the purpose of the study and provides a basis of recruitment (Sokolow, 1988; Chambers and Smith, 1991; Krueger, 1994).

The discussion is led, but not dominated by a qualified moderator familiar with the principles of group dynamics and interviewing techniques, and sufficiently knowledgeable in the category under discussion. The moderator guides the group using a questioning route designed to ensure full coverage of all pertinent issues. Although questions appear to be spontaneous, they have actually been carefully phrased and selected in advance to maximize the amount and quality of information that can be gained. Since the main format of the discussion is open-ended, the moderator can probe to clarify and follow up on comments made by participants (Sokolow, 1988; Krueger, 1994).

The focus group method is based on people's natural tendencies to form groups and engage in collective interaction. One limitation of interviews and surveys is the assumption that individuals form opinions in isolation when, in reality, attitudes and perceptions are developed in part by interaction with others. By reacting to other people's responses to questions or issues raised, participants find new frames of reference for assessing their own responses. Focus groups also have the added advantage of being less costly in terms of time and money on a per respondent basis when compared to interviews (Sokolow, 1988; Krueger, 1994).

The focus group interview has a narrow purpose for which it is particularly effective. It should be used to understand the perceptions, feelings, and attitudes of consumers about products, services, or opportunities. It should not be intended to arrive at a consensus (Krueger, 1994), determine trends and percentages as in a survey (Kraft, 1981), or estimate population attitudes by increasing the number of groups (Chambers and Smith, 1991). The uses of focus group interviews in food-product development and research have been extensively discussed by Kraft (1981) and Sokolow (1988).

23.6 Consumer Attitudes Should Be Translated into Objective Measures of Quality

Information gathered from focus group interviews should be treated as preliminary indicators and interpreted only in a broad sense (Katz, 1994). Verbal statements from consumers can sometimes be unreliable and a poor reflection of actual behavior (Koster, 1990). Focus group data must, therefore, be verified, preferably using quantitative methods (Kraft, 1981).

Translating information on consumer preferences into product properties that could be measured objectively is crucial in consumer-driven quality research (Koster, 1990). Established relationships between objective measures of quality and consumer preference can be used to (1) understand the effects of physicochemical properties on market acceptance of a commodity, (2) identify which attributes to emphasize when developing or improving a product, (3) establish guidelines for quality assurance and control, and (4) aid in product optimization (Schutz, 1990). For some objective methods, direct casual and indirect correlations with human perceptions have been determined (Hildrum et al., 1984; Monnot, 1990). Instrumental and chemical measures should be calibrated with respect to not only subjective, but objective sensory responses as well, since a measured change in product may not necessarily be perceived by the human senses (Martens and van der Burg, 1984).

23.7 Alternative Methods for Measuring Consumer Responses Are Timely

The interface of various quality methods can be promoted by developing reliable calibrations (i.e., mathematical relationships) among different objective and subjective measurements. Accurate predictions of consumer responses from objective measures may then be possible using mathematical relationships (Hildrum et al., 1984). To this end, quantifying consumer responses using valid techniques is critical (Sidel and Stone, 1993).

Various testing methods for measuring consumer response are available (Meilgaard et al., 1991; Stone and Sidel, 1993). The method of choice should depend on the objectives of the study and the type of information needed (Lawless and Claassen, 1993; O'Mahony, 1995). Ideally, the test should reproduce normal conditions of product consumption and minimize, if not eliminate, factors that bias results (Sidel and Stone, 1993; O'Mahony, 1995). Mall-intercepts and central location tests (CLTs) are preferred over affective tests conducted under carefully controlled laboratory conditions, as the latter creates an artificial situation that may cause consumers to respond unnaturally (O'Mahony, 1995; van Trijp and Schifferstein, 1995). Keeping test instructions simple and the avoidance of asking questions that consumers would not normally ask themselves can increase the validity of test results (Koster, 1990).

Of all measuring tools currently available for consumer testing, the nine-point hedonic scale is considered by some to be the most valid and reliable (Stone and Sidel, 1993). However, contrary to popular belief, the overly used scale is not without limitations (O'Mahony, 1991). Because of the "end-effects" bias observed in some consumer test participants, the nine-point scale violates basic assumptions of parametric statistics that are often used to analyze generated data. In addition, due to the relativeness of the scale, numbers used do not necessarily translate into information that is of immediate practical value in a business environment. Alternative scales for measuring consumer response are timely. Moskowitz (1993) suggested using a "probability of purchase" as an index of acceptability.

Traditional methods for consumer testing have been criticized for (1) often failing to produce actionable data that could be used as a base for marketing and business decisions, and (2) focusing too much on internal, and completely ignoring external, validity. Failure to address these criticisms has led to conflict between the marketing and sensory groups in most consumer-driven business establishments (van Trijp and Schifferstein, 1995). Scientists in sensory evaluation must continually challenge current theories and develop new techniques in the field (O'Mahony, 1991). These techniques should be adaptable to the diverse nature of consumers and must generate data useful in a business setting by not only being actionable but externally valid (Sidel and Stone, 1993). To this end, sensory scientists must move away from a product-centered approach to consumer research (Nwankwo, 1993) and study sensory experience more explicitly within the broader perspective as being influenced by both product properties and extrinsic cues. Extrinsic cues are attributes not physically related to but which still form part of the product. Obviously, these can be discrete variables (e.g., name brand, packaging) capable of influencing consumer perception (van Trijp and Schifferstein, 1995).

23.8 Statistical Methods Currently Applied to Consumer Data Have Limitations

Statistical analysis techniques are often applied to relate objective measures of quality to consumer response (Powers, 1981; Gacula and Singh, 1984; Danzart, 1986; Martens and Martens, 1986; Pike, 1986; Powers and Ware, 1986; Schiffman and Beeker, 1986; Mullet, 1988; Resurreccion, 1988; Schutz, 1988). Limitations of some statistical techniques in relating consumer data with objective quality measures have been discussed. Using linear correlation to relate hedonic with objective data has been considered inappropriate since, relative to stimulus concentration, the former usually exhibits a multimodal or parabolic response while the latter is often a linear function (Trant et al., 1981; MacFie and Hedderly, 1993). On the other hand, although regression analysis offers more information than correlation analysis by generating a mathematical model that relates subjective and objective data, the technique requires that variables be continuous. This requirement obviously limits the application of regression when discrete variables such as extrinsic product attributes are involved (Moskowitz, 1993). An alternative to regression analysis when independent variables are discrete would be logistic regression.

23.9 Logistic Regression Can Be Used in Consumer Research

Logistic regression is a statistical analysis method that can simultaneously relate both continuous and discrete product variables to any binary or ordinal response (Afifi and Clark, 1990; SAS, 1990; Hand, 1992) and, therefore, has potential for use in consumer research. This technique has found applications in the medical profession where it has been used in risk assessment and clinical diagnosis (Gordon et al., 1979; Chambless et al., 1990; Gilpin et al., 1990; Hand, 1992), but it has potential for application to food-product evaluation (Afifi and Clark, 1990).

Logistic regression can be used whenever an observation has to be classified into one of two or more alternative groups (or populations) on the basis of a set of measurements. The dependent response is expressed in terms of probability of occurrence, and the independent variables may or may not assume a multivariate normal distribution (Afifi and Clark, 1990; SAS, 1990; Hand, 1992). Therefore, logistic regression may be used to model the consumer index "probability of purchase" suggested by Moskowitz (1993) to any combination of discrete and continuous variables. The logistic function may be expressed as

$$P(1/x) = \exp(\alpha + \beta_1 x_1 + \beta_2 x_2 + \ldots + \beta_n x_n) / \{1 + \exp(\alpha + \beta_1 x_1 + \beta_2 x_2 + \ldots + \beta_n x_n)\} \qquad \textbf{(23.2)}$$

where $P(1/x)$ is the probability of purchase and the parameter estimates α, β_1, β_2, . . . , β_n are generated using the method of maximum likelihood (Afifi and Clark, 1990; Hand, 1992).

23.10 Examples Employing the Quality Enhancement Concept

Because the Quality Enhancement concept is quite new and still evolving, published examples are not available. However, a few investigations have involved consumer input and computer modeling in product development.

23.10.1 Dark Chicken Meat Products

The popularity of boneless, skinless chicken breasts in the United States has resulted in a surplus of dark meat, much of which is exported to other countries. The availability of raw, convenient-to-use products from dark chicken meat that satisfy consumer expectations could make a significant impact on domestic consumption of chicken. Oftentimes raw chicken products are presented in the supermarket display case in a frozen rather than refrigerated state. Although the example described below was limited to fresh and cooked chicken, the principles are directly applicable to frozen product. A five-part study based on a consumer-driven approach was conducted to determine consumer attitudes and preferences for raw dark chicken meat products and packaging of those products (Elsner, 1995). Although this study used a hedonic scale to measure willingness-to-purchase rather than the yes/no measure and probability of purchase associated with the Quality Enhancement approach, its integration of the consumer into the product development process is an appropriate model to follow.

Operating in the strategic mode of the QE model (fig. 23.1), focus group sessions were conducted in four different locations of metro-Atlanta, Georgia. These discussions were held to learn about consumers' poultry selection and purchase behavior, to develop product concepts for dark chicken meat, and to identify the most important attributes of chicken products and their packaging. Attributes that focus group participants considered most important in selection of raw poultry were price, quality, and color. Though price was important, participants said they would not sacrifice quality for lower price. Consumer confidence in the brand of chicken played an important role in purchase decisions, and participants sometimes paid more for a specific brand.

Focus group participants expressed a willingness to purchase new, convenient retail products made from dark chicken meat. These products were leg cuts; boneless, skinless thighs; kabob chunks; stir-fry strips; scallopini; and unbreaded nuggets. They also said they would consider using ground chicken as an alternative to ground beef or ground pork.

Current packaging and liners for raw chicken were considered inadequate by the participants. Consumers said they would like packaging that was clear and allowed a better view of the meat. The majority said they would consider purchasing larger packages of chicken if the packages were resealable.

The next phase of the study involved both the developmental and strategic modes of the QE model (fig. 23.1) and used a mailed survey to obtain information from consumers concerning purchase behavior, new products from dark chicken, and product packaging. The questionnaire consisted of 48 questions and was designed to be self-administered. Of 160 questionnaires mailed out, 72% were returned. Important information obtained from the mailed survey concerned the size preference for kabob and stir-fry pieces, desired serving sizes of various dark chicken meat products, the desire to have other menu items included in the package, and the type of packaging desired.

The third phase of the study involved the developmental mode of the QE model (fig. 23.1), in which consumer ranking was used to evaluate the effects of external product characteristics on acceptability. Packaging concepts that were suggested by the focus group participants were evaluated in a central location test with 103 consumers. As people entered either a Kmart store in Griffin, Georgia or a Sam's Club store in Athens, Georgia, they were intercepted and asked if they would like to participate. Those who wished to participate viewed five photographic representations of packaging for raw chicken and ranked their preference from least desirable to most desirable, considering both the package type and relative cost. Results showed that the resealable pouch was the most preferred, the clear tray was second, the clear box was third, the microwavable-ovenable box was fourth, and the traditional styrofoam tray was the least preferred. These results indicate that modification in the packaging currently used for raw chicken could promote increased purchase of retail products.

The fourth phase of the study involved the applied mode of the QE model (fig. 23.1). In this phase, information obtained from the mailed survey was used in the design of treatments to process new dark chicken meat products for sensory evaluation by consumers. Four potential products from dark chicken meat that were identified by the focus groups and from the mailed survey were processed, cooked, and evaluated for acceptability by untrained consumers. A nine-point hedonic scale ranging from 1 = dislike extremely to 9 = like extremely was used. After evaluating the cooked products, consumers were asked to evaluate packages of the raw product as it would be packaged for a retail setting and then answer questions concerning their likelihood of purchasing the product, their preference for white or dark meat, pricing suggestions, and any comments they wished to make.

Products that had the highest hedonic ratings for the sensory attributes and that consumers were the most willing to buy were as follows:

Scallopini	4 mm-thickness and 2% STPP
Kabobs	2″ chunk size and 2.6% seasoning
Stir-fry strips	Size not significant; 2.6% seasoning was preferred
Ground chicken	Sensory ratings too low to consider this an acceptable product

The final phase of this study involved evaluation of consumers' purchase intent of various raw products from dark chicken meat, using a simulated supermarket test. A large room on the University of Georgia campus was decorated with multicolored streamers and food-oriented posters to give the overall effect of being in a supermarket. Four new products (scallopini, kabobs, stir-fry strips, and ground chicken) processed from boneless, skinless thighs were used in the test. They were packaged in the preferred materials identified from the mailed survey. These products were compared in a retail display case against three traditionally packaged controls of boneless, skinless breasts; boneless, skinless thighs; and bone-in, skin-on thighs. In order to determine preference and compare perceived value, each package was weighed to give $2.00 worth of product with the packaging cost included.

Participants were each given $2.00 and asked to go to the retail display case to select one package of chicken. They brought their selection to a checkout counter where payment was accepted and the purchase was recorded. After each purchase, consumers were given a postshopping questionnaire asking why they bought the particular product instead of the others available, how they liked the packaging of the product, whether they thought the price was fair, and their plans for preparing the product at home.

Participants made four separate trips to the display case and were instructed to purchase a different product during each trip. This resulted in a forced choice ranking of four out of seven products. Consumers were paid a small cash honorarium for their participation and were allowed to take their selections home with them. They had been instructed to bring a cooler with them for transporting the products to their homes. Ice was provided for the coolers.

Results from the consumer responses were converted into a desirability index, which is a weighted average calculated from the choices and number of responses for each product. The desirability index values for each of the products are as follows: boneless, skinless breasts = 391; kabobs = 181; stir-fry = 174; scallopini = 134; boneless, skinless thighs = 138; bone-in, skin-on thighs = 86; and ground chicken = 82. Results showed that breasts were preferred over all other products. No significance was found among the dark chicken meat products. The control bone-in, skin-on thighs were less preferred than the other dark meat products except for ground chicken.

Sensory evaluation showed that three of the four products developed using a consumer-driven approach were very acceptable to consumers and that these convenience products have potential for expanding the use of dark chicken meat. Similar products may also be developed for the frozen foods market. These products should have a high probability of being successful in the marketplace because consumers were involved in their development from the very beginning. Development of the dark chicken meat products has illustrated use of the strategic, developmental, and applied modes of the QE model (fig. 23.1). Development of other products could involve these as well as other modes, which may operate simultaneously, once work in the strategic mode and the evaluation phase have been completed.

23.10.2 Akara (Prefried Frozen Veggie Nugget)

Akara is a popular West African food made from whipped cowpea (*Vigna unguiculata*) paste. The paste is whipped to achieve a light, spongy texture and is seasoned with chopped onion, peppers (either hot or mild), and salt. Deep-fat frying of spoonful portions of the seasoned paste produces flavorful nuggets (balls) with a light brown, crispy crust and a soft, breadlike interior. Although akara is relatively unknown in the United States, consumer acceptance tests conducted with Americans representing a wide range of ages (12 to 65+) and demographic characteristics indicate that akara has potential for certain segments of the population (McWatters et al., 1990, 1992, 1993).

The traditional process for preparing akara is extremely time consuming, labor intensive, and impractical for present-day lifestyles with family members working outside the home. Tan et al. (1995) conducted a study following the applied mode of the QE model to establish process conditions for prefried, frozen, reheatable akara that would closely resemble freshly fried akara. This convenient form of akara, which could be reheated by either frying, microwaving, or baking, would have potential for use in either home or food-service settings. Tan et al. (1995) used a 3 × 3 × 3 full factorial experimental design to evaluate the effects of initial cooking time, reheating oil temperature, and reheating time on consumer acceptance of frozen, reheated akara.

To evaluate the quality and acceptability of frozen akara reheated by different methods and conditions, both a consumer acceptance test and a trained panel test were employed. The trained panel identified six important quality attributes (color intensity, moistness, beany flavor, oiliness, crispness, and sponginess) of akara and measured their intensity when akara was reheated by frying (using both optimum and nonoptimum conditions), microwaving, and conventional oven baking. Results from the trained panel evaluation verified that akara reheated by the recommended (optimum) frying conditions was better than that reheated by non-recommended conditions. Using baking as the method of reheating produced a dryer, firmer-textured product than other methods, and microwaving resulted in a darker, less spongy product than freshly prepared akara. Instrumental and chemical measurements of proximate composition, color, and texture supported the trained panel results, suggesting that further adjustments in reheating conditions using baking and microwaving were warranted.

To develop a consumer acceptability function, data from the study by Tan et al. (1995) were analyzed by logistic regression to determine a probability of occurrence (e.g., probability of purchase). Consumers' yes or no responses to an intent-to-purchase question were decoded by assigning a 1 to every positive response and a 0 to each negative answer. The backward selection option of PROC LOGISTIC was used to relate consumer response, expressed as a probability of purchase, to several sensory properties. Results of logistic analysis showed that the logistic function fit the consumer data very well with only three sensory properties (flavor, texture, and oiliness) and had an index of discrimination c equal to 0.88. The c index is a measure of the predictability of the logistic regression equation. In instances where the outcome is dichotomous, as in this study, this index represents the probability of correctly classifying between two response classes (Afifi and Clark, 1990). An index of discrimination of 0.88 indicates 88% correct classification with the proposed equation

$$P(1/x) = \exp(\alpha + \beta_1 x_1 + \beta_2 x_2 + \beta_3 x_3) / \{1 + \exp(\alpha + \beta_1 x_1 + \beta_2 x_2 + \beta_3 x_3)\} \qquad \textbf{(23.3)}$$

where $P(1/x)$ was the probability of purchase; x_1, x_2, and x_3 were intensities of flavor, texture, and oiliness, respectively; and α (-10.08), β_1 (0.97), β_2 (0.34), and β_3 (0.45) were the parameter estimates associated with the model terms (Afifi and Clark, 1990; Miller et al., 1991; Hand, 1992).

To demonstrate the use of the consumer acceptability function involving the three critical quality attributes that were identified as important to the acceptability of prefried/frozen/reheated akara for a particular process condition, measured quality attribute values for flavor, texture, and oiliness can be put into equation 23.3 directly to calculate the probability of purchase. If a product with the quality attributes of flavor equals 5.5 (x_1), texture equals 4.5 (x_2), and oiliness equals 2.0 (x_3), $P(1/x)$ would equal 0.47, indicating that 47% of the target population would be willing to purchase the product resulting from this process condition. For further improvement in development of the product, the consumer acceptability equation employed in the QE approach may be used to assess product quality and predict the probability of purchase without repeated consumer tests.

The use of logistic regression in the analysis of consumer data has been applied only on a limited scale, therefore, few references are available for comparison with results obtained from the present study. A recent study by Malundo (1996) on fresh mango used logistic regression to relate consumer response to product properties and reported an index of discrimination $c = 0.70$. Therefore, findings from the present analysis of akara data, which resulted in a higher c index (0.88), support the application of this alternative approach to traditional methods often used in consumer research and data analysis. Such an approach will be helpful in relating acceptability with significant factors and in understanding the attributes of foods that are most important to consumers. The QE approach, with its emphasis on meeting the needs and desires of the consumer and involving the consumer from the first stage of the developmental process, has great potential for improving the success rate of new frozen-food introductions as well as other consumer-based products.

23.11 A Final Perspective

While there is much discussion on shifting from a product-based definition of food quality to one that is consumer based, there are few quantitative methods available to incorporate this concept into product-development or quality-control functions in the food industry, or even academic-based research programs. The answer does not lie in increasing the accuracy and precision of the measurement of specific quality characteristics. Rather it lies in the development of means to better quantify consumer acceptability and to predict acceptability as a function of critical quality characteristics.

This chapter offers a new approach, quality enhancement, to quality research and provides two examples of its application to fresh and frozen products. QE defines quality in terms of consumer acceptability. QE partitions acceptability into purchase and consumption aspects; refocuses acceptability measurement from strictly one of nine-point hedonic scaling to also incorporating percentage distributions of populations finding the product acceptable; emphasizes the measurement of only those quality characteristics that drive acceptability rather than completely characterizing sensory perception of that product or limiting evaluation to only those measurements that meet strict predetermined guidelines of accuracy and precision; needs effective mathematical techniques to best relate new measures of acceptability to those quality characteristics that drive acceptability. Logistic regression is a promising technique to achieve this goal.

QE may prove not to be the ultimate answer to a quest to transform quality research and practice from a product to a consumer orientation, but is an important step in the right direction.

References

Aaker, D.A. and Keller, K.L. 1990. Consumer evaluations of brand extensions. *J. Marketing* **54**:27–41.

Afifi, A.A. and Clark, V. 1990. *Computer-Aided Multivariate Analysis,* 2d ed. Lifetime Learning Publications, Belmont, Calif.

Brumfield, R.G., Adelaja, A.O., and Lininger, K. 1993. Consumer tastes, preferences, and behavior in purchasing fresh tomatoes. *J. Am. Soc. Hort. Sci.* **118**(3):433–438.

Chambers, E. and Smith, E.A. 1991. The uses of qualitative research in product research and development. In *Sensory Science Theory and Applications.* (Lawless, H.T. and Klein, B.P., eds.) pp. 395–412, Marcel Dekker, New York.

Chambless, L.E., Dobson, A.J., Patterson, C.C., and Raines, B. 1990. On the use of a logistic risk score in predicting risk of coronary heart disease. *Stat. Med.* **9**:385–396.

Crothers, D. 1994. ECR: what it means to you. *Produce Business* **10**(4):10–14.

Danzart, M. 1986. Univariate procedures. In *Statistical Procedures in Food Research.* (Piggott, J.R., ed.) pp. 19–60, Elsevier Applied Science, London.

Diaz-Knauf, K., Lopez, M., Ivankovitch, C., Aguilar, F., Bruhn, C., and Schutz, H. 1993. Hispanic consumer acculturation and food safety concerns. *J. Consumer Stud. Home Econ.* **17**:233–243.

Elsner, R.J.F. 1995. Consumer attitudes and preferences for raw dark chicken meat products and their packaging. M.S. Thesis, University of Georgia, Athens.

Fishken, D. 1983. Consumer-oriented product optimization. *Food Technol.* **37**(11):49–52.

Gacula, M.C., Jr. and Singh, J. 1984. *Statistical Methods in Food and Consumer Research.* Academic Press, Orlando, Fla.

Galvez, F.C.F. and Resurreccion, A.V.A. 1992. Reliability of the focus group technique in determining the quality characteristics of mungbean (*Vigna radiata* [L.] Wilczec) noodles. *J. Sens. Stud.* **7**:315–326.

George, R.J. 1993. A comparison of American and Irish consumers' perceptions of the quality of food products and supermarket service. *J. Food Prod. Marketing* **1**:73–81.

Gilpin, E.A., Olshen, R.A., Chatterjee, K., Kjekshus, J., Moss, A.J., Henning, R.E., Blacky, A.R., Dittrich, H., and Ross, J., Jr. 1990. Predicting 1-year outcome following acute and myocardial infraction: physicians versus computers. *Comput. Biomed. Res.* **23**:46–63.

Golomski, W.A. 1993. Total quality management and the food industry: why is it important? *Food Technol.* **47**(5):74–79.

Gordon, T., Kannel, W.B., and Halperin, M. 1979. Predictability of coronary heart disease. *J. Chron. Dis.* **32**:427–440.

Grijspaardt-Vink, C. 1993. TQM should include environmental management. *Food Technol.* **47**(5):30.

Hand, D.J. 1992. Statistical methods in diagnosis. *Stat. Methods Med. Res.* **1**:49–67.

Hatchwell, L.C. 1994. Overcoming flavor challenges in low-fat frozen desserts. *Food Technol.* **48**(2):98–100, 102.

Hildrum, K.I., Martens, M., and Martens, H. 1984. Research on analysis of food quality. In *Control of Food Quality and Food Analysis.* (Birch, G.G. and Parker, K.J., eds.) pp. 65–80, Elsevier Applied Science Publishers, New York.

Ingram, B. 1994. ECR and W. Edwards Deming. *Supermarket Business* **49**(3):13.

Katz, P.H. 1994. Sensory signals of success. *Food Process.* **55**(2):53–55.

Koster, E.P. 1990. Sensory analysis and consumer research in product development. *Acta Horticult.* **259**:13–23.

Kraft, L. 1981. Focus groups: letting consumers think about your new product idea. *Food Technol.* **35**(11):70–72.

Kramer, A. and Twigg, B.A. 1970. *Quality Control for the Food Industry,* vol. 3, 3d ed. AVI/van Nostrand, New York.

Krueger, R.A. 1994. *Focus Groups: A Practical Guide for Applied Research,* 2d ed. SAGE Publications, Newbury Park, Calif.

Lawless, H.T. and Claassen, M.R. 1993. Application of the central dogma in sensory evaluation. *Food Technol.* **47**(6):139–145.

Lidror, A. and Prussia, S.E. 1990. Improving quality assurance techniques for producing and handling agricultural crops. *J. Food Qual.* **13**:171–184.

Lyon, D.H. 1990. Factors affecting the sensory quality of vegetables. *Food Sci. Technol. Today* **4**(1):40–44.

MacFie, H.J.H. and Hedderley, D. 1993. Current practice in relating sensory perception to instrumental measurements. *Food Qual. Pref.* **4**:41–49.

Malundo, T.M.M. 1996. Application of the quality enhancement (QE) approach to mango (*Mangiferaindica* L.) flavor research. Ph.D. Dissertation, University of Georgia, Athens.

Martens, M. and Martens, H. 1986. Partial least squares regression. In *Statistical Procedures in Food Research.* (Piggott, J.R., ed.) pp. 293–360, Elsevier Applied Science, London.

Martens, M. and van der Burg, E. 1984. Relating sensory and instrumental data from vegetables using different multivariate techniques. In *Progress in Flavour Research 1984.* (Adda, J., ed.) pp. 131–148, Elsevier Science Publishers B.V., Amsterdam.

McWatters, K.H., Enwere, N.J., and Fletcher, S.M. 1992. Consumer response to akara (fried cowpea paste) served plain or with various sauces. *Food Technol.* **46**(2):111–114.

McWatters, K.H., Resurreccion, A.V.A., and Fletcher, S.M. 1990. Response of American consumers to akara, a traditional West African food made from cowpea paste. *Int. J. Food Sci. Technol.* **25**:551–557.

McWatters, K.H., Resurreccion, A.V.A., Fletcher, S.M., Peisher, A.V., and Andress, E.L. 1993. Physical and sensory characteristics of akara (fried cowpea paste) made from whole and decorticated cowpeas (*Vigna unguiculata*). *Lebensm. Wiss. Technol.* **26**:157–161.

Meilgaard, M., Civille, G.V., and Carr, B.T. 1991. Affective tests: consumer tests and in-house panel acceptance tests. In *Sensory Evaluation Techniques,* 2d ed. (Meilgaard, M., Civille, G.V., and Carr, B.T., eds.) pp. 201–235, CRC Press, Boca Raton, Fla.

Miller, M.E., Hui, S.L., and Tierney, W.M. 1991. Validation techniques for logistic regression models. *Stat. Med.* **10**:1213–1226.

Monnot, P. 1990. Internal product quality as a marketing instrument. *Acta Horticult.* **259**:25–29.

Moskowitz, H.R. 1993. Sensory analysis procedures and viewpoints: intellectual history, current debates, future outlooks. *J. Sens. Stud.* **8**:241–256.

Mullet, G.M. 1988. Applications of multivariate methods in strategic approaches to product marketing and promotion. *Food Technol.* **42**:145–156.

Nwankwo, S.A. 1993. A framework for conducting customer orientation audit. *J. Consumer Stud. Home Econ.* **17**:1–11.

O'Mahony, M. 1991. Techniques and problems in measuring sweet taste. *Food Technol.* **45**:128–133.

O'Mahony, M. 1995. Sensory measurement in food science: fitting methods to goals. *Food Technol.* **49**:72–82.

Pangborn, R.M. 1979. Physiological and psychological misadventures in sensory measurement or the crocodiles are coming. In *Sensory Evaluation Methods for the Practicing Food Technologist.* (Johnston, M.R., ed.) 1979 IFT short course sponsored by Institute of Food Technologists, Chicago.

Pike, D.J. 1986. A practical approach to regression. In *Statistical Procedures in Food Research.* (Piggott, J.R., ed.) pp. 61–100, Elsevier Applied Science, London.

Powers, J.J. 1981. Multivariate procedures in sensory research: scope and limitations. *MBAA Tech. Q.* **18**(1):11–21.

Powers, J.J. and Ware, G.O. 1986. Discriminant analysis. In *Statistical Procedures in Food Research.* (Piggott, J.R., ed.) pp. 125–180, Elsevier Applied Science, London.

Resurreccion, A.V.A. 1988. Application of multivariate methods in food quality evaluation. *Food Technol.* **42**(11):128, 130, 132–134, 136.

Sansolo, M. 1994. The more things change. *Prog. Grocer* suppl. **April**:4.

SAS. 1990. *SAS User's Guide: Statistics,* 6th ed. SAS Institute, Cary, N.C.

Schiffman, S.S. and Beeker, T.G. 1986. Multidimensional scaling and its interpretation. In *Statistical Procedures in Food Research.* (Piggott, J.R., ed.) pp. 255–292, Elsevier Applied Science, London.

Schutz, H.G. 1983. Multiple regression approach to optimization. *Food Technol.* **37**(11):46–48, 62.

Schutz, H.G. 1988. Multivariate analysis and the measurement of consumer attitudes and perceptions. *Food Technol.* **42**:141–144.

Schutz, H.G. 1990. Measuring the relative importance of sensory attributes and analytical measurements of consumer acceptance. *Acta Horticult.* **259**:173–174.

Shewfelt, R.L. 1993. Measuring quality and maturity. In *Postharvest Handling: A Systems Approach.* (Shewfelt, R.L. and Prussia, S.E., eds.) pp. 99–123, Academic Press, San Diego.

Shewfelt, R.L. 1994. Quality characteristics of fruits and vegetables. In *Minimal Processing of Foods and Process Optimization—An Interface.* (Singh, R.P. and Olivera, F.A.R., eds.) pp. 171–187, CRC Press, Boca Raton, Fla.

Shewfelt, R.L. and Prussia, S.E. 1993. Interdisciplinary solutions to challenges in postharvest handling. In *Postharvest Handling: A Systems Approach.* (Shewfelt, R.L. and Prussia, S.E., eds.) pp. 327–337, Academic Press, San Diego.

Shewfelt, R.L., Erickson, M.C., Hung, Y.-C., and Malundo, T.M.M. 1997. Applying quality concepts in frozen food development. *Food Technol.* **51**(2):56–59.

Sidel, J.L. and Stone, H. 1993. The role of sensory evaluation in the food industry. *Food Qual. Pref.* **4**:65–73.

Sokolow, H. 1988. Qualitative methods for language development. In *Applied Sensory Analysis of Foods,* vol. 1. (Moskowitz, H., ed.) pp. 3–19, CRC Press, Boca Raton, Fla.

Stone, H. and Sidel, J.L. 1993. *Sensory Evaluation Practices,* 2d ed. Academic Press, San Diego.

Tan, P., Hung, Y.-C., and McWatters, K.H. 1995. Akara (fried cowpea paste) quality as affected by frying/reheating conditions. *J. Food Sci.* **60**:1301–1306.

Trant, A.S., Pangborn, R.M., and Little, A.C. 1981. Potential fallacy of correlating hedonic responses with physical and chemical measurements. *J. Food Sci.* **46**:583–588.

van Trijp, H.C.M. and Schifferstein, H.N.J. 1995. Sensory analysis in marketing practice: comparison and integration. *J. Sens. Stud.* **10**:127–147.

von Alvensleben, R. and Meier, T. 1990. The influence of origin and variety on consumer perception. *Acta Horticult.* **259**:151–161.

Index